Quantitative Methods in Supply Chain Management

Ioannis T. Christou

Quantitative Methods in Supply Chain Management

Models and Algorithms

Springer

Prof. Ioannis T. Christou
Athens Information Technology
19Km Markopoulou Ave.
P.O. Box 68
19002 Paiania
Greece
e-mail: ichr@ait.edu.gr

ISBN 978-0-85729-765-5 e-ISBN 978-0-85729-766-2
DOI 10.1007/978-0-85729-766-2
Springer London Dordrecht Heidelberg New York

British Library Cataloguing in Publication Data
A catalogue record for this book is available from the British Library

Cover design: eStudio Calamar S.L.

Printed on acid-free paper

Springer is part of Springer Science+Business Media (www.springer.com)

Preface

This book presents some of the most important methods and tools available for modeling and solving problems arising in the context of supply chain management; in the context of this book, "solving problems" usually means "designing efficient algorithms for obtaining high-quality solutions". Modeling a real-world problem so that it becomes amenable to analysis and the later design of algorithms for actually solving it is a fascinating mixture of art, science, and engineering. The major purpose of this book is therefore to show what modeling techniques can be expected to work for a given situation, as well as what kinds of constraints or objective functions can render models intractable, and what to do when omitting them is not an option; above all, how to apply existing proven exact or heuristic methods, or even design a hybrid or completely new algorithm for a particular model.

As is often the case with textbooks, the material in this book grew out of a set of lectures that I gave to M.Sc. students of Carnegie-Mellon University's Master's in Information Networking program, and Ph.D. level graduate students at Aalborg University, Aalborg, Denmark on the topics of Business Management for Engineers, Supply Chain Management and Logistics, and Network Optimization. The enthusiasm of my students encouraged me to carefully write my lecture notes in book form, and the end result is this book.

The content of the book is organized as follows: the first chapter is a review chapter on methods for continuous as well as combinatorial optimization. It covers most areas of modern optimization:

- Unconstrained non-linear optimization, where the main focus is on methods that converge to a local optimum or at least a saddle point, including Newton-like methods, conjugate-gradient methods, and trust-region methods, but there is also a discussion on successful meta-heuristics for global optimization: simulated annealing, evolutionary algorithms, genetic algorithms, the differential evolution method. Theoretical results are given to show any guarantee that a method has for convergence to a local optimum or a saddle point.

- Constrained linear optimization: the revised simplex method for linear programming is covered in some detail, as is the revised network simplex method for linear network optimization. Advanced topics in network optimization including auction algorithms for the linear assignment problem are also discussed.
- Constrained non-linear optimization: the first-order necessary conditions for mathematical programming are given using the standard theorems of the alternative, and first-order sufficient conditions are presented for convex functions. From an algorithmic point of view, penalty methods and Lagrangean multiplier methods are discussed.
- Combinatorial and mixed-integer optimization, where the focus is on the framework of the Branch-and-Bound method and its variants including Branch-and-Price, Branch-and-Cut-and-Price etc. Successful meta-heuristics including Tabu search and the more recent nested partitions method are also covered.

The chapter also includes an introduction to dynamic programming, which plays an important role on many problems in planning, scheduling, and inventory control. All the material in this chapter can be considered *classical*, with the exception of the recent introduction of the nested partitions method in the arsenal of people working on NP-hard combinatorial optimization problems. As such, it can be skipped by readers familiar with the general (finite-dimensional) optimization techniques and serve only as a reference when the need arises.

The second chapter is an introduction to (short and medium term) demand forecasting using mostly time-series analysis methods. Demand forecasting is a tactical problem that is however of great importance to supply chain management as it forms the basis for setting sales targets, production plans, and consequently and even more seriously, lead-times, personnel levels and so on. Besides the classical exponential smoothing methods and their many variants, and time-series regression methods, and decomposition methods, there is a detailed derivation of fast order-recursive methods (i.e. the Levinson-Durbin method) for solving the Yule-Walker equations arising in auto-regressive based forecasting which is not the standard material in such manuscripts. This is also the case for prediction markets and their information aggregation capabilities, presented in that chapter as well. The material on ensemble forecasts on the other hand, is based mostly on the author's own research, and some computational results and conclusions are presented for the first time.

The third chapter is an introduction to tactical and operational level planning and scheduling problems, seen from the point of view of the interface between Operations Research and Computer Science. In this case, the focus is on formulating accurate models that are at the same time amenable to efficient algorithms for solving them to optimality or at least to near-optimality. Hierarchical production planning is introduced as a vehicle for reducing problem complexity, which then allows one to formulate optimization models at each level of the hierarchy that can be solved exactly or for which fast and efficient heuristics exist, even for large-scale problems. The algorithms for crew assignment scheduling problems provided here were developed in the context of my research on advanced decision support systems at Lucent Bell Labs, Transquest, and Delta Technology. Finally, the modeling of problems related to available-to-promise and order admission control and corresponding solution techniques are comprehensively presented here for the first time.

The fourth chapter deals with inventory control, a purely operational problem. The focus is mostly on single-echelon systems, but a brief discussion of the multi-echelon (serial) case is also presented. Starting with the simple case of deterministic and constant demand and the EOQ model, the text quickly turns to the much more challenging stochastic demand case. The material in this section requires a good understanding of probability theory and statistics. The modeling and analysis of such systems was completed more than forty years ago, but algorithms –exact or heuristic– for determining optimal policy parameters of some such systems have not appeared until very recently. For example, at the time of this writing, I have not been able to find any exact or heuristic algorithm for the (s,S,T) policy optimization under stationary demand and linear holding and backorder costs in the literature. In this chapter, both exact and fast heuristic algorithms for all major inventory control policies are discussed in detail, and computational results are provided.

The fifth chapter deals with the most strategic-level decision problems to be made in supply chain management, which are however intimately linked to operational-level decision problems: location theory and distribution management problems. A number of related location problems including the p-median problem, the uncapacitated and the capacitated facility location problem, as well as multi-echelon multi-commodity location/allocation problems are presented, modeled, and analyzed in this chapter, and efficient exact and heuristic methods are given for their solution. Some methods are again presented here for the first time (the cluster ensemble-based methods for the p-median and uncapacitated facility location problem in particular). Regarding distribution management, some of the most important techniques for vehicle routing problems under the general case of resource constraints and time windows are discussed. Both exact algorithms relying on column generation as well as carefully crafted heuristics are presented for this type of problem.

The last chapter (epilogue) presents a list of some problem areas that I deem will be of great importance in supply chain management in the near future. Some of these problems should be tackled via rigorous methods whereas other problems are purely information technology problems that should be attacked via rigorous software development techniques for building secure and dependable computing systems.

The intended audience of this book is advanced undergraduate and graduate students and researchers working on the interfaces of operations research and computer science; such persons are often affiliated with operations research, electrical and computer engineering, computer science, and industrial and systems engineering departments or graduate business schools. The prerequisites for understanding the material in this book are fairly standard: a two-semester undergraduate-level course on calculus and linear algebra should be enough to follow the mathematical developments in the manuscript. A first course on programming and data structures is also necessary to be able to implement most of the algorithms in this book. The material in Chap. 4 also requires a good background (i.e. a one-semester undergraduate-level course) of probability; some results however are developed using the notions of stochastic convexity and its applications. Assuming the students have also had some exposure to optimization methods, the material in Chaps. 2–5 can be presented in one full semester course. Otherwise, appropriate sections from the optimization review first chapter must be presented in a two- or three-week time span, and then selected topics from Chaps. 2–5 can be presented in the remaining time, probably skipping the material on auto-regressive methods from Chap. 2, and the material on inventory control under stochastic demand in Chap. 4.

At this point, I would like to thank all my students who carefully read hard-to-read portions of unfinished manuscripts of this book and made very useful suggestions. I would like to especially thank my colleague, Dr. Sofoklis Efremidis for proof-reading the first drafts of chapter one of this book and suggesting many corrections and improvements, my student Mr. Yongming Luo for carefully reading the second chapter of the book and preparing some of the figures in it, and finally, Mr. Panagiotis Apostolopoulos for reading parts of the third chapter and making useful suggestions. And of course, I would like to thank the editorial team at Springer for excellent job editing, formatting, and typesetting the book.

Last but not the least, I would like to dedicate this book to my daughter, Anna, and to guarantee to her that I will make up for all the time that she did not get to play with me while I was preparing this book.

Athens, February 2011

Contents

Notation

$\mathbb{N}, \mathbb{N}^*$	The set of natural numbers 0,1,2,…, and the set of natural numbers excluding 0, respectively
$\mathbb{Z}$	The set of all integers
$\mathbb{R}$	The set of all real numbers
$A \cup B, A \cap B, A - B, 2^A, \emptyset$	For sets A and B, the union of A and B, the intersection of A & B, the set containing all elements of A *not contained in* B, the set of all subsets of A including A, the empty set
$\forall, \exists$	"For all", "there exists"
$\lvert S \rvert$	When S is a set, the cardinality of S. When S is a real number, the absolute value of S. When S is a square matrix, the determinant of S
x_i, x^T, A^T, A^{-1}	The ith component of the n-dimensional vector $x \in \mathbb{R}^{\ltimes}$, the transpose of the vector x, the transpose of a matrix A, the inverse of a square invertible matrix A
$\det A$	The determinant of the square matrix A
x_+, x_-	The numbers $\max\{0, x\}$ and $\max\{0, -x\}$, respectively
$\lfloor x \rfloor, \lceil x \rceil$	The largest integer less than or equal to x (floor), the smallest integer greater than or equal to x (ceiling)
$\lim x_n$, $\limsup x_n$, $\liminf x_n$	The limit of the sequence $\{x_n\}$, the limit superior of the sequence, the limit inferior of the sequence
$\lim_{x \to x_0} f(x)$	The limit of the function $f(x)$ as $x \to x_0, x_0 \in \bar{\mathbb{R}} = \mathbb{R} \cup \{-\infty, +\infty\}$
$f'(x), df(x)/dx$	The first derivative of the uni-variable function $f : \mathbb{R} \to \mathbb{R}$
$f^{(n)}(x)$	The nth order derivative of the uni-variable function $f : \mathbb{R} \to \mathbb{R}$, $n \in \mathbb{N}^*$

$\frac{\partial f(x)}{\partial x_i}$	For a multi-variable function $f:\mathbb{R}^n \to \mathbb{R}$, the first partial derivatives of f with respect to its ith variable, $i=1\ldots n>1$
$\nabla f(x)$	For a multi-variable function $f:\mathbb{R}^n \to \mathbb{R}$, the gradient vector $\nabla f(x) = [\,\partial f(x)/\partial x_1 \;\ldots\; \partial f(x)/\partial x_n]^T$ evaluated at the point x
$\nabla_y f(x,y)$, $\nabla_{[t]} f(x)$	For a multi-variable function $f:\mathbb{R}^n \times \mathbb{R}^m \to \mathbb{R}$, the vector of partial derivatives $\partial f(x,y)/\partial y_i$, $i=1\ldots m$ For a multi-variable function $f:\mathbb{R}^n \to \mathbb{R}$, the vector of all first partial derivatives $\partial f(x)/\partial x_i,\;\; i \in t$
$\nabla^2 f(x)$	For a multi-variable function $f:\mathbb{R}^n \to \mathbb{R}$, the Hessian matrix $\nabla^2 f(x) = \begin{bmatrix} \partial^2 f(x)/\partial x_1^2 & \ldots & \partial^2 f(x)/\partial x_1 \partial x_n \\ \vdots & \ddots & \vdots \\ \partial^2 f(x)/\partial x_n \partial x_1 & \ldots & \partial^2 f(x)/\partial x_n^2 \end{bmatrix}$
$\int_a^b f(x)dx$	The Riemann integral of the uni-variable function $f(x)$ over the interval $[a,b] \subseteq \bar{\mathbb{R}}$
ℓ_k	The k-norm
$\lVert x \rVert_k$	The quantity $\left[\sum_{j=1}^{n} (x_j)^k\right]^{\frac{1}{k}}$ where $x \in \mathbb{R}^n$ is an n-dimensional vector. If the subscript k is omitted, k defaults to two
pdf, cdf	Probability density function, cumulative distribution function
$E[X]$, $Var[X]$	Expected value and variance of a random variable X
$o(x_n)$(little-o)	A sequence y_n is said to be $o(x_n)$ where x_n is another sequence if $\lim \frac{y_n}{x_n} = 0$
$o(x)$	A function $g = o(x)$ if $\lim_{x \to 0} \frac{g(x)}{x} = 0$ i.e. the function g tends to zero faster than x
$O(x_n)$(big-O)	A sequence y_n is said to be $O(x_n)$ where x_n is another sequence if $\exists c > 0, k_0 \geq 0 : \forall n \geq k_0, 0 \leq \lvert y_n \rvert \leq c x_n$

Chapter 1
A Review of Optimization Methods

This introductory chapter reviews a few selected major methods for optimization and mathematical modeling available today. The topic is extremely broad, so the focus is on methods that not only have sound theoretical ground, but more importantly, exhibit good performance in practice. Results from optimization theory and convex analysis—such as theorems of the alternative—are only mentioned and used to prove the necessary and sufficient conditions for mathematical programming. On the other hand, algorithms are described in enough detail so that the reader can understand under what circumstances and why they should be expected to work well in practice and also implement them from scratch if the need arises.

The objective of a generic optimization problem is to find the minimum (or maximum) value of a real-valued function $f(x)$ subject to a set of constraints that restrict the values that the variables x can take on. The set of values S with the property that any and only those $x \in S$ obey the constraints of the problem is known as the *feasible set* of the problem. The variables x are known as "decision variables" of the problem, and the problem is denoted as $\min_{x \in S} f(x)$. When there are no constraints on the values of the variables, the problem $\min f(x)$ belongs to the class of *unconstrained optimization* problems, whereas in the general case, the problem belongs to the class of *constrained optimization problems*. It should not be surprising that most engineering problems—and even most scientific problems—turn out to be optimization problems. Certainly, a large part of the problems encountered in Supply Chain Management are best modeled as optimization problems, and for this reason, optimization plays a key and central role in this field. As real-life teaches, in general, optimization problems constrained or otherwise, are intractable problems, wherein there is likely no way to ever completely solve them efficiently.

Despite the fact that in general, solving to optimality an optimization problem can be extremely difficult, there are classes of such problems for which efficient algorithms exist that guarantee to find the globally optimal solution. Such classes

I. T. Christou, *Quantitative Methods in Supply Chain Management*,
DOI: 10.1007/978-0-85729-766-2_1,

include *Linear Programming* and a more general class, known as *Convex Optimization*. We review some of the basic results for these classes.

For more general problems, such as *Discrete and Integer Optimization* problems where some or all the decision variables are constrained to take on values from a countable set C, finding the *globally optimal solution* and proving it is usually out of reach with the current state of computer and information technology. As a result, one is usually content with finding a "good", "acceptable" *local* optimum.

1.1 Continuous Optimization Methods

1.1.1 Unconstrained Optimization

One of the most strategic decisions that a manufacturing executive has to make is to decide on the location of a company's manufacturing plant. This decision is of high importance, because once made, it essentially cannot be "undone" without incurring extremely high costs that can drive the company even to bankruptcy. In an idealized and very simple model for such a case, the costs that matter the most would be the fixed cost of purchasing a plot and the transportation costs of transferring the products of the company to the markets it is going to serve. Such transportation costs can be considered roughly proportional to the sum of the distances from the plant to its markets. If the n markets to be served by the plant are located on a two-dimensional map in positions (x_i, y_i) $i = 1,\ldots,n$ and if the cost of a plot in location (x, y) is given by a cost function $c(x,y)$, the executive should locate the plant at coordinates (x^*, y^*) that *minimizes* the function $f(x,y) = c(x,y) + a\sum_{i=1}^{n}\left((x-x_i)^2+(y-y_i)^2\right)^{1/2}$ where a is the average fuel cost per unit distance over the time-horizon the plant will be operational. The subject of this section is to review the conditions, methods, and algorithms for solving such *unconstrained* problems.

Consider the generic problem of minimizing a multi-variable single-valued differentiable function $f : \mathbb{R}^n \to \mathbb{R}$. We start with a few basic definitions. A point $x^* \in \mathbb{R}^n$ is said to be a *local minimizer* (respectively maximizer) for f iff for any point x in an appropriately small neighborhood of x^*, $N(x^*) = \{x \in \mathbb{R}^n \,|\, \|x - x^*\| < \delta\}$, $\delta > 0$, the inequality $f(x^*) \leq f(x)$ (respectively $f(x^*) \geq f(x)$) holds. The $\|.\|$ notation indicates the ℓ_2 norm of a vector unless otherwise specified. A point x^* is said to be a *strict* local minimizer (respectively maximizer) for f iff for any point $x \neq x^*$ in an appropriately small neighborhood of the point x^*, $N(x^*) = \{x \in \mathbb{R}^n \,|\, \|x - x^*\| < \delta\}$, $\delta > 0$ the inequality $f(x^*) < f(x)$ (respectively $f(x^*) > f(x)$) holds. Finally, the point x^* is an *isolated* local minimizer (respectively maximizer) for f if it is the only local minimizer (respectively maximizer) in an appropriately small neighborhood of x^*.

Fermat's (little) theorem guarantees that if x^* is a local minimum (or maximum) for f, then $\nabla f(x^*) = 0$, thus providing a necessary—but not sufficient—condition to check for a candidate optimizer of a function.

Theorem 1.1 (Fermat's Theorem) *Let* $f: \mathbb{R}^n \to \mathbb{R}$ *be a differentiable function in a neighborhood of a point* x^*. *If f has a local minimum or local maximum at* x^*, *then* $\nabla f(x^*) = 0$.

Proof Without loss of generality, assume that x^* is a local minimum for f. Assume also, by contradiction that $\nabla f(x^*) \neq 0$. Then, at least one component of the gradient of $f(x)$ at x^* is non-zero, say component i. Then, depending on the sign of $\partial f(x^*)/\partial x_i$ being positive or negative, the function $g(t) = f(x_1^*, \ldots, x_{i-1}^*, t, x_{i+1}^*, \ldots, x_n^*)$ is increasing or decreasing in an appropriately small neighborhood of x_i^*, which means that for every $\varepsilon > 0$ small enough, the inequality $f(x_1^*, \ldots, x_{i-1}^*, x_i^* - \varepsilon, x_{i+1}^*, \ldots, x_n^*) < f(x^*)$ must hold, or else (for negative partial derivative sign), we must have that $f(x_1^*, \ldots, x_{i-1}^*, x_i^* + \varepsilon, x_{i+1}^*, \ldots, x_n^*) < f(x^*)$, therefore x^* cannot be a local minimum for f, which is a contradiction. So, if x^* is a local minimum for f, $\nabla f(x^*) = 0$. QED.

Fermat's theorem, as mentioned above provides only a necessary condition that any point must satisfy in order to qualify as a local optimizer for a function. In general, any point where the gradient of a function vanishes is called a "*stationary point*" for the function (sometimes also referred to as "*saddle* point"). A second-order sufficiency condition can guarantee that the point in question is indeed a local minimizer or local maximizer for a function. Second-order conditions require the notion of matrix Positive Definiteness and Positive Semi-Definiteness. In particular, we have the following

Definition 1.2 An $n \times n$ matrix A is Positive Definite (P.D.) if for any n-dimensional column vector $x \neq 0$, the quantity $x^T A x > 0$. The matrix A is called Positive Semi - Definite (P.S.D.) when for any n-dimensional column vector x, $x^T A x \geq 0$.

Theorem 1.3 (Second-Order Sufficient Conditions for Unconstrained Local Minimizer) *Let* $f : \mathbb{R}^n \to \mathbb{R}$ *be a twice continuously differentiable function in a neighborhood of a point* x^*. *If* $\nabla f(x^*) = 0$ *and the Hessian matrix* $\nabla^2 f(x^*) = \left[\frac{\partial^2 f}{\partial x_i \partial x_j}(x^*)\right]_{i=1\ldots n, j=1\ldots n}$ *of the function f is Positive Definite (P.D.) at* x^* *then f has an isolated strict local minimum at* x^*.

Proof Consider any n-dimensional vector $s \neq 0$. Expanding the function around x^* along s in a second-order Taylor series (Apostol 1962), we get

$$\begin{aligned} f(x^* + hs) &= f(x^*) + hs^T \nabla f(x^*) + \frac{1}{2} h^2 s^T \nabla^2 f(x^*) s + o(h^2) \\ &= f(x^*) + \frac{1}{2} h^2 s^T \nabla^2 f(x^*) s + o(h^2) \end{aligned}$$

where $h \in \mathbb{R}$ is any real number, and $o(x)$ (known as the *little-o notation*) is a function with the property that $\lim_{x \to 0} o(x)/x = 0$ (i.e. goes to zero faster than its argument). Therefore,

$$f(x^* + hs) - f(x^*) = \frac{1}{2} h^2 s^T \nabla^2 f(x^*) s + o(h^2)$$

Dividing both sides of the above equation by h^2 and taking the limit as $h \to 0$, we get $[f(x^* + hs) - f(x^*)]/h^2 \to \frac{1}{2} s^T \nabla^2 f(x^*) s > 0$ as the Hessian is P.D. at x^*. This implies that $f(x^* + hs) - f(x^*)$ remains positive for all s and all appropriately small step-sizes h, thus x^* is a strict local minimum for f.

To prove that x^* is an *isolated* local minimizer, assume by contradiction that it is not. Then, there exists a sequence of points $x_{(n)} = x^* + h_n s_{(n)}$, where $h_n \in \mathbb{R}, s_{(n)} \in \mathbb{R}^n, \|s_{(n)}\| = 1, h_n \to 0$ and whose limit is x^*, of local minimizers of f. Expanding the gradient of f around x^*, we obtain

$$\nabla f(x_{(n)}) = \nabla f(x^* + h_n s_{(n)}) = \nabla f(x^*) + h_n \nabla^2 f(x^*) s_{(n)} + o(h_n)$$

from which we obtain after multiplying both sides of the equality by $h_n s_{(n)}$ from the left,

$$h_n s_{(n)}^T \nabla f(x_{(n)}) = h_n s_{(n)}^T \nabla f(x^*) + h_n^2 s_{(n)}^T \nabla^2 f(x^*) s_{(n)} + o(h_n^2)$$

Because $x_{(n)}$ and x^* are local minimizers of f, we have that $\nabla f(x^*) = \nabla f(x_{(n)}) = 0$, so the above equality becomes $0 = h_n^2 s_{(n)}^T \nabla^2 f(x^*) s_{(n)} + o(h_n^2)$, and dividing by h_n^2 and taking the limit as $n \to \infty$ we get $\lim s_{(n)}^T \nabla^2 f(x^*) s_{(n)} = 0$ which is a contradiction because $\|s_{(n)}\| = 1$ for all n, and the Hessian matrix of f at x^* is P.D. (because according to Theorem 1.5, the Hessian matrix can be written as $P \Lambda Q$ where Q is the transpose of matrix P and Λ is a diagonal matrix, so the sequence of the direction vectors $s_{(n)}$ multiplied from the left by Q tends to zero implies that the direction vectors $s_{(n)}$ must also tend to zero, a contradiction) QED.

The conditions require the Hessian to be positive definite. Indeed, if the matrix is only positive semi-definite, then for some direction s, it is possible that the product $s^T \nabla^2 f(x^*) s = 0$ and the difference $f(x^* + hs) - f(x^*)$ are not guaranteed to be bounded above zero for small enough h. Yet, positive semi-definiteness is a second-order necessary condition for the existence of a local minimum at x^*.

Theorem 1.4 (Second-Order Necessary Conditions for Unconstrained Local Minimizer). *Let $f : \mathbb{R}^n \to \mathbb{R}$ be a twice continuously differentiable function in a neighborhood of a point x^*. If f has a local minimum at x^*, then $\nabla f(x^*) = 0$ and $\nabla^2 f(x^*)$ is positive semi-definite.*

Proof From Fermat's theorem we know that $\nabla f(x^*) = 0$. To see that $\nabla^2 f(x^*)$ is P.S.D., expand the function around x^* along any direction s with a step-size h in a second-order Taylor series to get as before,

$$f(x^* + hs) - f(x^*) = \frac{1}{2}h^2 s^T \nabla^2 f(x^*) s + o(h^2)$$

Now, since x^* is a local minimizer for f, this implies that the left-hand side of the above equation remains non-negative for all h small enough. Dividing both sides of the equation by h^2 and letting $h \to 0$, the left-hand side of the equation remains non-negative, and the right-hand side becomes the quantity $\frac{1}{2}s^T \nabla^2 f(x^*) s$ which must therefore be non-negative for all s. Therefore, $\nabla^2 f(x^*)$ is P.S.D. QED.

It is worth pointing out at this point that one can determine whether a matrix is positive definite by examining the matrix's eigenvalues. In particular, the following theorem establishes the Positive definiteness of a Hessian matrix in terms of the sign of the matrix's eigenvalues.

Theorem 1.5 *A symmetric $n \times n$ real matrix A is P.D. iff all its eigenvalues are positive.*

Proof From the standard linear algebra, we can decompose the matrix A as $A = P\Delta P^T$ where P is an orthonormal matrix whose columns are eigenvectors of A, and Δ is the diagonal matrix whose diagonal elements are the eigenvalues of A. Assuming that all eigenvalues of A are positive implies that for every column vector $s \neq 0$, $s^T A s = s^T P \Delta P^T s = p^T \Delta p = \sum_{i=1}^n \lambda_i p_i^2 > 0$ where $p = P^T s$, and λ_i are the eigenvalues of A. Therefore, matrix A is P.D.

Vice versa, if matrix A is P.D., then for every column vector $s \neq 0$, $s^T A s = s^T P \Delta P^T s = p^T \Delta p = \sum_{i=1}^n \lambda_i p_i^2 > 0$ which implies that all λ_i are positive, as otherwise, if say the jth eigenvalue was non-positive, by choosing $s = Pe_j$, where e_j is the unit vector in the jth dimension, we would have $s^T A s \leq 0$, a contradiction. Thus if matrix A is symmetric P.D., its eigenvalues are all positive. QED.

The first and second-order conditions determine whether a candidate point is a local minimizer for a multi-variable function f. One technique to find such a point then could be to solve the system of the n nonlinear equations with n variables $x_1, x_2, \ldots, x_n$ formed by "setting the derivative to zero": $\nabla f(x) = 0$, and then check whether the second-order conditions hold at a solution. Unfortunately, solving a system of n nonlinear equations of n variables is no less a complex task than finding directly a local optimum of a function. Nevertheless, one easy choice for solving a nonlinear system of equations of the form

$$\left[\frac{\partial f}{\partial x_1} \quad \cdots \quad \frac{\partial f}{\partial x_n}\right]^T = 0$$

could be to employ a Newton-type method, assuming second-order derivatives of the function f are available: let $H_{(k)}$ denote the Hessian of f, $\nabla^2 f(x_{(k)})$, evaluated at the kth iterate point $x_{(k)}$; then, in the next iteration, solve the $n \times n$ linear system $H_{(k)} s = -\nabla f(x_{(k)})$ to obtain $s_{(k)}$, and set $x_{(k+1)} = x_{(k)} + s_{(k)}$, and repeat until some convergence criterion is satisfied. Such a method (known as a Gauss–Newton method) is used in the generalized least squares method (Fletcher 1987). Usually, it is required that the initial estimate $x_{(1)}$ is a good estimate, i.e. it is sufficiently close to the optimizer.

The remainder of this section examines some of the most successful methods that directly attempt to optimize (at least locally) a function. Usually these methods converge at least to a stationary point of f, and often to a point satisfying the second-order necessary conditions of Theorem 1.4. The methods we shall review first are iterative methods which generate a sequence of points $\{x_{(k)}\ k = 1,2, \ldots\}$ that converge to a point x^* satisfying some order of necessary conditions.

1.1.1.1 Line-Search Type Methods

A line-search type algorithm for the unconstrained optimization of a multi-variable function $f(x)$, is an iterative algorithm where at each iteration a direction of search is determined $s_{(k)}$ (the line) and the program moves from the current iterate point $x_{(k)}$ along the search direction by an appropriately selected step-size $h_{(k)} \in \mathbb{R}$ so as to (approximately) minimize the *one variable* function $f(t) = f(x_{(k)} + ts_{(k)})$. Thus a generic line-search type algorithm looks like this:

Algorithm Line-Search Optimization

Inputs: Function $f(x)$, initial point $x_{(1)}$, criterion for termination
Outputs: A point x^* satisfying termination criteria

Begin

1. Set $k = 1$, $x = x_{(k)}$.
2. while x does not satisfy the termination criteria do
    a. Determine a line-search direction $s_{(k)}$
    b. Determine a minimizer $h_{(k)}$ of the uni-variable function $f(t) = f(x + ts_{(k)})$.
    c. Set $x = x + h_{(k)}s_{(k)}$, $k = k+1$.
3. end-while

End

The algorithm above of course is rather generic and in fact represents a whole *family* of algorithms; the most important choices to be made in a particular instantiation of a concrete algorithm from this family are in the method used to determine the direction of search (step 2.a), the method used to determine the step-size at each iteration (step 2.b), and to a lesser extent the termination criterion for the algorithm. Termination criteria usually include a test of change on the components of the iterate points $x_{(k)}$, such as a requirement that $\left|\left[x_{(k)}\right]_i - \left[x_{(k+1)}\right]_i\right| \leq \varepsilon,\ \forall i = 1, \ldots, n$ and $f\left(x_{(k)}\right) - f\left(x_{(k+1)}\right) \leq \varepsilon'$ for some user-defined tolerances ε & ε'. If the derivative is known—or can be well estimated using for example a first-order central difference scheme—the test $\left\|g\left(x_{(k)}\right)\right\| \leq \varepsilon$ is sometimes used, but it lacks some nice invariance properties that the above tests have.

Methods that choose the search direction s so that the algorithm is guaranteed to improve the objective function value at each iteration (and subsequently, terminates when *no* such improvement in objective function value is possible) are known as *iterative improvement* methods. For minimization problems, the name iterative *descent* method is often used for such algorithms. From now on, unless otherwise stated, *we will be concerned with minimization problems only*, because any problem of maximizing a function $f(x)$ can be stated in terms of the equivalent minimization problem of minimizing the function $-f(x)$: max. $f(x) = -\text{min.}\{-f(x)\}$. When a function f is differentiable determining whether a search direction is a descent direction is easy. First, we need to determine what we mean by a descent direction.

Definition 1.6 A column vector $s \in \mathbb{R}^n$, $\|s\| = 1$ is called a descent direction for a function f at a point x iff $\exists \varepsilon > 0 : \forall 0 < h < \varepsilon, \quad f(x + hs) < f(x)$.

Now, we can state the following lemma:

Lemma 1.7 *A unit-norm column vector s is a descent direction for a differentiable function f at a point x if* $s^T \nabla f(x) < 0$.

Proof By expanding the function f along direction s in a first-order Taylor series we get $f(x + hs) - f(x) = hs^T \nabla f(x) + o(h)$. Assume that$s^T \nabla f(x) < 0$. Dividing the equation by $h > 0$, for all h small enough, the right-hand side is negative, and therefore the left-hand side must be negative too, therefore s is a descent direction. QED.

In light of the above lemma, any direction that forms a negative inner product with the gradient vector $\nabla f(x_{(k)})$ at the current iterate point $x_{(k)}$ is a descent direction, and can potentially be used in a descent type algorithm. The obvious choice $s_{(k)} = -\nabla f(x_{(k)})$—which is guaranteed to be a descent direction, since $-\nabla f(x_{(k)})^T \nabla f(x_{(k)}) = -\|\nabla f(x_{(k)})\|^2 < 0$—leads to the *steepest descent* method. The name "steepest descent" arises from the fact that at the *extreme vicinity of the point* $x_{(k)}$, the direction opposite to the gradient vector at that point provides the maximum decrease rate for the objective function f. Unfortunately, this fact does not mean that the steepest descent direction is the best direction to move along so as to minimize the function. In fact, it is very often a very poor choice for a descent direction, and the steepest descent method has a reputation of being both inefficient and unreliable (Fletcher 1987). Nevertheless, it should be pointed out that this unreliable and ill-reputed method of steepest descent has often been successfully applied to large-scale complex problems whose structure rendered more robust methods unsuitable for a variety of reasons (Bertsekas 1995).

The line-search sub-problem that forms step 2.b in the line-search algorithm above is the second major issue in line-search type algorithms. In some cases (and in many theoretical results in the early optimization literature) the problem can be solved *exactly* so that the *global minimizer* of the function $f(t) = f(x_{(k)} + ts_{(k)})$ is returned, in which case the line-search algorithm is known as an "*exact line-search*" method. Note that when exact line-search is possible, and is employed in conjunction with a steepest descent choice for the search direction, the search direction $s_{(k+1)}$ is orthogonal to the previous search direction $s_{(k)}$ for all k. This is obvious because since an exact line-search is employed, by Fermat's theorem we must have that

$$0 = f'(a_{(k)}) = f'(x_{(k)} + ts_{(k)})|_{t=a_{(k)}}$$
$$= \lim_{h \to 0} \frac{f(x_{(k)} + (a_{(k)} + h)s_{(k)}) - f(x_{(k)} + a_{(k)}s_{(k)})}{h} \triangleq f'(x_{(k)} + a_{(k)}s_{(k)} : s_{(k)})$$

where $f'(x{:}y)$ is the notation for the directional derivative of f at x along the direction y and is equal to $y^T \nabla f(x)$ (from standard many-variable differential calculus, e.g. Apostol 1962). Therefore, $s_{(k)}^T \nabla f(x_{(k)} + a_{(k)}s_{(k)}) = s_{(k)}^T \nabla f(x_{(k+1)}) = 0$, and since in the steepest descent method, the search direction at any iterate point $x_{(k)}$ is the negative of the gradient at that point, we have $s_{(k)}^T s_{(k+1)} = 0$.

Since it is very rarely possible to solve the line-search problem exactly—because in the general case, it amounts to finding all the roots of a nonlinear equation which cannot be implemented with a finite number of operations—the problem is usually solved approximately. One easy implementation of an "approximate line-search" method would be to sample the function f along the line $x_{(k)} + ts_{(k)}$ for a number of progressively increasing t values, and select the one with the smallest function value. However, much better approaches exist for selecting a reasonable step-size value $a_{(k)}$. The best methods for solving the approximate line-search sub-problem are based on the idea that a good step-size should enforce reasonable progress to be made at each step. Such progress cannot be guaranteed by simply requiring the step-size to be such that $f(x_{(k+1)}) < f(x_{(k)})$ because this allows the possibility for arbitrarily slow rate of convergence, or even the non-convergence to a stationary point. But consider the least positive value $\bar{a}_{(k)}$ for which $f(x_{(k)}) = f(x_{(k)} + \bar{a}_{(k)}s_{(k)})$. The sequence of values $\{a_{(k)}\}$ should then be chosen so as to be bounded away from zero as well as $\bar{a}_{(k)}$. This can be guaranteed by the *Wolfe–Powell conditions* that require the chosen value $a_{(k)}$ to satisfy the inequalities

$$\begin{aligned} &f(x_{(k)} + a_{(k)}s_{(k)}) \le f(x_{(k)}) + \rho a_{(k)} s_{(k)}^T \nabla f(x_{(k)}) \\ &s_{(k)}^T \nabla f(x_{(k)} + a_{(k)}s_{(k)}) \ge \sigma s_{(k)}^T \nabla f(x_{(k)}) \\ &\rho \in (0,1), \quad \sigma \in (\rho, 1) \end{aligned} \tag{1.1}$$

where $s_{(k)}$ is a descent direction for f at $x_{(k)}$ and ρ and σ are user-supplied parameters to the algorithm employing the approximate line-search. Any number $a > 0$ that satisfies the above inequalities is a valid candidate point for an approximate line-search scheme.

The Wolfe–Powell conditions (or other similar ones, such as those set forth by Goldstein (1965)) allow the following major global convergence result to be established, which guarantees the convergence of any gradient descent type algorithm employing an approximate line-search to a stationary point of the function f. By global convergence, we mean that the choice of the initial iterate point $x_{(1)}$ is irrelevant for the convergence of the algorithm to a point satisfying the first-order necessary conditions.

Theorem 1.8 (Global Convergence Theorem) *Every iterative descent type algorithm for the unconstrained minimization of a differentiable function f with uniformly continuous derivative $g(x) = \nabla f(x)$ that employs an approximate line-search that obeys the Wolfe–Powel conditions (1.1) and is such that it picks its search directions $s_{(k)}$ so that they are always uniformly bounded away from becoming orthogonal to the gradient vector $\nabla f(x_{(k)})$ at the corresponding iterate point $x_{(k)}$ is globally convergent so that either there exists a $k > 0$ such that $\nabla f(x_{(k)}) = 0$, or $\nabla f(x_{(k)}) \to 0$, or else the sequence $f(x_{(k)}) \to -\infty$.*

Proof Assume that $\forall k, g_{(k)} = \nabla f(x_{(k)}) \neq 0$ and also that the sequence $f(x_{(k)})$ is bounded from below, so that it cannot go to negative infinity. We shall show that the sequence $g_{(k)} \to 0$. First, observe that the sequence of values $\{f(x_{(k)}):k = 1,2,\ldots\}$ is indeed convergent: by the first Wolfe–Powell condition $f(x_{(k+1)}) \leq f(x_{(k)}) + \rho a_{(k)} s_{(k)}^T g_{(k)}$ since $\rho > 0$ and $s_{(k)}$ is a descent direction, we have that $f(x_{(k+1)}) \leq f(x_{(k)})$ so that the sequence is decreasing and bounded from below and therefore converges to a point f^*. Also from the first Wolfe–Powell condition and the fact the direction $s_{(k)}$ is a descent direction we have that $0 \leq -\rho a_{(k)} g_{(k)}^T s_{(k)} \leq f(x_{(k)}) - f(x_{(k+1)}) \to 0$ so that either $g_{(k)} \to 0$, or $a_{(k)} \to 0$ or else $g_{(k)}$ becomes orthogonal to $s_{(k)}$ in the limit. Since by hypothesis of the theorem the latter cannot happen, either $g_{(k)} \to 0$, or $a_{(k)} \to 0$. Assume by contradiction that $g_{(k)} \not\to 0$. This implies (i) that $a_{(k)} \to 0$ and therefore the sequence of iterate points $x_{(k)}$ which satisfies $x_{(k+1)} = x_{(k)} + a_{(k)} s_{(k)}$ also satisfies $x_{(k+1)} - x_{(k)} = a_{(k)} s_{(k)} \to 0$ and (ii) that there must exist an $\varepsilon > 0$ and a subsequence L of indices k such that $\|g_{(k)}\| \geq \varepsilon \quad \forall k \in L$. Now, from the descent property of the search directions and the second Wolfe–Powell condition, we have the inequalities $0 \leq -g_{(k)}^T s_{(k)} \leq \frac{(g_{(k+1)} - g_{(k)})^T s_{(k)}}{1-\sigma}$.

From the Cauchy–Swartz inequality ($x^T y \leq \|x\| \|y\| \quad \forall x, y \in \mathbb{R}^n$) we then obtain $0 \leq \frac{-g_{(k)}^T s_{(k)}}{\|s_{(k)}\|} \leq \frac{\|g_{(k+1)} - g_{(k)}\|}{1-\sigma}$(with $\|s_{(k)}\| = 1$ since the s's are direction vectors). Dividing by $\|g_{(k)}\|$ for those indices k in the subsequence L, for which $\|g_{(k)}\| \geq \varepsilon$ we get $0 \leq -\frac{g_{(k)}^T s_{(k)}}{\|s_{(k)}\| \|g_{(k)}\|} \leq \frac{\|g_{(k+1)} - g_{(k)}\|}{(1-\sigma)\varepsilon}$.

But the norm $\|g_{(k+1)} - g_{(k)}\| \to 0$ because $x_{(k+1)} - x_{(k)} \to 0$ and $g(x)$ is a uniformly continuous function. Therefore, for the indices k in the subsequence L, the quantity $\frac{-g_{(k)}^T s_{(k)}}{\|g_{(k)}\| \|s_{(k)}\|} = \cos \vartheta_{(k)} \to 0$ where $\vartheta_{(k)}$ is the angle between the vectors $s_{(k)}$ and $g_{(k)}$. This is a contradiction because it is a hypothesis of the theorem that the algorithm picks its search directions in such a way that they are always bounded away from becoming orthogonal to the gradient vector $g_{(k)} = \nabla f(x_{(k)})$. QED.

Many different versions of the above global convergence theorem have been presented in the literature. Using the same proof technique as above, it is possible to show that if the derivative $g(x)$ of the function $f(x)$ is a contraction mapping on the level set $\{x : f(x) < f(x_{(1)})\}$, and if the search directions are chosen so that the angles $\theta_{(k)}$ they form with the gradient vectors at the iterate points satisfy the condition $\sum_{k=1}^{\infty} \cos^2 \theta_{(k)} = \infty$, and if f is bounded from below, then $g(x_{(k)}) = 0$ for some k, or else there exists a sub-sequence L of indices k on which $g(x_{(k)})$ converges to zero. The condition that the series of the square of the cosines of the angles that form between the search directions and the gradients diverges is clearly a much more relaxed condition than the requirement that the angles satisfy a condition such as $\theta_{(k)} \leq \frac{\pi}{2} - \varepsilon$ for all $k > 0$ for some positive ε.

One rather elegant and successful algorithm for finding such an acceptable step-size h that satisfies (1.1), is a two-phase algorithm, where a bracketing phase in which an interval known to contain an acceptable point is determined, is succeeded by a sectioning phase in which the algorithm zooms in inside the bracket, creating smaller and smaller intervals containing the acceptable point it seeks to find. The algorithm is guaranteed to converge to an acceptable point (Fletcher 1987) that satisfies the first Wolfe–Powell condition and the *stronger* two-sided condition $\left|s_{(k)}^T \nabla f(x_{(k)} + a_{(k)} s_{(k)})\right| \leq -\sigma s_{(k)}^T \nabla f(x_{(k)})$.

Algorithm Approximate Line-Search Optimization

Inputs: A differentiable function $f(x)$ and its derivative $g(x)$, a current iterate point x, a descent search direction s, and a threshold value $f_{\min}$ that the user is prepared to accept as value for $f(x_{(k+1)})$, and parameters $\tau_1 > 1$, τ_2, τ_3 satisfying $0 < \tau_2 < \tau_3 \leq 0.5$, and ρ in (0,1) and σ in $(\rho,1)$.

Outputs: An acceptable step-size h.
/* A. Bracketing Phase */

Begin

1. Set $\mu = (f_{min} - f(x))/(\rho g(x)^T s)$, $a_0=0$, $a_1 = \mu$.
2. for i=1,2,... do

   a. Set $f = f(x+a_i s)$.
   b. if $f \leq f_{min}$ then return $h = a_i$.
   c. if $f > f(x)+\rho a_i g(x)^T s$ OR $f \geq f(x+a_{i-1}s)$ then
      i. Set $A_i = a_{i-1}$, $B_i = a_i$.
      ii. break.
   d. end-if

e. Set $f' = g(x+a_i s)^T s$.
f. if $|f'| \leq -\sigma g(x)^T s$ then return $h = a_i$.
g. if $f' \geq 0$ then
   i. Set $A_i = a_i$, $B_i = a_{i-1}$.
   ii. break.

h. end-if
i. if $\mu \leq 2a_i - a_{i-1}$ then $a_{i+1} = \mu$ else $a_{i+1} = [2a_i - a_{i-1} + \min\{\mu, a_i + \tau_1(a_i - a_{i-1})\}]/2$.

3. end-for

/* B. Sectioning Phase */

4. for $j = i, i+1, \ldots$ do

a. Set $a_j = [A_j + B_j + (B_j - A_j)(\tau_2 - \tau_3)]/2$.
b. Set $f = f(x+a_j s)$.
c. if $f > f(x) + \rho a_j g(x)^T s$ OR $f \geq f(x+A_j s)$ then
   i. Set $A_{j+1} = A_j$, $B_{j+1} = a_j$
d. else
   i. Set $f' = g(x+a_j s)^T s$.
   ii. if $|f'| \leq -\sigma g(x)^T s$ return $h = a_j$.
   iii. Set $A_{j+1} = A_j$.
   iv. if $(B_j - A_j) f' \geq 0$ then Set $B_{j+1} = A_j$ else Set $B_{j+1} = B_j$.

e. end-if

5. end-for

End.

Al-Baali and Fletcher (1986) proved that the above algorithm must necessarily terminate in a finite number of iterations with an acceptable step-size h if the parameters σ and ρ satisfy $\sigma > \rho$.

An easier algorithm to implement for choosing an appropriate step-size h is known with the name "Armijo Rule" (Armijo 1966). The rule is in line with the ideas discussed above for adequate objective function value reduction at each step. It requires two user-supplied values: a positive base value β in (0,1) and a scaling factor γ. The rule then chooses as step-size h the largest value $\beta^m \gamma$ where m is a non-negative integer such that

$$f\left(x_{(k)} + \beta^m \gamma s_{(k)}\right) \leq f\left(x_{(k)}\right) + \rho \beta^m \gamma s_{(k)}^T \nabla f\left(x_{(k)}\right)$$

(which is the first Wolfe–Powell condition with $h = \beta^m \gamma$). Default (recommended) values for the parameters are as follows: ρ in $[10^{-5}, 10^{-1}]$, β in [0.1, 0.5], and γ—the initial step-size trial—is usually set to 1. The algorithm although trivial, is shown below in pseudo-code.

Algorithm Armijo Rule

Inputs: A differentiable function $f(x)$, its derivative $g(x)$, current iterate point x, chosen search direction s, user-defined values $\rho > 0$, $\beta > 0$, $\gamma > 0$.
Outputs: An appropriate step-size h.

Begin

1. Set $r_{\text{prev}} = \rho\gamma s^T g(x)$.
2. for $m = 0,1,2\ldots$ do
   a. Set $f = f(x + \beta^m \gamma s)$.
   b. if $f \le f(x) + r_{\text{prev}}$ then return $h = \beta^m \gamma$.
   c. Set $r_{\text{prev}} = \beta r_{\text{prev}}$.
3. end-for

End.

Armijo's Rule is clearly a *successive step-size reduction rule* that starts with an initial step-size γ and reduces it geometrically by multiplying the initial estimate by the factor β^m (with $\beta < 1$) in the mth iteration, until the first Wolfe–Powell condition is satisfied. The following global convergence result based on the same ideas used to prove Theorem 1.8, establishes the validity of any iterative descent line-search algorithm using it.

Theorem 1.9 (Armijo Rule Global Convergence Theorem). *Any descent algorithm for the unconstrained minimization of a differentiable function $f(x)$ that is bounded below, uses the Armijo Rule for selecting the step-size, and picks its search directions $s_{(k)}$ so that for any sub-sequence L of the sequence $\{x_{(k)}\ k = 1,2,\ldots\}$ that converges to a non-stationary point $x' : \nabla f(x') \neq 0$ the corresponding subsequence $\{s_{(k)}\}_{k\in L}$ is bounded and satisfies $\limsup_{k\in L, k\to\infty} s_{(k)}^T \nabla f(x_{(k)}) < 0$, is globally convergent in the sense that every limit point x^* of $\{x_{(k)}\ k=1,2,\ldots\}$ satisfies $\nabla f(x^*) = 0$.*

Proof The proof is by contradiction. Assume that there exists a limit point of the sequence $\{x_{(k)}\ k = 1,2,\ldots\}$, say $\bar{x}$ which is non-stationary, i.e. $\nabla f(\bar{x}) \neq 0$, and let L be a subsequence of that sequence converging to $\bar{x}$. Since the algorithm is an iterative descent algorithm, the sequence $\{f(x_{(k)})\ k = 1,2,\ldots\}$ is non-increasing, and since $f(x)$ is bounded below, the sequence converges, which from the Armijo Rule implies that $0 \le -\rho h_{(k)} s_{(k)}^T \nabla f(x_{(k)}) \le f(x_{(k)}) - f(x_{(k+1)}) \to 0$ where $h_{(k)} = \beta^m \gamma$ for some non-negative integer m, so that $h_{(k)} s_{(k)}^T \nabla f(x_{(k)}) \to 0$ and since by hypothesis $\limsup_{k\in L, k\to\infty} s_{(k)}^T \nabla f(x_{(k)}) < 0$, we must have that the subsequence $\{h_{(k)}\}_{k\in L}$ converges to 0. But since the step-sizes in the subsequence L converge to zero, there must exist an index k_0 such that for every k in L, $k \ge k_0$: $f\left(x_{(k)} + \frac{h_{(k)}}{\beta} s_{(k)}\right) > f(x_{(k)}) + \rho \frac{h_{(k)}}{\beta} s_{(k)}^T \nabla f(x_{(k)})$ because since $h_{(k)} \to 0$ for k in L, the step-size must be reduced at least once for all k in L. Now, since the vectors $s_{(k)}$

are direction vectors that satisfy $\|s_{(k)}\| = 1, \quad \forall k = 1, 2, \ldots$, we note that there exists a sub-subsequence L' of the subsequence L such that $s_{(k)} \underset{k \in L'}{\longrightarrow} s'$ for some direction vector $s', \quad \|s'\| = 1$. This is because *every* bounded (above *and* below) sequence of n-dimensional vectors $y_{(k)}$ has at least one limit point in $\mathbb{R}^n$, and the sequence of unit-length direction vectors $s_{(k)}$ is necessarily bounded both above and below. To see this, take any component of the vectors $s_{(k)}$, say component i, and observe that the sequence $\{s_{(k)_i}, \quad k = 1, 2, \ldots\}$ is bounded from below and above. If there is no non-increasing subsequence, then there must exist an increasing sub-sequence, because there must exist an index k_0 such that for every point $s_{(k)_i}$ with $k \geq k_0$ the points in the sequence after it must all be greater (otherwise there would exist a non-increasing subsequence). This subsequence being increasing and bounded from above must converge to its least upper bound, say s'_i. Otherwise, the sequence has a non-increasing subsequence which (since the whole sequence is bounded from below) converges, and thus there again exists a limit point s'_i of the subsequence. From the inequality above we have $\frac{f\left(x_{(k)} + \frac{h_{(k)}}{\beta} s_{(k)}\right) - f\left(x_{(k)}\right)}{\frac{h_{(k)}}{\beta}} > \rho s_{(k)}^T \nabla f(x_{(k)}), \ \forall k \in L', k \geq k_0$.

From the mean-value theorem of differential calculus (Apostol 1962) we have that there exists an $\tilde{h}_{(k)} \in [0, h_{(k)}/\beta]$ such that $s_{(k)}^T \nabla f\left(x_{(k)} + \tilde{h}_{(k)} s_{(k)}\right)$ equals the left-hand side of the above inequality, and therefore we get the inequality $s_{(k)}^T \nabla f\left(x_{(k)} + \tilde{h}_{(k)} s_{(k)}\right) > \rho s_{(k)}^T \nabla f(x_{(k)}), \quad \forall k \in L', k \geq k_0$.

Taking the limit within the subsequence L', from the fact that $h_{(k)} \to 0$ and that $s_{(k)} \underset{k \in L'}{\longrightarrow} s'$ and $x_{(k)} \underset{k \in L}{\longrightarrow} \bar{x}$ we get the inequality $(1 - \rho) s'^T \nabla f(\bar{x}) > 0 \Leftrightarrow s'^T \nabla f(\bar{x}) > 0$.

But from the hypothesis of the theorem we have that for any sub-sequence L of indices k indexing a sub-sequence of the sequence$\{x_{(k)}\ k=1,2,\ldots\}$ that converges to a non-stationary point $x' : \nabla f(x') \neq 0$ the corresponding subsequence $\{s_{(k)}\}_{k \in L}$ is bounded and satisfies $\lim\sup_{k \in L, k \to \infty} s_{(k)}^T \nabla f(x_{(k)}) < 0$. Therefore, we have that $s'^T \nabla f(\bar{x}) = \lim\sup_{k \in L', k \to \infty} s_{(k)}^T \nabla f(x_{(k)}) < 0$ contradicting the previous inequality. QED.

Newton-Like Methods

One of the most important and successful methods for unconstrained optimization is the family of Newton-like methods, which take advantage of second-order information regarding the variation of a function $f(x)$. Consider the second-order Taylor approximation of a function f with gradient $g(x)$ and Hessian matrix $H(x) = \nabla^2 f(x)$ around a point $x_{(k)}$, which can be written as

$$f(x) \approx f\left(x_{(k)}\right) + \left(x - x_{(k)}\right)^T g\left(x_{(k)}\right) + \left(x - x_{(k)}\right)^T H\left(x_{(k)}\right)\left(x - x_{(k)}\right)/2$$

for x sufficiently close to $x_{(k)}$. To minimize the right-hand side of the above approximation, one must minimize the quadratic function $q(s) = s^T g(x_{(k)}) + \frac{1}{2} s^T H(x_{(k)}) s$ which, assuming $H(x_{(k)})$ is Positive Definite, has a unique minimizer s that solves the linear system of equations $H(x_{(k)})s = -g(x_{(k)})$. A rather intuitive algorithm would then be to solve the linear system of equations for s, set $x_{(k+1)} = x_{(k)} + s$ and continue with the next iteration until some of the convergence criteria set forth in the previous pages are satisfied. Thus the generic Newton algorithm can be stated as follows.

Algorithm Newton Optimization

Inputs: A twice continuously differentiable function $f \in C^2$ with positive definite hessian everywhere, its first and second derivatives $g(x)$ and $H(x)$, an initial iterate point $x_{(1)}$, and a convergence criterion.
Outputs: The point x^* is the global minimizer of the function f.

Begin

1. Set $k = 1$.
2. while convergence criteria is not satisfied do
   a. Set $H_{(k)} = H(x_{(k)})$, $g_{(k)} = g(x_{(k)})$.
   b. Solve the linear system $H_{(k)}s = -g_{(k)}$.
   c. Set $x_{(k+1)} = x_{(k)}+s$.
   d. Set $k = k+1$.
3. end-while
4. return $x^* = x_{(k)}$.

End.

If Newton's method is applied to a quadratic function of the form $q(x) = x^T Ax/2 + b^T x$ where A is P.D., the algorithm will converge to the global minimum in a single iteration, as the minimization of a convex quadratic function is equivalent to solving a single linear system of equations. However, when minimizing a generic twice differentiable function $f(x)$ with not necessarily positive definite Hessian matrix at every iterate point $x_{(k)}$, the Newton optimization algorithm can easily fail to converge to a minimizer of the function. The method essentially tries to solve the system $\nabla f(x) = 0$ and because the solution of the system $H_{(k)}s = -g_{(k)}$ may not be a descent direction, so that it may happen that $f(x_{(k+1)}) > f(x_{(k)})$, the method may terminate in the general case with a local maximizer instead! For these reasons, a Newton-based line-search optimization algorithm has been proposed that is more robust and efficient than the pure Newton optimization algorithm.

Algorithm Newton Optimization with Line-Search

Inputs: Function $f(x)$, first derivative $g(x)$, second derivative $H(x)$, initial point $x_{(1)}$, criterion for termination.

Outputs: A point x^* satisfying termination criteria.

Begin

1. Set $k = 1$, $x = x_{(k)}$..
2. while x does not satisfy the termination criteria do:

   a. Solve the system $H(x_{(k)})s_{(k)} = -g(x_{(k)})$
   b. if $g(x_{(k)})^T s_{(k)}>0$ Set $s_{(k)} = -s_{(k)}$.
   c. Determine a minimizer $h_{(k)}$ of the uni-variable function $f(t) = f(x+ts_{(k)})$ (using e.g. Search Optimization or the Armijo Rule Algorithm).
   d. Set $x = x + h_{(k)}s_{(k)}$, $x_{(k+1)} = x$, $k = k+1$.

3. end-while
4. return $x^* = x_{(k)}$.

End.

The above algorithm, although an improvement over the "pure" Newton optimization algorithm, is not without problems. Difficulties arise when the Hessian $H_{(k)}$ is not invertible during an iteration (which will happen if the function f at the vicinity of $x_{(k)}$ is linear). Also, the Hessian matrix $H(x) = \nabla^2 f(x)$ must be available via a user-defined subroutine, which may sometimes pose a significant burden on the user—especially for high-dimensional highly complex functions.

Newton optimization with line-search would then be significantly enhanced if the difficulties with computing the Hessian and solving the linear system $H(x_{(k)})s_{(k)} = -g(x_{(k)})$ could be avoided. *Quasi-Newton methods* were developed for exactly this purpose. The basic idea of Quasi-Newton methods is then to approximate the inverse of the Hessian at any iteration so that determining the search direction at each iteration becomes a matrix–vector multiplication, and in such a way that fast convergence to a local minimizer is guaranteed to occur. The best approximation schemes require that the Hessian *inverse* is approximated in the kth iteration by a positive definite matrix $G_{(k)}$. Such schemes guarantee the descent property of the search direction since $s_{(k)}^T g(x_{(k)}) = -g_{(k)}^T G_{(k)} g_{(k)} < 0$. The most successful updating scheme for the approximation matrix $G_{(k)}$ is known as the *BFGS formula* (the acronym arising from the first letter of the four authors who independently proposed it Broyden (1970), Fletcher (1970), Goldfarb (1970), and Shanno (1970). Let $\delta_{(k)} = x_{(k+1)} - x_{(k)}$, $\gamma_{(k)} = \nabla f(x_{(k+1)}) - \nabla f(x_{(k)})\quad \forall k \geq 1$. Then, the approximation matrix $G_{(k+1)}$ is computed as

$$\begin{aligned} G_{(k+1)} &= G_{(k)} \\ &+ \left(1 + \frac{\gamma_{(k)}^T G_{(k)} \gamma_{(k)}}{\delta_{(k)}^T \gamma_{(k)}}\right) \frac{1}{\delta_{(k)}^T \gamma_{(k)}} \delta_{(k)} \delta_{(k)}^T - \frac{1}{\delta_{(k)}^T \gamma_{(k)}} \left[\left(\delta_{(k)} \gamma_{(k)}^T\right) G_{(k)} + G_{(k)} \left(\gamma_{(k)} \delta_{(k)}^T\right)\right] \end{aligned}$$

Initial estimates for $G_{(1)}$ usually set it to the identity matrix I. The BFGS method for unconstrained optimization of a continuously differentiable function $f(x)$ is outlined below.

Algorithm BFGS Method

Inputs: Function $f(x)$, first derivative $g(x)$, initial point $x_{(1)}$, criterion for termination.
Outputs: A point x^* satisfying termination criteria.

Begin

1. Set $k = 1$, $G = I$, $x = x_{(k)}$.
2. while x does not satisfy the termination criteria do:
   a. Set $s_{(k)} = -G\, g(x_{(k)})$.
   b. Determine a minimizer $h_{(k)}$ of the uni-variable function $f(t) = f(x+ts_{(k)})$ (using e.g. Algorithm Line-Search Optimization or the Armijo Rule Algorithm).
   c. Set $\delta = h_{(k)}s_{(k)}$, $x = x + h_{(k)}s_{(k)}$, $x_{(k+1)} = x$, $\gamma = g(x) - g(x_{(k)})$.
   d. Set $G = G + \left(1 + \frac{\gamma^T G\gamma}{\gamma^T\delta}\right)\frac{\delta\delta^T}{\gamma^T\delta} - \left[\frac{\delta\gamma^T G + G\gamma\delta^T}{\gamma^T\delta}\right]$
   e. Set $k = k+1$.
3. end-while
4. return $x^* = x_{(k)}$.

End.

There has been substantial theoretical and practical evidence that the BFGS method is the best possible Quasi-Newton method for unconstrained optimization. For example, it has been shown that the BFGS method using inexact line searches (i.e. line searches obeying the Wolfe–Powell conditions with σ less than but close to 1) is globally converging to a stationary point x^* for any general differentiable function $f(x)$ and in fact it converges to x^* *super-linearly* so that $\left\|x_{(k+1)} - x^*\right\| / \left\|x_{(k)} - x^*\right\| \to 0$. Also, many numerical studies on widely different test functions arising from completely different domains have shown the superiority of the method against any other line-search method developed so far. The way the method proceeds is shown graphically in Fig. 1.1 where the method is tested against the well-known Rosenbrock's function $f(x) = 100\left(x_2 - x_1^2\right)^2 + (1 - x_1)^2$. The method starts from $x_0 = [0\,0]^T$ and takes 25 iterations (each iteration point is marked with a blue asterisk in the figure) to reach the global minimum x^* located at $[1\ 1]^T$, using the Armijo Rule for line-search minimization (with default parameter values for β, γ and ρ). The termination condition $\left\|\nabla f(x_{(k)})\right\| < 10^{-8}$ was used, and the exact minimizer was located to an accuracy of 10^{-14}. The implementation of the full method in MATLAB (without any error checking conditions and safeguards necessary for a robust code) requires less than 65 lines of code.

Nevertheless, the method is not a panacea. Consider for example the function $f(x) = \frac{1}{2}\left(x_1^2 + x_2^2\right)e^{x_1^2 - x_2^2}$. Since the gradient of the function is $g(x) = \nabla f(x) = \left[x_1 e^{x_1^2 - x_2^2}\left(1 + x_1^2 + x_2^2\right) \quad x_2 e^{x_1^2 - x_2^2}\left(1 - x_1^2 - x_2^2\right)\right]^T$ it is not difficult to verify that the function has a *unique* minimizer at $x_0 = [0\ 0]^T$ and the minimum value is 0.

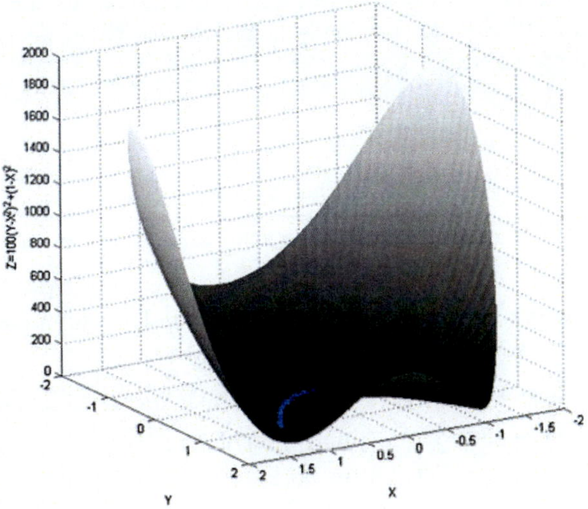

Fig. 1.1 3D plot of Rosenbrock's test function and the iterations of the BFGS method with Armijo Rule for step-size selection

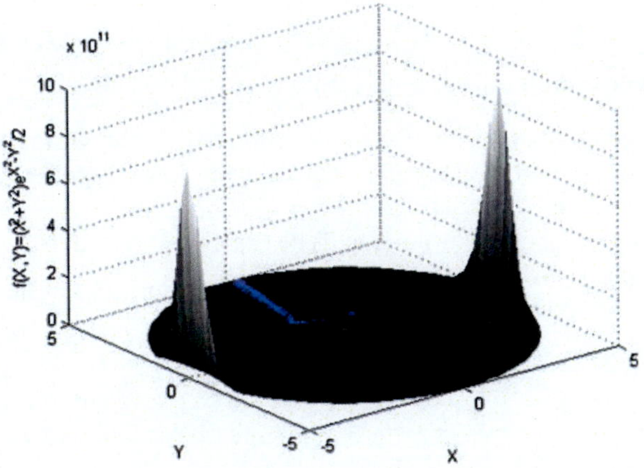

Fig. 1.2 3D plot of test function $f(x) = \frac{1}{2}\left(x_1^2 + x_2^2\right)e^{x_1^2 - x_2^2}$ and the 26 iterations of the BFGS method with Armijo Rule for step-size selection. The unique minimizer is shown in the figure as a *red circle*. It is easy to see how the method follows a trajectory taking it far from the true minimizer

However, applying the BFGS algorithm with default values, starting from the initial point $[1\ 1]^T$, the algorithm terminates far from the true minimizer, albeit with a point satisfying the termination condition $\left\|\nabla f(x_{(k)}\right\| < 10^{-8}$. The iterations of the algorithm are shown in Fig. 1.2. One of the main difficulties in this example is the fact that the derivative tends to zero rather fast along the direction $s = [0\ 1]^T$, which is approximately the direction the BFGS method moves along as can be seen in the figure. Other difficulties are the very flat landscape of the function, as well as the fact that the Hessian of the function is not positive definite.

The Conjugate-Gradient Method

The most successful non-Newton-like method that fits the line-search framework for optimization is perhaps the Conjugate-Gradient method, which combines the concepts of conjugacy of a set of directions with the simplest line-search method, the steepest descent method. Before describing the method, a definition of conjugacy is needed.

Definition 1.10 *A set of n n-dimensional non-zero column vectors* $S = \{s^1, s^2, \ldots, s^n\}$ *is conjugate with respect to a given Positive Definite matrix H if and only if* $(s^i)^T H s^j = 0, \quad \forall i \neq j$.

It is easy to establish that all the vectors in a conjugate direction set S are linearly independent of each other.

Lemma 1.11 *All vectors in a conjugate directions set S with respect to a P.D. matrix H are linearly independent.*

Proof Assume by contradiction that the set $S = \{s^1, s^2, \ldots, s^n\}$ is not linearly independent. Then, without loss of generality, we can write $s^n = \sum_{j=1}^{n-1} \lambda_j s^j, \quad \lambda_j \in \mathbb{R}, j = 1 \ldots n-1$, where not all $\lambda_j = 0$. From the conjugacy property of the set S for H and from the fact that H is P.D. we have that for all $j = 1 \ldots n-1$: $0 = (s^j)^T H s^n = (s^j)^T H \left(\sum_{i=1}^{n-1} \lambda_i s^i \right) = \lambda_j (s^j)^T H s^j \Leftrightarrow \lambda_j = 0$ which is a contradiction because not all $\lambda_j = 0$. QED.

Conjugate directions arise naturally in the Gram–Schmidt decomposition process of a linear system of equations in linear algebra and have a direct relationship with this process. The Gram–Schmidt process transforms a set of k linearly independent vectors $S = \{v^{[1]} \ldots v^{[k]}\}$ in $\mathbb{R}^n$ $(k \leq n)$ into an *orthogonal* set of vectors $U = \{u^{[1]} \ldots u^{[k]}\}$ that are orthonormal (so that

$$\left(u^{[i]}\right)^T u^{[j]} = \delta_{i,j} = \begin{cases} 1, & i = j \\ 0, & i \neq j \end{cases}$$

where $\delta_{i,j}$ are the Kronecker delta) spanning the same set as the vectors in S . By defining the projection of v onto u, $P_u(v)$ to be the vector

$$P_u(v) = \frac{u^T v}{\|u\|^2} u$$

the vectors $u^{[1]}, u^{[2]} \ldots$ are constructed as follows:

$$\begin{aligned}
u^{[1]} &= v^{[1]} \\
u^{[2]} &= v^{[2]} - P_{u^{[1]}}\left(v^{[2]}\right). \\
u^{[3]} &= v^{[3]} - P_{u^{[1]}}\left(v^{[3]}\right) - P_{u^{[2]}}\left(v^{[3]}\right) \\
&\ldots \\
u^{[k]} &= v^{[k]} - \sum_{i=1}^{k-1} P_{u^{[i]}}\left(v^{[k]}\right)
\end{aligned}$$

The above vectors are orthogonal to each other since we can show from the definition that $\left(u^{[2]}\right)^T u_1 = 0$, and using induction, if we assume that for all $j < i$ $u^{[i]}$ and $u^{[j]}$ are orthogonal to each other, then the vector $u^{[i+1]}$ is also orthogonal to all vectors $u^{[l]}$ $l = 1,\ldots, i$, since

$$\begin{aligned}\left(u^{[l]}\right)^T u^{[i+1]} &= \left(u^{[l]}\right)^T v^{[i+1]} - \sum_{j=1}^{i} \left(u^{[l]}\right)^T u^{[j]} \frac{\left(u^{[j]}\right)^T v^{[i+1]}}{\|u^{[j]}\|^2} \\ &= \left(u^{[l]}\right)^T v^{[i+1]} - \left(u^{[l]}\right)^T v^{[i+1]} = 0\end{aligned}$$

The set of orthonormal vectors U is now simply $\left\{u^{[1]}/\left\|u^{[1]}\right\|, \ldots, u^{[k]}/\left\|u^{[k]}\right\|\right\}$. That the span of the set U is the same as that of S follows from the fact that vectors $v^{[j]}$ can be written as linear combinations of vectors $u^{[i]}$, thus any vector z that is a linear combination of the vectors in S is also a linear combination of the vectors in U (and vice versa).

We can now state the Conjugate-Gradient method for the unconstrained minimization of a general differentiable function f.

Algorithm Conjugate Gradient Method (Polak-Ribiere)

Inputs: Function $f(x)$, first derivative $g(x)$, initial point $x_{(1)}$, criterion for termination.
Outputs: A point x^* satisfying termination criteria.

Begin

1. Set $k = 1$, $x = x_{(1)}$, $s_{(0)} = 0$, $s_{(1)} = -g(x)$, $\beta_{(0)} = 0$.
2. while x does not satisfy termination criteria do
   a. Set $s_{(k)} = -g(x_{(k)}) + \beta_{(k-1)}s_{(k-1)}$.
   b. Determine a minimizer $h_{(k)}$ of the uni-variable function $f(t) = f(x+ts_{(k)})$ using Algorithm Line-Search Optimization with small σ (e.g. ~0.1).
   c. Set $x_{(k+1)} = x + h_{(k)}s_{(k)}$, $x = x_{(k+1)}$.
   d. Set $\beta_{(k)} = [g(x_{(k+1)}) - g(x_{(k)})]^T g(x_{(k+1)}) \,/\, \|g(x_{(k)})\|^2$.
   e. Set $k = k+1$.
3. end-while
4. return $x^* = x_{(k)}$.

End

Comparing the Conjugate-Gradient method with the Polak–Ribiere formula for the search direction update with the BFGS method on the Rosenbrock benchmark test, the Conjugate-Gradient method is inferior to the BFGS method. Although both methods locate the global minimizer starting from the same initial point, it takes the Conjugate-Gradient method more than 2,000 iterations to establish the termination condition $\left\|\nabla f(x_{(k)})\right\| < 10^{-8}$. The trajectories of the two methods however are almost the same, as the graph of Fig. 1.3 clearly shows.

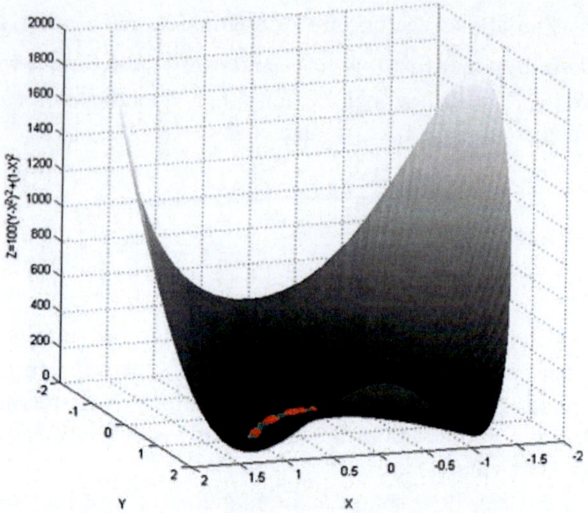

Fig. 1.3 3D plot of Rosenbrock's test function and the trajectory of iterations of the BFGS method with Armijo Rule for step-size selection versus the conjugate-gradient method with Polak–Ribiere formula. The trajectory of BFGS is shown in *green*, whereas the iterate points of conjugate-gradient methods are shown in *red*

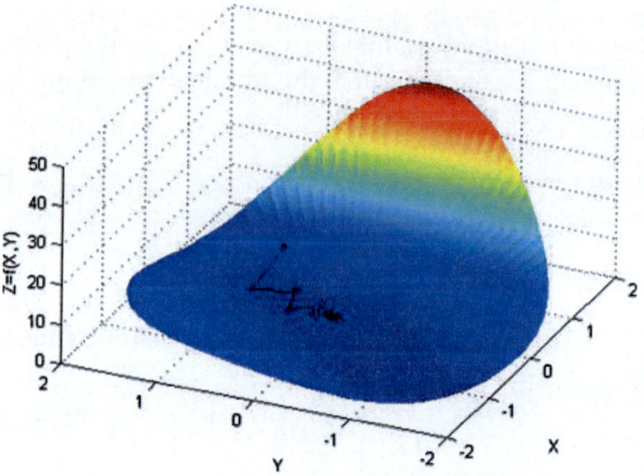

Fig. 1.4 3D plot of the test function $f(x) = 2x_1^2 + x_2^2 - 2x_1x_2 + 2x_1^3 + x_1^4$ and the trajectory of iterations of the conjugate-gradient method with Polak–Ribiere formula

On other test functions, the Conjugate-Gradient method performs much better. Consider for example the test function $f(x) = 2x_1^2 + x_2^2 - 2x_1x_2 + 2x_1^3 + x_1^4$. It should be rather easy for the reader to verify that the function has a unique minimizer at $[0\ 0]^T$. Starting from $[1\ 1]^T$ the Conjugate-Gradient method satisfies the termination condition $\left\|\nabla f(x_{(k)}\right\| < 10^{-8}$ and reaches the global minimizer in 56 iterations, as shown in Fig. 1.4.

Also, for the test function $f(x) = -10x_1^2 + 10x_2^2 + 4\sin(x_1x_2) - 2x_1 + x_1^4$ in Fig. 1.5 (vanden Berghen 2004), the Conjugate-Gradient method finds the minimum of the function in 19 iterations with a (reduced) accuracy of 10^{-6}.

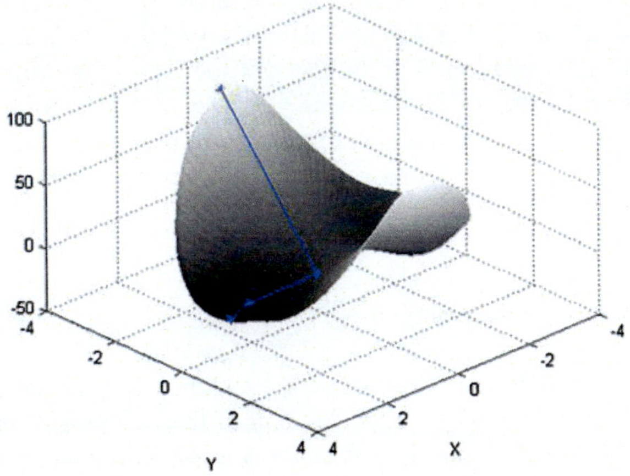

Fig. 1.5 3D plot of the test function $f(x) = -10x_1^2 + 10x_2^2 + 4\sin(x_1x_2) - 2x_1 + x_1^4$ and the trajectory of iterations of the Conjugate-Gradient method with Polak–Ribiere formula. In the area shown in the diagram there are two local minima

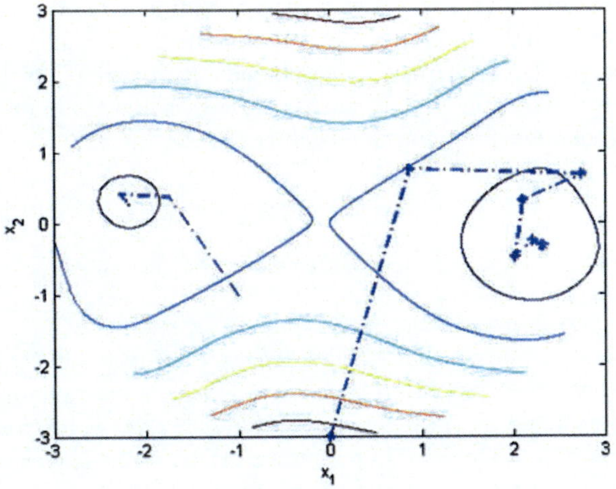

Fig. 1.6 Contours-plot of the test function $f(x) = -10x_1^2 + 10x_2^2 + 4\sin(x_1x_2) - 2x_1 + x_1^4$ and the trajectory of iterations of the BFGS method starting from two different initial points. Curves of colors approaching blue correspond to small function values compared to curves of near-red colors. In the area shown in the diagram there are two local minima, and the method will reach any of them depending on the initial iterate point

Figure 1.6 shows the contours (*curves along which the function value is the same*) of the previous function and the path the BFGS method takes starting from two different initial starting points to reach the two local minima of the function. The BFGS method converges in 14 and 19 iterations respectively, starting from the point $[0\ -3]^T$ and $[-1\ -1]^T$, illustrating the point to be always kept in mind that all descent methods for unconstrained optimization can at best guarantee to locate only a local minimum of the function to be minimized.

From the limited examples above, it might seem that Newton-like methods should be preferable to Conjugate-Gradient methods for unconstrained optimization. But in fact, the Conjugate-Gradient method has a *very important advantage* over Newton-like methods: it does not require the Hessian matrix or any approximation of the inverse of the matrix, and therefore has much less memory footprint in practical algorithmic implementations. For large-scale unconstrained optimization problems therefore, with many thousands of variables, the Conjugate-Gradient method has the distinct advantage that it may be the only feasible method of optimization even on today's modern computing systems, because of its very small memory requirements. Further, it has been observed in practice that many real-world applications exhibit certain structural properties that make the application of the Conjugate-Gradient method very fast as well (Fletcher 1987; Bertsekas 1995).

1.1.1.2 Trust-Region Methods

One of the problems the pure Newton's method faces as discussed in the previous section is that when the Hessian of the function at an iterate point $x_{(k)}$ is not P.D., the second-order Taylor series approximation may not have a unique minimizer and the iteration is no longer well-defined. However, if the area of the search for the new iterate point was small enough and concentrated around $x_{(k)}$ so that the second-order Taylor series approximation agreed with the value of the function to an appropriate degree, then the method could make reasonable progress at each step. This is the basic idea behind "trust-region" methods that derive their name from the fact that at each iteration, a new minimizer is sought by solving a Quadratic Programming problem of the form

$$\min q_{(k)}(\delta) = f(x_{(k)}) + \nabla f(x_{(k)})^T \delta + \frac{1}{2}\delta^T \nabla^2 f(x_{(k)})\delta$$
$$\textit{subject to the constraint} \|\delta\| \leq \eta_{(k)}$$

where the scalar $\eta_{(k)}$ is chosen so that $q_{(k)}(\delta) \approx f(x_{(k)} + \delta)$ for all δ having norm less than this scalar. In other words, a "trust region" algorithm restricts the choice of the next iterate point to a place where the function value is guaranteed to be very close to the *second-order Taylor series approximation of the function* at the current iterate point (a region we can trust the Taylor series to be valid). Since this is accomplished by restricting the step-size, the name "restricted step method" is also used sometimes to denote the family of algorithms based on this idea.

The generic family of trust region algorithms using quadratic approximations to the objective function is as follows.

Algorithm Trust Region-based Unconstrained Optimization

Inputs: Functions $f(x)$, first derivative $g(x)$, Hessian $H(x)$, initial point $x_{(1)}$, initial step-size $\eta_{(1)}$, parameters $0 < r_{\min} < r_{\max} < 1$, criterion for termination.
Outputs: A point x^* satisfying termination criteria.

Begin

1. Set $k = 1$, $x = x_{(1)}$.
2. while termination criteria are not satisfied at x do

a. Set $f_{(k)} = f(x)$, $g_{(k)} = g(x)$, $H_{(k)} = H(x)$.
b. Solve $\delta_{(k)} = \operatorname{argmin}_\delta \{ \delta^T g_{(k)} + (1/2)\delta^T H_{(k)}\delta \mid \|\delta\| \leq \eta_{(k)} \}$
c. Set $f_{\text{try}} = f(x + \delta_{(k)})$.
d. Set $Df_{(k)} = f_{(k)} - f_{\text{try}}$, $Dq_{(k)} = -(\delta^T_{(k)} g_{(k)} + \frac{1}{2}\delta^T_{(k)} H_{(k)} \delta_{(k)})$.
e. if $Dq_{(k)} = 0$ return $x^* = x$.
f. Set $r_{(k)} = Df_{(k)} / Dq_{(k)}$.
g. if $r_{(k)} < r_{\min}$ then

   i. Set $\eta_{(k+1)} = \|\delta_{(k)}\| / 4$

h. else if $r_{(k)} > r_{\max}$ AND $\|\delta_{(k)}\| = \eta_{(k)}$ then

   i. Set $\eta_{(k+1)} = 2\eta_{(k)}$

i. else Set $\eta_{(k+1)} = \eta_{(k)}$
j. end-if
k. if $r_{(k)} > 0$ then

   i. Set $x = x + \delta_{(k)}$.

l. end-if.
m. Set $k = k+1$.

3. end-while
4. return $x^* = x$.

End.

Default recommended values for the parameters $r_{\min}$ and $r_{\max}$ are 0.25 and 0.75 respectively. The ratio $r_{(k)}$ measures the deviation of the actual reduction on the objective value from the predicted reduction based on the quadratic approximation of the function via the second-order Taylor series around the current iterate point $x_{(k)}$. The logic of the algorithm should be clear—albeit the question of how to perform step 2.b of the algorithm asks for *the solution of a constrained quadratic optimization problem*. Within every iteration, the algorithm finds the minimum of the quadratic approximation of the function arising from the second-order Taylor series subject to a constraint on the distance that this solution is allowed to have from the current iterate point. If the actual reduction on the objective function agrees to a good degree or is even larger than the predicted reduction of the objective function the maximum allowed step-size is doubled in the next iteration, whereas if the actual reduction is significantly less than predicted, the maximum allowed step-size is halved into two for the next iteration to indicate that the trust region around that point should be made smaller. In that case, if the reduction is in fact negative, so that the solution of step 2.b leads to a point with a worse objective function value than the current iterate point, the new iterate point remains the same as that of the previous iteration—but now the trust region has been cut into half for the next iteration. Therefore, the sequence of iterations $\{x_{(k)}\}$ generates a *decreasing sequence of objective function values* $f_{(k)} = f(x_{(k)})$.

Importantly, trust-region algorithms have a very strong proof of global convergence, stating that any sequence of iterate points generated from a generic "trust-region" algorithm is globally convergent wherein under some mild conditions there exists a limit point of the sequence satisfying both the first and second-order conditions for unconstrained optimization stated in Theorems 1.1 and 1.4. The speed of convergence can also be shown to be at least super-linear.

Theorem 1.12 (Global Convergence of Trust Region Methods for Unconstrained Optimization). *If the sequence of iterate points* $\{x_{(k)}\}$ *generated by the Trust Region Algorithm for Unconstrained Optimization is bounded and if the objective function* $f : \mathbb{R}^n \to \mathbb{R}$ *is twice continuously differentiable then there exists a limit point* x^* *of the sequence that satisfies first and second-order necessary conditions for unconstrained optimization.*

Proof First, assume that the infimum of the sequence of step-sizes $\{\eta_{(k)}\}$ is zero. Now, there exists a subsequence L of indices k such that $x_{(k)}$ is convergent for k in L. Let $\lim_{k \in L} x_{(k)} = x^*$. For the indices $k \in L$ we have by definition of L that $r_{(k)} < r_{\min}$ and thus $\eta_{(k+1)}$ and $||\delta_{(k)}||$ both tending to zero in the subsequence L. We now claim that the point x^* defined above satisfies $\nabla f(x^*) = 0$, $s^T \nabla^2 f(x^*) s \geq 0, \quad \forall s \in \mathbb{R}^n$. Indeed, if the point x^* did not satisfy the first-order necessary condition, then the direction $s = -\frac{\nabla f(x^*)}{\|\nabla f(x^*)\|}$ would be a descent direction satisfying $s^T \nabla f(x^*) = -\|\nabla f(x^*)\| < 0$. Defining the sequences $f_{(k)} = f\left(x_{(k)}\right)$, $g_{(k)} = \nabla f(x_{(k)})$ the sequence $\Delta f_{(k)} = f\left(x_{(k)}\right) - f\left(x_{(k)} + \delta_{(k)}\right)$, and the sequence $\Delta q_{(k)} = f_{(k)} - q_{(k)}(\delta_{(k)})$, from the Taylor series we have $\Delta f_{(k)} = \Delta q_{(k)} + o\left(\|\delta_{(k)}\|^2\right)$. Now, at any iterate point $x_{(k)}$ with $k \in L$ consider taking a step of size $\theta_{(k)} = ||\delta_{(k)}||$ along the direction s. Since $\delta_{(k)}$ solves the Quadratic Programming problem $\min_\delta q_{(k)}(\delta) = \nabla f(x_{(k)})^T \delta + \frac{1}{2} \delta^T \nabla^2 f(x_{(k)}) \delta$ subject to $\|\delta\| \leq \eta_{(k)}$ we have that

$$
\begin{aligned}
\Delta q_{(k)} &= f_{(k)} - q_{(k)}(\delta_{(k)}) \geq f_{(k)} - q_{(k)}(\theta_{(k)} s) = -\theta_{(k)} s^T g_{(k)} + o(\theta_{(k)}) \\
&= \theta_{(k)} \|\nabla f(x^*)\| + o(\theta_{(k)})
\end{aligned}
$$

where the last equality follows from the fact that the function f is twice continuously differentiable, therefore $g_{(k)} \underset{k \in L}{\to} \nabla f(x^*) = g^*$. Now, from the way $\theta_{(k)}$ is defined, we have that it tends to zero for k in L, and therefore, dividing the previous equation $\Delta f_{(k)} = \Delta q_{(k)} + o\left(\theta_{(k)}^2\right)$ by $\Delta q_{(k)}$ (which, using the previous inequality, we get that $\lim_{k \in L} \frac{\Delta q_{(k)}}{\theta_{(k)}} \geq \|\nabla f(x^*)\|$, so that if $\Delta q_{(k)}$ goes to zero, it goes at a slower rate than $\theta_{(k)}$) we get $r_{(k)} = 1 + o\left(\theta_{(k)}^2\right) / \Delta q_{(k)} \underset{k \in L}{\to} 1$. The latest limit is a contradiction because $r_{(k)} < r_{\min} < 1$ for all k in L, and therefore, we must have $\nabla f(x^*) = 0$. To prove that $\nabla^2 f(x^*)$ is P.S.D., assume that there exists a direction s with $\|s\| = 1$ such that $s^T \nabla^2 f(x^*) s = -d, \quad d > 0$. For $k \in L$ consider taking a step of size $\theta_{(k)}$ along the direction $\pi_{(k)} s$ where we define $\pi_{(k)} = -sgn\left(s^T g_{(k)}\right)$. By the continuity of the Hessian and the fact that $\delta_{(k)}$ is the global minimizer of the problem $\min q_{(k)}(\delta) = \nabla f(x_{(k)})^T \delta + \frac{1}{2} \delta^T \nabla^2 f(x_{(k)}) \delta$ subject to $\|\delta\| \leq \eta_{(k)}$ we have that

$$\begin{aligned}\Delta q_{(k)} &= f_{(k)} - q_{(k)}(\delta_{(k)}) \geq f_{(k)} - q_{(k)}(\theta_{(k)} \pi_{(k)} s) \geq -\frac{1}{2} \theta_{(k)}^2 s^T \nabla^2 f(x_{(k)}) s \\ &= \frac{1}{2} \theta_{(k)}^2 d + o\left(\theta_{(k)}^2\right)\end{aligned}$$

and now dividing the equation $\Delta f_{(k)} = \Delta q_{(k)} + o\left(\theta_{(k)}^2\right)$ by $\Delta q_{(k)}$ gives again $r_{(k)} = 1 + o\left(\theta_{(k)}^2\right) / \Delta q_{(k)} \underset{k \in L}{\rightarrow} 1$ contradicting that $r_{(k)} < r_{\min} < 1$ for all k in L. Therefore, the Hessian at x^* is P.S.D.

Second and last, consider the case where the infimum of the sequence of step-sizes $\{\eta_{(k)}\}$ is greater than zero. Now, again, there exists a subsequence $L = (k_1\ k_2, \ldots, k_n, \ldots)$ of indices k in increasing order such that $\{x_{(k)}\}$ is convergent for k in L. Let $\lim_{k \in L} x_{(k)} = x^*$. In this second case, for all k in L, $r_{(k)} \geq r_{\min}$. Further, since the sequence $f_{(k)}$ is decreasing for all k = 1,2,3,..., we have that the series $\sum_{k \in L} \Delta f_{(k)} = f_{(k_1)} - f_{(k_1+1)} + f_{(k_2)} - f_{(k_2+1)} + \ldots$ (because $-f_{(k_i+1)} + f_{(k_{i+1})} < 0$ since $k_{i+1} \geq k_i + 1$) satisfies $\sum_{k \in L} \Delta f_{(k)} \leq f_{(k1)} - f_{(\infty)} \leq f_{(1)} - f*$ (that the series is convergent is obvious since for all k, $\Delta f_{(k)} > 0$ and any partial sum of the series is bounded by the term $f_{(1)}$–f^*). Therefore, $\Delta f_{(k)} \rightarrow 0$ for $k \in L$. As $r_{(k)} = \Delta f_{(k)} / \Delta q_{(k)} \geq r_{\min} > 0$, this implies that $\Delta q_{(k)} \rightarrow 0$ for $k \in L$. Now define the "limit point" quadratic function

$$q^*(\delta) = f^* + \delta^T \nabla f(x^*) + \frac{1}{2} \delta^T \nabla^2 f(x^*) \delta.$$

Choose any $\bar{\eta} \in (0, \inf \eta_{(k)})$. Let $\bar{\delta}$ be the solution of the Quadratic Programming problem $\min_\delta q^*(\delta)$ subject to $\|\delta\| \leq \bar{\eta}$ and define $\bar{x} = x^* + \bar{\delta}$. For all large enough $k \in L$, $\left\|\bar{x} - x_{(k)}\right\| \leq \left\|\bar{\delta}\right\| + \left\|x_{(k)} - x^*\right\| \leq \eta_{(k)}$ because $\lim_{k \in L} x_{(k)} = x^*$. Therefore, for all such k, the point $\bar{x} - x_{(k)}$ is a feasible point for the Quadratic Programming problem of step 2.b of the generic trust region algorithm, and since $\delta_{(k)}$ is its solution , we have $q_{(k)}(\bar{x} - x_{(k)}) \geq q_{(k)}(\delta_{(k)}) = f_{(k)} - \Delta q_{(k)}$. As $k \rightarrow \infty$ for $k \in L$, from the twice continuous differentiability of $f(x)$ we have that $f_{(k)} \rightarrow f^*, g_{(k)} \rightarrow \nabla f(x^*), \nabla^2 f(x_{(k)}) \rightarrow \nabla^2 f(x^*)$ and thus $\bar{x} - x_{(k)} \rightarrow \bar{x} - x^* = \bar{\delta}$. Since we already showed that $\Delta q_{(k)} \rightarrow 0$ in the subsequence, then $q^*(\bar{\delta}) \geq f^* = q^*(0)$. But $\bar{\delta}$ is the global minimizer of $q^*(\delta)$ subject to $\|\delta\| \leq \bar{\eta}$, therefore 0 is also a global minimizer of the same problem, and at zero, the constraint is inactive, therefore the first and second-order necessary conditions for unconstrained minimization of

$q^*(x)$ must be satisfied, thus $0 = \nabla q^*(0) = \nabla f(x^*)$ *and* $\nabla^2 q^*(0) = \nabla^2 f(x^*)$ is P.S.D. QED.

The proof presented above follows Fletcher (1987), expanding on those subtle points that are not immediately clear in that text. The proof is valid if the region of twice continuous differentiability of $f(x)$ is restricted to any set B that contains the sequence $\{x_{(k)} k = 1, 2, \ldots\}$. The proof is also valid if for any index k^* it happens that $\Delta q_{(k^*)} = 0$ in which case the algorithm terminates returning the current iterate point $x_{(k)}$ satisfying the first and second-order necessary conditions as can be easily checked by the reader.

The generic trust region-based algorithm for Unconstrained Optimization cannot be fully implemented without the description of a method for carrying out step 2.b, which asks for the solution of a Quadratic Programming problem subject to a distance bound constraint. Let us define the problem to solve as

$$\begin{array}{ll} (\mathrm{QP}) & \min\limits_{x} q(x) = \frac{1}{2} x^T H x + g^T x \\ \text{s.t.} & x^T x \leq h^2 \end{array} \tag{1.2}$$

where H is a symmetric square real matrix (the Hessian of the original objective function at the current iterate point $x_{(k)}$), g is a non-zero vector (the gradient of the objective function at the point $x_{(k)}$ which cannot be zero as the algorithm would stop at such a point), and $h > 0$ is a scalar parameter value. Consider the vector $s(\lambda) = -(H + \lambda I)^{-1} g$. We show that the ℓ_2 norm $\|s(\lambda)\|$ is decreasing in λ for $\lambda \geq 0$. To see this, observe that since H is a symmetric square real matrix, by the linear algebra matrix decomposition theory we have that $H = S \Lambda S^T$ where S is an orthonormal matrix satisfying $S^{-1} = S^T$ and Λ is a diagonal matrix with diagonal elements λ_i, $i = 1, \ldots, n$. Then, $\|s(\lambda)\| = \|(H + \lambda I)^{-1} g\| = \|(S(\Lambda + \lambda I) S^{-1})^{-1} g\| = \|S(\Lambda + \lambda I)^{-1} S^{-1} g\|$. Letting s_{ij} denote the (i,j) element of S, we have that the (i,j) element of the matrix $(H + \lambda \mathrm{I})^{-1} = S(\Lambda + \lambda I)^{-1} S^{-1}$ is equal to $\sum_{k=1}^{n} s_{ki} (\lambda_k + \lambda)^{-1} s_{kj}$ and therefore, the mth component of the vector $s(\lambda) = S(\Lambda + \lambda I)^{-1} S^{-1} g$ is equal to $\sum_{j=1}^{n} \left[\sum_{k=1}^{n} \frac{s_{km} s_{kj}}{\lambda_k + \lambda} \right] g_j \quad \forall \lambda \geq 0$. So, every component of the vector $s(\lambda)$ is decreasing in λ and therefore the whole norm of the vector is decreasing in λ *and in fact converges to zero as* λ *goes to infinity*. Further, it is now obvious that even if the matrix H is not P.D., the matrix $H + \lambda I$ will be P.D. for all choices of λ such that $\lambda > \max\{-\lambda_i\, i = 1, \ldots, n\}$.

Suppose s^* satisfies $(H + \lambda^* I) s^* = -g$ for some $\lambda^* \geq 0$ such that $H + \lambda^* I$ is at least P.S.D. and $\lambda^* (s *^T s^* - h^2) = 0$. From the positive semi-definiteness of the matrix $H + \lambda^* I$ we have that s^* is a global minimizer of the quadratic function $q_{\lambda*}(x) = \frac{1}{2} x^T (H + \lambda^* I) x + g^T x$ which implies that $q_{\lambda^*}(x) \geq q_{\lambda^*}(s^*)$ for all x which in turn, after some algebra, implies that $q(x) \geq q(s^*) + \frac{\lambda^*}{2} (s^{*T} s^* - x^T x)$. So, for all x such that $\|x\| \leq h$ we have that $q(x) \geq q(x^*)$ and therefore x^* is a global minimizer for (QP).

Otherwise, the solution of the problem (QP) will have to satisfy the Karush–Kuhn–Tucker First-Order Necessary Conditions (FONC) for mathematical programming (discussed in Sect. 1.1.2 on constrained optimization), which state that if s^* is a solution to (QP), then there exists a scalar $\lambda^* \geq 0$ (Lagrange multiplier) such that $Hs^* + g + 2\lambda^* s^* = 0$ and $\lambda^*(s*^T s^* - h^2) = 0$. If the constraint associated with the Lagrange multiplier λ^* is inactive at the optimal solution s^*, then $\lambda^* = 0$, and the solution satisfies $Hs^* = -g$. Otherwise, the constraint is active, implying that $s^{*T}s = h^2$, and also that s^* satisfies $(H + 2\lambda^* I)s^* = -g$. So, s^* will be the solution to the system of linear equations $(H + 2\lambda^* I)s^* = -g$ *such that* $s*^T s = h^2$ *and* $\lambda^* > 0$ Based on the above observations, an algorithm for solving the problem (QP) above and therefore solving step 2.b in the trust region algorithm, is the following

Algorithm Restricted Step-Size Quadratic Programming

Inputs: Non-zero vector g, symmetric square matrix H, step-size restriction $h > 0$, user-defined increment parameter $\tau > 0$.
Outputs: A point x^* that is a guaranteed local solution of the problem (QP).

Begin

1. Compute the matrix decomposition $H = S\Lambda S^T$.
2. Set $\lambda = \max\{0, (-\Lambda_{ii})\ i=1,\ldots,n\}$.
3. while true do
   a. Set $H_k = H + \lambda I$.
   b. Solve the linear system $H_k\, x = -g$.
   c. if $\|x\| \leq h$ then
      i. break.
   d. else Set $\lambda = \lambda + \tau$.
   e. end-if
4. end-while
5. return $x^* = x$.

End.

The above algorithm requires checking the solution of the system of linear equations, after having first computed the matrix decomposition $H = S\Lambda S^T$ which can best be done using a modified Cholesky factorization (Cheney and Kincaid 1994).

Running the trust region algorithm on the test function $f(x) = -10x_1^2 + 10x_2^2 + 4\sin(x_1 x_2) - 2x_1 + x_1^4$ having plugged-in the above algorithm for carrying out step 2.b finds the local minimum shown on Fig. 1.7 in only 8 iterations, showing the very good convergence speed of the trust region family of algorithms.

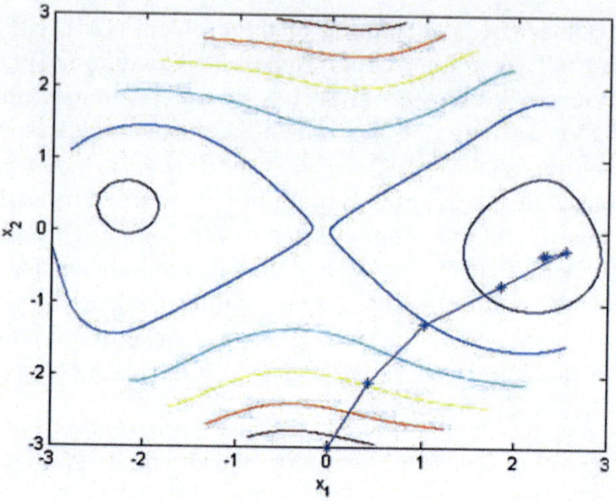

Fig. 1.7 Contourplots of the test function $f(x) = -10x_1^2 + 10x_2^2 + 4\sin(x_1x_2) - 2x_1 + x_1^4$ and the trajectory of iterations of the trust region algorithm starting from $[0\ -3]^T$. The norm of the gradient at the final iterate point is less than $10^{-13,}$ whereas the tolerance requested was set at 10^{-8}

1.1.1.3 Randomized Meta-Heuristics for (Global) Nonlinear Optimization

The methods discussed until now are essentially local search methods based on the idea of *iterative descent* as mentioned before. From the steepest descent method, to the trust region methods, all the methods have the common trait that they seek to improve the objective function value in every iteration until no more improvement can be made, often guaranteeing the existence of a local minimizer. This iterative improvement strategy obviously suffers from the problem that when multiple local minimizers exist, the starting point of the algorithm will determine to a large extent the solution of the local optimum that will be found.

To make matters worse, in nonlinear programming, the existence of multiple local minima is usually the norm and not the exception. This means that one should apply the algorithms described above several times, starting from several different initial points and choose the best local minimizer found. In order to implement this idea, one must choose how to select the starting points. The obvious choice of selecting initial points at random (from within a multi-dimensional cube or a ball of radius r centered around zero, $B(r) \in \mathbb{R}^n$) may actually work reasonably well in some circumstances (and the Wolpert and McReady "No Free Lunch Theorems" for optimization suggest that *theoretically*, when averaged over all problem instances, such a method will perform exactly the same as any other.) Nevertheless, in practice, it would likely be better to search the decision variable space in a more systematic manner so as to maximize the likelihood that a point with an objective function value close to that of the global minimum will be found.

A significant body of work on global optimization that was carried out during the past 40 years has resulted in a small number of optimization schemes for selecting points to evaluate a function . These schemes are generally applicable to any function and are independent of any local search iterative improvement procedure that may be applied to refine the selection target points that the scheme is able to find, but most often the schemes work best when an appropriate iterative improvement method from the list of methods described above is applied to an appropriate subsequence of points they generate. This also implies that *none of the schemes to be presented below requires gradient or curvature information*, which in practice can prove to be a significant advantage since often the function to optimize can hardly be written down as an analytical function whose derivatives can be symbolically computed. All these methods have proved successful in many and diverse fields of science and engineering, and have the added advantage that they are easy to understand and implement. They have been applied successfully in various sub-fields of Supply Chain Management and Optimization.

Simulated Annealing

Simulated annealing (SA) derives its name from an analogy in solid-state physics and metallurgy. More specifically, SA simulates the way metals restructure via the annealing process. At high temperatures T, when metals are in liquid condition the system is considered to be in a disordered state, where atoms of the metal can easily move around escaping possible configurations of high energy. As the metal goes through the annealing process, the temperature falls in a controlled manner adhering to an "*annealing schedule*" and the system self-reorganizes to approach the lowest possible energy state. Eventually, the system freezes to a halt when the temperature $T \approx 0$ is translated into a near-thermodynamic equilibrium point, where the atoms of the metal have rearranged themselves in very low energy configurations resulting in much stronger metallic structures. Similarly, an SA algorithm attempting to minimize a given function $f(x)$ starts from an initial point $x_{(1)}$—that is to a large degree irrelevant to the final point x^* to be returned by the process—and an initial very high temperature level T_0. The algorithm proceeds in phases, where within a phase the annealing temperature is kept constant, and at the end of each phase the temperature is decreased according to the annealing schedule mentioned above. Within a phase, for a given number of iterations, the algorithm evaluates the function f at various *neighboring* candidate positions x_{new} that result by moving from the current position $x_{(c)}$ along a randomly selected direction and (restricted) step-size. A candidate position is always accepted as the new current iterate point if $\Delta f = f(x_{\text{new}}) - f\left(x_{(c)}\right) < 0$, but is also accepted as new current iterate point even if $\Delta f \geq 0$, with probability proportional to $e^{-\Delta f/T}$, thus striking a strong difference from all algorithms we have seen so far (indeed, the trust region algorithm allows the generation of trial points and their evaluation in steps 2.b, 2.c that may be worse than the current iterate point function value, but such points are never accepted as the new iterate points; instead, the trust region is shrunk by shortening the trust-region step-size h.) This difference presumably allows for the essential characteristic of SA-based algorithms to escape local minima. During early phases (when T is large and thus $-\Delta f/T$ is a small negative number) moves leading to worse objective function values are more often accepted than in later stages when as T approaches zero, $\exp(-\Delta f/T)$ approaches zero exponentially fast.

In theory, when allowed to iterate forever with a slow enough cooling schedule, the algorithm converges to the global optimum. The speed of convergence however is unknown, and in fact, there are theoretical reasons why there can be no approximation guarantee to the global optimum for any SA procedure that terminates in a finite number of steps. The operational description of the process for unconstrained nonlinear optimization in pseudo-code is as follows.

Algorithm Simulated Annealing (SA) for Unconstrained Optimization

Inputs: Function $f(x)$, initial point $x_{(1)}$, step-size restriction h defining state neighborhood, temperature annealing schedule *sched* mapping iteration number to temperature, termination criteria.
Outputs: A point x^*.

Begin

1. Set $k = 1$, $x = x_{(k)}$, $f_x = f(x)$.
2. while termination criteria at x are not satisfied do
   a. Set $T = sched[k]$.
   b. Select at random a direction vector d with $\|d\| = 1$.
   c. Select at random a positive number $\eta \leq h$.
   d. Set $x_{try} = x + \eta d$.
   e. Set $f_{try} = f(x_{try})$.
   f. if $f_{try} \leq f_x$ then
      i. Set $x = x_{try}$, Set $f_x = f_{try}$.
   g. else
      i. Select at random a number $P \in [0, 1]$.
      ii. if $P \leq \exp(-(f_{try} - f_x)/T)$ then
         1. Set $x = x_{try}$, Set $f_x = f_{try}$.
      iii. end-if
   h. end-if
   i. Set $k = k + 1$.
3. end-while
4. return $x^* = x$.

End

It is easy to write a small matlab function implementing the SA algorithm (free open-source implementations are readily available on the web as well.) Running the algorithm with a maximum of $k_0 = 1{,}000$ iterations to the test function $f(x) = -10x_1^2 + 10x_2^2 + 4\sin(x_1x_2) - 2x_1 + x_1^4$ (same as in Figs. 1.5–1.7) with a

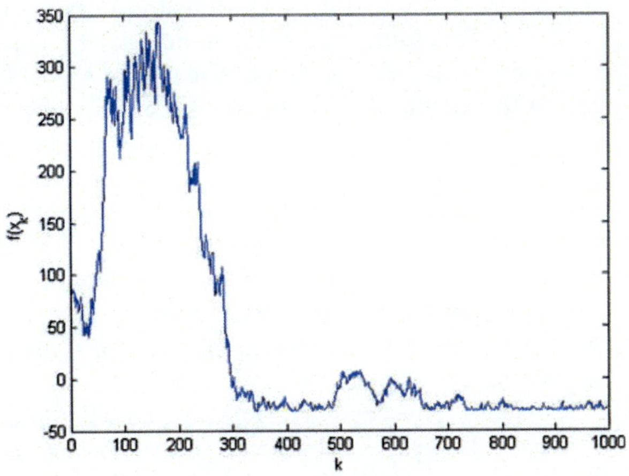

Fig. 1.8 Progress of an SA algorithm on test function $f(x) = -10x_1^2 + 10x_2^2 + 4\sin(x_1x_2) - 2x_1 + x_1^4$ starting from $[0\ -3]^T$. The best function value found was $f_{SA} = -31.1797$, whereas the trust region algorithm starting from the same point locates the optimal point with $f^* = -31.1807$. Widely different choices of the initial starting point lead to almost identical results

schedule that maps iteration number k to temperature T according to the equation $T = k_0\left(1 + \left\lfloor\frac{20k}{k_0}\right\rfloor\right)^{-2}$ and with a step-size restriction $h = 0.25$ we get a sequence of points whose function values are plotted against the sequence number of the iteration they appeared in Fig. 1.8. Observe how the algorithm is allowed to make "bad" moves in the beginning, whereas in the latter iterations "uphill" moves are rarely accepted as the threshold to be overcome is very large, allowing only improving moves. This comprises the main characteristic of SA type algorithms in general.

Note also that whereas the trust region algorithm starting from the same starting point locates the global minimum in only 8 iterations, the SA algorithm required more than 350 function evaluations to find a slightly worse solution. This apparent inefficiency of the SA method compared to trust region algorithms—and many of the other line-search iterative improvement type algorithms discussed above—is due to the fact that SA does not use any type of gradient or curvature information, as pointed out before. As we have seen, when a good guess of the optimizer of the function is available, most iterative improvement methods will find efficiently the true optimizer—except perhaps in pathological cases, see Fig. 1.2. The SA method, as well as the rest of the methods to be discussed below is most often used to find such a guess, rather than locate the true optimizer which is then a task left to the line-search and/or trust region methods discussed before.

Many variants of the SA algorithm are concerned with the choice *of the schedule* function that maps iteration number to temperature, as well as with optimizations for steps 2.b,c and 2.f,g,h that deal with choosing a next candidate point, and deciding whether to accept a "good" or "bad" move respectively. We discuss each briefly.

- The schedule map must be a function that allows the algorithm to remain at a (near-) constant temperature long enough so as to settle in a "promising" region of the search space. In general however, how long is long enough is strongly dependent on the nature of the function to be optimized and therefore default annealing scheduling strategies should not be expected to work very often. It seems necessary for the decision maker to experiment with different schedules until a reasonable balance between speed and solution quality is achieved.

- The way the next move is generated, must be such that the optimal solution can be located from any initial point in a small number of iterations. In implementations of Simulated Annealing for Discrete Optimization problems in a discrete search space, choosing the neighbor of a point x should usually be done in such a way that *reflexivity* of the concept of neighborhood is preserved, and if x_1 is a neighbor of x_2, then x_2 is a neighbor of x_1 (this condition is satisfied in the description of SA given above, but may not always be guaranteed if steps 2.b,c are implemented in a different manner.)
- The strategy for accepting or rejecting moves should clearly be such that "bad" moves are progressively harder to accept. The exponential function $e^{-\Delta f/T}$ comes from an analogy with statistical physics and happens to have the right properties for threshold purposes. However, issues of scale come into play; depending on the "ruggedness" of the functional landscape, or equivalently, depending on the gradient values, Δf may be orders of magnitude larger or smaller than the initial temperature T_0 and the subsequent temperatures, yielding the SA search ineffective, restricting it to either a blind "random greedy search" or simply a "random walk" of the search space. Scaling techniques have been used to either scale the term $-\Delta f/T$ or to redefine the initial temperature T_0 and perhaps the annealing schedule as well after a small number of initial iterations show this to be necessary.

Genetic Algorithms

Just as SA is analogous to a natural phenomenon—the hardening of metals when cooled in a controlled manner— Genetic Algorithms are inspired from the biological theory of evolution of the species (Darwin 1861), a main concept of which is that of *natural selection* that states that within a population of individuals, the ones that are the fittest under the current environmental conditions are the ones leaving the most offspring in the next generations. The idea behind the Genetic Algorithm method for function optimization (GA for short) is then to construct and breed a population of candidate solution points that will progress towards the optimum—or at least toward a near-optimal solution. Although the origins of this class of algorithms may be traced back to the early 1950s (according to Michalewicz (1994)) it was John Holland (Holland 1975) who pioneered the field and developed a theoretical framework for the use of GAs. GAs are another class of randomized algorithms that breed a whole population—usually fixed in size— of *individuals* represented by strings of an *alphabet*—traditionally the binary alphabet $B = \{0, 1\}$—which themselves represent a possible solution to a given problem. Each of these strings represents therefore an *encoding* of a possible solution to the given problem and is known as the individual's *chromosome* obviously borrowing the terminology from genetics. Crucial to the success of a GA is the existence of a *fitness function* that can be applied to any individual chromosome to produce a reasonable metric of the quality of the solution the individual represents for our problem. This metric is called the *fitness value* of the individual.

The relative fitness value of the individual compared to others in the current population comprises the *relative fitness value* of the individual. Obviously, when applied to unconstrained optimization problems, the fitness function is often just the function to be minimized multiplied by -1 (as the GA attempts to maximize the fitness of the individuals in successive generations.) When given such a fitness function, the algorithm bootstraps its computations by creating a random initial population of individuals by building strings of random letters from the chosen alphabet, and then proceeds in generations where at each generation some of the most fit individuals are selected for mating and form couples that exchange parts of their strings to form a new individual, a process called crossover. Other genetic operators such as mutation or inversion may be applied to the newly created individual and then this newborn offspring is evaluated using the fitness function. Then, according to a survival policy the individual may or may not replace one of its parents in the population. A pure survival policy will always replace one of the offspring's parents with the offspring no matter what the fitness value of the offspring is. More elitist strategies will discard the offspring in favor of the parents if they are fitter than their children. The algorithm stops when some termination criterion is met, such as a provably optimal solution found, or a maximum generation number is reached, or the population has converged to a point from which no further improvement can be made. The above can be summarized in the following description of the workings of a GA (graphically shown in Fig. 1.9).

1. Initialize randomly a population of individuals.
2. Evaluate the fitness of each individual in the population and their relative fitness values.
3. Select a number of individuals from the current population for mating.
4. Apply crossover operator to pairs of individuals producing pairs of offspring; apply any other genetic operators to the offspring.
5. Decide whether to replace original pairs of individuals with their offspring or not, according to a "survival of the fittest" concept.
6. Increment a generation counter.
7. If termination criteria are not satisfied, GOTO 2.

This generic procedure leaves many decisions to be made about the specific workings of the GA.

- Solution Representation: The string representation of the solution can be altered or generalized to include other more appropriate data structures for the original problem or simply to include more letters in the original alphabet; indeed as we shall see in later sections, this often results in serious performance

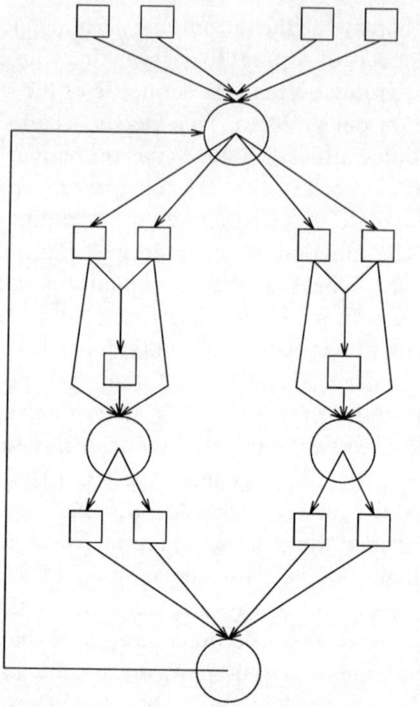

Fig. 1.9 Schematic presentation of the workings of a genetic algorithm

improvements in discrete optimization problems such as the Quadratic Assignment Problem or the famous Travelling Salesman Problem etc.

- Genetic Operators: Having decided on the representation of the solutions, the next task is to choose the genetic operators to be used for the creation of the next generation. The most widely used ones are one-point or two-point crossover (less often used is uniform crossover), and mutation. Less often used (and much less studied) is the inversion operator which, like the other operators, is also inspired from genetics. Besides the previous 3 "universal" operators, quite often special purpose operators might be needed to enhance the performance or to ensure the feasibility of the solutions represented by the offspring. It is common for GA designers/users to design special purpose solution representations in conjunction with special purpose genetic operators in an attempt to make it more likely that the Building Block Hypothesis holds good in that setting, and correspondingly to make it easier for the algorithm to locate good near-optimal solutions.
- Survival Policy: The survival policy is a set of rules that determines when an offspring will replace one of its parents in the next generation or it will be discarded. Elitist policies prefer to not replace the parents with their offspring if the offspring are less fit than their parents. The more elitist a policy is, the faster the whole population converges to a homogeneous state, where every individual is similar to every other, and with approximately the same fitness value.

The problem is that usually, fast convergence locates only a local optimum of moderate quality.

- Selection: Another decision that has to be made concerns the process that selects individuals for mating. Several such routines have been proposed. Among the most successful ones are "tournament selection", "roulette-wheel based", and "remainder stochastic sampling with replacement". These policies try to select individuals for mating based on their fitness values in such a way as to strike a good balance between selective pressure (the force that drives individuals towards promising regions of the search space) and population diversity (for the exploration of as much of the search space as possible).
- Steady-State versus Generational Replacement: Members of the current population might not be selected for mating but may still continue their existence in the next generation. This policy is known as the steady-state approach where in each generation only a small percentage of the population is selected for mating and their offspring replace the parent individuals. The traditional GA follows the generational replacement model where the whole population is replaced by offspring in the next generation.
- Parameters Setting: Finally, when all the above decisions have been made, the appropriate parameters must be set (crossover and mutation rate, population size etc.). Choosing a set of parameters is usually done via extensive experimentation—a common feature in most of the methods in this section.

In the following, we present in detail the pseudo-code for a GA that follows the steady-state approach, uses one-point crossover, uniform mutation and two-point inversion genetic operators, selects individuals for mating using the roulette-wheel method, and implements the elitist survival approach.

Algorithm Genetic Algorithm (GA) for Unconstrained Optimization

Inputs: Individual length l, alphabet B, fitness function $f : B^l \rightarrow \mathbb{R}$, crossover probability px, mutation probability pm, inversion probability pi, population size s, termination criteria.
Outputs: A point x^*.
Function select($cpf[n]$, r).
Inputs: cpf: array double[n], r: double.
Output: index j in $\{1, 2, \ldots, n\}$.

Begin

1. for $i = 1$ to n do
   a. if $cpf[i] \geq r$ return i.
2. end-for
3. return -1.

End

Begin

1. Generate at random a set $S=\{v_1, v_2, \ldots, v_s\}$ of s individuals by generating random strings of length l with letters from the alphabet B.
2. Set $f^* = -\infty$.
3. while termination criteria are not satisfied do
    a. Set $pf_{tot} = \sum_{j=1}^{s} f(v_i)$, Set $cpf_0 = 0$.
    b. for $i = 1$ to s do
        i. Set $pf_i = f(v_i)/pf_{tot}$.
        ii. Set $cpf_i = cpf_{i-1} + pf_i$.
    c. end-for
    d. Set $P = \{\}$.
    e. for $i = 1$ to $\lceil px \times s \rceil$ do
        i. Select at random a number $r \in [0, 1]$.
        ii. Set j = select(cpf, r).
        iii. Set $P = P \cup \{v_j\}$.
    f. end-for
    g. Set $ps = |P|$.
    h. for $i = 1$ to ps do
        i. Select at random a number $r \in [0, 1]$.
        ii. if $r \leq px$ then
            1. Select at random a number $m \in [1, ps + 1)$.
            2. Set $j = \lfloor m \rfloor$.
            3. Select at random a number $c \in \{1, 2, \ldots, l-1\}$.
            4. Set $child_1 = v_i[1\ldots c] + v_j[c+1\ldots l]$.
            5. Set $child_2 = v_j[1\ldots c] + v_i\ [c+1\ldots l]$.
            6. for $k = 1$ to l do
                a. Select at random a number $r \in [0, 1]$.
                b. if $r \leq pm$ then
                    i. Select at random a letter $e \in B$.
                    ii. Set $child_1[k] = e$.
                c. end-if
                d. Select at random a number $r \in [0, 1]$.
                e. if $r \leq pm$ then
                    i. Select at random a letter $e \in B$.
                    ii. Set $child_2[k] = e$.
                f. end-if
            7. end-for
            8. Select at random a number $r \in [0, 1]$.
            9. if $r \leq pi$ then
                a. Select at random two numbers $i_1, i_2 \in \{1, \ldots l\}$, so that $i_1 < i_2$.
                b. Invert the sub-string $child_1[i_1 \ldots i_2]$.
            10. end-if
            11. Select at random a number $r \in [0, 1]$.

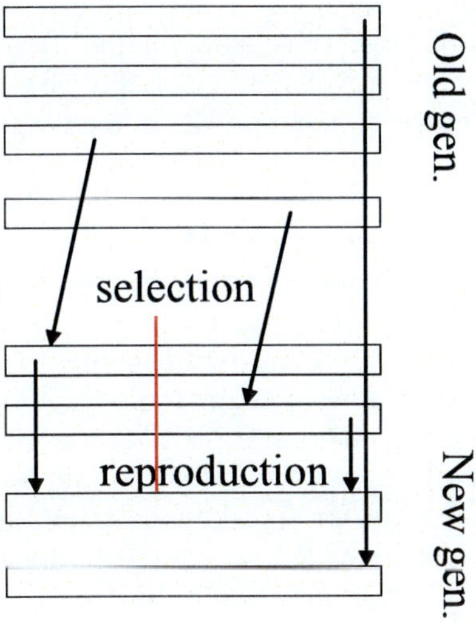

Fig. 1.10 Generational progress of a GA using one-point crossover operator

12. if $r \leq pi$ then
    a. Select at random two numbers $i_1, i_2 \in \{1, \ldots l\}$, so that $i_1 < i_2$.
    b. Invert the sub-string child$_2[i_1 \ldots i_2]$.
13. end-if
14. Set v_i = child$_1$, v_j = child$_2$.

iii. end-if

i. end-for
j. Select $x_{\text{best}} = \text{argmax}\{ f(v_i)\ i = 1 \ldots s\}$.
k. if $f(x_{\text{best}}) > f^*$ then
   i. Set $x^* = x_{\text{best}}$, Set $f^* = f(x^*)$.
l. end-if.

4. end-while
5. return x^*.

End.

The one-point crossover genetic operator works, as its name suggests, by selecting a point at random in the sequence of letters (also known as alleles) comprising the individual chromosomes and cutting the chromosomes of the two parents along that position, and then pasting together the first part of the first parent's chromosome with the second part of the second parent's chromosome to form the first offspring, and then pasting the remaining two parts from the two parents together to form the second offspring (see Fig. 1.10).

Genetic Algorithms (originally called Genetic Plans in Holland's terminology) presumably work when the so-called "Building Block Hypothesis" holds. This hypothesis states that the problem to be solved by a Genetic Algorithm has a representation encoding structure so that any sufficiently large sub-strings of strings representing solutions that are near-optimal, when contained in other strings will tend to have fitness values higher than strings that do not contain such large sub-strings of the optimal solutions. This condition is made more formal in Holland's "Schema Theorem" which states under certain conditions, a Genetic Algorithm will in the long run improve the average fitness value of the population in successive generations. However, the precise meaning of the theorem came under attack in the last decade of the 20th century and the question regarding the theoretical convergence of Genetic Algorithms towards optimal or near-optimal solutions remains an open issue. In fact, although GAs have proved to be powerful optimization tools in many domains that have the added advantage of being easily amenable to parallelization, the practical performance of GAs in other domains remains an open issue. It seems that out-of-the-box GA implementations for discrete and combinatorial optimization (to be discussed later) are often incapable of solving real-world problems, but when used as a blue-print for algorithmic design, together with engineering intuition they often lead to excellent problem-solving tools that perform as well as or better than most other optimization methods.

The GA paradigm has offered an excellent compromise between the themes of exploration and exploitation of a generic search space. Problems with poor structure (such as the NP-hard problems) have to be tackled with search methods capable of exploiting promising regions of the search space as well as exploring in some reasonable way the rest of the space and if they are to be of any practical use, they have to manage to accomplish both tasks without facing combinatorial explosion (else brute force will explore enough of the search space to come up with the provably optimal solution to the problem—the key is in obtaining high-quality solutions within an acceptable time frame.) GAs, whether in their traditional form, or modified so as to take advantage of any problem-specific structure and when combined with iterative improvement methods of the previous sections have often been able to strike such an exploration/exploitation balance and to find good solutions to otherwise intractable problems.

The Evolutionary Strategy Method

Whereas GAs place a lot of emphasis on the crossover recombination operator for exploring the search space, the *Evolutionary Strategy* (also known as Evolutionary Algorithm, EA for short) *relies solely on random mutations* of the solution and an elitist, greedy strategy to evolve new solutions. The original EA predates both GAs and SA and is an extremely simple algorithm that starts with an initial point $x_{(1)}$ and in the kth iteration, applies random changes to all the components of the vector $x_{(k)}$ where the changes are random variables drawn from the Normal distribution $N(0,\sigma)$ with zero mean and user-defined standard deviations $\sigma = [\sigma_1 \ldots \sigma_n]^T$. The new "mutated" point x' is accepted as the new iterate point $x_{(k+1)}$ iff $f(x') < f(x_{(k)})$, otherwise $x_{(k+1)} = x_{(k)}$. The algorithm in pseudo-code is therefore as follows.

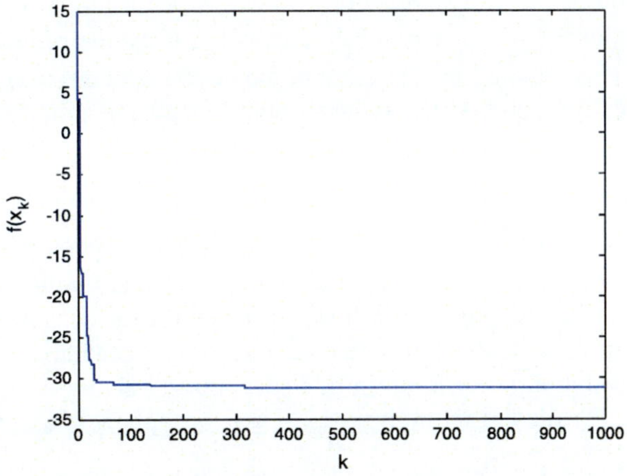

Fig. 1.11 Generational progress of the EA for 1,000 iterations. As the algorithm rejects worsening moves, the graph of the values $f(x_{(k)})$ is necessarily a non-increasing sequence. Nevertheless, the random nature of the directions to move to and the step-sizes taken leads the algorithm to overcome local minima and locate a solution that is very close to the true global optimum in about 300 function evaluations

Evolutionary Algorithm (EA) for Unconstrained Optimization

Inputs: Function $f : \mathbb{R}^n \to \mathbb{R}$, initial point $x_{(1)}$, standard deviations vector σ, termination criteria.
Outputs: A point x^* satisfying termination criteria.

Begin

1. Set $x = x_{(1)}$, $fx = f(x)$, $x^* = x$, $f^* = fx$.
2. Set $k = 1$.
3. while termination criteria at x are not satisfied do
   a. Select a random vector $\delta \in \mathbb{R}^n$, $\delta_i \sim N(0, \sigma_i)\ \forall i = 1\ldots n$.
   b. Set $x_{try} = x + \delta$.
   c. If $f(x_{try}) < fx$ then
      i. Set $x = x_{try}$, $fx = f(x_{try})$.
      ii. if $fx < f^*$ then
         1. Set $x^* = x$, $f^* = fx$.
      iii. end-if
   d. end-if
   e. Set $k = k+1$.
4. end-while
5. return x^*.

End.

The termination criteria are often just an upper bound on the number of iterations to perform. Applying the above very simple algorithm to the test function $f(x) = -10x_1^2 + 10x_2^2 + 4\sin(x_1x_2) - 2x_1 + x_1^4$ with an s.d. vector $\sigma = [1\ 1]^T$ we get a series of points with function values shown in Fig. 1.11.

Just as SA is guaranteed in theory to converge to the global optimum (in infinite number of iterations), so is the EA method guaranteed to converge to the global optimum of any function $f(x)$ for which a global minimum $x^* \in \mathbb{R}^n$ exists when allowed to run forever. In particular, it can be proved that $P\{\lim x_{(k)} = x^*\} = 1$. The meaning of this theorem however, is in essence that a random walk over the entire search space will eventually locate the global minimum.

Differential Evolution

Particularly well-suited to continuous unconstrained nonlinear optimization is a specialized form of Genetic Algorithms (and Evolutionary Computation) known as Differential Evolution (DE). The main ideas of combining individual solutions to produce better solutions in successive generations by random crossover and mutation are present in this specialized version of the classical GA method. The DE structure is intuitive and easy to implement as well.

Algorithm Differential Evolution (DE) for Unconstrained Optimization

Inputs: Function $f : \mathbb{R}^n \to \mathbb{R}$, crossover probability px, differential weight w, population size s, termination criteria.
Outputs: A point x^*.

Begin

1. Set $x^* = 0$, $f^* = f(x^*)$.
2. Initialize an array p of size s with s random vectors in $\mathbb{R}^n$.
3. while termination criteria for p are not met do
   a. for $i = 1$ to s do
      i. Set $x = p[i]$.
      ii. Select random vectors a, b, c from p so that $a \neq b, a \neq c, b \neq c$, $x \notin \{a, b, c\}$.
      iii. Select at random an index $r \in \{1, 2, \ldots, n\}$.
      iv. Set $x_{\text{try}} = x$.
      v. for $j = 1$ to n do
         1. Select at random a number r_j in (0,1).
         2. if $j = r$ OR $r_j < px$ Set $(x_{\text{try}})_j = a_j + w(b_j - c_j)$
      vi. end-for
      vii. if $f(x_{\text{try}}) < f(x)$ Set $p[i] = x_{\text{try}}$.
      viii. if $f(p[i]) < f^*$ then
         1. Set $x^* = p[i], f^* = f(p[i])$.
      ix. end-if
   b. end-for
      1. end-while
      2. return x^*.

End.

The evolution of the best solution found as iterations progress in the course of a run of the DE algorithm is depicted in Fig. 1.12, together with the graph of the average value of each generation's candidate solutions.

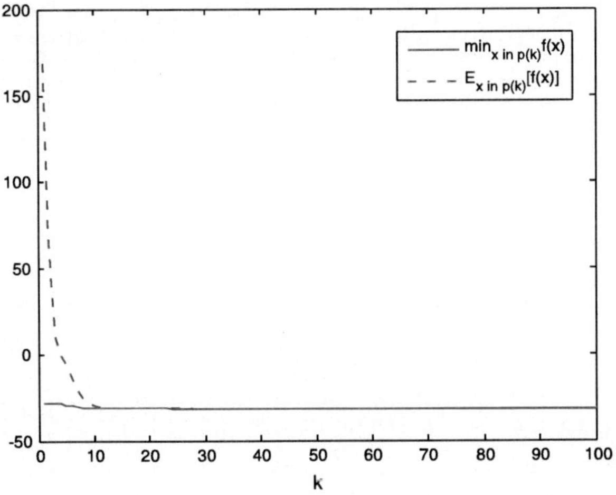

Fig. 1.12 Generational progress of the DE method for 100 iterations. The test function as before is $f(x) = -10x_1^2 + 10x_2^2 + 4\sin(x_1x_2) - 2x_1 + x_1^4$. The population size was set at 20, $px = 0.7$, $w = 0.1$. The graph shows both the average function value of the population as well as the minimum value among the population members at each generation. The best function value found was -30.9944, close to the global minimum value of the function. It is found within less than 10 generations of the algorithm run

Several variations and improvements of the above randomized methods have been proposed in the literature, but cannot be described in the small space of a review chapter. Parallel versions of all the above methods have been implemented and tested on a variety of parallel computers, as well as hybrid methods combining almost any of the above methods. It seems that in practice, often the above algorithms have to be fine-tuned and/or modified to yield near-optimal and satisfying results depending on the problem to be solved. Nevertheless, the algorithms above form a substantial arsenal of methods that can be used to attack an extremely wide variety of problems in industrial settings, and in Supply Chain Management in particular. The ideas behind the algorithms above are also often carried with little or no modification to the context of constrained continuous optimization discussed next.

1.1.2 Constrained Optimization

The objective of smooth constrained optimization—also known as nonlinear programming—is to solve the following generic problem (*NLP*)

$$(NLP)\ \min_x f(x)$$
$$s.t. \begin{cases} c_i(x) \geq 0, & \forall i \in I \\ c_j(x) = 0, & \forall j \in E \end{cases} \tag{1.3}$$

The sets I and E are index sets of inequality and equality constraints respectively, and any one may be empty, but not both. The real function f and the vector function

$$c(x) = \begin{bmatrix} c_1(x) \\ \vdots \\ c_m(x) \end{bmatrix}$$

containing all nonlinear constraint functions in the definition of (*NLP*) are usually assumed to be continuously differentiable functions. Before going in depth into important classes of the (*NLP*) such as standard Linear Programming and Network Flows, and Quadratic Programming (*QP*), we prove some theorems leading to the First-Order Necessary Conditions for mathematical programming (*FONC*) that are at the heart of most algorithmic developments in the field. These conditions are natural extensions to the classical (and much easier to prove) Fermat's theorem. The notion of convexity is necessary at this point.

Definition 1.13 A set $S \subseteq \mathbb{R}^n$ is said to be convex iff $\forall x,\ y \in S, \forall \lambda \in [0,1] \quad \lambda x + (1-\lambda)y \in S$. A function $f : \mathbb{R}^n \to \mathbb{R}$, defined over a convex domain S is convex iff for each x and y in S, and for every $\lambda \in [0,1]$ it holds that $f(\lambda x + (1-\lambda)y) \le \lambda f(x) + (1-\lambda)f(y)$. A function g defined over the same domain S but satisfying the opposite inequality, i.e. $g(\lambda x + (1-\lambda)y) \ge \lambda g(x) + (1-\lambda)g(y)$ is called a concave function.

Lemma 1.14 *Consider a function $f : \mathbb{R}^n \to \mathbb{R}$ defined and differentiable over a convex open set $S \subseteq \mathbb{R}^n$. The function f is convex iff $f(x) \ge f(y) + (x-y)^T \nabla f(y), \quad \forall x, y \in S$.*

Proof Suppose f is convex on S. Then, for all x, y in S and λ in [0,1] define $x_\lambda = \lambda x + (1-\lambda)y$. Then, $f(x_\lambda) - f(y) \le \lambda(f(x) - f(y))$ and by dividing this inequality by λ for $\lambda > 0$ we have

$$\frac{f(y + \lambda(x-y)) - f(y)}{\lambda} \le f(x) - f(y)$$

and by taking the limit as $\lambda \to 0+$, the left-hand side of the inequality becomes the directional derivative of f at y along the direction $x - y$, which is equal to $(x-y)^T \nabla f(y)$, so we get $f(x) \ge f(y) + (x-y)^T \nabla f(y)$. Vice versa, if $f(x) \ge f(y) + (x-y)^T \nabla f(y)$ holds for all x,y in the convex open set S, then, for the vectors y, and x_λ the inequality holds $f(y) \ge f(x_\lambda) + (y - x_\lambda)^T \nabla f(x_\lambda)$, and for the vectors x and x_λ the inequality $f(x) \ge f(x_\lambda) + (x - x_\lambda)^T \nabla f(x_\lambda)$ holds. Multiplying the first inequality by $(1-\lambda)$ and the second by λ, and adding the two, we get $(1-\lambda)f(y) + \lambda f(x) \ge f(x_\lambda) + [(1-\lambda)(y - x_\lambda) + \lambda(x - x_\lambda)]^T \nabla f(x_\lambda)$. Since $x - x_\lambda = (\lambda - 1)(y - x), y - x_\lambda = \lambda(y - x)$ we have that $[(1-\lambda)(y - x_\lambda) + \lambda(x - x_\lambda)] = 0$, so that $(1-\lambda)f(y) + \lambda f(x) \ge f(x_\lambda)$. QED.

Equivalently, from the definition of functional convexity, we have that a function is convex if its epigraph is a convex set.

The following lemmas and theorems, known as *Lemmas and Theorems of the Alternative*, have played a major role and lead directly to the development of necessary and sufficient conditions for nonlinear optimization.

Lemma 1.15 (Tucker's Lemma). *For every $p \times n$ real matrix A having as rows the n-dimensional row vectors $a_1, \ldots a_p$, the systems $Ax \geq 0$ and $A^T y = 0, y \geq 0$ have solutions x, y satisfying $a_1 x + y_1 > 0$.*

Proof By induction on the number of rows of matrix A, p assume $p = 1$. If $a_1 = 0$, then $x = 0$ and $y_1 = 1$ satisfies the conditions of the lemma. Otherwise, take $y_1 = 0$ and $x = a_1$ and the conditions are again satisfied. Now, assume that the lemma holds for any matrix A of up to p rows. We shall prove that the theorem holds for any matrix of the form $\hat{A} = \begin{bmatrix} A \\ a_{p+1} \end{bmatrix}$. From the induction hypothesis we have that there exist vectors x & y satisfying $Ax \geq 0$, $A^T y = 0, y \geq 0$ and $a_1 x + y_1 > 0$. If x satisfies $a_{p+1} x \geq 0$, then the vectors x and $y' = [y^T\ 0]^T$ satisfy the conditions of the lemma for the new matrix $\hat{A}$, otherwise $a_{p+1} x < 0$. Now, consider the real $p \times n$ matrix B whose ith row is equal to $b_i = a_i - (a_i x / a_{p+1} x) a_{p+1}$ so that $Bx = 0$. From the induction hypothesis we have the existence of two vectors u, v such that $Bu \geq 0$, $B^T v = 0$, $v \geq 0$ *and* $b_1 u + v_1 > 0$.

Define

$$\hat{v} = \begin{bmatrix} v^T & \dfrac{-\sum_{j=1}^{p} v_j a_j x}{a_{p+1} x} \end{bmatrix}^T$$

Obviously $\hat{v} \geq 0$, and furthermore,

$$\begin{aligned} \hat{A}^T \hat{v} &= A^T v - \left(\sum_{j=1}^{p} \frac{a_j x}{a_{p+1} x} v_j \right) a_{p+1}^T \\ &= B^T v + \left(\sum_{j=1}^{p} \frac{a_j x}{a_{p+1} x} v_j \right) a_{p+1}^T - \left(\sum_{j=1}^{p} \left(\frac{a_j x}{a_{p+1} x} \right) v_j \right) a_{p+1}^T = 0 \end{aligned}$$

Now consider the vector $w = u - (a_{p+1} u / a_{p+1} x) x$. This vector satisfies

$$a_{p+1} w = a_{p+1} u - a_{p+1} u = 0$$

and also

$$\begin{aligned} \hat{A} w &= \begin{bmatrix} A \\ a_{p+1} \end{bmatrix} w = \begin{bmatrix} Aw \\ 0 \end{bmatrix} = \begin{bmatrix} Bw + \frac{Ax}{a_{p+1} x} a_{p+1} w \\ 0 \end{bmatrix} = \begin{bmatrix} Bw \\ 0 \end{bmatrix} = \begin{bmatrix} Bu - \frac{a_{p+1} u}{a_{p+1} x} Bx \\ 0 \end{bmatrix} \\ &= \begin{bmatrix} Bu \\ 0 \end{bmatrix} \geq 0 \end{aligned}$$

Now,

$$a_1 w + v_1 = \left(b_1 + \left(a_1 x / a_{p+1} x\right) a_{p+1}\right) w + v_1 =$$
$$b_1 w + v_1 = b_1 u - \left(a_{p+1} u / a_{p+1} x\right) b_1 x + v_1 = b_1 u + v_1 > 0$$

completing the proof. QED.

Theorem 1.16 (Tucker's First Theorem of the Alternative) *For every real $p \times n$ matrix A there exist vectors x, y satisfying $Ax \geq 0$ and $A^T y = 0, y \geq 0$ such that $Ax + y > 0$.*

Proof By permuting pairs of rows a_1 and a_i $i = 2,\ldots p$ of matrix A and applying Lemma 1.15 on the permuted matrix, we guarantee the existence of pairs of vectors $x^{[i]}$ and $y^{[i]}$ $i = 1\ldots p$ such that $Ax^{[i]} \geq 0$, $A^T y^{[i]} = 0$, $y^{[i]} \geq 0$, $a_i x^{[i]} + y_i^{[i]} > 0$. Define $x = x^{[1]} + \cdots + x^{[p]}$ *and* $y = y^{[1]} + \cdots + y^{[p]}$. It is easy to verify that $Ax \geq 0$ and that $A^T y = 0$, $y \geq 0$. Also,

$$a_i x + y_i = a_i x^{[i]} + y_i^{[i]} + \sum_{\substack{j=1 \\ j \neq i}}^{p} \left(a_i x^{[j]} + y_i^{[j]}\right) > 0 \forall i = 1\ldots p$$

since the first two terms of this sum are positive and the rest of the summation is non-negative. The last p inequalities can be written in matrix form as $Ax + y > 0$. QED.

Theorem 1.17 (Tucker's Second Theorem of the Alternative). *For every $p \times n$ real matrix $A \neq 0$ and every $q \times n$ real matrix B there exist vectors x, y and z such that $Ax \geq 0$, $Bx = 0$, $A^T y + B^T z = 0$, $y \geq 0$ that satisfy $Ax + y > 0$.*

Proof Consider the matrix $W = \begin{bmatrix} A^T & B^T & -B^T \end{bmatrix}^T$. By Tucker's first theorem of the alternative (Theorem 1.16), there must exist vectors u *and* v such that $Wu \geq 0$, $W^T v = 0$, $v \geq 0$, $Wu + v > 0$. Consider the partition of vector v induced by the matrices A B and $-B$ in the definition of W so that $v = \left[y^T r^T s^T\right]^T$. Then, we must have $A^T y + B^T r - B^T s = 0, y, r, s \geq 0$ and by setting $z = r - s$ this can be rewritten as $A^T y + B^T z = 0$ $y \geq 0$. Setting $x = u$, we have $Ax \geq 0$, $Bx = 0$, $A^T y + B^T z = 0$, $y \geq 0$ and from the first p rows of the inequality $Wu + v > 0$ we get $Ax + y > 0$. QED.

As a corollary of Tucker's second theorem of the alternative, we have

Corollary 1.18 *For every real matrices A, B, C, and D all having n columns and p, q, r, and s rows respectively, there are n-dimensional vector x, p-dimensional vector $u \geq 0$, q-dimensional vector $v \geq 0$, r-dimensional vector $w \geq 0$ and s-dimensional vector z that satisfy $Ax \geq 0$, $Bx \geq 0$, $Cx \geq 0$, $Dx = 0$ and $A^T u + B^T v + C^T w + D^T z = 0$ and $Ax + u > 0$, $Bx + v > 0$, and $Cx + w > 0$.*

The above corollary leads immediately to Motzkin's theorem of the alternative.

Theorem 1.19 (Motzkin's Theorem of the Alternative). *For every real $p \times n, q \times n, s \times n$ matrices A, C, and D, one and only one of the following conditions is always true:*

$$\begin{cases} (I) & \exists x \in \mathbb{R}^n : Ax > 0, Cx \geq 0, Dx = 0 \\ (II) & \exists u \in \mathbb{R}^p, v \in \mathbb{R}^q, w \in \mathbb{R}^s : A^T u + C^T v + D^T w = 0, \quad u \geq 0, u \neq 0, v \geq 0 \end{cases}$$

Proof Showing $(I) \Rightarrow not(II)$ is easy. If *(II)* holds while *(I)* holds, by multiplying the equation $A^T u + C^T v + D^T w = 0$ by x^T from the left, we would obtain on the left-hand side the quantity $x^T A^T u + x^T C^T v + x^T D^T w > 0$ since $Ax > 0$ $u \geq 0$ and not all its components are zero, and the last two terms of the sum are non-negative.

Vice versa, in order to show that $not(I) \Rightarrow (II)$ note that if *(I)* does not hold then if a vector x satisfies $Ax \geq 0$, $Cx \geq 0$, and $Dx = 0$ (such a vector for example can be $x = 0$), then Ax cannot be strictly greater than zero. But then, from Corollary 1.18, (with $B = 0$), we get x, $u \geq 0$, $v \geq 0$, and w satisfy $Ax \geq 0$, $Cx \geq 0$, and $Dx = 0$, $A^T u + C^T v + D^T w = 0$, with u non-zero (the non-zero components will be at least those of Ax that are zero). QED.

Lemma 1.20 (Farkas' Lemma) *For every real $m \times n$ matrix A and every $m \times 1$ column vector $a \in \mathbb{R}^n$, one and only one of the following two conditions (I) or (II) is always true:*

$$\begin{cases} (I) & \exists x \in \mathbb{R}^n : x \geq 0 : \quad Ax = a \\ (II) & \exists u \in \mathbb{R}^m : A^T u \geq 0, \quad u^T a < 0 \end{cases}$$

Proof Showing $(I) \Rightarrow not(II)$ is easy again. Assume the existence of a vector $x \geq 0$ such that $Ax = a$. Now, if there were a vector u satisfying condition *(II)*, by multiplying the equation $Ax = a$ by u^T from the left we would have $u^T Ax = u^T a < 0$. But $u^T Ax = (A^T u)^T x \geq 0$ since $A^T u \geq 0$, $x \geq 0$, and their inner product would have to be non-negative, which is a contradiction.

To prove that $not(I) \Rightarrow (II)$ we make use of Motzkin's theorem with $D = 0$, $A = -a^T$ and $C = A^T$. The second condition of Motzkin's theorem states that there exists a non-negative vector v and a positive scalar u such that $-au + Av = 0$, or equivalently, (by considering $x = u^{-1} v$) that $Ax = a$ for some $x \geq 0$. If no such x exists, then there must exist an x such that $-a^T x > 0$ and $A^T x \geq 0$. QED.

Farkas' Lemma can also be shown by observing that if condition *(I)* does not hold, then the polyhedral closed convex cone $K = \{Ax | x \geq 0\}$ does not contain a and therefore the vector a can be strongly separated from K, which by definition means that there exists a vector u such that $u^T k \geq 0 \forall k \in K, u^T a < 0$, and this implies that $A^T u \geq 0$, because otherwise, if the ith component of $A^T u$ was negative, the inner product $u^T\left(Ae_{[i]}\right) = \left(A^T u\right)^T e_{[i]} < 0$ (where $e_{[i]}$ is the unit vector in the ith dimension) which is a contradiction. The proof as presented, avoids the need to show that the set K is closed (showing that it is a polyhedral convex cone is easy).

The next theorem known also as the famous *Karush–Kuhn–Tucker theorem* establishes the conditions that must be satisfied by any local minimizer of an (NLP) problem of the form (1.3).

Definition 1.21 A vector $x \in \mathbb{R}^n$ is a feasible point for (NLP) iff it satisfies all constraints defined by (1.3). A column vector $s \in \mathbb{R}^n$ of unit-norm is said to be a feasible direction for problem (NLP) at a feasible point $x \in \mathbb{R}^n$ iff there exists an $\varepsilon > 0$ such that $x + hs$ is feasible for all $0 \le h < \varepsilon$.

Theorem 1.22 (First-Order Necessary Conditions for NLP). *Assume x^* is a local minimizer for the (NLP) problem (1.3), where the functions f and c_i are differentiable in a region of x^*, and assume further that the Kuhn–Tucker constraint qualification holds at x^* so that the set $\Im^*$ of feasible directions for (NLP) at x^* is equal to the set $F* = \left\{s \in \mathbb{R}^n | A_{I*}^T s \ge 0, A_E^T s = 0, \|s\| = 1\right\}$ where A_{I*} is the matrix whose columns are the gradient vectors $\nabla c_i(x^*)$ of all inequality constraints that are active at x^* (evaluated at that point), and A_E is the matrix whose columns are the gradient vectors $\nabla c_i(x^*)$ of all equality constraints $i \in E$ (that are active everywhere). Then, there exist scalar values λ_i for each constraint $i \in I \cup E$ known as Lagrange multipliers, such that $\lambda_i \ge 0 \forall i \in I$ with the property that $\lambda_i = 0 \forall i \in (I - I^*)$ so that the following holds*

$$\nabla f(x^*) = A_I \tilde{\lambda}_I + A_E \tilde{\lambda}_E, \quad \sum_{i \in I \cup E} \lambda_i c_i(x^*) = 0$$
$$c_i(x^*) \ge 0 \forall i \in I, \quad c_i(x^*) = 0 \forall i \in E$$

where the vectors $\tilde{\lambda}_I, \tilde{\lambda}_E$ contain as their components the Lagrange multipliers λ_i of the inequality and equality constraints of (NLP) respectively.

Proof As x^* is a local minimizer for (*NLP*), by definition x^* is a feasible point for (*NLP*) and thus the conditions $c_i(x^*) \ge 0 \, \forall i \in I$, $c_i(x^*) = 0 \, \forall i \in E$ are satisfied. Further, since x^* is a local minimizer for (NLP), there can be no descent direction for f at x^* that is also a feasible direction for (NLP)—otherwise x^* would not be a local minimizer. A descent direction s for f at x^* is a direction vector that satisfies $s^T \nabla f(x^*) < 0$, while a feasible direction s for (NLP) from the hypothesis is a direction vector that satisfies $A_{I*}^T s \ge 0, \quad A_E^T s = 0$. Since no descent direction for f at x^* can also be a feasible direction for NLP at x^*, the following system of equalities and inequalities has to be inconsistent:

$$\begin{cases} s^T \nabla f(x*) < 0 \\ A_{I*}^T s \ge 0 \\ A_E^T s = 0 \end{cases}$$

(note that the constraint $\|s\| = 1$ is redundant in the system above). But this system can be rewritten as

$$\begin{cases} \nabla f(x*)^T s < 0 \\ \begin{bmatrix} A_{I*}^T \\ A_E^T \\ -A_E^T \end{bmatrix} s \geq 0 \end{cases}$$

and since the above system is inconsistent, by Farkas' Lemma (1.20) there must exist a vector $\tilde{\lambda} \geq 0$ such that $[A_{I^*} \quad A_E \quad -A_E]\tilde{\lambda} = \nabla f(x^*)$ which, after partitioning the vector $\tilde{\lambda} = \begin{bmatrix} \tilde{\lambda}_{I*}^T & \tilde{\lambda}_{pE}^T & \tilde{\lambda}_{nE}^T \end{bmatrix}^T \geq 0$ according to the columns of the matrix $[A_{I*} \quad A_E \quad -A_E]$, can be written as $\nabla f(x^*) = A_{I^*}\tilde{\lambda}_{I^*} + A_E\left(\tilde{\lambda}_{pE} - \tilde{\lambda}_{nE}\right) = A_{I^*}\tilde{\lambda}_{I^*} + A_E\tilde{\lambda}_E, \quad \tilde{\lambda}_{I^*} \geq 0$. Setting to zero Lagrange multipliers for the inactive inequality constraints we get $\nabla f(x^*) = A_I\tilde{\lambda}_I + A_E\tilde{\lambda}_E, \quad \tilde{\lambda}_I \geq 0, \lambda_i = 0 \, \forall i \in I - I^*$, and $\sum_{i \in I \cup E} \lambda_i c_i(x^*) = 0$ which completes the proof. QED.

In general, $\Im^* \subseteq F^*$. This is easy to establish by considering an *active* inequality constraint i at x^* from the set I^*. Letting s be a feasible direction for (*NLP*) at x^*, expanding the function $c_i(x^* + hs)$ in a first-order Taylor series around x^* we obtain

$$c_i(x* + hs) = c_i(x^*) + hs^T \nabla c_i(x^*) + o(h) = hs^T \nabla c_i(x^*) + o(h) \geq 0 \forall h \in (0, \varepsilon)$$

for appropriately small positive ε. Dividing over by h and letting h tend to zero we obtain the inequality $s^T \nabla c_i(x^*) \geq 0 \forall i \in I^*$. A similar argument shows that $s^T \nabla c_i(x^*) = 0 \forall i \in E$, proving $\Im^* \subseteq F^*$.

There are some important cases where the Kuhn-Tucker constraint qualification is guaranteed to hold: in particular, if the active inequality constraints at x^* are linear constraints, then the qualification is guaranteed to hold. Also, if the columns of the matrix A_{I*} are all linearly independent, then the qualification holds. Proof of this claim can be found in Fletcher (1987).

We conclude this presentation of the basic theoretical results for smooth constrained optimization with the following first-order sufficient criterion for *global* optimality of a point for the particular class of NLP problems known as *convex optimization problems*.

Theorem 1.23 (First-Order Sufficient Conditions for Constrained Convex Optimization). *Assuming that for the* (*NLP*) *problem* (1.3) *the functions f and c_i are all continuously differentiable for all $i \in I, E = \emptyset$, f is convex and c_i are concave, and assuming further that for a point x^* there exist scalar values $\lambda_i \geq 0$ for each constraint $i \in I$ known as Lagrange multipliers so that the following holds*

$$\nabla f(x^*) = \sum_{i \in I} \lambda_i \nabla c_i(x^*)$$
$$c_i(x^*) \geq 0, \lambda_i \geq 0 \quad \forall i \in I$$
$$\sum_{i \in I} \lambda_i c_i(x^*) = 0$$

then x^* *is the global minimizer for that problem.*

Proof Let x be any feasible point, i.e. any point satisfying $c_i(x) \geq 0$ for all i in I. From the non-negativity of λ_i and $c_i(x)$ for all i in I we have

$$\begin{aligned} f(x) \geq & \ f(x) - \sum_{i \in I} \lambda_i c_i(x) \\ & \geq f(x*) + (x - x*)^T \nabla f(x*) - \sum_{i \in I} \lambda_i \left(c_i(x*) + (x - x*)^T \nabla c_i(x*)\right) \end{aligned}$$

using the easily verifiable fact that the function $L(x) = f(x) - \sum_{i \in I} \lambda_i c_i(x)$ (known as the Lagrangean of the problem) is convex, and Lemma 1.14. Since by hypothesis, $\sum_{i \in I} \lambda_i c_i(x^*) = 0, \quad \nabla f(x^*) = \sum_{i \in I} \lambda_i \nabla c_i(x^*)$, the right-hand side of the second inequality is exactly equal to $f(x^*)$, proving that x^* is the global solution to the problem. QED.

Corollary 1.24 *Any local minimizer* x^* *of a convex function* f *defined over a convex domain* S *is a global minimizer for the function* f.

The fact that any local minimizer for a convex function is also a global minimizer is of the greatest importance as it provides a certificate of optimality using only local information—which is usually much easier to obtain. For this reason, optimization problems that turn out to be instances of convex optimization are considered "well behaved" problems. We shall encounter a number of such problems when we discuss *Inventory Management*.

1.1.2.1 Linear Programming

Consider the problem that a small pharmaceutical company faces in choosing the optimal production quantities for two different drugs so that the company profits are maximized. In particular, assume that company F has a budget B of €20,000 for renting processing time on two machines A and B that are needed to make two different drugs a and b. The cost of renting 1 h of processing on machine A c_A = €1,000 while that of renting 1 h on machine B is c_B = €2,000. Making 1 metric ton of drug "a" requires $p_{A,a} = 2$ h of processing on machine A and $p_{B,a} = 3$ h of processing on machine B and returns a profit of r_a = €1,500. Making 1 metric ton of drug "b" requires $p_{A,b} = 4$ h of processing on machine A and $p_{B,b} = 1$ h of processing on machine B, for a profit of r_b = €2,000. How much of drug a and how much of drug b should company F produce in order to maximize its profits?

Let x_a denote the quantity (in metric tons) of drug a that the company will produce, and x_b the quantity (in metric tons again) of drug b that the company will produce. The objective is to maximize the total profit $r_a x_a + r_b x_b$. However, the quantities produced are constrained by the fact that the company has a limited available budget for renting processing time on both machines. In particular, we know that the amount of money the company will have to pay for producing x_a tons of drug a will be $(c_A p_{A,a} + c_B p_{B,a})x_a$ and similarly, the amount of money to be paid for producing x_b tons of drug b will be $(c_A p_{A,b} + c_B p_{B,b})x_b$. Therefore, the quantities x_a and x_b must be such that they satisfy the constraint

$$\left(c_A p_{A,a} + c_B p_{B,a}\right)x_a + \left(c_A p_{A,b} + c_B p_{B,b}\right)x_b \leq B.$$

The problem that the pharmaceutical company faces can now be written down in numbers as a Linear Programming maximization problem:

$$\max_{x_a, x_b} \quad 1500x_a + 2000x_b$$
$$\text{s.t.} \begin{cases} 8000x_a + 6000x_b \leq 20000 \\ x_a, x_b \geq 0 \end{cases}$$

The term Linear Programming (LP) refers to the fact that the objective function $f(x_a, x_b) = r_a x_a + r_b x_b$ of the optimization problem the company faces is linear in the decision variables $x_{a,b}$, as are the constraints of the problem, which can be expressed as inequality constraints on the values of linear functions of the (decision) variables of the form $c(x) \geq 0$ or equality constraints of the form $g(x) = 0$ where the functions $c(.)$ and $g(.)$ are linear in x.

The reader can easily spot the optimal solution to the above problem. Economically, it makes sense to invest the company's entire budget in producing only drug b because it has a higher profit margin and costs less to make. Therefore, the optimal solution should have $x_a^* = 0, x_b^* = 20/6 = 3.33$ tons of drub b. However, the solution is less obvious if there is a quota on the time use of machine A . Suppose the owner of machine A only allows the pharmaceutical company to rent up to $q_A = 5$ h of processing time on machine A. Now the previous schedule becomes infeasible because in order to make 3.33 tons of drug b, 13.33 h of processing on machine A is needed, which is clearly impossible (infeasible.) The new constraint must be taken into account. The constraint can be easily written down as $P_{A,a}\, x_a + P_{A,b}\, x_b \leq q_A$, and the new more constrained problem now becomes

$$\max. \quad 1500x_a + 2000x_b$$
$$\text{s.t.} \begin{cases} 8000x_a + 6000x_b \leq 20000 \\ 2x_a + 4x_b \leq 5 \\ x_a, x_b \geq 0 \end{cases}$$

The solution to this problem is no longer trivial to find; it requires the use of a systematic method that will guarantee to find the global optimum for all optimization problems having the special structure that their objective function is linear, and their feasible region can be described as a set of linear equality and inequality constraints.

The *SIMPLEX* method for Linear Programming developed in the late 1940s by George Dantzig and others represents the most popular method for solving such problems and in practice happens to be one of the most efficient—if not the most efficient—method to date. Due to its extremely wide range of applicability, the development of the simplex method is considered today as one of the greatest scientific and technological achievements of the twentieth century.

In the following we will provide a brief description of the steps the simplex method follows when solving an LP in its so-called *standard form* (Chvatal 1983). Every LP problem can be stated in standard form which is as follows:

$$(LP) \quad \max c^T x \\ \text{s.t.} \begin{cases} Ax \leq b \\ x \geq 0 \end{cases} \tag{1.4}$$

where b and c are n-dimensional column real vectors, A is an $m \times n$ real matrix, and x is the n-dimensional column vector of continuous decision variables. Note how our example production mix problem above is an LP problem already in standard form. (Parenthetically, we note that some authors prefer to call as the standard, an LP problem of the form $\min_x\{c^T x | Ax = b,\ x \geq 0\}$; this is nothing but a matter of convention).

Converting a general LP into standard form is pretty straightforward. The conversion of an inequality constraint of the form $a^T x \geq b$ to a lesser constraint, can be done by rewriting the inequality as $-a^T x \leq -b$. An equality constraint of the form $a^T x = b$ can be written as the following two inequalities: $a^T x \leq b$ & $-a^T x \leq -b$. Obviously, the same trick can be used to handle upper bound constraints on individual variables of the form $x_i \geq u$. In case the original LP formulation poses no lower or upper bound box constraints on a variable x_i, so that the variable is allowed to take on any value in $(-\infty, +\infty)$ the variable can be replaced in the objective function as well as the constraints in which it appears by two new non-negative variables y_i and z_i obeying $x_i = y_i - z_i\ y_i,\ z_i \geq 0$. And finally, if the original problem is stated as a minimization problem of the function $f(x) = c^T x$, as we noted in the beginning of the chapter, the problem is equivalent to minimizing the function $g(x) = -c^T x$.

As an example, consider the following LP that is not in standard form:

$$\max_x 3x_1 - 2x_2 + 5x_3 \\ \text{s.t.} \begin{cases} x_1 + x_3 = 5 \\ x_2 \geq 3 \\ x_1 \geq 0 \end{cases}$$

The problem's first constraint is an equality constraint, but as mentioned above, it is equivalent to the following two constraints: $x_1 + x_3 \leq 5$, $-x_1 - x_3 \leq -5$. The second constraint can be written as $-x_2 \leq -3$. Finally, variable x_3 is not constrained below, so we replace it by two non-negative variables y_3 and z_3 as follows $x_3 = y_3 - z_3$. The original LP now becomes

$$\max_{x_1,x_2,y_3,z_3} 3x_1 - 2x_2 + 5y_3 - 5z_3$$
$$\text{s.t.} \begin{cases} x_1 + y_3 - z_3 \leq 5 \\ -x_1 - y_3 + z_3 \leq -5 \\ -x_2 \leq -3 \\ x_1 \geq 0, x_2 \geq 0, y_3 \geq 0, z_3 \geq 0 \end{cases}$$

and this problem now is in standard form.

The SIMPLEX Method for LP

We will describe the (revised) simplex method following Chvatal (1983). The first step the method takes when solving an LP in standard form is to convert the $m > 0$ inequality constraints $Ax \leq b$ into equality constraints by introducing new non-negative variables $x_{n+i} \geq 0$ for each constraint $i = 1\ldots m$ called slack variables. In algebraic form, we obtain the following set of equations, called a *dictionary*:

$$\begin{aligned} z &= \sum_{j=1}^{n} c_j x_j \\ x_{n+i} &= b_i - \sum_{j=1}^{n} a_{ij} x_j, \quad i = 1\ldots m \\ x_i &\geq 0, \quad i = 1\ldots n+m \end{aligned} \tag{1.5}$$

Definition 1.25 A dictionary of the form (1.5) that has m of the $n + m$ variables on the left-hand side of m equations expressing the left-hand side variables as affine combinations of the remaining n variables, and sets their value to non-negative quantities when the right-hand side variables are set to zero is called a feasible dictionary. The m variables that appear on the left-hand size of a dictionary are called basic and constitute a basis, while the n variables that appear on the right-hand side of the dictionary are called non-basic.

Associated with a feasible dictionary is the notion of *basic feasible solutions* (b.f.s.). A b.f.s. is any feasible point described by a feasible dictionary, where the n non-basic variables assume the value zero.

The simplex method proceeds by finding a feasible dictionary (if one exists), and then in successive iterations modifying the previous dictionary to yield a new dictionary with a better objective value z'. A dictionary is modified by rearranging the system of equations (1.5) by moving one of the right-hand side variables to the left-hand side, and correspondingly, moving one of the left-hand side variables to the right-hand side. When the right-hand side variables of the *new* dictionary are set to zero, the left-hand side variables assume such values so that the objective function value z' of the problem is improved. The variable that enters the left-hand side in the new dictionary is called the *entering variable*, and the variable that leaves the left-hand side is called the *leaving variable* (and modifies its status from basic to non-basic.) Each simplex iteration is called a *pivot* iteration in LP terminology.

In matrix notation, the problem can be written down as

$$\max . \quad \begin{bmatrix} c^T & 0_m^T \end{bmatrix} \begin{bmatrix} x_1 \\ \vdots \\ x_{n+m} \end{bmatrix}$$

$$\text{s.t.} \begin{cases} \begin{bmatrix} A & I_m \end{bmatrix} \begin{bmatrix} x_1 \\ \vdots \\ x_{n+m} \end{bmatrix} = b \\ \begin{bmatrix} x_1 \\ \vdots \\ x_{n+m} \end{bmatrix} \geq 0 \end{cases}$$

where 0_m is an m-dimensional column vector of all zeros, and I_m is the $m \times m$ identity matrix. At the beginning of an iteration, the simplex method has an associated dictionary that expresses the m variables that constitute the current basis as linear functions of the remaining n (non-basic) variables. The basis forms a partition therefore of the variables into basic and non-basic. This partition can be used to partition also the matrix equation $[A\ I_m]x = b$ in the form $A_B x_B + A_N x_N = b$ where the columns of the matrix $[A\ I_m]$ and the rows of the vector x have been permuted so that the m basic variables of the dictionary (denoted x_B) appear first in the vector x, and the n non-basic variables (denoted x_N) appear last. The permuted matrix $[A\ I_m]$ is now written as $A = [A_B |\ A_N]$ and from now on, we shall consider the expanded matrix $[A\ I_m]$ as the problem's matrix. The $m \times m$ square matrix A_B is called the *basis matrix* of the dictionary.

The equation $A_B x_B + A_N x_N = b$ can be uniquely solved for x_B if the $m \times m$ square matrix A_B is non-singular—and we shall prove below that this is always the case.

Lemma 1.26 *The matrix A_B associated with a b.f.s. $\bar{x} = (\bar{x}_B, \bar{x}_N)$ is non-singular.*

Proof From standard linear algebra, it is enough to prove that the system $A_B x_B = b$ has a unique solution. The existence of a solution for this system is given simply as $\bar{x}_B$. Uniqueness of the solution follows by considering any other vector x'_B satisfying $A_B x'_B = b$. Considering the vector $x' = \left(x'_B, x'_N = 0\right)$ we see that it satisfies $Ax' = A_B x'_B + A_N x'_N = b$ so it must satisfy the bottom m equations (the constraints) of the dictionary describing $\bar{x}$, and therefore, since $x'_N = 0$, it follows that $x'_B = \bar{x}_B$. QED.

By multiplying the equation $A_B x_B + A_N x_N = b$ by A_B^{-1}, we see that the vector x_B and the objective function satisfy:

$$\begin{aligned} x_B &= A_B^{-1} b - A_B^{-1} A_N x_N \\ z &= c_B^T A_B^{-1} b + \left(c_N^T - c_B^T A_B^{-1} A_N\right) x_N \end{aligned} \tag{1.6}$$

where c_B represents the m-dimensional column vector of the coefficients in the objective function of the basic variables x_B and c_N represents the n-dimensional column vector of the coefficients of the non-basic variables x_N.

From (1.6) we see that if we can increase the value of any of the non-basic variables, say the jth variable, which has positive coefficient $(c_N - (A_B^{-1} A_N)^T c_B)_j > 0$ without driving any of the basic variables to assume negative value, we can find a *better* new feasible solution that satisfies the constraints $Ax = b$ and the non-negativity constraints $x \geq 0$. This is because the current dictionary represents the solution $x_B = A_B^{-1} b, \quad x_N = 0$ with objective function value $z = c_B^T A_B^{-1} b$, whereas the new solution will have $z' = c_B^T A_B^{-1} b + \left(c_N - \left(A_B^{-1} A_N\right)^T c_B\right)_j x'_j, \quad x'_j > 0$. The variable x_j that can be increased without violating the non-negativity constraints of the x_B variables forms a *candidate pivot variable for entering the basis.*

If there is no variable among the non-basic variables that has a positive coefficient $(c_N - (A_B^{-1} A_N)^T c_B)_j$ the simplex method stops its iterations, as the current dictionary is optimal.

If on the other hand there exists a non-basic variable with positive coefficient which can be increased without bound without violating the non-negativity constraints of x_B, the problem is unbounded (we can find feasible solutions of the problem driving the objective function value to $+\infty$). This can only happen in the current iteration if there is a column l of the matrix $A_B^{-1} A_N$ whose entries are all non-positive numbers and $(c_N - (A_B^{-1} A_N)^T c_B)_l > 0$.

Otherwise, we may choose among the non-basic variables any x_j that satisfies $(c_N - (A_B^{-1} A_N)^T c_B)_j > 0$. This variable can be increased until any one of the x_B variables becomes zero. The first basic variable to become zero—and therefore the one to become the *leaving* variable—will be the one whose index i in B minimizes the quantity

$$\frac{\left(A_B^{-1} b\right)_i}{\left(A_B^{-1} A_N\right)_{ij}},$$

as the variable corresponding to this index is the first (or at least one of the first) to go to zero as x_j is increased. In other words,

$$i = \arg\min_{k=1\ldots n} \frac{\left(A_B^{-1}b\right)_k}{\left(A_B^{-1}A_N\right)_{kj}}.$$

At this point we can state the revised simplex method in pseudo-code.

Algorithm Revised SIMPLEX Method for Linear Programming

Inputs: Objective coefficients column vector c, $m \times n$ constraints matrix A, m-dimensional constraints right-hand side column vector b.
Outputs: A point x^* that solves the standard form LP $\max. c^T x, \quad s.t.\ Ax \le b, x \ge 0$, or an indication that the problem is unbounded or infeasible.

Begin

1. Add m slack variables $x_{n+1}, \ldots, x_{n+m}$ to the decision variables to create an $n+m$-dimensional column vector $x = [x_1 \ \ldots \ x_{n+m}]^T$ and convert the inequality constraints to equality constraints $Ax = b$ by setting $A = [A \mid I_m]$, $c = [c^T \mid 0_m^T]^T$.
2. Determine an initial feasible dictionary (x_B, x_N) with associated basis matrix B and non-basic matrix A_N. If no dictionary can be found then ERROR('INFEASIBLE').
3. Set h=1.
4. while true do
   a. Set c_B to the m-dimensional column vector containing the coefficients of the basic variables in the objective function.
   b. Solve the linear system of m equations in m unknowns $B^T y = c_B$.
   c. Choose an index j in $\{1, 2, \ldots, n\}$ such that the jth column of $A_N a_j$ satisfies $y^T a_j < (c_N)_j$. If no such index exists then return the solution $x^* = (x_B = B^{-1}b,\ x_N = 0)$.
   d. Solve the linear system of m equations in m unknowns $Bd = a_j$.
   e. Set $i = \arg\min_{k=1\ldots n} \frac{\left(B^{-1}b\right)_k}{d_k}, u = \frac{\left(B^{-1}b\right)_i}{d_i}$.
   f. if $u < 0$ ERROR('UNBOUNDED').
   g. Move x_j from the non-basic variables to the basic variables set and set its value equal to u.
   h. Decrease the value of the rest of the basic variables' by ud.
   i. Move x_i from the basic variables to the non-basic variables set.
   j. Set $h = h + 1$.
5. end-while

End.

It remains open how to determine an initial feasible dictionary as required in step 2 of the revised simplex method. Clearly, if $b \geq 0$, setting all original variables to zero is a feasible solution and therefore the dictionary where the basic variables consist of only the slack variables is a feasible initial dictionary. Otherwise, a separate procedure has to be followed that will terminate in a finite number of steps either with a feasible dictionary, or with an indication that no such dictionary exists.

A procedure that allows us to determine an initial feasible dictionary for a given problem if one exists, or else declare the problem infeasible makes use of the very same revised simplex method we just described, operating this time on a special auxiliary LP for which we know how to obtain an initial feasible dictionary.

$$(ALP)\ \max -x_0$$
$$\text{s.t.} \begin{cases} \sum_{j=1}^{n} a_{ij}x_j - x_0 \leq b_i, & i = 1\ldots m \\ x_j \geq 0, & j = 0, 1, \ldots n \end{cases} \tag{1.7}$$

From the definition of problem (ALP), the reader should easily verify that the problem has an optimal solution of value zero, if and only if the original LP

$$\max .c^T x$$
$$\text{s.t.} \begin{cases} Ax \leq b \\ x \geq 0 \end{cases}$$

has a feasible solution. Otherwise, the optimal solution value of ALP will be strictly less than zero. Clearly ALP cannot be unbounded, as the maximum possible value its objective function can take is the value 0.

Finding a feasible solution to ALP is trivial: the solution $x_j = 0$ $j = 1,\ldots,n$. $x_0 = \max(0, \{-b_i | i = 1,\ldots,m\})$ is feasible. Finding a feasible dictionary (and thus a feasible basis) for ALP is also very easy. As before, we introduce the non-negative slack variables x_{n+i} $i = 1\ldots m$ that convert the inequality constraints into equality constraints, and use them to form a dictionary

$$z = -x_0$$
$$x_{n+i} = b_i - \sum_{j=1}^{n} a_{ij}x_j + x_0, \quad i = 1\ldots m$$
$$x_i \geq 0, \quad i = 0, 1, \ldots n + m$$

where the basis is the m slack variables, and the non-basic variables are the original ALP variables $x_0 \ldots x_n$. This dictionary is infeasible unless $b_i \geq 0$ $i = 1,\ldots, m$ (in which case we would not have to use the auxiliary problem at all.) But if we pivot the x_0 variable in the basis, then the resulting dictionary becomes feasible, so that we may apply the revised simplex method to it, without doing anything in step 2 this time. Indeed, let x_0 be the pivot variable, and without loss of generality, assume b_m satisfies $b_m \leq b_i$ $i = 1,\ldots,m-1$, and let x_{n+m} be the leaving variable. The new dictionary becomes

$$z = b_m - \sum_{j=1}^{n} a_{mj}x_j - x_{n+m}$$

$$\underset{-1}{x_{n+i}} = b_i - b_m + \sum_{j=1}^{n} \left(a_{ij} + a_{mj}\right)x_j + x_{n+m}, \quad i = 1\ldots m - 1$$

$$x_0 = -b_m + \sum_{j=1}^{n} a_{mj}x_j + x_{n+m}$$

$$x_i \geq 0, \quad i = 0, 1\ldots m$$

This dictionary is feasible because $b_i - b_m \geq 0$ for all $i = 1\ldots m{-}1$ and $b_m \leq 0$ as b_m is the smallest of all b_i $i = 1\ldots m$ and not all b_i are greater than or equal to zero. So the dictionary is feasible, and the revised simplex method can begin from step 3 above. If the method terminates with an optimal solution that is zero, the last dictionary is a feasible dictionary for the original problem, otherwise the original problem is infeasible and the algorithm issues the appropriate error message. Step 2 of the revised simplex method is also known as Phase 1 of the simplex method.

The reason that step 4.c is valid as an exit condition with optimality, needs some explanation. A dictionary in matrix form (1.6) at an iteration in which the conditional expression of step 4.c is true, expresses the objective function as an affine combination of n variables, none of which can be incremented without reducing the objective function value, since they all appear with negative coefficients. Therefore, the objective function cannot be greater than the value $c_B^T A_B^{-1} b$ which is obtained at the feasible point $x^* = (x_B = A_B^{-1}b, x_N = 0)$ which is precisely the solution returned in step 4.c, when the conditional expression becomes true.

One final comment regards the termination of the revised simplex method. The method proceeds in pivots, in which dictionaries are modified and produce new basic feasible solutions. There are finitely many different basic feasible solutions, and their number is bounded by the total number of ways by which we can choose m elements from a set of cardinality $n + m > m$. This number is $\binom{n+m}{m}$ and is only an upper bound because not every dictionary represents a basic feasible solution. Nevertheless, unless care is taken, the revised simplex method may cycle in the presence of degenerate iterations, that is, iterations in which the basic feasible solutions have the same objective function value, which in turn occurs when in an iteration, the value of one or more *basic* variables becomes zero. (By cycling we mean that starting from a certain basis, the simplex method proceeds in two or more iterations, after which at some point the same basis is encountered again.) It can be easily proven (Chvatal 1983) that if the choice of the leaving variable in step 4.e is done via the *lexicographic rule*, no pivot iteration will be degenerate, and thus the revised simplex method will not cycle, and therefore *will always terminate*. The lexicographic rule introduces "symbolic" perturbations $0 < \varepsilon_m \ll \varepsilon_{m-1} \ldots \ll \varepsilon_1 \ll 1$ to each of the original constraints

$\sum_{j=1}^{n} a_{ij}x_j \le b_i, \quad i = 1\ldots m$ so that they become $\sum_{j=1}^{n} a_{ij}x_j \le b_i + \varepsilon_i$. Now, in the course of simplex iterations, each of the equations describing the m constraints of the problem in the current dictionary will be of the form

$$x_k = r_k + \sum_{i=1}^{m} r_{ik}\varepsilon_i + \sum_{j \in N} d_j x_j, \quad k \in B$$

The lexicographic rule states that the leaving variable will be the variable x_k which has the smallest vector $r = [r_k/d_j \; r_{1k} \; r_{2k} \; \ldots \; r_{mk}]^T$ in the lexicographic sense among all variables in the basis, where x_j is the entering variable.

Theorem 1.27 (Fundamental Theorem of Linear Programming) *Every LP that has no optimal solution must be either infeasible or unbounded, and every LP that has a feasible solution must also have a basic feasible solution. Further if an LP has an optimal solution, then it must have a basic feasible solution that is an optimal solution.*

Proof Every LP can be converted into standard form. Applying the revised simplex method on the converted LP in standard form, we see that step 2 of the revised simplex method either discovers that the problem is infeasible or constructs a basic feasible solution. In steps 3–5, the method either discovers that the problem is unbounded or terminates with an optimal solution that is in the form of a basic feasible solution. QED.

The Duality Theorem of Linear Programming

Besides the development of the simplex method, the duality theorem stands as the most important development in the theory of LP, with tremendous implications that resulted in many significant theoretical and algorithmic innovations in the field of optimization, such as dual methods, primal–dual methods, and so on. It is not an exaggeration to state that most of the modern duality theories in general optimization and control stem from the fundamental observations making up the duality theorem of LP.

Consider an LP in standard form

$$(PLP)\ \max . \sum_{j=1}^{n} c_j x_j$$

$$\text{s.t.} \begin{cases} \sum_{j=1}^{n} a_{ij}x_j \le b_i, & i = 1\ldots m \\ x_i \ge 0, & i = 1\ldots n \end{cases}$$

Such an LP is called the primal LP. The dual problem of the primal LP has m decision variables y_i $i = 1\ldots$ m associated with the m constraints of the primal problem and is stated as follows

$$(DLP)\ \min.\sum_{i=1}^{m} b_i y_i$$
$$\text{s.t.}\begin{cases}\sum_{i=1}^{m} a_{ij} y_i \geq c_j, & j = 1\ldots n\\ y_i \geq 0, & i = 1\ldots m\end{cases} \tag{1.8}$$

Now, the dual LP (1.8) has a number of very interesting properties. First of all, observe that the dual is constructed by using the column vector b that corresponds to the right-hand side of the primal constraints as the vector of coefficients in the objective function of the dual, using the vector of coefficients c in the objective function of the primal as the right-hand side of the constraints of the dual, transposing the constraints matrix A of the primal to obtain the constraints matrix of the dual, and finally reversing the direction of optimization and the constraints of the primal problem for the dual. Clearly, if we repeat this process starting from the dual problem, what we end up with is the original primal problem. Therefore, *the dual of the dual is the primal problem.*

Furthermore, observe that if $(\hat{y}_1, \ldots, \hat{y}_m)$ is a feasible solution of (DLP), and if $(\hat{x}_1, \ldots, \hat{x}_n)$ is a feasible solution for (PLP), *the objective function value of the dual is greater than or equal to the objective function value of the primal.* This is because

$$\sum_{j=1}^{n} c_j \hat{x}_j \leq \sum_{j=1}^{n}\left(\sum_{i=1}^{m} a_{ij}\hat{y}_i\right)\hat{x}_j = \sum_{i=1}^{m}\left(\sum_{j=1}^{n} a_{ij}\hat{x}_j\right)\hat{y}_i \leq \sum_{i=1}^{m} b_i \hat{y}_i$$

This inequality is at the heart of the proof of the duality theorem stated below.

Theorem 1.28 (Duality Theorem of Linear Programming). *If the primal LP (PLP) has an optimal solution x^* then the dual problem (DLP) has an optimal solution y^* and the objective function value of the two problems at their optimum is the same. If the primal problem (PLP) is unbounded then the dual problem is infeasible, and if the dual is unbounded then the primal is infeasible.*

Proof Given that a solution of the dual is always greater than or equal to any solution of the primal problem, it is enough to construct a feasible solution for (DLP) whose objective function value is equal to the optimal objective function value of the primal problem (PLP). Such a solution is immediately available from the final dictionary obtained by the revised simplex method, and corresponds to setting the variables y_i equal to the negative of the coefficients with which the slack variables x_{n+i} $i = 1\ldots m$ appear in the final expression of the objective function in the last (optimal) dictionary constructed by the simplex method. Indeed, let the final dictionary obtained by applying the simplex method to (PLP) be the following

$$x_k = x_k^* + \sum_{j \in N} d_{kj} x_j, \quad k \in B$$

$$z = z^* + \sum_{j=1}^{n+m} \bar{c}_j x_j$$

where the coefficients $\bar{c}_j$ are all non-positive (and in fact $\bar{c}_j = 0 \, \forall j \in B$), and z^* is obviously the optimal value of the primal problem (*PLP*). We shall show that the solution $y_i^* = -\bar{c}_{n+i}, \quad i = 1 \ldots m$ is a feasible solution for the dual whose objective function value equals $c^T x^*$. Indeed the primal objective function $z = c^T x$ from the last row of the final dictionary above, can be written as

$$z = \sum_{j=1}^{n} c_j x_j = z^* + \sum_{j=1}^{n} \bar{c}_j x_j + \sum_{i=1}^{m} \bar{c}_{n+i} x_{n+i} = z^* + \sum_{j=1}^{n} \bar{c}_j x_j + \sum_{i=1}^{m} \bar{c}_{n+i} \left(b_i - \sum_{j=1}^{n} a_{ij} x_j \right)$$

$$= z^* + \sum_{j=1}^{n} \bar{c}_j x_j - \sum_{i=1}^{m} y_i^* \left(b_i - \sum_{j=1}^{n} a_{ij} x_j \right) = z^* - \sum_{i=1}^{m} b_i y_i^* + \sum_{j=1}^{n} \left(\bar{c}_j + \sum_{i=1}^{m} a_{ij} y_i^* \right) x_j$$

Now, since this equation has been obtained by algebraic transformations of the initial dictionary, it must hold for every possible set of values that the variables $x_j \; j = 1 \ldots n$ may assume. Therefore, we must have $z^* - \sum_{i=1}^{m} b_i y_i^* = 0 \Leftrightarrow z^* = \sum_{i=1}^{m} b_i y_i^*$, and also $\left(\bar{c}_j + \sum_{i=1}^{m} a_{ij} y_i^*\right) = c_j \forall j = 1 \ldots n$. But this equation implies that $\sum_{i=1}^{m} a_{ij} y_i^* \geq c_j \forall j = 1 \ldots n$ since the coefficients $\bar{c}_j \leq 0 \forall j = 1 \ldots n+m$. The two last facts imply that the solution $y_i^* = -\bar{c}_{n+i}, \; i = 1 \ldots m$ is indeed a feasible solution for the dual problem (DLP) and since the objective function value of the dual at this solution equals the optimal value of the primal problem, it is the optimal value of the dual.

If the primal is unbounded, since any solution of the dual would have to be greater than or equal to any solution of the primal, the dual must be infeasible. By the same reasoning, if the dual is unbounded (the objective function of the dual goes to minus infinity) the primal must be infeasible. QED.

Now, consider an LP problem in standard form (the primal problem) and the corresponding dual in the form (1.8), with corresponding feasible solutions x^* and y^*. Since y^* is a feasible solution for (DLP), it satisfies $\sum_{i=1}^{m} a_{ij} y_i^* \geq c_j \forall j = 1 \ldots n$. Multiplying this inequality by x_j^* we get $c_j x_j^* \leq x_j^* \sum_{i=1}^{m} a_{ij} y_i^*, \quad \forall j = 1 \ldots n$ and by applying the same argument on the inequalities of the primal problem, we get $y_i^* \sum_{i=1}^{m} a_{ij} x_j^* \leq b_i y_i^*, \quad \forall i = 1 \ldots m$. From the duality theorem, we know that if x^* and y^* are the optimal solutions of the primal and dual problem respectively, then $c^T x^* = b^T y^*$ which in turn means that the above stated $n + m$ inequalities must all hold as strict equalities. Therefore, if x^* and y^* are the optimal solutions of the primal and dual problem respectively, then the following $n + m$ equations are satisfied

$$\left(c_j - \sum_{i=1}^{m} a_{ij} y_i^*\right) x_j^* = 0, \quad \forall j = 1 \ldots n$$

$$\left(b_i - \sum_{j=1}^{n} a_{ij} x_j^*\right) y_i^* = 0, \quad \forall i = 1 \ldots m$$

These equalities can be stated as the following complementary slackness theorem of LP.

Theorem 1.29 (Complementary Slackness Theorem) *A feasible solution x^* of the primal problem (PLP) and a feasible solution y^* of the dual problem (DLP) are simultaneously optimal if and only if the following equations hold*

$$\left(c_j - \sum_{i=1}^{m} a_{ij} y_i^*\right) x_j^* = 0, \quad \forall j = 1 \ldots n$$

$$\left(b_i - \sum_{j=1}^{n} a_{ij} x_j^*\right) y_i^* = 0, \quad \forall i = 1 \ldots m$$

Proof By the argument above, we have already proved that if x^* and y^* are optimal and dual-optimal respectively, the equations hold. Vice versa, if the equations hold, then the objective function value of the primal solution x^* equals the objective function value of the dual, and since the objective function value of the primal will always be less than or equal to the objective function of the dual, x^* and y^* must be optimal for the primal and the dual respectively. QED.

The importance of the duality theorem and the complementary slackness theorem cannot be stressed enough. It has found many applications and admitted interpretations in economic theory, production theory, and many other fields of engineering and science in general. It has been the starting point for the development of very fast algorithms and heuristics for the solution of linear and quadratic programming problems, as well as the much more difficult mixed integer programming problems.

The prevalence of linear programming both as a theory and algorithmic tool, as well as modeling tool will be made clear in the subsequent chapters of this book in which a large set of problems will be modeled as linear programming problems. In practice, the running time of the simplex method as measured by the number of pivot iterations needed to terminate, is usually in the order of $3m\log n$ where m is the number of constraints and n the number of variables. (This number is of course not a strict upper bound on the complexity of the algorithm, as the simplex method using any myopic rule for choosing the entering variable can require a number of steps to terminate what is exponential in the number of variables n, as the famous Klee–Minty examples demonstrated (Chvatal 1983). However, this result only says that *in theory*, the simplex method is not a satisfactory method; *in practice*, as we mentioned before, the simplex method is still the best method known today for solving large-scale LP problems and *is extremely satisfactory*.) In the next section, we present one of the most important practical applications of linear programming, namely linear network flows, for which a special variant of the simplex method called the network simplex method, relying on special data structures, can be made to run at least one order of magnitude faster than the revised simplex method.

1.1.2.2 Linear Network Optimization

In this section, we review the results of optimization problems defined over a network or graph, that happen to have a great many applications in the field of Supply Chain Management. Before proceeding, we need the definition of a directed graph.

Definition 1.30 A directed graph $G(V,E)$ consists of a finite set of nodes V and a set of edges E that are ordered pairs (u,v) of nodes u, v from V defining one-way connections between nodes. An edge-weighted directed graph is a directed graph $G(V,E,W)$ that has a scalar weight value associated with every edge so that $W : E \to \mathbb{R}$.

Clearly, since the set of nodes is finite, so is the set of edges E connecting the nodes of the graph. Nodes are also called *vertices*, and edges are sometimes referred to as *arcs* or *nets*. A directed graph is also known as a *network*.

The Shortest Path Problem

Perhaps the most well-known optimization problem defined over an edge-weighted directed graph G with non-negative edge weights is to find the shortest path between a source node s and a target node t in V.

Definition 1.31 A path $\langle s, t\rangle$ from a node s to a node t in a graph $G\ (V,\ E)$ is a sequence of edges (s,n_1), $(n_1, n_2)\ldots(n_k, t)$ all of them belonging to E such that the ending node of one edge is the starting node of the next edge in the sequence, and the starting node of the first edge in the sequence is s and the ending node of the last edge in the sequence is t. The weight of a path is the sum of all weights of the edges in the path.

The Dijkstra method (Dijkstra 1959) is a label-setting method for solving the Shortest Path Problem (SPP) which according to its inventor, took less than 20 min to design—on the back of a napkin. The algorithm is an example of application of the optimality principle in Dynamic Programming that will be briefly discussed later. The idea is to maintain a set of distance labels for each node that are initially set to infinity except for the source node whose label is set to zero. At each iteration of the algorithm the so far unvisited neighbor of the current node with least distance from the source node is selected to be "visited". The labels of all neighbors of the newly visited neighbor are then updated if the sum of the "visited" node's label and the weight of the edge connecting the node to the neighbor is smaller than the neighbor's current label. The algorithm stops when the target node is visited. A formal description of the algorithm in pseudo-code follows:

Algorithm Dijkstra

Inputs: Network $G(V,E,W)$ with non-negative edge weights, source node $s \in V$, target node $t \in V$.
Outputs: A shortest path p connecting node s to node t, or an indication node t is not reachable from s in G.

Begin

1. if $s = t$ return {}.
2. Create an empty mapping $label : V \to \mathbb{R}_+$.
3. Create an empty mapping $prev : V \to V$.
4. for each $v \in V$do:
   a. if $v = s$ then Set $label[v] = 0$.
   b. else Set $label[v] = +\infty$.
   c. end-if
   d. Set $prev[v]$ = nil.
5. end-for
6. Create a priority queue Q storing each node v in V in ascending order of label value.
7. while $Q \neq \{\}$ do:
   a. Set n = smallest label node in Q.
   b. Set $Q = Q - \{n\}$.
   c. if $label[n] = +\infty$ then
      i. ERROR('target node unreachable').
   d. end-if
   e. if $n = t$ then
      i. Set $p = \{\}$.
      ii. while $prev[n] \neq$ nil do:
         1. Set $p = p \cup \{(prev[n],n)\}$.
         2. Set $n = prev[n]$.
      iii. end-while
      iv. return p.
   f. end-if
   g. for each m in V such that (n,m) is in E do:
      i. Set $d_{new} = label[n]+w_{(n,m)}$.
      ii. if $label[m] < d_{new}$ then
         1. Set $label[m] = d_{new}$.
         2. Set $prev[m] = n$.

iii. end-if

h. end-for

8. end-while

End.

If a binary heap data structure is used to implement the priority queue Q, the complexity of the algorithm in terms of running time is $O((|V|+|E|)\log|V|)$. The correctness of the algorithm is also easy to establish. Correctness is immediately evident from the fact that nodes are visited in ascending order of shortest distance from s *and labels of visited nodes contain the shortest distance from* s. To see this, observe first that any node label that is set or updated at any time represents a valid distance of that node from s. Now observe that the property holds in the first iteration trivially. If it holds for the kth iteration when the algorithm visits say node n_k, then in the $(k+1)$-st iteration, the algorithm picks, say node n_{k+1} with smallest label in Q. If there was another node m in Q with strictly smaller shortest distance from s than n_{k+1}, let the shortest path from s to m have as last edge an edge (l,m). Node m could not have been visited at the time node n_k was visited, because in the $(k+1)$-st iteration node m would not be in Q. Further, node l could not have been visited within the first k iterations, because if it had, by the induction hypothesis, its label would have been correctly set, and the label of node m would also have been correctly updated in step 7.g to reflect the shortest distance from s to m and then in the $(k+1)$-st iteration node m would have been selected instead. But then, node l cannot be among the first k shortest distance nodes and therefore node m cannot be among the first $k+1$ shortest distance nodes either, proving the induction step, and completing the proof of correctness of Dijkstra's algorithm.

It is important to note that algorithm Dijkstra as given above may never terminate if there exists a cycle with negative cost—since the more one repeats traveling along the cycle, the smaller their total path weight becomes. For this reason, the assumption of non-negative edge weights was introduced.

The SPP has a tremendous number of applications in the real-world, and in Supply Chain Management in particular. In logistics, when deliveries must be made from one source to one destination, fuel costs often dominate the total transportation costs and solving the SPP becomes a requirement.

There are several very important extensions to the SPP encountered in Supply Chain Management domains. For instance, in routing and other logistics problems, one must find a shortest path in a network between a source node and a target node that are subject to resource constraints: traveling along an edge (i,j) in the network not only incurs a cost $w_{(i,j)}$ to be added to the objective function, but also incurs the consumption of a resource by an amount $r_{(i,j)}$. The constraint is that the path to be selected must not incur more than a pre-specified amount of resource consumption. Examples of a resource could be time spent traveling along an edge, whereas the cost of an edge could indicate the actual distance between nodes i and j. The objective then would be to find the shortest distance path

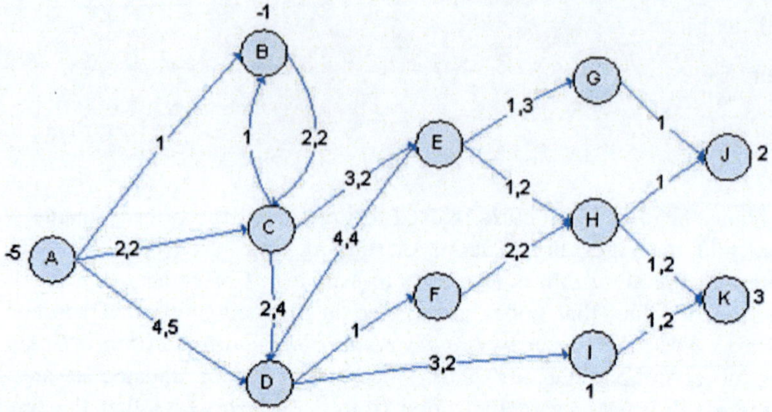

Fig. 1.13 Example Transshipment problem graph

between two nodes s and t so that the path can be traveled in less than one hour. The model for the Shortest Path Problem with Resource Constraints then becomes the following optimization problem defined over a network $G(V,E)$ with costs c_{ij} and resources r_{ij} defined over the edges (i,j) in E.

$$(SPPRC)\ \min_x \sum_{(i,j)\in E} c_{ij}x_{ij}$$

$$\text{s.t.} \begin{cases} \sum_{(i,j)\in E} r_{ij}x_{ij} \le R \\ \sum_{j:(s,j)\in E} x_{sj} = \sum_{i:(i,t)\in E} x_{it} = 1 \\ \sum_{i:(i,k)\in E} x_{ij} - \sum_{j:(k,j)\in E} x_{kj} = 0, \quad \forall k \in V - \{s,t\} \\ x_{ij} \in \{0,1\}, \quad \forall (i,j) \in E \end{cases}$$

Contrary to the SPP, there are no polynomial time algorithms for solving the SPPRC but there are some very elegant pseudo-polynomial time algorithms for solving it. We shall describe such algorithms for a more constrained variant of the problem in the chapter on distribution management.

The Minimum Cost Network Flow Problem

Consider the network in Fig. 1.13, representing a *single-commodity* transshipment problem. The network contains 11 nodes and 17 arcs. Some nodes have an associated supply or demand. For example node A is a supply node that supplies to the network 5 units of a single commodity, whereas node I is a demand node requesting 1 unit of the same single-commodity. Arcs have an associated unit commodity cost incurred for every unit of the commodity that flows along it. For example, the cost incurred when one unit of the commodity flows along arc (B,C) is 1, whereas the cost incurred when two units of the commodity flow along arc (D,E) is 8(=2 × 4). When a second number is associated with an arc in the figure, it denotes the *capacity of the arc*, namely it represents a constraint on the maximum value the flow along an arc can take. In the figure, the capacity of arc (E,G) is 3, whereas arc (A,B) has infinite capacity (no constraint on the value of the flow along this arc exists).

The (linear, single-commodity) minimum cost network flow problem asks to move all quantities produced at the supply nodes from the supply nodes to the demand nodes meeting the demand exactly without violating any arc capacity constraints, and with minimum cost incurred. Arcs may also have non-zero lower bounds on the amount of flow they must carry, but one can always convert a problem with non-zero lower bounds into an equivalent minimum cost network flow problem where all arcs have zero lower bounds. The details of this conversion are left as an exercise for the reader.

Let $G(V,E)$ be a network whose nodes $n \in V$ have demand d_n and its arcs $(i,j) \in E$ having cost c_{ij} and capacity u_{ij}. We define the Linear Minimum Cost Network Flow Problem (*LMCNF*) to be the following problem:

$$(LMCNF)\ \min.\ \sum_{(i,j)\in E} c_{ij}x_{ij}$$

$$\text{s.t.}\begin{cases} \sum\limits_{i:(i,j)\in E} x_{ij} - \sum\limits_{k:(j,k)\in E} x_{jk} = d_j, & \forall j \in V \\ x_{ij} \le u_{ij}, & \forall (i,j) \in E \\ x_{ij} \ge 0, & \forall (i,j) \in E \end{cases} \quad (1.9)$$

The equality constraints in (1.9) represent *flow balance constraints* at every node, and by adding them all up, we immediately see that any instance of the (*LMCNF*) problem is feasible only if the right-hand sides of the equality constraints add up to zero (flow preservation).

The *LMCNF* represents the most common form of LP problems encountered in practice. It is of course an LP (not in standard form), with $|E|$ variables and $|V|+|E|$ constraints, and can be solved by the revised simplex method. However, inspecting the problem reveals a very special structure that gives rise to many interesting properties. The most important property of the problem's structure is that the constraints of the problem form a matrix A whose entries are elements of the set $\{-1, 0, 1\}$, and in fact, each column of the matrix A corresponds to an edge in the network and has all elements zero except the two entries corresponding to the starting node of the edge and the ending node of the edge who have as entries the numbers 1 and −1 correspondingly. This, of course, is a direct consequence of the underlying network structure of the problem in the first place.

We shall describe a specially tailored version of the simplex method that applies to the LMCNF problem only, which in practice is usually an order of magnitude faster than the standard revised simplex method.

Similar to the revised simplex method, the network simplex method proceeds in pivot iterations, moving from one basic feasible solution to another until no further improvement can be made, at which point the method terminates. However, dictionaries and systems of linear equations are abandoned in favor of much more efficient data structures that are applicable when the network structure is present. A b.f.s. in the network simplex method is represented by a *feasible spanning tree T*—rooted at a certain node n_r in V—that consists of a subset of arcs of the network that have the property that all nodes in the network are covered by at least one arc that is in T and that the sub-graph induced by the restriction of the original network to the arcs of T is still connected, the arcs in T form no cycle and finally, it is only the arcs in T that may have positive flow less than the arc's capacity. In a b.f.s., the arcs that are not part of the spanning tree will have flow that will be either zero, or at the arc's capacity, so that the balance constraints are preserved for every node. The network simplex method also maintains a vector π of *prices for every node essentially representing the dual variables of the* LP (one variable for each node which is represented by one constraint in the balance flow constraints set.) These price—also known as node potentials or as Lagrange or simplex multipliers—have the property that $\pi_r = 0$ for the root node of the tree T representing the current basis, and all other nodes have prices satisfying the equation $c_{ij} - \pi_i + \pi_j = 0\ \forall(i,j) \in T$

In a pivot iteration of the network simplex method, the entering variable is any variable x_{ij} of an arc $(i,j) \notin T$ that has zero flow and *negative reduced cost* $r = c_{ij} - \pi_i + \pi_j < 0$ *or* has flow at capacity ($x_{ij} = u_{ij}$) and *positive reduced cost* $r = c_{ij} - \pi_i + \pi_j > 0$. If one of the above conditions is satisfied, we can push (or pull) flow along the cycle formed by the introduction of the non-basic arc (i,j) in T so as to reduce the objective function value of the problem and still maintain all flow balance at the nodes and variable "box" constraints, in the same spirit as that of the standard revised simplex method. Such a variable then becomes a legitimate candidate entering variable. In an exact analogy with the revised simplex method, the leaving variable would be that basic arc (p,q) in the cycle whose flow would reach the upper or lower bound first. (If no such arc can be found, the problem is unbounded.) This arc (p,q) is removed from $T \cup \{(i,j)\}$ resulting in a new basic feasible spanning tree T'. The price vector π is then updated so that reduced costs along each arc of T' are zero, and the tree is re-rooted if needed.

When no entering variable can be found in a pivot iteration, the optimality conditions (1.10) for x and π below are satisfied, and the current b.f.s. represented by the spanning tree T and the remaining non-basic variables set at their lower or upper bounds is an optimal solution for *LMCNF*.

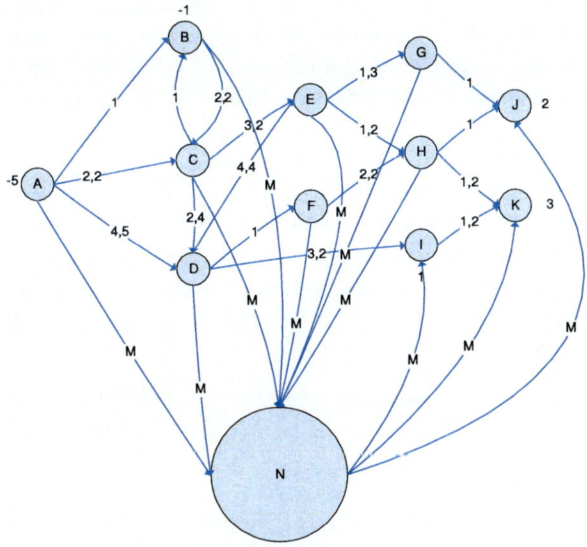

Fig. 1.14 The auxiliary network for the Big-M method created for the example network

$$
\begin{aligned}
&\sum_{i:(i,j)\in E} x_{ij} - \sum_{k:(j,k)\in E} x_{jk} = d_j, \quad \forall j \in V \\
&0 \le x \le u \\
&c_{ij} - \pi_i + \pi_j = 0, \quad \forall (i,j) \in T \\
&c_{ij} - \pi_i + \pi_j > 0 \Rightarrow x_{ij} = 0, \quad \forall (i,j) \notin T \\
&c_{ij} - \pi_i + \pi_j < 0 \Rightarrow x_{ij} = u_{ij}, \quad \forall (i,j) \notin T
\end{aligned} \tag{1.10}
$$

In a clear analogy with the revised simplex method for LP, the issues of initialization and degeneracy arise here as well. To obtain an initial basic feasible solution—or to prove that the problem is infeasible—an *auxiliary* network flow problem for which an initial basic feasible solution is readily available is solved; this new problem is constructed in such a way that if it is unbounded, the original problem is also unbounded; and if the original problem has an optimal solution, then the auxiliary problem's optimal solution is also an optimal solution for the original problem, whereas if the original problem is infeasible, then the optimal solution of the auxiliary problem will provide an appropriate indication for the original problem's infeasibility.

This technique is known as the *big-M* method, and consists of adding a *dummy* node N with zero demand $d_N = 0$ to the original network and $|V|$ *dummy* arcs so as to connect the new dummy node to each of the original network nodes. The new arcs have infinite capacity each, and connect supply nodes s to the dummy node, and the dummy node is connected to each demand node t with arcs (N,t) (see Fig. 1.14). Clearly, *assuming the total demands of the nodes equal the total nodes supply which is simply a test on the condition* $\sum_{j\in V} d_j = 0$, this "augmented" network admits an initial basic feasible solution, whereby the root of the basic feasible spanning tree is the new dummy node N, and the spanning tree arcs are all the new arcs that were inserted to connect N with the rest of the network nodes. Along the arcs connecting N to a demand node t flows a flow equal to the node's demand $d_t > 0$ whereas along the arcs connecting a supply node s to N (s,N) flows a flow equal to the node's supply $-d_s > 0$. Zero flow flows among arcs connecting N to a zero demand node n having $d_n = 0$. This flow vector is a b.f.s. for the auxiliary problem that the network simplex method can now solve. By imposing costs along the new dummy arcs high enough so that in an optimal solution there should never be flow along any of these dummy arcs, we can guarantee that an optimal solution of the auxiliary problem will never have any flow along the dummy arcs unless the original problem is infeasible. Such a cost value for the dummy arcs can be

$$M = \frac{|V| \max\limits_{(i,j)\in E} c_{ij}}{2} + 1$$

hence the name *big-M method*.

It is easy to see why the auxiliary problem of the big-M method cannot be unbounded if the original problem is not unbounded. If the network flow problem for the auxiliary network is unbounded, there must exist a simple (forward) cycle along which one can push any amount of flow and the total cost along this cycle must be negative. If this cycle does not contain any of the dummy arcs, then this cycle is valid for the original network as well, and the original problem is unbounded as well. If on the other hand, the cycle contains any one dummy arc, then it must contain actually two dummy arcs (since the dummy node N connects to the rest of the network through dummy arcs only), and letting $k \leq |V|$ of the original arcs participate in the cycle, the cost along it has to be greater than $(|V| - k) \max\limits_{(i,j)\in E} c_{ij} \geq 0$ contradicting that the cycle has negative cost.

The possibility of cycling in the network simplex method needs to be addressed as well. In the revised simplex method, cycling can be avoided by selecting the leaving variable via the lexicographic rule. In the context of the network simplex method, degeneracy that can lead to cycling is avoided via the use of *Cunningham's Rule*, which states that the out-arc e_o should be the arc in the cycle C formed by the current tree T and the in-arc $e_i = (s_i, t_i)$ is the arc with minimum flow among the arcs in the cycle C that are oriented in the opposite direction as we travel the cycle in the direction specified by the in-arc e_i that is encountered first when we start the traversing of the cycle from the node that is the first node of C that lies on the unique path of T starting from the root node of T and ending at s_i.

The network simplex method is given below in pseudo-code.

Algorithm Network SIMPLEX Method for Linear Minimum Cost Network Flow

Inputs: Network $G(V,E)$ with associated scalar costs c_{ij} and non-negative (possibly infinite) capacities u_{ij} on the network arcs, and associated demands d_n on the network nodes.

Outputs: A point x^* that solves the (*LMCNF*) problem (1.9), or an indication that the problem is unbounded or infeasible.

Begin

1. Set $d = \sum_{i \in V} d_i$.
2. if $d \neq 0$ then
   a. ERROR('INFEASIBLE').
3. end-if
4. Add dummy node and dummy arcs to the original graph according to the big-M method.
5. Create an initial b.f.s. by considering the spanning tree T hanging from the dummy node with arcs all the dummy arcs with appropriate flows according to the big-M method.
6. Select the entering arc a_e to enter the tree.
7. while no termination condition has been detected do
   a. Select the leaving arc a_o according to Cunningham's Rule.
   b. if a_o = nil then
   i. ERROR('UNBOUNDED').
   c. end-if
   d. Update the flows along the simple cycle formed by the current tree T and the arc a_e.
   e. Rebuild the tree T so that it includes the arc a_e and excludes the arc a_o.
   f. Update the depth at which each node is from the root of the new tree.
   g. Update the prices of each node.
   h. Set a_e equal to a new entering arc.
   i. if a_e = nil then
      i. if any dummy arc has non-zero flow then
         1. ERROR('INFEASIBLE').
      ii. else
         1. Set x^* equal to the current flows.
         2. Set termination condition for optimality to true.
      iii. end-if
   j. end-if
8. end-while
9. return x^*.

End

Perhaps the most remarkable result regarding the single-commodity linear minimum cost network flow problem is that if an instance has an optimal solution, *the optimal flow is guaranteed to have integral values when the node demands are integer quantities independent of the cost coefficients along the network arcs.*

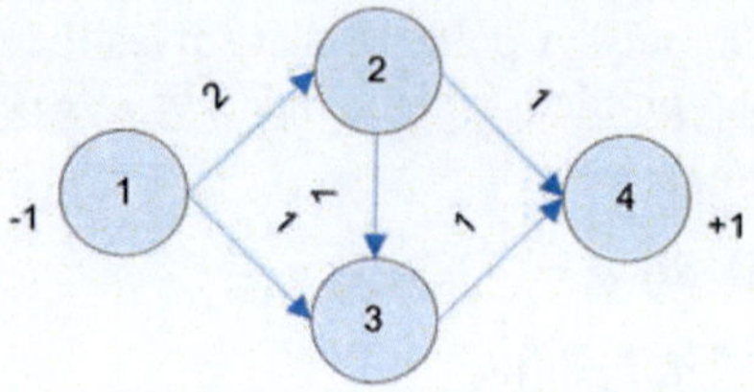

Fig. 1.15 An example network problem for the network simplex method

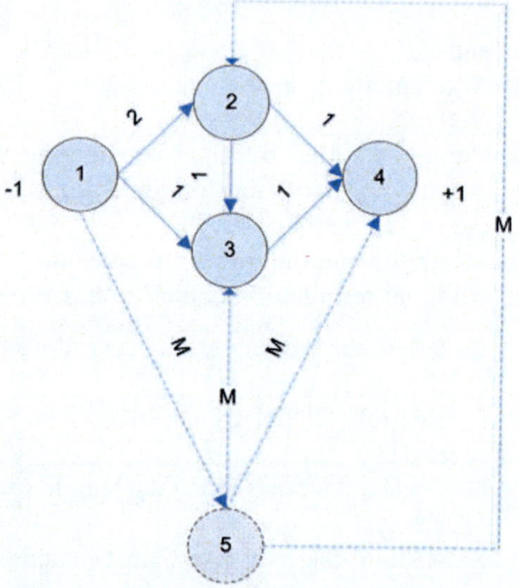

Fig. 1.16 The auxiliary network

This property is easy to establish by observing that in the algorithm the flows along each arc start at integer quantities, and maintain their integrality since whenever flow is pushed along a cycle, an integer quantity is added or subtracted from the flow of the arcs participating in that cycle.

As a small example, consider the network flow problem shown in Fig. 1.15. The only nodes with non-zero demand are node 1 (with supply equal to 1) and node 4 (with demand equal to 1). Arc costs are shown in the figure, and there are no capacities on the arcs.

Applying the network simplex method to this problem first yields the auxiliary network in Fig. 1.16 below.

1. In the beginning, the value of M is determined to be 4. The basis consists of arcs (1,5), (5,2), (5,3), and (5,4). The root of the tree is of course node 5. Flow along arcs (1,5) and (5,4) is 1. Node prices are set to obey $c_{ij} = \pi_i - \pi_j$ for all arcs (i,j) in the tree T, and since $\pi_5 = 0$, we have $\pi_1 = 4$, and $\pi_2 = \pi_3 = \pi_4 = -4$.
2. In the first iteration, the entering arc is determined to be arc (1,3), as it has the greatest reduced cost. The leaving arc is determined to be arc (5,3). The price of node 3 becomes 3 (from −4). The basis is now (1,5), (1,3), (5,2) and (5,4). The total cost of this solution is 8.

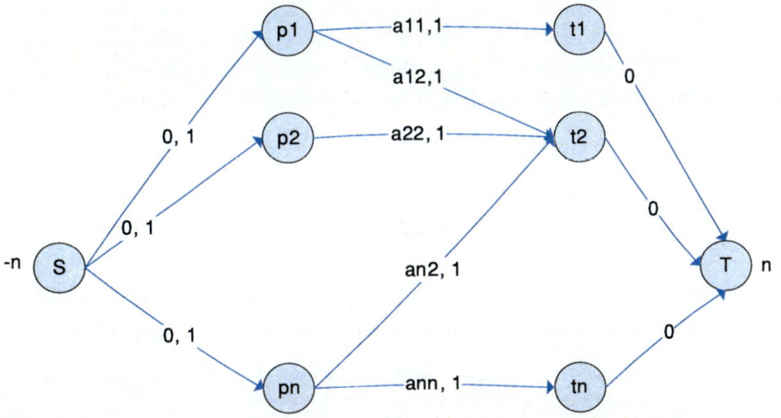

Fig. 1.17 Linear assignment as a linear minimum cost network flow problem

3. The new incoming arc is determined to be arc (3,4). The leaving arc is determined to be arc (1,5). The price of node 3 becomes −3 (from 3), and the price of node 1 is also updated to −2 (from a value of 4). The basis is now (1,3), (3,4), (5,2) and (5,4). A flow of 1 flows along arcs (1,3) and (3,4), with a total cost of 2.
4. At this point, there is no entering arc and the algorithm terminates with the optimal solution which is pushing a flow of one unit along arcs (1,3) and (3,4). The node prices for nodes 1 through 4 are −2, −4, −3, & −4 respectively.

The Linear Assignment Problem

A very important special case for the LMCNF problem is the *Linear Assignment* problem, that can be stated as follows: consider a set of n persons $p = \{p_1, p_2, \ldots, p_n\}$ that must be assigned to n tasks from a set $T = \{t_1, t_2, \ldots, t_n\}$, so as to maximize the "total value" of the assignment measured as the sum of known values a_{ij} that are incurred when person p_i is assigned to task t_j, while each person p_i can only be assigned to any task from a given set $T_i \subseteq T$. In other words, assuming the values a_{ij} represent the "preference" of person p_i for the task $t_j \in T_i$, we seek the optimal set of pairs (p_i, t_j) that maximizes the total sum of preferences of persons for tasks so that each person is assigned exactly one task.

The Linear Assignment Problem (LAP) can be formulated as an instance of the *LMCNF* problem, as shown in Fig. 1.17. Node S has a supply $d_S = -n$ and node T has a demand $d_T = n$. All arc capacities are at 1.

Since the problem data (except arc costs) are all integers, the integrality of the solution is guaranteed and for this reason, the solution of the *LMCNF* problem represents the optimal solution to the LAP as well. Therefore, the network simplex method described above can be used to solve any linear assignment problem as well. However, another idea eventually based on non-smooth optimization techniques lead to a different and much faster family of algorithms for solving LAP, namely *auction algorithms* (Bertsekas 1988). The generic auction algorithm for LAP proceeds in iterations during which a set of previously unassigned persons are each assigned to an object so that a property called *ε-complementary slackness* (ε-CS) is maintained. ε-CS is strongly related to the standard complementary slackness conditions of linear programming (Theorem 1.29) which guarantee optimality of a proposed solution x for an LP instance and a proposed solution y for its dual. In the context of the LAP, the complementary slackness conditions mandate that an assignment of the n persons from to the n tasks from T specified by the set $A = \{(p_1, t_{j_1}), (p_2, t_{j_2}), \ldots (p_n, t_{j_n})\}$ and a set of prices π_j for the tasks t_j in T satisfy $a_{ij_i} - \pi_{j_i} = \max_{j \in T_i}\{a_{ij} - \pi_j\}, \quad \forall i = 1 \ldots n$. The assignment A is then optimal, and even more, the price vector π is the global solution to the optimization problem $\min_y d(y) = \sum_{i=1}^n \max_{j \in T_i}\{a_{ij} - y_j\} + \sum_{j=1}^n y_j$. To see why, observe that the cost of *any* feasible assignment $\hat{A} = \{(p_1, t_{k_1}), (p_2, t_{k_2}), \ldots (p_n, t_{k_n})\}$ must necessarily satisfy

$$\sum_{i=1}^{n} a_{ik_i} \leq \sum_{i=1}^{n} \left\{ \max_{j \in T_i}\{a_{ij} - y_j\} + y_i \right\} \tag{1.11}$$

for *any* set of price values y_i, $i = 1 \ldots n$ since $a_{ik_i} - y_{k_i} \leq \max_{j \in T_i}\{a_{ij} - y_j\}, \quad \forall i = 1 \ldots n$. So, if the assignment A and price vector π satisfy complementary slackness conditions, then by adding up the n equations, we immediately see that inequality (1.11) is satisfied with equality, and then, the assignment must be optimal because no other assignment can exceed in value the right-hand side of (1.11). On the other hand, the price vector π must solve the problem of minimizing the function $d(y)$ because no price vector $\hat{\pi}$ can attain a value of the objective function $d(\hat{\pi})$ lower than the left-hand side of (1.11) which is attained by the specific price vector.

Now, define a partial assignment to be a (possibly empty) set A of pairs (p_i, t_j) where $t_j \in T_i$. Let y denote any vector of "*current prices*" for the tasks in T. In the context of the LAP, the *ε-CS* conditions are as follows:

$$(\varepsilon - CS) \quad a_{ij} - y_j \geq \max_{k \in T_i}\{a_{ik} - y_k\} - \varepsilon, \quad \forall (p_i, t_j) \in A$$

where $\varepsilon > 0$ is a user-defined parameter. By extending the argument given above for the complementary slackness conditions' optimality, so as to include the perturbation ε, it can easily be proven that if a vector of current prices y can be found so that *ε-CS* hold and set A represents a complete assignment (i.e. no person is unassigned by A), then the assignment represented by A *is within nε of being optimal*. Since ε can be chosen arbitrarily small by the user, an algorithm that maintains *ε-CS* for the assignments A it produces, and is guaranteed to eventually find a complete assignment is also guaranteed to find an solution to the LAP that is within $n\varepsilon$ of being optimal; if the problem data are integer, then by choosing $\varepsilon < 1/n$ the final assignment will be necessarily optimal.

As mentioned before, the auction algorithm maintains at each iteration a partial assignment A, and a price vector y so that ε-CS holds. For the first iteration, the price vector is initialized with zeros, and A is set to $\emptyset$ (the empty set). Each iteration then consists of two phases that execute in sequence, the bidding and the assignment phase. In the bidding phase, a set I of currently unassigned persons from the set $p{-}A_p$ where $A_p = \{p \in P | \exists t \in T : (p, t) \in A\}$ is the set of all currently assigned persons is chosen for possible assignment. For each person $p_i \in I$ then, the algorithm computes the currently "best" choice for that person, where the best choice is not the task t_j with maximal value a_{ij} among all tasks in T_i but rather, the task t_j with maximal value $a_{ij} - y_j$ that the person will gain if they had to "pay" the amount y_j of the price of the task t_j. In other words, the algorithm computes the task t_j that maximizes the quantity $v_j = a_{ij} - y_j$ among all $t_j \in T_i$. The algorithm also computes t_k, the second-best choice for person p_i among the tasks in T_i and offers a bid for the best choice in the amount of $y_j + (v_j - v_k) + \varepsilon$. In other words, the person p_i *raises* the amount he/she offers to pay for their best choice task t_j above the current price y_j by an amount equal to the difference of the current value of their best and second-best choice plus a tiny increment ε. After every person in I has offered a bid for their best choice task, in the second phase, namely the assignment phase, for each task t_j that has received a bid in the previous phase the algorithm raises the price of the task to the maximum bid previously offered, and assigns the task to the person that gave the highest bid, un-assigning the task from any previous assignment it might have had. The algorithm in pseudo-code is as follows:

Algorithm Forward Auction for Linear Assignment

Inputs: A set of n persons P and a set of n tasks T, and a user-defined parameter $\varepsilon > 0$. For each person p in P, a set T_p with elements from T is given, together with values a_{ij} indicating the "value" of task t_j in T_i for person p_i.
Outputs: A set of n pairs (p_i, t_j) that assign each person in P to exactly one task in T that is within ε of being optimal.

Begin

1. Set A={}, Set π=0.
2. Set I=P.
3. while $I \neq \{\}$ do
   a. Create a new empty mapping $Q : T \rightarrow 2^P$ mapping tasks in T to sets of persons in P.
   b. for each p_i in I do
      i. Set $j_i = \arg\max_{j \in T_i} \{a_{ij} - \pi_j\}$, Set $v_i = \max_{j \in T_i} \{a_{ij} - \pi_j\}$.
      ii. if $T_i = \{j_i\}$ then Set $w_i = -\infty$.
      iii. else Set $w_i = \max_{j \in T_i, j \neq j_i} \{a_{ij} - \pi_j\}$.
      iv. end-if

v. Set $b_{ij_i} = a_{ij_i} - w_i + \varepsilon$.
vi. Set $Q(t_{j_i}) = Q(t_{j_i}) \cup \{p_i\}$.
c. end-for
d. for each task t_j in T do
i. if $Q(t_j) \neq \{\}$ then
1. Set $\pi_j = \max_{p_i \in Q(t_j)} b_{ij}$, Set $p_j = \arg\max_{p_i \in Q(t_j)} b_{ij}$.
2. Remove from A any assignment to t, (p, tj), Add p to I.
3. Set $A = A \cup \{(p_j, t_j)\}$, Remove p_j from I.
ii. end-if
e. end-for
4. end-while
5. return A.

End

The auction family of algorithms is in sharp contrast to most optimization algorithms wherein it guarantees finite termination with guaranteed optimality *without* necessarily improving either the primal cost or the dual cost in a given iteration. Its complexity when appropriate data structures are utilized is $O(n|E|log(nC))$ where C is the maximum of the absolute value of the quantities a_{ij}, but its practical performance has surpassed most other assignment algorithms to date, and is routinely used (usually in the combined form of forward/backward auction) for large-scale assignment problems in many fields.

The Multi-Commodity Linear Network Flow Problem

It is important to stress that the integrality of the optimal solution of *LMCNF* is lost when more than one commodity is required to flow along the same network (the *multi-commodity* linear minimum cost network flow problem), which is a special category of LP problems that generalize the (*LMCNF*) problem. In a multi-commodity network flow problem, set as a multi-commodity transshipment problem, there are $K > 1$ different commodities produced at some nodes of a network $G(V,E)$ that need to be transferred to nodes that demand each commodity so that the demand for each commodity at each network node is completely satisfied. If amount x_{ek} of commodity k flows along arc $e = (i,j)$, a cost $c_{ek}x_{ek}$ is incurred. The integrality properties of the optimal solution are destroyed by the fact that arcs may have a capacity $u_e > 0$ so that the total flow of all commodities flowing along each arc is below this threshold $\sum_{k=1}^{K} x_{ek} \leq u_e, \quad \forall e \in E$. This fact introduces a *coupling* of the flow variables between commodities in the constraints. Indeed, consider the formulation of the Multi-Commodity Linear Network Flow Problem (*MCLNF*):

$$(MCLNF)\ \min .\sum_{k=1}^{K}\sum_{(i,j)\in E} c_{ijk}x_{ijk}$$

$$\text{s.t.}\begin{cases}\sum\limits_{i:(i,j)\in E} x_{ijk} - \sum\limits_{k:(j,k)\in E} x_{ijk} = d_{jk}, & \forall j\in V, k=1\ldots K\\ \sum\limits_{k=1}^{K} x_{ijk}\le u_{ij}, & \forall (i,j)\in E\\ x_{ijk}\ge 0, & \forall (i,j)\in E\end{cases} \tag{1.12}$$

If the flows of the different commodities along each arc did not have to be *together adding to a sum smaller than the arcs' capacity*, then the multi-commodity problem would be a super-position on the same network of K different single-commodity network flow problems that could be solved independently of each other and in parallel, and would be guaranteed to have integral solutions. In fact, if solving the K independent single-commodity problems leads to a solution such that the coupling constraints are satisfied, the overall solution is guaranteed to be optimal. If, however, any of the coupling constraints is violated, then the integrality of the solution will likely not exist.

The MCLNF is also known as a block-angular system (or, alternatively, as a stair-case system). The reason for this name comes from writing the system in matrix format as follows. Consider a partitioning of the decision variables so that $x = \left[\left(x^{[1]}\right)^T \left(x^{[2]}\right)^T \ldots \left(x^{[K]}\right)^T\right]^T$ where each vector $x^{[i]}$ contains the decision variables x_{ei} of the flows for the ith commodity, and consider the induced partitioning of the cost coefficients vector c into $c^{[1]}$, …, $c^{[K]}$, and similarly for node demands $d^{[i]}$ i = 1…K. Letting A denote the network edge incidence matrix of the problem, (1.12) can be rewritten as

$$(MCLNF)\ \min\sum_{k=1}^{K}\left(c^{[k]}\right)^T x^{[k]}$$

$$\text{s.t.}\begin{cases}Ax^{[1]} & = d^{[1]}\\ Ax^{[2]} & = d^{[2]}\\ \ldots & \\ Ax^{[K]} & = d^{[K]}\\ x^{[1]}+\ldots+x^{[K]} & \le u\\ x^{[i]}\ge 0, \quad \forall i=1\ldots K & \end{cases}$$

The problem received significant attention from many research teams due to its very large range of practical applicability and also due to its significant potential for decomposition and subsequent speedups obtained from parallel execution of the independent sub-problems.

One approach for solving large-scale MCLNF problem instances is based on the *Dantzig–Wolfe decomposition* principle, known also as (delayed) *column generation*, a method that extended the range of applicability of LP to instances with up to an order of magnitude larger than the previously largest size of solvable instances with the simplex method. The technique is intuitively easy to understand, and is theoretically applicable to any LP, but we shall restrict our attention to the MCLNF because it is only in problems of such special structures that the technique is worth applying. The MCLNF problem can be formulated as

$$(MCLNF - CG)\ \min_x c^T x$$

$$\text{s.t.} \begin{cases} A_M x = u \\ A_s x = d \\ x \geq 0 \end{cases}$$

where we have defined non-negative slack variables s_e for each arc e of the network and collect them in a slack vector s and define $x = \left[x^{[1]^T} \ldots x^{[K]^T} s^T\right]^T$, while the matrices A_M and A_s are defined as follows:

$$A_M = [I_m I_m \cdots I_m], \quad A_s = \begin{bmatrix} A & 0 & 0 & 0 \\ 0 & \ddots & 0 & \vdots \\ 0 & 0 & A & 0 \end{bmatrix}, \quad m = |E|$$

and d is defined as the vector $\left[d^{[1]^T} \ldots d^{[K]^T}\right]^T$.

The dimensions of matrix A_M are obviously $|E| \times ((K+1)|E|)$ whereas the dimensions of matrix A_s are $(K|V|) \times ((K+1)|E|)$. Now, from the fundamental theorem of polyhedral theory (see for example Rockafellar 1970), the set of feasible points of *any* polyhedral set $P = \{x \in \mathbb{R}^n | Ax \leq b\}$ is equal to the set of points that can be expressed as a direct sum of the non-negative combinations of the extreme directions (rays) $r^{[k]}$ $k = 1 \ldots p$ of the set and the convex combinations of its extreme points, $q^{[j]}$ $j = 1 \ldots q$ i.e. $P = \{x | \exists \mu_k \in \mathbb{R}_+\ k = 1 \ldots p, \exists \lambda_j \in \mathbb{R}_+,\ j = 1 \ldots q : x = \sum_{k=1}^p \mu_k r^{[k]} + \sum_{j=1}^q \lambda_j q^{[j]}, \sum_{j=1}^q \lambda_j = 1\}$ where the numbers p and q are finite natural numbers. Therefore, the set of non-negative points $x \geq 0$ satisfying the (network flow balance) equations $A_s x = d$ can be expressed as

$$x = \sum_{k=1}^p \mu_k r^{[k]} + \sum_{j=1}^q \lambda_j q^{[j]}, \mu \geq 0, \lambda \geq 0, \lambda^T e = 1$$

where e is a q-dimensional column vector of all ones. The (*MCLNF-CG*) problem then can be expressed as

$$(MCLNF-MP)\ \min_{\mu,\lambda} c^T\left(\sum_{k=1}^{p}\mu_k r^{[k]}+\sum_{j=1}^{q}\lambda_j q^{[j]}\right)$$

$$\text{s.t.}\begin{cases} A_M\left(\sum_{k=1}^{p}\mu_k r^{[k]}+\sum_{j=1}^{q}\lambda_j q^{[j]}\right)=u \\ e^T\lambda=1 \\ \mu\geq 0, \lambda\geq 0 \end{cases}$$

This formulation is known as the "*Master Problem*" for column generation. Note that at this point, it is not possible to apply the revised simplex method to work on this formulation since the extreme points $q^{[j]}$ and directions $r^{[k]}$ are unknown data still—and in fact, the numbers p and q can become prohibitively large. Fortunately, it is not necessary to compute the whole matrix $\tilde{A}=\left[A_M r^{[1]}\ \ldots\ A_M r^{[p]}\ \ A_M q^{[1]}\ \ldots\ A_M q^{[q]}\right]$ or compute *all* the extreme points and directions of the set $\{x \mid A_s x=d,\ x\geq 0\}$. Indeed, each iteration of the revised simplex method applied to an LP in standard form, as described in the beginning of this section, starts with an invertible basis matrix $\tilde{A}_B$ (so that after some permutation of the columns of $\tilde{A}$ we have $\tilde{A}=\left[\tilde{A}_B|\tilde{A}_N\right]$), a corresponding cost vector $\tilde{c}_B$ and a basic variables vector $\tilde{x}_B\geq 0$ satisfying $\tilde{A}_B\tilde{x}_B=\tilde{b}$. Solving the system $\tilde{A}_B^T y=\tilde{c}_B$ is still possible in the new formulation given the data $\tilde{A}_B$, (whose dimensions are $((|E|+1)\times(|E|+1))$ and $\tilde{c}_B$. The non-trivial step is to find an entering column $\tilde{a}_i$ from the matrix $\tilde{A}_N$ which satisfies $\tilde{a}_i^T y>\tilde{c}_i$. As it turns out, one such column can be found by solving the *column generation subproblem*

$$(MCLNF-SP)\ \min_{x}\left(c-(A_M)^T Ly\right)^T x$$

$$\text{s.t.}\begin{cases} A_s x=d \\ x\geq 0 \end{cases}$$

where L is the component removal operator, and is a matrix of dimensions $|E|\times(|E|+1)$ defined as $L=[I_{|E|}|0]$. This subproblem however is nothing more than the super-position of K independent network flow problems, which are directly solvable by applying the network simplex method K times *in parallel.* Indeed, every feasible solution of the master problem, describing the original problem, is also feasible for the subproblem which only contains a subset of the constraints of the original problem. The b.f.s. $\tilde{x}_B$ then guarantees that the subproblem has a solution. Now, if the problem (*MCLNF-SP*) has an optimal solution x^* such that $\left(c-A_M^T Ly\right)^T x^*<y_{|E|+1}$ the revised simplex method applied to the subproblem will find a b.f.s. v such that $\left(c-A_M^T Ly\right)^T v<y_{|E|+1}$ that will be one of the extreme points $q^{[j]}$ $j=1\ldots q$ of the set $\{x \mid A_s x=d,\ x\geq 0\}$. This extreme point then can form a valid entering column of the matrix $\tilde{A}$ as follows:

$$\tilde{a}_i = \begin{bmatrix} A_M v \\ 1 \end{bmatrix}$$

with a corresponding component of $\tilde{c}$ equal to $c^T v$. The reader should verify that the proposed column is indeed a valid column to enter the basis. Otherwise, the subproblem may be unbounded or have an optimal solution such that $(c - A_M^T Ly)^T v \geq y_{|E|+1}$. In the former case, the revised simplex method will instead locate an extreme direction r such that $(c - A_M^T Ly)^T r < 0$ which will be one of the vectors $r^{[k]}$ $k = 1 \ldots p$ and therefore the column

$$\tilde{a}_i = \begin{bmatrix} A_M r \\ 0 \end{bmatrix}$$

is a valid entering column of the matrix $\tilde{A}$ with corresponding cost component equal to $c^T r$. In the latter case $\left((c - A_M^T Ly)^T v \geq y_{|E|+1}\right)$ it is easy to prove that the current b.f.s. of the master problem is optimal for the original problem and the method may terminate (the reader should verify this claim as well).

Another approach, based on penalty methods to be discussed in detail immediately after, is to delete the coupling constraints $x^{[1]} + \cdots + x^{[K]} \leq u$ from the set of constraints (a process called *relaxation* of the original problem) and add a "*penalty*" to the objective function proportional to the amount of violation of any coupling constraint, the idea being that for large proportionality weights, the solution of the relaxed optimization problem would be in the feasible region of the original problem—since solutions outside the feasible region of the original problem would have very high objective cost for the relaxed problem. Such a penalty function can be the function $p(w) = w^T \left(\sum_{i=1}^K x^{[i]} - u\right)$ which will be strictly positive if the weights w_{ij} on an arc (i,j) are positive when the coupling constraint is violated on that arc, and zero otherwise. The new objective function for the relaxed problem then becomes $L(x, w) = \sum_{k=1}^K \left(c^{[k]} + w\right)^T x^{[i]} - u^T w$, and the full relaxed problem is written down as

$$(RMCLNF) \min_x \sum_{k=1}^K \left(c^{[k]} + w\right)^T x^{[i]} - u^T w$$

$$\text{s.t.} \begin{cases} Ax^{[k]} = d^{[k]}, & \forall k = 1 \ldots K \\ x^{[k]} \geq 0, & \forall k = 1 \ldots K \end{cases}$$

Note that for any given choice of "penalty" *non-negative* weights w, the (*RMCLNF*) problem is a super-position of K single-commodity linear minimum cost network flow problems of the form $\min_{x^{[k]}}\left\{\left(c^{[k]} + w\right)^T x^{[k]} | Ax^{[k]} = d^{[k]}, x^{[k]} \geq 0\right\}$ that can be solved in parallel and provide a solution, so RMCLNF is particularly easy to solve, being fully amenable to decomposition. However, the solution of the relaxed problem is by no means guaranteed to be the optimal solution to the original problem. If however, the solution is such that $\sum_{k=1}^{K} x_{ijk} < u_{ij}$ only for those arcs for which $w_{ij} = 0$ and happens to be a feasible solution for MCLNF, then the solution must be optimal for the original problem because at such a point the penalty term simply vanishes. So, the penalty-based approach iteratively modifies the weights w of the penalty function and solves the K independent sub-problems constituting the solution to *RMCLNF* until it satisfies this condition. The weights are updated in the $(q + 1)$st iteration according to the formula $w_{(q+1)} = [w_{(q)} + \theta_{(q)}(x - u)]_+$ where $\theta_{(q)}$ are scalars that satisfy $\theta_{(q)} \to 0$ and $\sum_{q=1}^{\infty} \theta_{(q)} = \infty$, and x_+ denotes the vector whose components are the quantities $(x_i)_+ = \max\{x_i, 0\}$. This updating formula guarantees convergence to the optimal solution.

The algorithm in pseudo-code is shown

Algorithm Lagrangian Relaxation for Multi-Commodity Linear Network Flow Problem

Inputs: A graph $G(V,E)$, with arc incidence matrix A, costs along each arc c_{ijk} for each commodity $k = 1\ldots K$, and arc capacities u_{ij}, node demands for each commodity $k = 1\ldots K$, d_{ik} for each node i in V.
Outputs: Optimal flow vector x^* satisfying all constraints of (*MCLNF*) and minimizing the objective function.

Begin

1. Set $q = 1$.
2. Set $|E|$-dimensional column vector $w = 0$.
3. while true do
   a. for $k = 1$ to K do (in parallel)
      i. Call the network simplex method to solve the problem $\min\left\{ \left(c^{[k]} + w\right)^T x^{[k]} | Ax^{[k]} = d^{[k]}, x^{[k]} \geq 0\right\}$
   b. end-for
   c. Set $x = x^{[1]} + \ldots + x^{[K]}$.
   d. if $x \leq u$ then
      i. return $x^* = x$.
   e. end-if
   f. Set $\theta = 1/q$.
   g. Set $w = (w + \theta(x - u))_+$.
   h. Set $q = q+1$.
4. end-while

End.

1.1.2.3 Nonlinear Constrained Optimization

When the constraints and/or the objective function of the (*NLP*) problem (1.3) are not linear, the algorithms of the previous paragraph are not applicable any more, and other approaches are needed to be able to discover points that satisfy for example the first-order necessary conditions for mathematical programming. In the special case where the constraints are linear equality constraints of the form $Ax = b$ and the objective function is a quadratic of the form

$$q(x) = \frac{1}{2}x^T Bx + c^T x$$

where B is a $n \times n$ symmetric real matrix and c is an n-dimensional column vector, the problem is known as a (equality constrained) *Quadratic Programming* problem (EQP). In order for the problem to be non-trivial, the row rank of the matrix A must be less than n, the number of variables in the problem, otherwise, solving the linear system $Ax = b$ will provide the unique point in the feasible region of the problem. In the non-trivial case where all the m rows of the matrix A are linearly independent and $m < n$, it is straightforward at least conceptually, to express m of the x variables as linear combinations of the remaining $n - m$ variables and then substituting in the quadratic to obtain an unconstrained quadratic optimization problem in $n - m$ variables of the form $\min\{x^T Qx/2 + q^T x\}$, which in turn can be solved by either directly solving the system $Qx = -q$ or by applying Algorithm Restricted Step-Size Quadratic Programming described in Sect. 1.1.1.2.

Penalty Methods for *NLP*

In the following, we discuss equality constrained problems of the form $\min_x\{f(x)|c(x) = 0\}$ where f and c are both differentiable functions $f : \mathbb{R}^n \to \mathbb{R}$, $c : \mathbb{R}^n \to \mathbb{R}^m$. This does not cause any loss of generality, because any general constrained *NLP* as formulated in (1.3) is equivalent to the following equality constrained problem containing $|I|$ new slack variables y:

$$\min_{x,y} f(x)$$
$$\text{s.t.} \begin{cases} c_i(x) = 0, & \forall i \in E \\ c_i(x) - y_i^2 = 0, & \forall i \in I \end{cases}$$

The idea behind penalty methods is to *relax* the constraints of the problem and *solve an unconstrained optimization problem* in which the objective function is modified and a term is added that is increasing with an appropriate measure of the violation of the constraints at any point. Without loss of generality therefore, consider the equality constrained problem

$$(P)\ \min_x\{f(x)|c(x) = 0\}$$

where f and c are simply required to be continuous functions, and denote by x^* the *global* minimizer for (P). Assuming M is a known lower bound for the value of $f(x^*)$, so that $M \leq f(x^*)$, let us define the penalty function $v(M, x) =$

$(f(x) - M)^2 + \|c(x)\|^2$. Now, it is easy to see that if M is tight, so that $M = f(x^*)$, then minimizing the function $v(M,x)$ without any constraints with respect to x, solves the problem (P); this is because the function $v(M,x)$ is non-negative for all $x \in \mathbb{R}^n$ and at $x = x^*$, $v(M,x^*) = 0$, so x^* is a solution for $\min_x v(M,x)$, and vice versa, if $v(M,y) = 0$, then both terms of the function $v()$ must be zero, so that $f(y) = M$ and $c(y) = 0$, and thus y is then a solution to (P). Further, for any M, if $x(M)$ is a solution to the problem $\min_x v(M,x)$ then $(f(x(M)) - M)^2 \leq (f(x(M)) - M)^2 + \|c(x(M))\|^2$ and the right-hand side of this inequality is in turn less than or equal to $(f(x^*) - M)^2 + \|c(x^*)\|^2 = (f(x^*) - M)^2$. Therefore, $(f(x(M)) - M)^2 \leq (f(x^*)) - M)^2$ and since $f(x^*) - M \geq 0$, we get $f(x(M)) \leq f(x^*))$ which implies that as long as M is a lower bound to the optimum value of the original problem, the optimum values of the unconstrained problems are less than or equal to the optimum value of the original problem (P). Consider now the following naïve algorithm for solving the equality constrained problem (P):

Algorithm Naïve Penalty Method for Equality Constrained (*NLP*) Problem

Inputs: Continuous objective function $f : \mathbb{R}^n \to \mathbb{R}$, continuous equality constrained vector function $c : \mathbb{R}^n \to \mathbb{R}^m$, initial lower bound M on the global minimum of (P), termination criteria at any point x.
Outputs: A point x^* that is a global minimizer of (P).

Begin

1. Set $k = 1$, $M_k = M$.
2. Solve the unconstrained optimization problem $(RP_k)\ \min_x \left\{ (f(x) - M_k)^2 + \|c(x)\|^2 \right\}$ to obtain the global minimizer $x_{(k)}$.
3. if termination criteria are satisfied at $x_{(k)}$ then
   a. return $x^* = x_{(k)}$.
4. end-if
5. Set $M_{k+1} = M_k + \sqrt{v(M_k, x_{(k)})}$.
6. Set $k=k+1$.
7. GOTO 2.

End.

Assuming the criteria for termination is that $v(M,x)$ is sufficiently close to zero (e.g. less than 10^{-8}), the above algorithm is guaranteed to converge to the global optimum of the original problem (P). This is because the sequence of values $\{M_k\ k = 1,2,\ldots\}$ remain lower bound estimates of the value $f(x^*)$. Indeed, for $M_1 = M$ this holds by hypothesis. Assuming it holds for k, then from the update formula in step 5 of the algorithm, $M_{k+1} = M_k + \sqrt{v(M_k, x_{(k)})}$, we have that

$$M_{k+1} = M_k + \sqrt{\left(f(x_{(k)}) - M_k\right)^2 + \left\|c(x_{(k)})\right\|^2}$$
$$\leq M_k + \sqrt{(f(x^*) - M_k)^2 + \|c(x^*)\|^2} = M_k + \sqrt{(f(x^*) - M_k)^2} = f(x^*)$$

where the inequality follows from the fact that $x_{(k)}$ is the global minimizer for $v(M_k, x)$. So, the induction step has been proved and the proof by induction is complete. In fact, it is now easy to see that *the sequence of values M_k converges to the value $f(x^*)$*. Indeed, $\{M_k\ k = 1,2\ldots\}$ is obviously an increasing sequence that is bounded from above by $f(x^*)$, so it converges to a limit, M^*. At the limit, $M^* = M^* + \sqrt{(v(M^*, x'))}$, so $v(M^*, x') = 0$ and by our previous argument, $M^* = f(x^*)$, and x' is a solution to (P).

The above described algorithm however is accurately described as *naïve* because carrying out the *global* optimization in step 2, as discussed in Sect. 1.1.1 cannot be done in a finite number of steps in general, except in very particular circumstances (e.g. when the function $v(M,x)$ is convex), and therefore there can be no general guarantees about the performance of the algorithm under more realistic assumptions. Numerical stability issues can arise during the execution of this algorithm in difficult problems. Nevertheless, the performance of the algorithm can be tested even when step 2 is not carried out to its completion, but rather when an algorithm for unconstrained minimization to a local optimum from Sect. 1.1.1 is carried out embedded in a heuristic search for global optimization as the ones described in Sect. 1.1.1.3. The performance of the naïve algorithm for equality constrained (*NLP*) for the problem of minimizing the test function $f(x) = -10x_1^2 + 10x_2^2 + 4\sin(x_1x_2) - 2x_1 + x_1^4$ subject to the constraint $x_1 + x_2 \leq 0$ is shown in Fig. 1.18 (the constraint is converted into the equality constraint $x_1 + x_2 + z^2 = 0$ first, where z is a new slack variable). From an initial choice $M = -30$, the algorithm terminates in only 4 major iterations with a function value accuracy of 10^{-6}.

More sophisticated penalty methods for solving the equality constrained nonlinear programming problem $\min_x\{f(x)|c(x) = 0\}$, as already mentioned, consist of transforming the original objective function to include a penalty term for the constraints violation of any vector x, as in the Courant penalty function, $\varphi(x,p) = f(x) + pc(x)^Tc(x)/2$. The unconstrained minimizer of the function $\varphi(x,p)$ with respect to x for increasingly larger values of p eventually leads to a solution that is feasible for the original problem. A generic penalty-based method for equality constrained nonlinear programming problems is therefore the following:

Algorithm Penalty Method for Equality Constrained (*NLP*) Problem

Inputs: Objective function $f : \mathbb{R}^n \rightarrow \mathbb{R}$, equality constrained vector function $c : \mathbb{R}^n \rightarrow \mathbb{R}^m$, termination criteria to be satisfied at x^*, increasing penalty sequence $\{p_{(k)}, k = 1,2,\ldots\}$.
Outputs: A point x^* satisfying the FONC for mathematical programming.

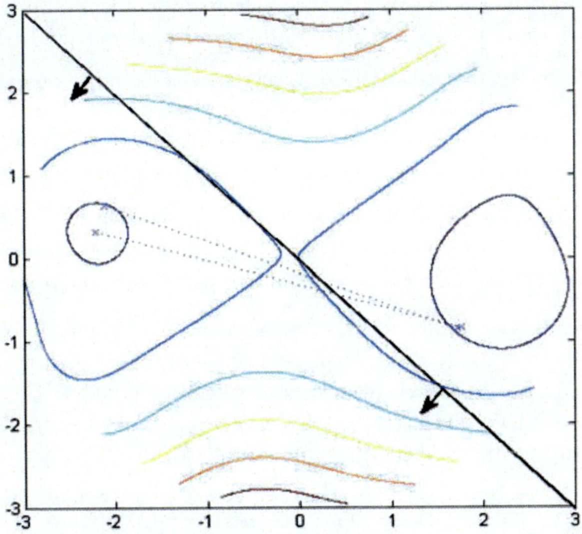

Fig. 1.18 Path of Na algorithm for constrained (NLP) shown in dotted line. The contours of the objective function to be optimized, along with the boundary of the constraint $x_1 + x_2 \leq 0$ is shown. The constraint is inactive at the global minimum. Step 2 of the algorithm is solved by calling the EA strategy 100 times to locate a reasonable starting point for the BFGS method which is called subsequently to minimize the penalty function $v(M,x)$. Note how the algorithm, starting from a feasible point, crosses the boundary of feasibility before locating the optimal solution for the constrained problem

Begin

1. Set k=1.
2. while true do:
   a. Set x equal to the solution of the unconstrained nonlinear optimization problem $\min_x \{\varphi(x, p_{(k)}) = f(x) + \frac{p_{(k)}}{2} c(x)^T c(x)\}$ using for example any of the methods in Sect. 1.1.1.
   b. if termination criteria are satisfied at x break.
   c. Set k=k+1.
3. end-while
4. Set x^*=x.
5. return x^*.

End

Usually the termination criteria is that $\|c(x)\| \leq \varepsilon$ where ε is a user-specified constraint violation tolerance, often set at 10^{-6} or 10^{-8}. The sequence of penalty values $p_{(k)}$ is often taken as {0, 1, 10, 100, 1,000,...}. The method is motivated by the following fact that, just as in the previous naïve algorithm case, it is only true in an idealized situation.

Theorem 1.32 (Penalty function convergence to global optimum) *Assuming the problem* $\min_x\{f(x)|c(x) = 0\}$ *is feasible and bounded so that* f^* *is the infimum of the function f over the feasible region, and that the penalty sequence* $\{p_{(k)}, k = 1,2,\ldots\}$ *is increasing without bound to infinity, and assuming the unconstrained minimization of the penalty cost function can be carried out to find out the global minimum* $x_{(k)}$ *of the function* $\varphi(x,p_{(k)})$ *then the sequence* $\varphi(x_{(k)},p_{(k)})$ *is non-decreasing, the norm of the constraints* $\|c(x_{(k)})\|^2$ *is non-increasing tending to zero, and the sequence of objective function values* $f(x_{(k)})$ *is non-decreasing and any limit point* x^* *of* $\{x_{(k)}\ k = 1,2,\ldots\}$ *is the global minimizer for the problem.*

Proof From the monotonicity of the sequence of $p_{(k)}$ values we have that $p_{(k)} < p_{(k+1)}$, and since $x_{(k)}$ is the global minimizer of $\varphi(x, p_{(k)})$ for all k, we have $\varphi(x_{(k)},p_{(k)}) \le \varphi(x_{(k+1)},p_{(k)})$ which in turn is less than or equal to $\varphi(x_{(k+1)},p_{(k+1)}) \le \varphi(x_{(k)},p_{(k+1)})$. So, we have that $\varphi(x_{(k)},p_{(k)}) \le \varphi(x_{(k)},p_{(k+1)})$ proving the monotonicity of $\varphi(x_{(k)},p_{(k)})$. Adding the two inequalities we obtain $\varphi(x_{(k)},p_{(k)})-\varphi(x_{(k)},p_{(k+1)}) \le \varphi(x_{(k+1)},p_{(k)})-\varphi(x_{(k+1)},p_{(k+1)})$ which is equivalent to $\|c(x_{(k)})\|^2(p_{(k)}-p_{(k+1)}) \le \|c(x_{(k+1)})\|^2(p_{(k)}-p_{(k+1)})$ and since $p_{(k)}$ are increasing, $\|c(x_{(k)})\|$ is indeed non-increasing. But then, from the first inequality $\varphi(x_{(k)},p_{(k)}) \le \varphi(x_{(k+1)},p_{(k)})$ we must have that $f(x_{(k)}) \le f(x_{(k+1)})$ so this sequence is indeed non-decreasing. Now, the sequence of values $\varphi(x_{(k)},p_{(k)})$ is non-decreasing and is bounded from above by the value $f^* = \inf_x \{f(x) \mid c(x) = 0\}$ so as $p_{(k)} \to \infty$ $\|c(x_{(k)})\|$ must tend to zero in order for the quantity $\varphi(x_{(k)},p_{(k)})$ to remain bounded. Finally, if x^* is a limit point of the sequence of points $x_{(k)}$ then $c(x^*) = 0$ from the previous argument, and $f(x^*) \ge f^*$. But $f(x_{(k)}) \le \varphi(x_{(k)},p_{(k)}) \le f^*$ for all k, so $f(x^*) = f^*$, and thus x^* is the global optimum of the problem $\min_x\{f(x)|c(x) = 0\}$. QED.

It is also possible to prove that if only a local minimum of the function $\varphi(x, p_{(k)})$ is computed each time in step 2.a of the penalty method for equality constrained (*NLP*), as long as the matrix $A = \nabla c(x^*)$of the constraints gradient at any limit point x^* of the sequence $\{x_{(k)}\ k = 1,2\ldots\}$ has full rank, then the point x^* satisfies the Karush–Kuhn–Tucker first-order necessary conditions for mathematical programming. Unfortunately, in many practical situations, the above algorithm and its variants do not behave well. The reason is that as the penalties $p_{(k)}$ get larger, the Hessian of the penalty function φ, $\nabla^2\varphi(x_{(k)}, p_{(k)})$ becomes more and more ill-conditioned, with some eigenvalues tending to infinity whereas others remaining bounded, and thus making the optimization step 2.a more and more difficult to solve even for the location of only a local minimum.

Multiplier Penalty Methods

Despite the numerical stability issues that may arise in the application of penalty methods for the solution of an NLP problem, the key idea of introducing a penalty term in the objective function proportional to the degree of violation of the constraints is of very high importance in the theoretical and algorithmic developments in constrained nonlinear optimization. In general, given the problem

$$(P)\ \min_x\{f(x)|c(x)=0\}$$

for continuously differentiable functions $f:\mathbb{R}^n\rightarrow\mathbb{R}$ and $c:\mathbb{R}^n\rightarrow\mathbb{R}^m$, the *Lagrangian function* for (P) is defined as $L(x,\lambda)=f(x)+\lambda^T c(x)$. Now, if x^* is a local minimum of f subject to the constraints of (P), and assuming that the columns of $\nabla c(x^*)$ are linearly independent, there exist unique Lagrange multipliers $\lambda_i^*,\quad i=1\ldots m$ such that $\nabla_x L(x^*,\lambda^*)=0$ or equivalently $\nabla f(x^*)=-\sum_{i=1}^m \lambda_i^*\nabla c_i(x^*)$. This formula is in very close analogy to the Karush–Kuhn–Tucker first-order necessary conditions for the case where there are no inequality constraints, so that $I=\{\}$. The proof of the following theorem follows the arguments in Bertsekas (1995).

Theorem 1.33 (Lagrange first-order multiplier theorem). *Let $f:\mathbb{R}^n\rightarrow\mathbb{R}$ and $c:\mathbb{R}^n\rightarrow\mathbb{R}^m$ be continuously differentiable functions, and assume x^* is a local solution for the problem (P) $\min_x\{f(x)|c(x)=0\}$. If the constraint gradient matrix $\nabla c(x^*)$ has full column rank, then there exists a unique Lagrange multiplier vector λ^* such that $\nabla_x L(x^*,\lambda^*)=0$ and $\nabla_\lambda L(x^*,\lambda^*)=0$.*

Proof If $n<m$, then the matrix $A^*=\nabla c(x^*)$ cannot have full rank, and the theorem is null (its conditions are not true). If $n=m$, then since A^* has full column rank, its columns are linearly independent and span all of $\mathbb{R}^n$ so clearly $\nabla f(x^*)$ is uniquely expressed as a linear combination of the columns of A^*. It remains to prove the validity of the theorem in case $m<n$. Because of the full column rank hypothesis, we can reorder the rows of A^* so that the first m rows of A^* form a matrix A_B^* that is invertible. This reordering of the rows induces a reordering and partitioning of the x variables so that x^* can be expressed as a pair of vectors $\left[(x_B^*)^T\quad(x_N^*)^T\right]^T$, and the constraint $c(x)=0$ can be written as $c(x_B,x_N)=0$ which has the solution $x^*=\left[(x_B^*)^T\quad(x_N^*)^T\right]^T$ and it also happens that c is a differentiable function with continuous partial derivatives, and also the $m\times m$ square matrix $A_B^*=\nabla_{[B]}c(x^*)$ is invertible so $\det A_B^*\neq 0$. Therefore, by the implicit function theorem of mathematical analysis (Apostol 1981), the equation $c(x_B,x_N)=0$ in a neighborhood of x^* defines a continuous and differentiable function $\psi(x_N)$ that satisfies $c(\psi(x_N),x_N)=0$ for all x_N in an appropriately small neighborhood of x_N^*. Applying the chain rule of differentiation in the previous equation we obtain $\nabla_{[B]}c(\psi(x_N),x_N)\nabla\psi(x_N)+\nabla_{[N]}c(\psi(x_N),x_N)=0$ and from the invertibility of A_B we get $\nabla\psi(x_N)=-\left(\nabla_{[B]}c(\psi(x_N),x_N)\right)^{-1}\nabla_{[N]}c(\psi(x_N),x_N)$. Now, observe that the vector x_N^* must satisfy the first-order conditions for unconstrained optimization for the function $\hat{f}(x_N)=f\left(\begin{bmatrix}\psi(x_N)\\ x_N\end{bmatrix}\right)$ because x_N^* must be an unconstrained minimum for the function $\hat{f}$ (otherwise, we could find another point $\bar{x}^*$ that is also

feasible and has smaller objective value than x^*). Therefore, we must have $\nabla\psi(x_N^*)\nabla_{[B]}f(x^*) + \nabla_{[N]}f(x^*) = 0$ and substituting for $\nabla\psi(x_N^*)$ from above, we get that the gradient $\nabla f(x^*)$ can indeed be expressed as a unique linear combination of the columns of $\nabla c(x^*)$. QED.

In fact, if f and c are twice continuously differentiable functions and the conditions of Theorem 1.33 hold, then any vector y belonging to the set $F^* = \{s|\nabla c(x*)^T s = 0\}$ satisfies $y^T L_{xx}(x^*, \lambda^*)y \geq 0$, where $L_{xx}(x,\lambda)$ is the Hessian of the function $L(x,\lambda)$ with respect to x.

Combining the idea of the Lagrangian with the *Courant* penalty method described above, leads to an *augmented Lagrangian function for solving the equality constrained* (*NLP*) problem (*P*) that has the form

$$L(x,\lambda,p) = f(x) + \lambda^T c(x) + \frac{p}{2}\|c(x)\|^2 \tag{1.13}$$

Now, if f and c are *twice continuously differentiable*, λ is close to the optimal λ^*, and p is above a certain threshold, the unconstrained minimization of $L(x,\lambda,p)$ will lead to a point close to the optimal x^*, as also when p is *sufficiently* large—as in the case of the standard penalty method. To understand why assuming λ is close to λ^*, minimizing subject to no constraints the augmented Lagrangian function will lead to a point close to x^*, we use the fact that in such a case and assuming also that the *strict* inequality $y^T L_{xx}(x^*, \lambda^*)y > 0$ holds for all y in F^*, there exists a positive scalar $\gamma > 0$ such that $L(x,\lambda^*,p) \geq L(x^*,\lambda^*,p) + \gamma\|x-x^*\|^2/2$ for all feasible x in an appropriately small neighborhood of x^*. This fact can be easily proved by contradiction, arguing that if the last inequality was not true, there would exist a sequence of *feasible* points $\{x_{(k)}\ k = 1,2,\ldots\}$ converging to x^* satisfying $x_{(k)} \neq x^*$ for all k, such that the bounded sequence $y_{(k)} = (x_{(k)}-x^*)/\|x_{(k)}-x^*\|$ would have a limit point, say y. This point would then satisfy $\nabla c(x^*)^T y = 0$ by application of the mean-value theorem of differential calculus and some algebra, and we would obtain $y^T L_{xx}(x^*, \lambda^*)y \leq 0$ which would contradict the *strict* inequality assumed at the beginning.

The multiplier penalty method essentially is yet another sequential minimization method similar to the two previous algorithms presented, but the special structure of the function it minimizes as well as the update formula for the Lagrange multipliers it computes at each iteration allow the penalty sequence $p_{(k)}$ to remain bounded, and thus avoids all ill-conditioning issues associated with the standard penalty methods. The Lagrange multipliers update formula is simply

$$\lambda_{(k+1)} = \lambda_{(k)} + p_{(k)}c(x_{(k)})$$

which essentially adds to the previous estimate of the Lagrange multipliers $\lambda_{(k)}$ the degree of violation of the constraints multiplied by the penalty for the previous iteration $p_{(k)}$. The multiplier penalty method in pseudo-code is as follows.

Algorithm Multiplier Penalty Method for Equality Constrained (*NLP*) Problem

Inputs: Objective function $f : \mathbb{R}^n \to \mathbb{R}$, equality constrained vector functionc : $\mathbb{R}^n \to \mathbb{R}^m$, termination criteria to be satisfied at x^*, increasing penalty sequence $\{p_{(k)}, k = 1,2,\ldots\}$.
Outputs: A point x^* satisfying the (FONC) for mathematical programming.

Begin

1. Set k=1.
2. Set $\lambda_{(k)} = 0$.
3. while true do:

   a. Set x equal to the solution of the unconstrained nonlinear optimization problem $\min_x \left\{ L(x, \lambda_{(k)}, p_{(k)}) = f(x) + \lambda_{(k)}^T c(x) + \frac{p_{(k)}}{2} \|c(x)\|^2 \right\}$using for example any of the methods in Sect. 1.1.1.
   b. if termination criteria are satisfied at x break.
   c. Set $\lambda_{(k+1)} = \lambda_{(k)} + p_{(k)} c(x)$.
   d. Set $k = k+1$.

1. end-while
2. Set $x^* = x$.
3. return x^*.

End

The termination criteria for the above method are usually as before, namely that $\|c(x)\| \le \varepsilon$ where ε is a user-specified constraint violation tolerance, often set at 10^{-6} or 10^{-8}. However, even if step 3.a only returns a near-local minimum of the augmented Lagrangian $L(x,\lambda,p)$, under mild assumptions, the algorithm can be guaranteed to converge to a local solution of the original problem (*P*). More importantly, the penalty terms $p_{(k)}$ do not have to be increased to infinity, and in fact there is a finite threshold, above which step 3.a will return a feasible point for (*P*), and for this reason, the ill-conditioning problems associated with the standard penalty methods essentially disappear in the multiplier penalty method. For a detailed analysis of the properties of this algorithm, the reader is referred to (Bertsekas 1982).

Implementing the algorithm above is rather straightforward. An application of the multiplier penalty method on the problem

$$\begin{aligned} &\min f(x) = -10x_1^2 + 10x_2^2 + 4\sin(x_1 x_2) - 2x_1 + x_1^4 \\ &\text{s.t.} \quad x_1^2 + x_2^2 = 1 \end{aligned}$$

is shown in Fig. 1.19, where we show the path the method follows superimposed on the contours of the objective function, along with the feasible region (in heavy black line).

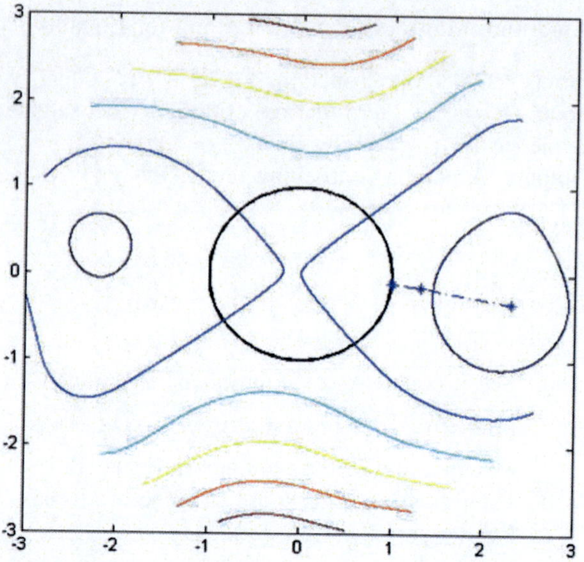

Fig. 1.19 Path of multiplier penalty algorithm for equality constrained NLP shown in dotted line. The algorithm computes the global optimum of the problem with a value $f^* = -11.2074$ at $[0.9947–0.1029]^T$. The path requires only 5 major iterations. The minimization of step 3.a is carried out by running the EA method for 100 iterations to produce a point x_1 and then applying the BFGS method starting from x_1

1.1.3 Dynamic Programming

Dynamic programming is the name given to a method for finding the optimal controls to apply to a dynamical system that evolves from an initial state $x_{(0)}$ according to a recursive equation of the form

$$x_{(k+1)} = f_k(x_{(k)}, u_{(k)}, r_{(k)}) \tag{1.14}$$

where $x_{(k)}$ is a point in a space S_k (usually a finite-dimensional norm vector space), $u_{(k)}$ correspond to the decision or control variables that are constrained to take values from a nonempty set $U_k(x_k)$ that is a subset of a space U_k and $r_{(k)}$ is a—possibly non-existent—random variable from a space Q_k whose probability distribution $P(r|x_{(k)},u_{(k)})$ is assumed to be known and may only depend on $x_{(k)}$ and $u_{(k)}$. Clearly, when the random variables are present, the system state $x_{(k+1)}$ at times 1,2,... is in general a stochastic variable as well, because its value depends on other stochastic variables ($x_{(k)}$ and $r_{(k)}$). The *finite-horizon Dynamic Programming problem* is to choose the controls $u_{(k)}$ at each time $k = 0,1,\ldots N-1$ over a finite length horizon consisting of N periods so as to minimize the expectation of a cost function that is additive over time, so that the expected cost has the form

$$C(u) = E_r\left[\sum_{k=0}^{N-1} c_k\left(x_{(k)}, u_{(k)}, r_{(k)}\right) + c_N\left(x_{(N)}\right)\right] \tag{1.15}$$

where the expectation operator $E_r[.]$ is with respect to the joint probability distribution of the random variables $r_{(0)} \ldots r_{(N-1)}$.

The controls $u_{(k)}$ must be the result of applying a *policy function* $\mu_k : S_k \to U_k$ that satisfies $\forall x_{(k)} \in S_k,\ \mu_k(x_{(k)}) \in U_k(x_{(k)})$. The objective is then to determine the policy functions $\mu_k^*,\ k = 0\ldots N-1$ that minimize the expectation

$$C(\mu_0, \ldots \mu_{N-1}) = E_r\left[\sum_{k=0}^{N-1} c_k(x_{(k)}, \mu_k(x_{(k)}), r_{(k)}) + c_N(x_{(N)})\right]$$

where the system now evolves according to the equation $x_{(k+1)} = f_k(x_{(k)}, \mu_k(x_{(k)}), r_{(k)})$.

The control vector $\mu = [\mu_0 \mu_1 \ldots \mu_{N-1}]^T$ is known as a *policy*, and the vector μ^* comprising policy functions that minimize the above expression is known as an *optimal policy* for the given problem. The Dynamic Programming (DP) basic algorithm is founded on the fundamental *principle of optimality* (Bellman 1957), which states if μ^* is the optimal policy for the problem of minimizing (1.15), and if at time i we are at state $x_{(i)}$, then to minimize the remainder cost from time i to time N given by the remainder cost function $C_i(\mu_i, \ldots \mu_{N-1}) = E_r\left[\sum_{k=i}^{N-1} c_k(x_{(k)}, \mu_k(x_{(k)}), r_{(k)}) + c_N(x_{(N)})\right]$ the optimal remainder policy vector we should apply is given by the vector $\tilde{\mu}^* = [\mu_i^* \ \ldots \ \mu_{N-1}^*]^T$. The proof of this claim should be rather obvious: if the proposed policy was not optimal for the remainder problem, then another set of policies $\kappa_i \ \ldots \ \kappa_{N-1}$ for the remainder problem would have a smaller expected cost from time i to time N; but then, the vector $\hat{\mu} = \left[\mu_0^* \cdots \mu_{i-1}^* \ \kappa_i \cdots \kappa_{N-1}\right]^T$ would have a total expected cost over the whole horizon $0\ldots N$ that would be strictly smaller than the cost of the optimal policy μ^*, a contradiction.

The principle of optimality implies that problems of the form (1.15) can be solved in stages: first, the last stage problem (corresponding to $k = N-1$), formulated as

$$\min_{\mu_{N-1}} E_{r_{(N-1)}, x_{(N-1)}} \left[c_N\left(f_{N-1}\left(x_{(N-1)}, \mu_{N-1}\left(x_{(N-1)}\right), r_{(N-1)}\right)\right) + c_{N-1}\left(x_{(N-1)}, \mu_{N-1}\left(x_{(N-1)}\right), r_{(N-1)}\right)\right]$$

is solved to obtain the optimal last stage policy function μ_{N-1}^*. Then, iteratively, the second last stage problem can be solved to obtain the optimal policy μ_{N-2}^* given that the optimal policy μ_{N-1}^* is known and must be applied at the last stage, and so on until the first stage $k = 0$. The problem to be solved at any stage $k = 0\ldots N-1$ is simply formulated as

$$C_k\left(x_{(k)}\right) = \min_{u_{(k)} \in U_k(x_{(k)})} E_{r_{(k)}}\left[c_k\left(x_{(k)}, u_{(k)}, r_{(k)}\right) + C_{k+1}\left(f_k\left(x_{(k)}, u_{(k)}, r_{(k)}\right)\right)\right], \quad k = N-1\ldots 0 \tag{1.16}$$

Fig. 1.20 A location problem in 1 dimension. The problem shows 16 markets along a highway that must be optimally served by $k > 1$ warehouses

The above equation (1.16) is known as the *Dynamic Programming equation*, and constitutes the basis for the backwards recursive algorithm discussed above. The *structure* of the problem determined by the state-transition functions f_k and the single-period cost functions c_k play a fundamental role in the complexity of the DP algorithm. At each stage k, the minimization dictated by the DP equation must occur for each possible state $x_{(k)}$ that the system might be in that allows state $x_{(k+1)}$ to be a state *reachable* by the application of the optimal policy μ^* up to time k, and the set of all such states must be recorded. In general therefore, to apply the DP algorithm, the set of states S_k must be a countable set with cardinality $|S_k|$. The algorithm must then carry out in the order of $\sum_{k=1}^{N-1} |S_k|$ minimizations, which is exponential in nature. However, as we shall see, in many practical optimization problems arising in Supply Chain Management such as Inventory Management, Production Planning and other fields as well, the *structure of the problem allows the application of a principle of decomposition* by which the number of states $x_{(k)}$ that must be considered in stage k remains bounded by a constant that is independent of the value of $|S_k|$, and then the problem can possibly be solved in polynomial time.

We illustrate the difference between a tractable application of DP and an intractable application of the same basic principle in the context of clustering data—a problem that arises in Location Theory within Supply Chain Management, as well as in other fields such as statistical pattern recognition. In this particular *deterministic* context, there are no random variables $r_{(k)}$ involved and the expectation of the costs degenerates to a simpler expression of the form

$$C_k\left(x_{(k)}\right) = \min_{u_{(k)} \in U_k\left(x_{(k)}\right)} \left[c_k\left(x_{(k)}, u_{(k)}\right) + C_{k+1}\left(f_k\left(x_{(k)}, u_{(k)}\right)\right)\right], \quad k = N-1\ldots0$$

Let n markets be placed along a highway represented by a straight line in Euclidean space, and consider the problem of locating k warehouses along the highway so as to minimize the sum of distances from each market to its nearest warehouse (see Fig. 1.20).

By selecting an arbitrary point along the highway to serve as the origin of the line—for example the location of the left-most market—, the position of each of the n markets can be uniquely determined by its distance from the origin, and the same holds for the positions of the warehouses that are the decision variables. In a formal model of the described problem therefore, consider a set of numbers $S = \{a_1, a_2, \ldots, a_n\}$, and without loss of generality assume that the sequence $a_i, i = 1, 2, \ldots, n$ is non-decreasing (otherwise sort it in $O(n \log n)$ time). We are interested in computing the optimal clustering of the numbers in the set among up to k components, so that we minimize the sum of the intra-cluster distances of points from their cluster centers. We shall derive a highly parallelizable DP algorithm that computes the optimal clustering in time that is $O(kn^2/E)$ where E is the number of available processing units. First, observe that *an optimal clustering of the set S has only contiguous clusters, i.e. in every cluster in the optimal partition, all points between the minimum and maximum element in the cluster belong to the same cluster.* To see why, consider any clustering with a non-contiguous cluster. Consider the element b that has the maximum distance from its cluster center among all non-contiguous clusters. Without loss of generality, assume element b is less than the center to which it belongs. Swap element b with the first element that is bigger than b and belongs to a cluster other than the cluster to which b belongs. The resulting partition is better, and thus there can be no non-contiguous clustering that is an optimal solution to our (1-dimensional) clustering problem. Define c_{ij} to be the cost of the cluster containing the points $\{a_i, a_{i+1}, \ldots, a_j\}$, or simply

$$c_{ij} = \sum_{r=i}^{j} \left| a_r - \frac{\sum_{m=i}^{j} a_m}{j-i+1} \right|.$$

Clearly $c_{ii} = 0 \quad \forall i = 1, \ldots, n$. Define $P_{i,m}$ to be the cost of the optimal clustering of the points $\{a_i, a_{i+1}, \ldots, a_n\}$ among up to m clusters. Then the following DP equation clearly holds, because of the contiguous property of the optimal clusterings:

$$P_{i,m} = \min_{j=i,\ldots,n-1} [c_{ij} + P_{j+1,m-1}] \quad i = 1, \ldots, n \quad m = 1, \ldots, k$$

Note how the above DP equation is indeed an instance of the generic equation for the deterministic DP problems derived above. There are a total of nk states in this formulation, and being in a state x_{ij} in the original formula corresponds to a clustering of the points $\{a_1 \ldots a_i\}$ among j clusters and the control u_{ij} corresponds to the decision whether to include the point a_{i+1} in the last cluster j of that state or whether to start a new cluster j+1 to include that point (in case $j = k$, the last choice is clearly not available as an option). The optimal clustering value is obviously $P_{1,k}$. Note the following boundary conditions: $P_{i,1} = c_{in} \quad \forall i = 1, \ldots, n$ and $P_{i,m} = +\infty$ if $m > n - i + 1$. Finally,

$$P_{n,m} = \begin{cases} 0 & m = 1 \\ +\infty & \text{else} \end{cases}$$

Having computed and stored the costs c_{ij} in time and space $O(n^2)$, we can now use the recursive equation for computing $P_{i,m}$ to construct the matrix $P[n \times k]$ containing the P values. If the construction of the matrix proceeds column-wise from bottom row to top row, i.e. the left-most column elements are computed first bottom-to-top, then the second column's elements are computed from bottom-to-top etc., then the construction of this matrix can be done in parallel with as many

Fig. 1.21 Computing the optimal costs in the 1-D clustering problem in parallel

$$\left[\begin{array}{cccc} P_{1,1}=c_{1,n} & & & P_{1,k} \\ \vdots & \vdots & \cdot^{\cdot^{\cdot}} & \\ P_{n-1,1}=c_{n-1,n} & 0 & \cdots & +\infty \\ P_{n,1}=0 & +\infty & \cdots & +\infty \end{array}\right] \begin{array}{c} \underline{Thread\_1} \\ \vdots \\ \hline \vdots \\ \hline Thread\_E \end{array}$$

as n processing units (e.g. cores in a multi-core CPU). Each processing unit is assigned to the task of computing an equal-sized part of the column under construction. So clearly computing the optimal clustering requires time that is $O\left(\frac{kn^2}{E}\right)$ where E is the number of processing units available. Figure 1.21 illustrates this parallel computation.

The actual cluster assignment can be easily found by maintaining an auxiliary matrix $I[n \times k]$ which is updated during the computation of matrix P in the same loop. The quantity $I_{i,m}$ is the index such that the optimal clustering of the data set $\{a_i, \ldots, a_n\}$ among m blocks is the cluster $\{a_i \ldots a_{I_{i,m}}\}$ plus the optimal clustering of the set $\{a_{I_{i,m}+1}, \ldots, a_n\}$ among $m-1$ clusters. The indices of the markets served by the jth warehouse are then given by the following recursive equations:

$$\begin{aligned} J_1 &= I_{1,k} \\ J_j &= I_{J_{j-1}+1,k-j+1}, \quad j = 2\ldots k \\ S_1 &= \{1\ldots J_1\} \\ S_j &= \{J_{j-1}+1\ldots J_j\}, \quad j = 2\ldots k \end{aligned}$$

and from this, the location of the jth warehouse can be computed as the number $\sum_{i\in S_j} a_i/|S_j|$ for all $j = 1, \ldots, k$.

Now, consider the problem of locating k warehouses on a 2D map so as to minimize the sum of the (Euclidean) distances of each market located in a given position (x_i, y_i) to its nearest warehouse, where the markets form a finite set S of two-dimensional vectors. The problem is known as *p-median* problem and is known to be NP-hard (Megiddo and Supowit 1984), therefore any DP algorithm for solving it, must necessarily have exponential complexity. The root cause for this is that the DP equation developed above for the 1-D case (clustering of points along a line) does not hold any more. In particular, the contiguous property of optimal clusterings that holds for the 1-D case is clearly not true anymore, and therefore, the DP equation for solving the p-median problem on the plane becomes

$$P_{I,m} = \min_{J\subseteq S-I}\left[c_J + P_{S-J,m-1}\right], \quad m = 2\ldots k, \quad I \subseteq S$$

where by c_J we denote the sum of distances of the points in J (subset of S) from its center of gravity. The number of states this problem has is in the order of $2^n k$ since essentially all subsets of the set S must be considered. From the above, it is now clear that *the structure of the problem plays a crucial role in the practical applicability of Dynamic Programming in its solution process.* We shall study

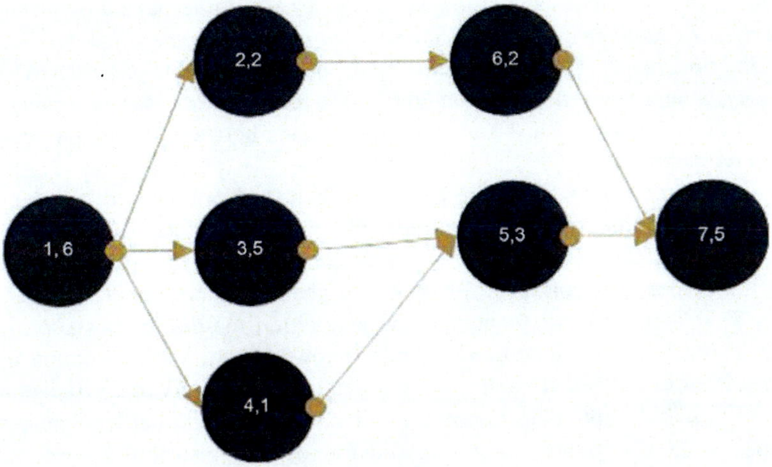

Fig. 1.22 An example of a precedence graph in a paced-assembly line-balancing problem. Each node in the graph has two numbers, the first is the task id and the second is the time the task requires for its completion

practical algorithms for the solution of the p-median problem when we discuss location theory and distribution management.

1.2 Mixed Integer and Combinatorial Optimization Methods

Mixed integer programming problems arise in problems that deal with physically discrete objects and/or problems that contain logical conditions. Also, many problems with an underlying geometry often can only be described as Mixed integer programming problems. Despite the superficial resemblance of such problems to the class of LP problems, their complexity is completely different from that of LP, and it is *almost* certain that *in general* there is no way to efficiently solve such problems to optimality.

As an introduction to discrete optimization, consider the following. In production and operations management, a classical problem that arises when a *paced-assembly line* (Hopp and Spearman 2008) is designed, is the so-called *line-balancing* problem. In order to manufacture a certain product, a number of (assembly) tasks J_1, J_2, ..., J_n must be performed, and each task takes a certain amount of time t_1, ..., t_n to execute. There is also a partial order imposed on the execution of the tasks, so that a *precedence graph* is defined determining for each task J_i what are the tasks that must be completed before execution of J_i can begin. An example of such a graph is shown in Fig. 1.22. Tasks are executed within a *workstation* that they are assigned to, and every workstation is typically operated by a human employee. Given a *target* number of final products D that the assembly line should produce every day (usually corresponding to daily demand) in a given total time-frame T (usually equal to a shift, e.g. 8 h), the line-balancing problem is to *minimize the number of assembly stations required to run the line while meeting the production target D and simultaneously minimizing any imbalances of the workloads between any two stations*, the latest requirement usually arising from personnel concerns over "just and fair" work-loads between the employees running the workstations.

Given the time-frame T and the target number D of products to be assembled within that time frame, assuming a number of k *sequential* stations is used to make up the totality of the *paced-assembly production line*, the time each station will have at its disposal to complete all of its assigned tasks will be $T_s = T/D$. The quantity T_s is then an upper bound on the available time that each station has in order to complete its assigned tasks, and obviously then, a lower bound on the number of stations needed to form the assembly line is given by the quantity $k_{\min} = \sum_{i=1}^{n} t_i/T_s$. This lower bound may or may not be attainable depending on the values of the task times t_i $i = 1\ldots n$. We are now in a position to formulate the line-balancing problem as an optimization problem.

The objective function to minimize can be expressed as a cost on the total number of workstations used, and the total "imbalance" of the load of the workstations in the line. The major decision variables should specify to which station each task is assigned. Clearly, there can be no more than n stations in the line (since there are a total of n tasks to be performed). Introducing therefore the *discrete (binary) variables* x_{ij} $i,j = 1,\ldots n$ that are set to 1 if task J_i is assigned to station j and zero otherwise, and auxiliary variables y_j $j = 1\ldots n$ that are set to one if $k \geq j$ stations are used (equivalently, if at least one task is assigned to station j), and zero otherwise, the cost function for the line-balancing problem becomes

$$c(x,y) = T_s \sum_{j=1}^{n} y_j + \max_{j=1\ldots n} \left[\left(T_s - \sum_{i=1}^{n} t_i x_{ij} \right) y_j \right] \tag{1.17}$$

This objective function, if minimized over all feasible (x,y) pairs, will produce an allocation of tasks to stations so that the minimum number of stations is utilized while among all allocations that require the minimum possible number of stations, minimizes the maximum imbalance of the line (why this is so, is left as an exercise for the reader). Of course, not every (x,y) vector pair will be feasible for the problem. The allocation of tasks to stations must be such that every station is assigned to a set of tasks whose total time is no greater than T_s, and a task can only be assigned to a *single* station j if and only if all of its preceding tasks are assigned on stations numbered $i \leq j$. To formulate these two constraints, let p_{ij} $i,j = 1,\ldots n$ be parameters set to one if task J_i precedes task J_j in the precedence graph, and zero otherwise. In the example graph of Fig. 1.22 therefore we have that $p_{1,3} = p_{1,5} = 1$ but $p_{5,6} = 0$. The time-limit constraint on each station can be expressed as $\sum_{i=1}^{n} t_i x_{ij} \leq T_s, \quad j = 1\ldots n$ whereas the precedence constraints can be expressed as the inequalities $p_{ij} x_{is} + \sum_{t=1}^{s-1} x_{jt} \leq 1, \quad \forall i,j,s = 1\ldots n$. These inequalities require that if task J_i precedes task J_j and task J_i is assigned to station s, then task J_j cannot be assigned to any of the stations $1\ldots s-1$. If however, task J_i is not assigned to station s, the inequality is simply inactive because the first term of the sum of the left-hand-side of the inequality is zero, and the second term is always less than or equal to 1, since a task must be assigned to exactly one station, expressed by the equality constraint $\sum_{j=1}^{n} x_{ij} = 1, \quad \forall i = 1\ldots n$. The full model for the line-balancing problem becomes the following:

$$\min_{x,y}\left[T_s \sum_{j=1}^{n} y_j + \max_{j=1\ldots n}\left[\left(T_s - \sum_{i=1}^{n} t_i x_{ij}\right) y_j\right]\right]$$

$$\text{s.t.}\begin{cases} p_{ij}x_{is} + \sum_{t=1}^{s-1} x_{jt} \le 1, & \forall i,j,s = 1\ldots n \\ \sum_{i=1}^{n} x_{ij} \le T_s, & \forall j = 1\ldots n \\ \sum_{j=1}^{n} x_{ij} = 1, & \forall i = 1\ldots n \\ y_i \ge x_{ji}, & \forall i,j = 1\ldots n \\ x_{ij} \in B = \{0,1\}, \quad y_i \in B, & \forall i,j = 1\ldots n \end{cases}$$

If we drop the desire for equally distributing the load between the minimum number of workstations, we arrive at a problem known as a Pure Binary Programming Problem:

$$\min_{x,y} \sum_{j=1}^{n} y_j$$

$$\text{s.t.}\begin{cases} p_{ij}x_{is} + \sum_{t=1}^{s-1} x_{jt} \le 1, & \forall i,j,s = 1\ldots n \\ \sum_{i=1}^{n} x_{ij} \le T_s, & \forall j = 1\ldots n \\ \sum_{j=1}^{n} x_{ij} = 1, & \forall i = 1\ldots n \\ y_i \ge x_{ji}, & \forall i,j = 1\ldots n \\ x_{ij} \in B = \{0,1\}, \quad y_i \in B & \forall i,j = 1\ldots n \end{cases}$$

This problem has the structure of a standard LP, wherein the objective function as well as each of the constraints is a linear combination of the decision variables, except that all its variables are constrained to take on the binary values 0 or 1, and so are *discretized*. In this section, we shall review a number of the most widespread available techniques for solving such problems involving a mix of continuous and integer variables, known as mixed integer programming problems.

The general Linear Mixed Integer Programming Problem (*LMIP*) is defined as follows (Nemhauser and Wolsey 1988):

Definition 1.34 Given real matrices $A_{m\times n}$, $G_{m\times p}$, *n-dimensional* cost vector c and *p-dimensional* cost vector h and an *m-dimensional* vector b representing the right-hand side of the constraints (where *m, n, and p* are natural numbers) the Linear Mixed Integer Programming Problem is defined to be the problem

$$(LMIP) \min_{x,y} .c^T x + h^T y$$
$$\text{s.t.} \begin{cases} Ax + Gy \le b \\ x \in \mathbb{Z}_+^n \\ y \in \mathbb{R}_+^p \end{cases} \tag{1.18}$$

The set $\Im = \{(x, y) : x \in \mathbb{Z}_+^n, y \in \mathbb{R}_+^p, Ax + Gy \le b\}$ is known as the feasible region of the problem, and a problem is called feasible if the set $\Im$ is non-empty. When $n = 0$ the problem reduces to the Linear Programming problem studied in the previous sections, and when $p = 0$, the problem is called a *Pure* Integer Programming Problem, sometimes denoted as IP. Equivalently, LMIP can be written as follows:

$$(LMIP) \min_{x} c^T x$$
$$\text{s.t.} \begin{cases} Ax \le b \\ x \ge 0 \\ x_j \in \mathbb{N}, \quad \forall j \in J \subseteq S = \{1, 2, \ldots n\} \end{cases}$$

where the matrices A and G have been merged in a single matrix A, as the variables x and y merged in a single vector x.

The case when some of the discrete variables x can only take on values from the binary set $B = \{0,1\}$ can be expressed explicitly in the LMIP formulation by adding the constraint $x_k \le 1$ for each variable x_k that is constrained to take values from the binary set B. More generally, when some variables x_k must take values from a finite set $C \subset \mathbb{Z}$ the LMIP formulation is still sufficient to express this fact, but some more variables must be introduced to the problem to express this constraint.

A naïve glance at the problem may suggest that one could solve the corresponding LP by "*forgetting*" about the integer constraints in (1.18), obtain a solution, and then somehow "*round about*" the values of those variables that are required to be integer and obtain the optimal solution. Although the idea of *relaxing* the integrality constraints and solving the "underlying" LP has turned out to be very fruitful and has led to very important classes of optimization algorithms for solving the (*LMIP*) problem, the prior naïve claim is false, and in fact rounding the solution of the LP may not provide the optimal solution, or even worse, may even lead to an infeasible point for the original problem. As a simple example, consider the problem $\{\max 6x_1 + 10x_2 \mid 2x_1 + 3x_2 \le 7, x_1, x_2 \ge 0, \textit{integer}\}$. The naïve approach of solving the corresponding LP $\{\max 6x_1 + 10x_2 \mid 2x_1 + 3x_2 \le 7, x_1, x_2 \ge 0\}$ produces the solution $x^* = [0\ 7/3]^T$ which when rounded produces the feasible point (for the original problem) $[0\ 2]^T$ with an objective function $z = 20$. However, the optimal value of the original problem is located at the point $x^{**} = [2\ 1]^T$ with value $z^{**} = 22$ which the reader can verify by enumerating all feasible points for this problem (all eight of them) and computing the objective function values at these points. The following easy to prove lemma provides further formal justification why the naïve approach cannot be expected to work in most cases.

Lemma 1.35 *Consider the (IP)* $\{\min c^T x \mid Ax = b,\ x \geq 0,\ x \text{ integer}\}$ *and its relaxation (LP)* $\{\min c^T x \mid Ax = b,\ x \geq 0\}$. *If the optimal solution* x^* *of the relaxed problem (LP) is unique, but it is not feasible for the original problem (IP), then the point* $y = round(x^*)$ *obtained by rounding each component of* x^* *to the nearest integer is not a feasible point for (IP).*

Proof The solution y can be written as $y = x^* + r,\quad r_i \in [-1/2, 1/2]\quad \forall i = 1, \ldots n$ where n is the dimensionality of the vectors x and y. Assuming by contradiction that y is feasible for the *(IP)* problem, it must satisfy $Ay = A(x^*+r) = b$ which implies $Ar = 0$. Now since x^* is the unique solution for the problem (*LP*), we must have that $c^T x^* < c^T x$ for all non-negative vectors x satisfying $Ax = b$, and therefore, by using $x = y = x^*+r$ in the previous strict inequality, we have that $c^T x^* < c^T(x^*+r)$ which in turn implies $c^T r > 0$. Now consider the vector $w = x^*-r$ that satisfies $Aw = b$ and is non-negative, because if any component i of x^* is greater than or equal to ½ then $(x^*)_i - r_i \geq 0$ and if otherwise, a component i of x^* is less than ½ then $r_i = -(x^*)_i$ and $w_i > 0$. Therefore, w is a feasible point for LP, and its objective value is $c^T w = c^T(x^*-r) = c^T x^* - c^T r < c^T x^*$ which is a contradiction as x^* is the optimal solution to the problem (*LP*). QED.

1.2.1 Mixed Integer Programming Modeling

In the following sections, Mixed Integer Programming (MIP) will refer to a *Linear* Mixed Integer Programming problem, unless otherwise specified.

As mentioned already, there are a large number of sources for problems that can be modeled as discrete optimization problems. In fact, throughout this book, we shall formulate and discuss a large number of such problems, model them as discrete optimization problems, and propose algorithms for their solution. At this point, we present a few classical discrete optimization problems and discuss a few modeling issues that arise.

1.2.1.1 The Knapsack Problem

The knapsack problem (KS) is the problem of selecting from a set of n items each having a profit p_i $i = 1\ldots n$ and weight w_i $i = 1\ldots n$ the best possible subset of items that maximize the total profit in the knapsack subject to a capacity constraint imposed by the size of the knapsack c. Introducing n decision variables x_i $i = 1 \ldots n$ where the variable x_j is set equal to the number of "j" items the decision maker will carry, the KS is modeled as follows:

$$(KS)\ \max_x z = \sum_{j=1}^{n} p_j x_j$$

$$\text{s.t.} \begin{cases} \sum_{j=1}^{n} w_j x_j \le c \\ x_j \in \mathbb{N}, \quad \forall j = 1 \ldots n \end{cases}$$

This particular discrete optimization problem has been the subject of many intense studies since the 1950s. In case the variables x_j are forced to be binary so that x_j is either 0 or 1, the problem is known as the 0–1 Knapsack Problem, *possibly the simplest* Integer Programming problem (single linear constraint, all variables binary). It also appears as a sub-problem in many general MIP procedures where its solution can be used to strengthen LP bounds obtained from LP relaxations of general LMIP problems.

1.2.1.2 Assignment and Network Flow Problems

We have already discussed the linear assignment problem (LAP) in Sect. 1.1.2.2, and we have seen that efficient algorithms exist for its solution. Modeling the LAP with binary variables x_{ij} that denote whether person p_i is assigned to task t_j is very easy:

$$(LAP)\ \max \sum_{i=1}^{n} \sum_{j \in T_i} a_{ij} x_{ij}$$

$$\text{s.t.} \begin{cases} \sum_{i=1}^{n} x_{ij} = 1, \quad \forall j = 1 \ldots n \\ \sum_{j \in T_i} x_{ij} = 1, \quad \forall i = 1 \ldots n \\ x_{ij} \in B = \{0, 1\}, \quad \forall i = 1 \ldots n, j \in T_i \end{cases}$$

We have already seen that the structure of the constraints is such that it allows the binary variables x_{ij} to be replaced by continuous variables $0 \le x_{ij} \le 1$ and the solution of the problem is guaranteed to be binary, which is the reason for the existence of efficient algorithms for this particular problem. Similarly, the linear network flow problems of Sect. 1.1.2.2 are discrete optimization problems, whose very special structure allows the linear relaxation to obtain valid integer solutions nevertheless. Unfortunately, very few discrete optimization problems share such properties that render them amenable to solution by polynomial time algorithms.

1.2.1.3 Facility Location Problem

In the introduction to Sect. 1.1.1 (on continuous unconstrained optimization) we discussed the problem of optimally determining the location of a plant to minimize the acquisition and transportation costs incurred by serving the markets that will be assigned to it. In a discrete version of the problem, there are n existing facilities with capacities c_i $i = 1\ldots n$ that have an associated acquisition price p_i. There are also given costs s_{ij} of shipping items from each plant location i to any one of m market places, say j that have demand d_j. The objective is to choose which of the n plants to acquire so that the acquisition and shipping costs are minimized while the demand at each of the m market places is fully met subject to the capacity constraints of the acquired plants. Introducing n binary decision variables x_i, as before, indicating acquisition or not of the respective plant, the optimization problem is formulated as follows:

$$\min_{x,y} . \sum_{j=1}^{n} p_i x_i + \sum_{j=1}^{n}\sum_{i=1}^{m} s_{ij} y_{ij}$$

$$\text{s.t.} \begin{cases} \sum_{i=1}^{m} y_{ji} \le c_j x_j, & \forall j = 1\ldots n \\ \sum_{j=1}^{n} y_{jk} \ge d_k, & \forall k = 1\ldots m \\ x_j \in B = \{0,1\}, & \forall j = 1\ldots n \\ y_{ij} \ge 0, & \forall i = 1\ldots n, j = 1\ldots m \end{cases}$$

The problem is a true Mixed-Integer Programming problem because in order to model the flow balance constraints regarding the satisfaction of market demands, we had to introduce mn continuous non-negative variables y_{ij} indicating the amount of goods that will be shipped from each acquired facility to each of the m markets. We shall discuss specialized algorithms for facility location problems in Chap. 5 on location theory.

1.2.1.4 Set Covering, Packing and Partitioning Problems

In many problems arising in location theory or clustering of data, the problem data include a finite—but usually very large—set S of n points s in some finite-dimensional space. A collection $\{S_j,\ j = 1\ldots m\}$ of subsets of S is implicitly or explicitly provided, and each subset S_j has a cost c_j.

The set covering problem (SCP) is the problem to find a collection of subsets $S_{j_k},\ \ j_k \in I \subseteq \{1,\ldots m\}$, not necessarily disjoint to each other, so that the sum of the costs c_{j_k} is minimized while the selected subsets completely cover the initial set

S, i.e. $\bigcup_{j_k \in I} S_{j_k} = S$. Consider the matrix A of dimensions $n \times m$ whose element A_{ij} is one if the element s_i is contained in the set S_j and zero otherwise. We can now formulate the model for SCP using binary decision variables x_j denoting whether the set S_j will be part of our collection or not based on whether the value of x_j is set to 1 or 0, which becomes

$$(SCP) \min_x \sum_{j=1}^{m} c_j x_j$$

$$\text{s.t.} \begin{cases} Ax \geq e \\ x \in B^m \end{cases}$$

where e is the n-dimensional column vector of all ones.

If there is an extra requirement that the subsets $S_{j_k}, j_k \in I \subseteq \{1, \ldots m\}$ we pick must be disjoint to each other so that $S_k \cap S_l = \emptyset, \quad \forall k,l \in I, k \neq l$ then the problem is known as the set partitioning problem (SPP), and is modeled as follows:

$$(SPP) \min_x \sum_{j=1}^{m} c_j x_j$$

$$\text{s.t.} \begin{cases} Ax = e \\ x \in B^m \end{cases}$$

Both the above problems will play a crucial role in the development of important algorithms for location problems discussed later in the book.

Finally, the set packing problem (PP) asks that we select a collection of subsets S_j so that all subsets in our selection are disjoint from each other—as in the SPP problem—but without the requirement that the resulting collection completely covers the set S, and with an objective to maximize the sum of the costs c_j of the selected subsets (otherwise, with non-negative costs the problem has the trivial solution $x = 0$, corresponding to selecting no subset at all). The model for the packing problem becomes

$$(PP) \max_x \sum_{j=1}^{m} c_j x_j$$

$$\text{s.t.} \begin{cases} Ax \leq e \\ x \in B^m \end{cases}$$

1.2.1.5 Traveling Salesman Problem

The traveling salesman problem (TSP) together with the knapsack problem is among the most intensely studied problems in discrete optimization. The problem is formulated over a graph $G(V,E,W)$ where nodes represent cities and arcs represent roads between those cities, whereas the arc weights w_e in W represent distances between cities. The objective is to find the *minimum distance tour* of a traveling salesman that begins from the salesman's home city s in V and ends at the same city so that all cities (nodes) in V are visited exactly once. By introducing binary variables x_{ij} that are set to 1 if city j immediately follows city i in a tour, and are set to zero otherwise, the objective function of the problem is clearly the linear function $c(x) = \sum_{(i,j)\in E} w_{ij}x_{ij}$. The constraint that requires each city to be visited exactly once can be expressed as $\sum_{j:(i,j)\in E} x_{ij} = \sum_{k:(k,i)\in E} x_{ki} = 1, \quad \forall i \in V$. In order to express the constraint that the resulting solution is a full tour however and not a collection of disconnected smaller sub-tours, we need to ensure that for each subset V_1 of V with a cardinality greater than 1, and its complement $V_2 = V - V_1$, there are at least two arcs in the tour that connect nodes from V_1 to V_2. This introduces a huge number of constraints as the total number of subsets of V is $2^{|V|}$ where $|V|$ is the cardinality of the set V. The constraints become

$$\sum_{i\in V_1, j\in V-V_1:(i,j)\in E} x_{ij} \geq 1, \quad \forall V_1 \subseteq V : |V_1| > 1, |V - V_1| > 1$$

and the full model for the traveling salesman problem becomes:

$$(TSP)\ \min \sum_{(i,j)\in E} w_{ij}x_{ij}$$

$$\text{s.t.} \begin{cases} \sum\limits_{j:(i,j)\in E} x_{ij} = \sum\limits_{k:(k,i)\in E} x_{ki} = 1, \quad \forall i \in V \\ \sum\limits_{i\in V_1, j\in V-V_1:(i,j)\in E} x_{ij} \geq 1, \quad \forall V_1 \subseteq V : |V_1| > 1, |V - V_1| > 1 \\ x_{ij} \in B, \quad \forall (i,j) \in E \end{cases}$$

There are other, more compact formulations for the TSP but they are usually inferior to the above formulation.

1.2.1.6 Modeling Disjunctive Constraints

In many domains, the problem model must explicitly express the logical condition that some decision variables x must take on such values so that at least one of two functions $g(x)$ and $h(x)$ must be non-positive, i.e. $g(x) \leq 0$ OR $h(x) \leq 0$. Assuming we have at our disposal an upper bound M on the values any of the two functions can assume for any x in the feasible set of the problem, we can express this logical or condition by introducing a new binary variable y as the following constraints that must be simultaneously satisfied:

$$g(x) \leq My$$
$$h(x) \leq M(1-y)$$
$$y \in \{0,1\}$$

Indeed, when $y = 1$, $h(x)$ must be non-positive whereas the constraint for $g(x)$ is inactive for all feasible x, while for $y = 0$ the opposite is true. In any case, one of the two functions will always be non-positive in the feasible set, and the disjunctive constraint is correctly expressed in the model. Note that the nature of the decision variables x is left unspecified, i.e. we do not care if they are continuous or discrete variables. Note also that the constraints will inevitably be nonlinear in the decision variables x if the functions $g(.)$ or $h(.)$ are nonlinear.

This can be particularly useful in models where a variable w must be expressed as the minimum of two other *non-negative* variables x_1, x_2, in which case, assuming we have an a priori known upper bound M on the values x_1 or x_2 may assume in the feasible set, we may use the above technique to model the equation $w = \min\{x_1, x_2\}$ as the following set of constraints:

$$w \leq x_1$$
$$w \leq x_2$$
$$x_1 - w \leq M(1-y)$$
$$x_2 - w \leq My$$
$$x_1, x_2 \geq 0$$
$$y \in B = \{0,1\}$$

Implication constraints of the form IF $g(x) > 0$ THEN $h(x) \geq 0$ can also be expressed in MIP models, assuming again that an upper bound of the values $g(x)$ and $-h(x)$ is a priori available for all feasible x. This is because the above logical implication can be equivalently expressed as the following constraints:

$$-h(x) \leq My$$
$$g(x) \leq M(1-y)$$
$$y \in \{0,1\}$$

In case $g(x) > 0$, the binary variable y must necessarily assume the value 0, and this will force the value of $h(x)$ to be non-negative as requested.

General disjunctive constraints can be formulated using the same ideas. Consider the problem of formulating a disjunctive constraint that requires at least k out of m different sets of linear inequalities of the form $A^{[i]}x \leq b^{[i]}$ $i = 1 \ldots m$ to be satisfied subject to some box bounding constraints on the decision variables x: $0 \leq x \leq u$. It is easy to see that there will always exist a column vector ω such that the inequalities $A^{[i]}x \leq b^{[i]} + \omega$ will hold for all x satisfying $0 \leq x \leq u$ for all

$i = 1\ldots m$ (simply consider the maximum value $A^{[i]}x - b^{[i]}$ can take for each $i = 1\ldots m$ in the interval $0 \le x \le u$.)

Now, the requirement that at least k out of the m systems of inequalities must be satisfied, can be expressed by introducing m binary variables y_j $j = 1\ldots m$ and adding the following set of constraints:

$$A^{[i]}x \le b^{[i]} + \omega(1 - y_i), \quad \forall i = 1, \ldots m$$

$$\sum_{i=1}^{m} y_i \ge k$$

$$0 \le x \le u$$

$$y \in B^m$$

In the first m sets of constraints, the ith constraint set will be inactive if y_i is set to zero, and will be enforced to hold when $y_i = 1$. The requirement $y_1 + \cdots + y_m \ge k$ then ensures that at least k of the constraint sets will be enforced.

1.2.2 Methods for Mixed Integer Programming

1.2.2.1 Preprocessing Models

Model preprocessing refers to the techniques applied by modern combinatorial optimization solvers to MIP models in order to transform the input model problem into another equivalent problem which is however easier for the solver algorithms to solve. Among other things, preprocessing attempts to eliminate variables from the input problem that must be necessarily fixed to some value (explicitly or implicitly fixed variables), remove redundant constraints (e.g. constraints that are linear combinations of other constraints); usually, both primal and dual problem information is used where feasibility reasoning techniques work on the primal problem, whereas objective function bounds and variable elimination techniques work with dual problem information.

Other preprocessing techniques attempt to strengthen the value of the LP relaxation of the input MIP problem so as to speed-up Branch-and-Bound methods employed later on by the solver. This can be done by tightening bounds on individual variables and/or by tightening constraint coefficients.

Finally, most state-of-the-art solvers utilize some Artificial Intelligence-based techniques whereby they attempt to extract implications and clique constraints from the problem constraints formulation that can be very useful when adding cuts in cutting plane techniques to be discussed further below. To illustrate the nature of pre-processing in MIP solvers we list a number of cases that solvers routinely check for in order to strengthen a problem formulation

Preprocessing Inequality Constraints of Binary Variables

Consider the binary programming problem

$$\min c^T x$$
$$\text{s.t.} \begin{cases} Ax \le b \\ x \in B^n \end{cases}$$

where $B = \{0,1\}$. Consider any of the linear inequality constraints of the above problem and write it as follows:

$$\sum_{j \in N_+} a_j x_j + \sum_{j \in N_-} a_j x_j \le b \quad N_+ = \{j : a_j > 0\}, N_- = \{j : a_j < 0\}$$

It is easy to see that if the inequality $\sum_{j \in N_-} a_j > b$ holds, then the constraint is infeasible, since b is obviously less than zero, and so a lower bound on the value of the left-hand side of the constraint, the value $\sum_{j \in N_-} a_j$ (attainable by setting $x_j = 0$ for all j in N_+ and setting $x_j = 1$ for all j in N_-) is greater than the right-hand side of the constraint.

It is equally easy to see that if the inequality $\sum_{j \in N_+} a_j \le b$ holds, then the constraint is always satisfied, since an upper bound on the left-hand side of the constraint given by $\sum_{j \in N_+} a_j$ is always less than or equal to the right-hand side of the constraint. Therefore, the constraint is redundant and can be safely removed from the model.

Further, the following implications always hold:

$$a_k > b - \sum_{j \in N_-} a_j, \quad k \in N_+ \Rightarrow x_k = 0$$

$$-a_k > b - \sum_{j \in N_-} a_j, \quad k \in N_- \Rightarrow x_k = 1$$

which can be used to fix variables to zero or one and immediately after to remove them from the problem formulation. Their justification is trivial: if the coefficient of a binary variable x_k in a certain constraint $a^T x \le b$ is positive and greater than the quantity $b - \sum_{j \in N_-} a_j$, then the constraint can never be satisfied if that variable is set to 1, since the smallest value the left-hand side of the constraint can take is $a_k + \sum_{j \in N_-} a_j$ which will still be greater than b. Therefore, in such a case, the variable x_k must be set to zero. A similar reasoning proves the second implication above.

Clique Preprocessing

Continuing with the previous setting, consider a constraint of the form

$$\sum_{j \in N} a_j x_j \le b$$

where all variables x_j j in N are binary. If $a_j \le b$ for all j in N, then for any subset C of N for which the inequality $\sum_{j \in C} a_j > b$ holds, the implication also holds that

$$\sum_{j \in C} x_j \le |C| - 1$$

which is a valid inequality that can be added to the constraint set of the problem to improve the bound obtained from the linear programming relaxation of the original problem (the validity of the implication is obvious, since if the inequality does not hold, all the variables x_j for j in C will be set to 1, and the original constraint will be violated).

General Inequalities Preprocessing

In general, inequality constraints in MIP problems have the form $l \le a^T x \le u$. Some constraints are bounding-box constraints on the values of some variables: $l_j \le x_j \le u_j$. Solvers usually first detect bounding-box constraints, and fix any variable that is explicitly fixed because $l_j = u_j$ to the constant value $x_j = l_j$ and remove the box constraints from the problem. Then, the variable is eliminated from the problem by modifying the remaining constraints of the form $l \le a^T x \le u$ by setting $l = l{-}a_j l_j e$ and $u = u{-}a_j l_j e$ and finally setting $a_j = 0$ where e is an n-dimensional column vector of all ones (where n is the number of variables in the original problem).

Constraint normalization in MIP preprocessing is performed by checking for integrality (or if possible, for rationality) of the coefficients of the constraint. If all constraint coefficients are integral, then the coefficients, including the right-hand side coefficient of the constraint, are divided by their greatest common divisor (GCD).

Finally, tightening of left- and right-hand side of constraints in a general MIP problem can be sometimes done by checking if all coefficients of vector a are integral and all variables x_j for which a_j are not zero are integer variables, then the left-hand side of the constraint can be raised to its ceiling whereas the right-hand side of the constraint may be lowered to its floor so that the original inequality constraint can be replaced by $\lceil l \rceil \le a^T x \le \lfloor u \rfloor$ which is tighter than the original constraint.

A technique known as Domain Propagation is also often successfully employed in MIP preprocessing that consider a set partitioning, or set packing constraints of the form

$$\sum_{j \in C} x_j = 1, \quad C \subseteq N = \{1, 2 \ldots n\}$$

or, respectively

$$\sum_{j \in C} x_j \le 1, \quad C \subseteq N = \{1, 2 \ldots n\}$$

Such constraints are added in a *clique table* for quickly drawing implications by domain propagation of the variables.

Now consider integer variables x_j for which the side of the constraint becomes redundant if the variable is not set to its lower or upper bounds respectively. Then, we can reduce the coefficients a_j and the bounds l or u in order to obtain the same redundancy effect if the variable is not set at its lower or upper bound respectively and the same restrictions on the other variables if the variable is set to one of its bounds. The above process can be repeated until the constraints no longer change.

Pre-Solving Equality Constraints

For constraints involving 2 continuous variables x_i and x_j, one variable can be expressed as an affine transformation of the other. In this case, one variable can be immediately deleted from the problem, with appropriate modifications on the coefficients and right-hand side parameters of the constraints of the problem (and by deleting the corresponding variable and coefficient from the objective function and adding/subtracting a constant term).

The same can be done *if* the two variables are integer, and their coefficients are divisible by each other so that a_j/a_k or a_k/a_j is integer. This process can also be repeated until there is no equality constraint involving only two variables in the way described above.

1.2.2.2 Branch & Bound

The Branch & Bound method (B&B) is a *framework* for the solution of combinatorial optimization problems that combines *relaxation* ideas discussed earlier to obtain *bounds* on the *optimal* value of a problem with the well-known divide-and-conquer method based on the *decomposition principle*. The decomposition principle states that a problem (P) of the form

$$\begin{aligned} &(P) \min_x f(x) \\ &\text{s.t.} \quad x \in S \end{aligned}$$

where $f : \mathbb{R}^n \to \mathbb{R}$, whose feasible set S can be decomposed into a finite number of subsets $S_j \subseteq S, j \in J$ where J is a finite nonempty index set such that $\underset{j \in J}{\cup} S_j = S$ can be optimized by optimizing $f(x)$ over each of the subsets S_j giving solutions $x^*_{[j]}$ with values z^*_j and choosing the best (i.e. with smallest objective function value) optimizer, where it is assumed that $z^*_j = -\infty$ if the corresponding problem $(P_j) \min_{x \in S_j} f(x)$ is unbounded, and $z^*_j = +\infty$ if (P_j) is infeasible. Indeed, observe that the optimal value will be $+\infty$ iff none of the $|J|$ sub-problems

(P_j) are feasible, in which case, the whole problem (P) is also infeasible. Otherwise, the solution will be $-\infty$ iff at least one sub-problem (P_j) is unbounded, in which case, the whole problem (P) is also unbounded. Otherwise, let $z^* = \min\left\{z_j^*, j \in J\right\}$, $j^* = \arg\min\left\{z_j^*, j \in J\right\}$. The value z^* is the optimal value for (P) since it is the optimal value $f()$ can assume in S_{j*} and the optimal value of the problem $\min\{f(x) \mid x \text{ in } S{-}S_{j*}\}$ is not smaller than the value z^*. Clearly then, the optimizer $x^*_{[j^*]}$ is a valid optimizer for (P). The result also holds if we replace the optimal values z_j^* with lower bounds $\underline{z}_j \le z_j^*$. In particular, we have the following

Lemma 1.36 *If* $\underline{z}_k = \min_{j \in J}\left\{\underline{z}_j\right\}$ *then* $\underline{z}_k$ *is a lower bound for the optimal value of* (P). *In addition, if it happens that* $\underline{z}_k = z_k^*$ *then the optimal value of* (P) *is* $z^* = z_k^*$ *and the optimizer* $x^*_{[k]}$ *is an optimizer for* (P).

Proof In all of the set $S{-}S_k$, the function $f(x)$ cannot take a value smaller than $\underline{z}_k$ so $\underline{z}_k$ is a valid lower bound for f in S. Now, if $\underline{z}_k = z_k^*$ is satisfied, then the function $f(x)$ cannot assume a value smaller than $f\left(x^*_{[k]}\right) = z_k^*$ in all of S and therefore, $z_k^*, x^*_{[k]}$ are respectively the optimal value and the optimizer point for (P). QED.

The above established facts provide a guideline for the B&B framework. Ideally, one would like to develop a method so that the collection of subsets S_j to consider does not grow too much and at the same time develops a method to relax the constraints $x \in S_j$ by creating a superset $T_j \supseteq S_j, \quad j \in J$ so that the solution of the *relaxed* problems $z_j = \min\{f(x) \mid x \text{ in } T_j\}$ is at the same time, easy to obtain, coincides with the optimal solution $x^*_{[k]}$ at the subset S_k in which an optimal solution lies, and does not produce values z_j (lower bounds) that are below the optimal solution value $f\left(x^*_{[k]}\right) = z_k^*$ for (P). Unfortunately, this ideal situation is very rarely possible.

The B&B method generates a tree that is augmented in successive iterations by selecting a node n with no children yet having certain properties, and partitioning the feasible region S_n corresponding to that node among a collection of some m_n (usually mutually disjoint) subsets $S_{nj}\, j = 1 \ldots m_n$ so that their union equals S_n. This collection of nodes is denoted as $D(n)$. The augmentation of the B&B tree continues until a solution to (P) obtained in some node n^* of the tree is provably optimal, or until infeasibility or unboundedness of the problem is proved. Initially, the tree consists of a single node, called the root, designated with the index 0, and $D(0) = \{\}$. The process by which a node n is selected for processing and a set of its immediate descendants, $D(n)$ is eventually added to the tree, is called "branching on the node n". A node with an empty set of children is called a terminal node, and the set of terminal nodes is denoted as T. The set of the rest of the nodes in the tree will be denoted as F.

Consider the relationship between the nodes of the B&B tree and the original problem (P). The root node 0 corresponds to an initial relaxation of the set $S_0 = S$, denoted by T_0. The relaxed problem at the root node is therefore the problem

$$
\begin{aligned}
&(R_0)\ \min_x f(x)\\
&\text{s.t.}\quad x \in T_0 \supseteq S
\end{aligned}
$$

Branching at a node n corresponds to adding a restriction K_n such that $S \subseteq K_n = \cup_{j\in D(n)} K_n(j)$ defining the relaxed set T_j at a node j in $D(n)$ as $T_j = T_n \cap K_n(j)$, then moving the now father node n from T to F and adding the children nodes $D(n)$ to the set of terminal nodes T. Clearly then, associated with any node n in the B&B tree are two problems:

1. the relaxed problem (R_n) $\min\{f(x) \mid x \text{ in } T_n\}$, and
2. the underlying problem $(\underline{P}_n)$ $\min\{f(x) \mid x \text{ in } S_n\}$ where $S \supseteq S_n = S \cap T_n$.

Note that the relaxed problem (R_n) is a relaxation of the underlying problem $(\underline{P}_n)$ but not necessarily a relaxation of the original problem (P).

In general, when a node is selected, one would ideally like to solve the underlying problem $(\underline{P}_n)$ but most of the time in the B&B process its solution will not be possible without further branching on that node. A successful B&B algorithm therefore, must construct relaxations (R_n) that are easy to solve and provide optimal solutions $\underline{z}_n$ that represent lower bounds on the optimal value z_n^* that are as tight as possible (meaning that they are not significantly lower than the value z_n^*). This is because if S_n does not contain an optimal solution of (P), then if the value $\underline{z}_n$ is above a known upper bound $\bar{z}$ on the optimal value of (P)—often computed as the best feasible solution constructed in the B&B process thus far—the node can be "*fathomed*" or "closed" in the sense that no children for this node need be created, and no further processing for the underlying problem $(\underline{P}_n)$ need be done (this can be accomplished in the B&B process by simply discarding the node from the B&B tree). Note that due to the branching process rules, for any child node j in $D(n)$ it holds that $T_j \subseteq T_n$ and therefore, any lower bound on z_n^* is also a lower bound on z_j^*.

The following essentially establishes a lower bound for the root node 0.

Lemma 1.37 *If T is the set of terminal nodes of a finite B&B tree, then* $\bigcup_{j\in T} T_j \supseteq S$.

Proof If the tree consists of only the root node, then the lemma trivially holds as $T_0 \supseteq S$ by definition. Otherwise, observe that the children $D(n)$ of a node n will always satisfy $\bigcup_{j\in D(n)} T_j \supseteq T_n$ so if a node n is not a terminal node, the union of all the sets T_j of nodes j that are descendants of n and are also terminal nodes will be a superset of T_n. Applying this fact to the root node 0, we get $\bigcup_{j\in T} T_j \supseteq T_0 \supseteq S$. QED.

This immediately implies that a lower bound on the optimal value of the original problem (P) will be the *minimum* of all lower bounds obtained at the terminal nodes of the B&B tree *at any point in the B&B tree construction process*. This minimum lower bound may of course be $\pm\infty$ at any point but it is non-decreasing throughout the tree construction process. There is also an associated upper bound on the optimal solution value of the original problem that is $+\infty$ until a feasible solution is discovered, and decreases thereafter as new better feasible solutions for the original problem are discovered (the best solution found at any point in time is known as the *incumbent solution*) and may become $-\infty$ if an underlying subproblem $(\underline{P}_j)$ is proved to be unbounded, in which case the algorithm terminates with an indication of unboundedness.

To describe a *Generic Branch&Bound* algorithm (GBB) for MIP, a few more definitions are needed. We define the set of active nodes A as the set of terminal nodes T that may require branching, and the set of closed nodes (the fathomed nodes) as C. The GBB algorithm in pseudo-code is described below:

Algorithm Generic Branch & Bound Algorithm for Mixed Integer Programming

Inputs: Cost vector c, matrix B, right-hand side vector b, index set J of integer variables.

Outputs: A point x^* that is the global minimizer for the problem

$$\min_x c^T x$$

$$s.t. \begin{cases} Bx \leq b \\ x_j \in \mathbb{Z}, \quad \forall j \in J \subseteq \{1, \ldots n\} \end{cases}$$

Begin

1. Select an initial relaxation $T_0 \supseteq S$, Set $T = A = \{0\}$, Set $z_l = -\infty, z_u = +\infty, k = 0$.
2. Select from A an active node k.
3. Compute a lower bound $\underline{z}_k$ for the value z_k^* by solving the relaxed problem (R_k) and the corresponding solution $\underline{x}_k$.
4. Set $z_l = \min\{\underline{z}_j | \ j \text{ in } T\}$, $j^* = \text{argmin}\{\underline{z}_j | \ j \text{ in } T\}$.
5. if $z_u \leq \underline{z}_k$ then
    a. Set $A = A - \{k\}, C = C \cup \{k\}$.
    b. GOTO 8.
6. end-if
7. if $\underline{z}_k \leq z_u$ then
    a. if $\underline{z}_k = z_k^*$ then
        i. Set $z_u = \underline{z}_k$.
        ii. for each j in A do
            1. if $\underline{z}_j > z_u$ then
                a. Set $A = A - \{j\}$.
                b. Set $C = C \cup \{j\}$.
            2. end-if

iii. end-for

b. end-if

8. if $A = \{\}$ then

a. Set $z^* = z_l$, $x^* = x_{j^*}$.
b. return x^*.

9. else

a. Select a node j from A.
b. if analysis not involving branching must be performed on j then

i. Set $k = j$.
ii. GOTO 3.

c. end-if
d. Set $D(j) = \{j_1, j_2\}$ where the nodes j_1 and j_2 are defined in terms of the smallest index i of variables in J such that the value of the optimal solution of problem (R_j) $\hat{x}$, is non-integer: $K_{j_1} = \{x|x_i \leq \lfloor \hat{x}_i \rfloor\}, K_{j_2} = \{x|x_i \geq \lceil \hat{x}_i \rceil\}$.
e. for each m in $D(j)$ do

i. Compute $K_j(m)$.
ii. Set $T_m = T_j \cap K_j(m)$.
iii. Set $\underline{z}_m = \underline{z}_j$.
iv. Set $T = T - \{j\}$.
v. Set $A = A \cup D(j)$.

f. end-for
g. Select k from $D(j)$.
h. GOTO 3.

10. end-if

End.

It should be emphasized that the above algorithm—as is the case with some of the previous algorithms described in this chapter—describes a family of algorithms rather than a concrete algorithm. Specific instantiations of B&B need to specify the following:

- Rules for selecting among active nodes a node for further analysis in steps 2, 9.a and 9.g
- Methods for computing lower bounds of problems (R_i) in step 3

Even step 9.d which is completely specified in the above B&B algorithm is often modified in order to enhance the algorithm performance. As specified, the strategy for branching in step 9.d is broadly known as "*branching on variables*", and it remains the most popular branching rule implemented and set as default in most successful commercial codes. For this reason, it is often called the "*standard branch*" (terminology introduced in the 1970s). Other successful branching rules will be briefly mentioned shortly.

In the sequence we shall discuss a number of successful strategies and heuristics for implementing each of the above generic steps. First, however, we present some important results that hold irrespective of the choices made for each of the steps above.

The above algorithm has the property that the lower and upper bounds generated are non-decreasing and decreasing respectively, and the optimal value of the original problem (*P*) always lies between them. In particular, we have

Theorem 1.38 *The values z_l and z_u generated by the algorithm GBB are respectively non-decreasing and decreasing, and always satisfy $z_l \le z^* \le z_u$.*

Proof Initially, z_l is negative infinity and z_u is positive infinity, so the inequality clearly holds. When a lower bound is generated in step 4 of the algorithm, the value is a lower bound on the optimal value of a relaxed problem (R_j) and it is a lower bound on (*P*) by virtue of Lemma 1.36. When the set *T* changes, at least one element $\underline{z}_m$ of the set $\{\underline{z}_J | j \text{ in } T\}$ is deleted, and if branching occurs then additional replicas of the value $\underline{z}_m$ are added to the set according to the algorithm. This process cannot decrease the value of the minimum element in this set. Some of the elements can only increase their value as a result of the analysis in step 3. Therefore, the value z_l is monotonically non-decreasing.

Regarding the upper bound z_u observe that an upper bound is generated only in step 7.a.i and is the optimal objective function value over a subset of the original feasible set *S*, hence it is an upper bound on the optimal value of the original problem (*P*). According to the algorithm, the value z_u only changes in order to be made smaller, so it is indeed monotonically decreasing. QED.

The most well-known B&B framework instantiation corresponds to LP-based B&B algorithms, whereby the underlying problems $(\underline{P}_i)$ at individual nodes in the B&B tree correspond to a particular MIP instance, and the relaxation (R_i) chosen for the underlying problem is obtained by deleting any and all integrality constraints on the problem variables that results in an LP that must be solved. In this case, if the LP is unbounded, then it can be proved that if the matrix *A* consists of rational data (i.e. each element is of the form $a_{ij} = q/r,\ q, r \in \mathbb{Z}$), the original problem (*P*) is either unbounded or infeasible (Meyer 1992). Alternatively, by relaxing the linear constraints $Ax \le b$ but maintaining the integrality constraints at any node a completely different algorithm, known as the *Additive Algorithm* takes shape (Balas 1965).

We can easily derive some more properties of the Generic B&B algorithm for MIP. For instance, assuming the LP relaxation set $T = \{x | Ax \le b\}$ is bounded, the algorithm must terminate in a finite number of branches. This follows immediately from the fact the standard branching in step 9.d limits the number of nodes at any depth level *k* to a maximum of at most 2^k nodes and since from the boundedness of the set *T* we have that there exists an appropriately large but finite integer *M* such that for all *x* in *T* it holds that for each component of *x*, $x_i \le M$, then the maximum number of levels in the tree must be $M|J| + 1$, since the range of each integer variable is initially at most M, and in each branch, the length of the range of the variable being branched on, is reduced by at least 1. The total number of branches therefore will be at most $2^{M|J|+1}$ (although finite, this bound is not at all satisfactory from a theoretical point of view, to say the least.) As a corollary, if x is constrained to be a binary vector, the number of levels in any resulting B&B tree cannot exceed $|J| + 1$.

Unfortunately, if the set $T = \{x|\ Ax \leq b\}$ is unbounded, the algorithm is not guaranteed to terminate. To show this is the case, consider the following counterexample:

$$\min_x -x_1$$
$$\text{s.t.} \begin{cases} 2x_1 - 2x_2 = 1 \\ x_1, x_2 \in \mathbb{Z} \end{cases}$$

The above problem is clearly infeasible since the constraint $2x_1 - 2x_2 = 1 \Leftrightarrow x_1 - x_2 = 1/2$ can never be satisfied by integer x_1 and x_2. But the resulting LP relaxation of this problem is unbounded and an LP-based B&B algorithm for MIP would leave z_l at $-\infty$ after solving the LP at the root node 0. Such an algorithm will produce an unbounded number of nodes because the standard branching of the form $x_i \leq \lfloor \hat{x}_i \rfloor, x_i \geq \lceil \hat{x}_i \rceil$ (where $\hat{x}$ is the solution of the relaxed LP) for any $i = 1{,}2$ will not improve the objective function bounds at all. With standard branching on variables then, the node corresponding to $x_i \geq m + 1$ for some m will have negative infinity as the optimal value of the LP relaxation and z_l will remain at negative infinity for ever, whereas z_u will also remain at positive infinity for ever. Note that if another type of branching was employed, so that instead of branching on x_i being not less than or not greater than the nearest integers of its current value, we branch on the restriction $x_1 - x_2 \leq 0$ OR $x_1 - x_2 \geq 1$, the resulting nodes' LP relaxations are both infeasible (the systems

$$\begin{array}{l} 2x_1 - 2x_2 = 1 \\ x_1 - x_2 \leq 0 \end{array}, \text{ and } \begin{array}{l} 2x_1 - 2x_2 = 1 \\ x_1 - x_2 \geq 1 \end{array}$$

are both infeasible) and therefore such an algorithm would terminate after generating 3 nodes in the B&B tree. This example illustrates the importance of branching rules in the successful application of the B&B framework on various problems.

In the context of LP-based B&B, closing a node in step 5.a, does not necessarily mean that the node does not contain an optimum of the original problem (P) but rather that the node does not contain a point that is strictly better than the current incumbent solution. The node selection in step 9.a can follow one of a number of strategies:

- The *Best Open Node Strategy* selects the node with largest index j from A that satisfies $\underline{z}_j = z_l$ which under some mild conditions is optimal in the sense that all the nodes of the tree generated by the application of this strategy will have to be generated by any other node selection strategy that attempts to locate the globally optimal solution to the problem (P). Unfortunately, this strategy usually requires an extraordinarily large storage space for storing open nodes in the set A, and so is rarely used in its pure form in practice.

- The *Depth-First Strategy* selects the node with largest index j from A regardless of bounds. Since nodes are indexed according to the order in which they were constructed, this strategy leads to quickly finding feasible solutions by "diving" into the B&B tree and so produces upper bounds z_u at the fastest time possible. However, no guarantees can be made about the quality of such solutions. Yet, the storage requirements for such a strategy are kept at a minimum and for this reason many practical implementations of B&B algorithms use this strategy at least for some part of the solution process (in conjunction with other rules giving rise to *hybrid* strategies).
- The *Breadth-First Strategy* selects from the list A the node j with the smallest index—essentially the opposite of the Depth-First Strategy. This strategy is rarely used as it has neither of the advantages of the previous two strategies but suffers from both strategies disadvantages (requires extremely heavy storage space and does not obtain feasible solutions early that can lead to upper bounds and node closures).
- *Hybrid strategies* that combine information from node analysis and other AI-based approaches—for example Constraint Programming techniques—to determine the "best" node to select for branching next. Such branching rules have been the subject of intense research in the 1980s and have also seen renewed interest in the past decade due to the integration of optimization-based and AI-based techniques for solving large and difficult combinatorial optimization problems (Achterberg et al. 2005; Achterberg and Berthold 2009; Achterberg 2007).

Related to hybrid strategies for node selection, one particularly successful technique was *node estimation*, by which an estimate ε_j of the optimal value of the underlying problem $(\underline{P}_j)$ at any node j is obtained via some heuristic which may be above, equal, or below the actual value z_j^*. In such a strategy, the node from A with the best (smallest) estimated value ε_j is selected in step 9.a. One approach to node estimation goes as follows. Let $\underline{x}_{[j]}$ denote the optimal solution for the relaxed problem (R_j) at node j, and let $f_{i,j}$ denote the fractional part of the value of the ith *basic* variable $\left(\underline{x}_{B_{[j]}}\right)_i$ in the optimal solution of (R_j) so that $f_{i,j} = \left(\underline{x}_{B_{[j]}}\right)_i - \left\lfloor\left(\underline{x}_{B_{[j]}}\right)_i\right\rfloor$ where $B_{[j]}$ is the basis of the optimal solution $\underline{x}_{[j]}$ of (R_j) at node j. Let $I_j = \sum_{i \in B_{[j]}} \min\{f_{i,j}, 1 - f_{i,j}\}$, so that I_j denotes the sum of integer infeasibilities of the optimal relaxed solution at node j. Assume that $I_0 > 0$ and let $\varepsilon^* > \underline{z}_0$ be an estimate for z^*, the optimal solution of the original problem (P). The quantity $W = (\varepsilon^* - \underline{z}_0)/I_0$ represents an estimate for the *rate of change of the objective function associated with forcing the sum of integer infeasibilities to zero*. Therefore, a "reasonable" estimate ε_j for the underlying problem at node j could be set to

$$\varepsilon_j = \underline{z}_j + WI_j$$

The quantity W may be adjusted *dynamically* in the course of the solution process, by modifying the quantity ε^* to be equal to the current incumbent solution value—which is improved whenever a new *better* incumbent solution is found.

Another strategy for open node selection involves the use of *pseudo-costs*. The *lower* pseudo-cost for an integer variable i that has fractional value in the solution $\underline{x}_{[j]}$ of the relaxed problem (R_j) at node j is defined as $s_{i,j}^- = \left(\underline{z}_k - \underline{z}_j\right)/f_{i,j}$ where $\underline{z}_k$ is the solution value of the relaxed problem (R_k) of the son node k of j corresponding to the restriction $x_i \le \left\lfloor \left(\underline{x}_{[j]}\right)_i \right\rfloor$—assuming of course that standard branching on variables is performed. The *upper* pseudo-cost for an integer variable i that has fractional value in the solution $\underline{x}_{[j]}$ of the relaxed problem (R_j) at node j is then similarly defined to be $s_{i,j}^+ = \left(\underline{z}_m - \underline{z}_j\right)/\left(1 - f_{i,j}\right)$ where $\underline{z}_m$ is the solution value of the relaxed problem (R_m) of the son node m of j corresponding to the restriction $x_i \ge \left\lfloor \left(\underline{x}_{[j]}\right)_i \right\rfloor + 1$. In the course of the solution process then, upper and lower pseudo-costs s_i^+, s_i^- for any integer variable x_i, i in J are computed as (possibly weighted) averages of the values $s_{i,j}^\pm$ over the nodes j created so far in the tree. The values s_i^+, s_i^- are then used to obtain the estimate ε_j for the value z_j^* according to the formula:

$$\varepsilon_j = \underline{z}_j + \sum_{k \in B_j} \min\left\{s_k^- f_{k,j}, s_k^+\left(1 - f_{k,j}\right)\right\}$$

Again, the node from A with the best (smallest) estimated value ε_j is selected in step 9.a. of the B&B Algorithm.

Using the pseudo-costs s_i^+, s_i^-, it is also reasonable to modify step 9.d and establish the following *pseudo-cost branching rule* for variable selection to branch on: *the integer variable x_i to branch on is the variable with the highest priority* $s_i = \max\left\{s_i^+, s_i^-\right\}$. Various computational studies in the past 30 years have established the superiority of this approach and its variants over less sophisticated approaches for performing node selection or variable selection as described above.

Branching Rules Revisited

Besides pseudo-cost branching described above (and the trivial *most-fractional* or *most-infeasible variable rule* that picks the fractional variable whose value's decimal part is closest to 0.5, a rule that has been shown to offer essentially no advantage over random selection of a fractional variable to branch on), other branching rules have been proposed in the literature.

The *full strong branching rule* requires that *for each fractional variable x_i* that emerges from the solution $\underline{x}_{[j]}$ of the LP relaxation of the underlying problem at a node

j with solution value $\underline{z}_j$, the variable to be set at its floor and ceiling values and the two LPs with the bounding-box inequalities $x_i \leq \left\lfloor \left(\underline{x}_{[j]} \right)_i \right\rfloor$ and $x_i \geq \left\lceil \left(\underline{x}_{[j]} \right)_i \right\rceil$ to be solved, so as to give new LP solutions $\underline{z}_{j,i}^-, \underline{z}_{j,i}^+$. The rule then selects as branching variable the (fractional) variable x_i that maximizes a score function that is a convex combination of the two values $\min\left\{ \underline{z}_{j,i}^+ - \underline{z}_j, \underline{z}_{j,i}^- - \underline{z}_j \right\}, \max\left\{ \underline{z}_{j,i}^+ - \underline{z}_j, \underline{z}_{j,i}^- - \underline{z}_j \right\}$, where the parameter λ in the convex combination is a user-defined parameter often set between 0.1 and 0.2. Computational experience with this rule shows that it leads to small number of nodes in the B&B tree when seeking optimality, but this advantage is often offset by the computational requirements per node. Indeed, in the worst-case, in the order of $2|J| + 1$ LP problems must be solved per node, which can be very demanding—where J is the set of indices of discrete variables in the problem.

A *hybrid combination of pseudo-cost and full strong branching* is possible, whereby full strong branching is only applied to nodes that are no deeper in the B&B tree than a certain (user-defined) threshold depth d, and pseudo-cost branching is applied for variable selection on nodes below that level. The rationale behind this approach is that at high levels of the tree, the pseudo-costs do not carry much information since variables have rarely been branched on at that point, so full strong branching is used for such nodes, but as the tree grows deeper, pseudo-cost branching rules have significant information available in the quantities they compute, so that they can be very effective while at the same time being much "cheaper" computationally than full strong branching.

Reliability branching is a recently introduced branching rule, which extends hybrid pseudo-cost/full strong branching by intelligently choosing whether to use the full strong branching rule or pseudo-cost branching rule to select the variable to branch on a node. The decision of which rule to use is based on the "reliability" of the variable pseudo-costs at a node: the pseudo-cost estimates s_i^+, s_i^- of variable x_i at any node j, are called reliable iff the number of problems where x_i was selected as the branching variable (after the corresponding relaxed problem produced a solution $\underline{x}$), and both the resulting sub-problems with the restrictions $x_i \leq \lfloor \underline{x}_i \rfloor$ and $x_i \geq \lceil \underline{x}_i \rceil$ were solved and found to be feasible exceed a reliability threshold t_{rel}. The rationale is that the pseudo-costs of a variable can only be reliable if there was a sufficient number of nodes in the B&B tree where the variable was branched on and the resulting sub-problems were solved and found to be feasible (otherwise the estimates do not make much sense). If the pseudo-cost estimates are reliable, it makes sense to use them; otherwise, strong branching should be used to select the variable to branch on this node.

1.2.2.3 Cutting Planes

A fundamentally different method for solving MIP, pioneered by Gomory (1958), works by solving a (finite) series of linear programs, starting with an initial relaxation of the original MIP, and incrementally adding linear constraints (called *cutting planes*) that "*cut-off*" a region of the LP feasible region but without cutting-off an optimal solution to the MIP. Eventually, the process results in an LP whose solution is integral in all variables that are required to be integer, and this solution is guaranteed to be an optimal solution to the original MIP. Unfortunately, the number of constraints that need to be added before the LP solution becomes a solution for the MIP grows exponentially with the input size of the problem—not surprisingly, since otherwise, the algorithm guaranteeing the optimality of the solution, would run in polynomial time and, as already mentioned, such an algorithm is extremely unlikely to exist. Nevertheless, the idea behind cutting plane methods has turned out to be very important, and it has been the source of hybrid methods known as "Cut & Branch"; it has also been very successfully utilized in other frameworks such as "Branch-Cut-and-Price" and so on.

For simplicity of presentation, the problem to be solved is assumed to be the following:

$$(P)\min_{x} c^T x$$

$$\text{s.t.}\begin{cases} Ax = b \\ x \geq 0 \\ x \in \mathbb{Z}^n \end{cases}$$

Gomory Cuts

The Gomory cut is easily computed from the final optimal dictionary generated by the simplex method applied to the LP relaxation of (P), $(R_0)\ \min_x\{c^T x \mid Ax = b, x \geq 0\}$. Recall that the optimal dictionary of problem (R_0) is of the form

$$x_B = A_B^{-1} b - A_B^{-1} A_N x_N$$
$$z = c_B^T A_B^{-1} b + \left(c_N^T - c_B^T A_B^{-1} A_N\right) x_N$$

Consider the *r*th row ($r = 1 \ldots m$, the number of constraints) in the above dictionary, which can be written in the form:

$$x_r = \tilde{b}_r - \sum_{j \in N} \tilde{a}_j x_j$$

where N is the set of non-basic variables in the optimal solution computed. The above equation can be equivalently expressed as

$$x_r + \sum_{j \in N} \tilde{a}_j x_j = \tilde{b}_r$$

Because x is required to be non-negative, the following inequality must always hold by all feasible solutions of (R_0):

$$x_r + \sum_{j \in N} \lfloor \tilde{a}_j \rfloor x_j \leq \tilde{b}_r$$

Now, if $\tilde{b}_r$ is not integer (so that the value of the variable x_r in the optimal solution of the relaxed problem is not integral), then by the requirement of integrality of all x, it follows that the following inequality must be satisfied by all feasible points of the original problem (P):

$$x_r + \sum_{j \in N} \lfloor \tilde{a}_j \rfloor x_j \leq \lfloor \tilde{b}_r \rfloor$$

Subtracting the above inequality from the final dictionary equation $x_r + \sum_{j \in N} \tilde{a}_j x_j = \tilde{b}_r$ yields the Gomory cut:

$$(GC) \sum_{j \in N(r)} (\tilde{a}_j - \lfloor \tilde{a}_j \rfloor) x_j \geq \tilde{b}_r - \lfloor \tilde{b}_r \rfloor$$

wherc $N(r)$ denotes the subset of indices of the non-basic variables in the optimal relaxed solution that have non-integer coefficients in row r. It is trivial to verify that the introduction of this linear inequality constraint to the relaxed problem (R_0) leaves out the previously computed optimal solution, and that therefore by introducing this *valid inequality* to the problem implies that a new relaxed solution will be generated whose solution will be greater than or equal to the solution of the problem (R_0) and will be less than or equal to the solution value of the problem (P).

0–1 Knapsack Cover Inequalities

Another class of valid inequalities for a special case of the knapsack problem can be derived as follows. The 0–1 knapsack problem—mentioned already in the beginning of Sect. 1.2.1—is the problem of selecting from a list of *unique* items a subset so that the total value of the items selected is maximized, whereas the total size of the items does not exceed the size of the knapsack s:

$$(0-1KS) \max_x \sum_{i=1}^{n} p_i x_i$$

$$\text{s.t.} \begin{cases} \sum_{i=1}^{n} s_i x_i \leq s \\ x_i \in B = \{0,1\}, \quad \forall i = 1 \ldots n \end{cases}$$

Now, consider a set $C \subseteq \{1 \ldots n\}$ such that $\sum_{i \in C} s_i > s$. The knsapsack cover (valid) inequality specifies that the inequality

$$\sum_{i \in C} x_i \leq |C| - 1$$

is a valid inequality for the problem ($0–1KS$), which should be clear from the fact that if the decision variables violate this inequality, then they will surely violate the knapsack constraint. A *minimal cover* for a cover knapsack valid inequality consists of a set C' of indices in $\{1 \ldots n\}$ such that $\sum_{i \in C'} x_i \leq |C'| - 1$ is a valid inequality, but any strict subset of C', $C'' \subset C'$ satisfies $\sum_{i \in C''} s_i \leq s$ and so is not a valid inequality for (0–1KS) anymore. Cover inequalities for the 0–1 knapsack problem have found important applications in the development of theoretical as well as practical algorithmic advances in combinatorial optimization.

1.2.2.4 Branch-and-Cut and Related Methods

Branch and Cut

Typically, the first few iterations of a generic cutting-plane method based on Gomory cuts produces the largest increase in the lower bound of the relaxed problem (R_0) but after a while, introducing new cuts leads to minimal or negligible progress in terms of new lower bounds for the problem at hand. For this reason, as already mentioned, cutting planes are usually incorporated as a sub-routine in B&B algorithms, whereby cuts are introduced to the root node 0 and perhaps at other nodes in the tree as long as these cuts lead to a substantial improvement of the lower bound for the node, and stop as soon as progress is slowed down, returning to the main routines of the calling B&B method. Such an approach is termed "*Branch & Cut*", although it should more appropriately be called "Cut & Branch" since cutting planes in the form of new constraints are introduced to the problem before any branching occurs, as a means to tighten the LP relaxation's computed optimal value to serve as a lower bound for the node.

Branch and Price

For MIP problems with a *special structure* however, another method combining column generation techniques for solving the LP relaxation at a node and standard B&B emerged in the 1990s that became known as "*Branch & Price*". The basic idea behind Branch & Price is that for certain MIP problems, it is possible to reformulate the problem in such a way that possibly much less constraints are needed—at the expense of introducing a very large number of variables (columns). The LP relaxation of this new formulation of the same problem provides tighter lower bounds than the original formulation's relaxation (otherwise there would not be much of an advantage in undertaking this new formulation's solution). The LP relaxation is solved via column generation methods mentioned when discussing the multi-commodity linear network flow problem (MCLNF) in Sect. 1.1.2.2 where only a few of the variables of the relaxed problem are initially considered giving rise to a *Restricted Master Problem* (RMP), and when an optimal solution of the RMP is found, a special sub-problem—often a shortest path problem or a knapsack problem—is solved whose solution determines whether the restricted master solution is optimal for the complete LP relaxation or whether there exist columns (variables) not considered in the RMP that price favorably and should be included in the RMP, in which case, all such columns with negative reduced costs enter the RMP formulation and the RMP is solved again, usually starting from the previous solution found. This process continues until the optimal value of the last RMP solved is provably the optimal solution of the LP relaxation, at which point a lower bound for the current node becomes available. We shall briefly describe this approach using as application domain the Airline Crew Scheduling Problem (Barnhart et al. 1998), the natural continuation of which, we shall encounter again in Chap. 3 on Planning & Scheduling.

In the airline industry, the crew scheduling problem, also known as the *pairing problem*, is to generate sequences of flights (known as *flight legs*), where the starting leg takes off from a home *base* city airport $d_1 = b$ and lands on another city airport l_1; the next leg in the pairing must start from airport $d_2 = l_1$ and will land in some other airport l_2 so that they can be assigned to crew stationed at that home base. In general, every feasible pairing must be a sequence of legs $\{(d_1 = b, l_1), \ldots (d_{i-1}, l_{i-1}), (d_i = l_{i-1}, l_i = d_{i+1}) \ldots (d_m, l_m = b)\}$ that also satisfies certain constraints set forth by the FAA that aim to ensure the safety of flights and specify for instance minimum rest of crew between consecutive flights and so on; extra regulations by pilot unions also often impose other constraints on what constitutes feasible pairings. The objective of the problem is to generate pairings so that all flight legs are covered by exactly one feasible pairing, and that the resulting set of pairings has minimum cost. The cost function is the sum of the cost of each individual pairing constructed, but the cost of an individual pairing often has a highly nonlinear functional form of several factors involving the duration of each flight leg, the rest interval between legs in the pairing and others. For this reason, the optimization problem corresponding to the pairing problem is rarely modeled using binary variables x_{ij} indicating whether leg i is assigned to pairing j or not. Instead, the problem is usually modeled as a set partitioning problem where pairings are implicitly or explicitly considered in the formulation and a set of pairings that completely partition the legs set is sought that has minimum cost. We have already seen earlier a generic formulation for the Set Partitioning Problem (*SPP*). Since the full matrix A in the (*SPP*) formulation is too large to work with even for modestly sized problems, some algorithms for the crew pairing problem directly generate a small number of low-cost pairings so that the resulting matrix A has a manageable size and then directly solve the resulting restricted (*SPP*) (Anbil et al. 1991). This was also the approach followed by AT&T's KORBX system suite (Kennington 1989).

Some other more recent algorithms depend on a reformulation of the problem to include the pairing generation problem in the formulation in a column generation methodology. Given n flight legs $f_1, f_2, \ldots, f_n$ to arrange into a set of feasible pairings with minimum total cost, a pairing p can be defined as a column vector of n dimensions, where the ith component of the vector is 1 if leg f_i is in the pairing and zero otherwise. The cost of a pairing $c = c(p)$ is defined via a function $c(.)$ which is a highly nonlinear function in the binary components of vector p. Now, initially the problem can be formulated as follows:

$$(CPP)\ \min_{p} \sum_{j=1}^{N} c(p^{[j]})$$

$$\text{s.t.} \begin{cases} \sum_{j=1}^{N} p_i^{[j]} = 1, & \forall i = 1\ldots n \\ Dp^{[j]} \le d, & \forall j = 1\ldots N \\ p_i^{[j]} \in \{0,1\}, & \forall i = 1\ldots n, j = 1\ldots N \end{cases}$$

where D is a given data matrix and d a given data vector so that the constraint $Dp \le d$ is a side-constraint that must be satisfied by each feasible pairing in the optimal solution. Note that in the above formulation the number N of pairings in the optimal solution is not known in advance, but is certainly less than n, the total number of legs to be scheduled.

It should be mentioned also that the partitioning constraint is not a strict requirement. Indeed, in practice, a set covering problem is usually solved instead, where the constraint $\sum_{j=1}^{N} p_i^{[j]} = 1$ for all $i = 1\ldots n$ is replaced with $\sum_{j=1}^{N} p_i^{[j]} \ge 1$. This is because any flight leg can be assigned to more than one pairing, but in all assigned pairings except one, the leg will be designated as "*deadheading leg*" meaning that all crews assigned to that flight leg will get aboard that flight, but only one—the one assigned to the pairing where the leg is not designated as deadheading—will actually operate the flight, whereas the other crews will fly as normal passengers, reducing the available seats for *real* passengers.

Now, consider the set F of all feasible pairings satisfying $Dp \le d,\ d \in B^n$ and assume the set is bounded, so that $F = \{y^{[1]},\ldots y^{[M]}\}$ for some number M. Define $c_j = c(y^{[j]})$ to be the cost of the jth pairing in the set F. The problem can be written in column generation form as

$$(CPP - CG)\ \min_{\lambda} \sum_{j=1}^{M} c_j \lambda_j$$

$$\text{s.t.} \begin{cases} \sum_{k=1}^{M} y_i^{[k]} \lambda_k = 1 \\ \lambda_k \in B = \{0,1\}, & \forall k = 1\ldots M \end{cases}$$

Again, the constraint $\sum_{k=1}^{M} y_i^{[k]} \lambda_k = 1$ in practice is usually replaced by $\sum_{k=1}^{M} y_i^{[k]} \lambda_k \ge 1$ for all $i = 1\ldots n$. The formulation (*CPP-CG*) has a possibly much larger number of variables (*M*) than the formulation CPP which has at most n^2 binary variables, but as it turns out it has significant advantages over CPP. The most significant advantage of CPP-CG is that it eliminates some of the inherent symmetry of CPP that causes B&B algorithms to perform rather poorly when attempting to solve CPP. The symmetry is caused by the fact that swapping the contents of any two vectors $p^{[i]}$ and $p^{[j]}$ in the optimal solution still results in an optimal solution, so that the number of optimal solutions in the B&B tree is exceptionally large, while all (or most) of them represent the same solution set. This however implies that nodes cannot be closed early enough in the B&B process for it to be effective and thus its poor performance in such problems. This symmetry is broken in the CPP-CG formulation of the problem and further LP relaxations of this model are tighter than the LP relaxations of the first problem formulation, i.e. provide optimal solution values that are higher than the solution of the corresponding LP in the CPP formulation, and thus allow closing nodes earlier in the B&B tree construction process. This then implies that meaningful progress can be made as the B&B tree expands by branching on the λ_k variables.

The Branch & Price algorithm works on the model CPP-CG, starting with a few pairing vectors $y^{[1]}\ldots y^{[K]}$ where K is a relatively small number that covers the problem's legs, known as we saw before as the RMP. Now, let π be an optimal dual solution to the *current* RMP. By applying the Dantzig–Wolfe decomposition technique described in Sect. 1.1.2.2, by solving the corresponding subproblem, either we identify new reduced cost columns to enter the basis and augment the pairings set to work with, or else we prove the optimality of the LP relaxation of the RMP. At this point, branching is performed, usually as follows: assuming the optimal solution $\bar{\lambda}$ of the LP relaxation of the RMP at a node is fractional, there will always exist two rows r and s in the data matrix $Y = [y^{[1]}\ldots y^{[M']}]$ such that $0 < \sum_{k:y_{rk}=1 \wedge y_{sk}=1} \bar{\lambda}_k < 1$ (Ryan and Foster 1981). Therefore, branching is performed on this node by finding two such rows r and s and adding the restriction $\sum_{k:y_{rk}=1 \wedge y_{sk}=1} \lambda_k = 0 \vee \sum_{k:y_{rk}=1 \wedge y_{sk}=1} \lambda_k = 1$ so that the left child node contains the extra constraint $\sum_{k:y_{rk}=1 \wedge y_{sk}=1} \lambda_k = 0$ whereas the right child node contains the extra constraint $\sum_{k:y_{rk}=1 \wedge y_{sk}=1} \lambda_k = 1$.

Branch, Price & Cut

The Branch, Price & Cut method is a natural combination of the Branch & Cut and Branch & Price methods, based on column generation to solve the relaxed LP at the nodes of the B&B tree, and enhancing the process by finding and adding—hopefully useful—valid inequalities after the LP solution of the relaxed problem at any node has been found and has been shown to violate some of the integrality constraints of the original problem. However, combining column generation (adding variables to the RMP) with row generation (adding constraints to improve the gap between the LP solution and the MIP solution) is usually highly non-trivial and domain-specific, because adding constrains to a problem can easily destroy the structure of the (pricing) sub-problem that has to be solved in the column generation method to determine whether more variables should be added in the restricted master problem LP. We shall describe the method in the context of an important transportation problem that arises in the distribution as well as the telecommunications industry, namely the *Origin–Destination Integer Multi-Commodity Network Flow Problem* (*OD-IMCNF*) (Barnhart et al. 2000).

The OD-IMCNF problem is clearly directly related to the MCLNF problem studied in Sect. 1.1.2.2. The problem is defined on a network $G(V,E,W)$ with weighted arcs that are capacitated so that the total flow along each arc (i,j) in E cannot exceed a quantity u_{ij} that may be $+\infty$ for some (or all) arcs. There are K different commodity types, and an integer quantity q_k of each commodity type $k = 1\ldots K$ must be wholly and indivisibly sent from an origin node s_k in V to another destination node t_k in V. Therefore, demand for each commodity type k along the nodes i of the network is

$$d_{i,k} = \begin{cases} q_k, & i = t_k \\ -q_k, & i = s_k \\ 0 & else \end{cases}$$

The unit flow cost along arc (i,j) for commodity type k is denoted c_{ijk}. We can now model this problem in a classical formulation using $|E|K$ binary arc-flow variables x_{ijk} denoting whether commodity k uses arc (i,j) in E to send the quantity q_k or not:

$$(OD-IMCNF)\ \min_x \sum_{k=1}^{K} \sum_{(i,j)\in E} c_{ijk} q_k x_{ijk}$$

$$\text{s.t.} \begin{cases} \sum_{k=1}^{K} q_k x_{ijk} \le u_{ij}, & \forall (i,j) \in E \\ \sum_{j:(j,i)\in E} x_{jik} - \sum_{j:(i,j)\in E} x_{ijk} = d_{i,k}, & \forall i \in V, \forall k = 1\ldots K \\ x_{ijk} \in B = \{0,1\}, & \forall (i,j) \in E, \forall k = 1\ldots K \end{cases}$$

The problem has $|E| + (|V| + |E|)K$ constraints. An equivalent path-based formulation for the problem that is amenable to column generation techniques uses the paths that exist in the network that connect each commodity type origin node s_k with the corresponding destination node t_k. Denote by $P(k)$ the set of all *distinct* feasible paths in the network G connecting s_k to t_k so that every arc e in a path p in $P(k)$ satisfies $q_k \le u_e$. Introducing binary variables y_{pk} denoting whether commodity k will follow path p in $P(k)$ or not, the problem can be formulated as

$$(OD-IMCNF-CG)\ \min_y \sum_{k=1}^{K} \sum_{p\in P(k)} c_{pk} q_k y_{pk}$$

$$\text{s.t.} \begin{cases} \sum_{k=1}^{K} \sum_{p\in P(k)} q_k \delta_{ijp} y_{pk} \le u_{ij}, & \forall (i,j) \in A \\ \sum_{p\in P(k)} y_{pk} = 1, & \forall k = 1\ldots K \\ y_{pk} \in B = \{0,1\}, & \forall k = 1\ldots K, \forall p \in P(k) \end{cases}$$

In the formulation, δ_{ijp} is one iff arc (*i,j*) is in path $p \in \bigcup_{k=1}^{K} P(k)$ and zero otherwise, and c_{pk} corresponds to the quantity $\sum_{e \in p} c_{ek}$. Now, the number of binary variables is much larger, and is equal to $\sum_{k=1}^{K} P(k)$. Nevertheless, column generation methods based on the Dantzig–Wolfe decomposition can be successfully used to solve the LP relaxation of this problem. Starting with a small set of origin–destination feasible paths for each commodity type *k*, the RMP contains a small set of columns to choose from for each *k* = 1…*K*. Deleting the integrality constraints and replacing them with the bounding-box constraints $0 \leq y_{pk} \leq 1$ where for each *k*, *p* belongs to a rather small set of paths, the resulting (restricted) LP can be easily solved to optimality. Next, we determine whether the resulting solution is optimal for the full LP relaxation of the problem OD-IMCNF-CG. Let $-\pi_{ij}$ denote the *nonnegative* dual variables in the optimal solution associated with the coupling capacity constraints $\sum_{k=1}^{K} \sum_{p \in P(k)} q_k \delta_{ijp} y_{pk} \leq u_{ij}$ and let σ_k represent the *unrestricted* dual variables in the optimal solution associated with the set partitioning constraints $\sum_{p \in P(k)} y_{pk} = 1$. As c_{pk} equals $\sum_{e \in E} c_{ek} \delta_{ep}$ the reduced cost of column for commodity type *k* can be written as

$$\bar{c}_{pk} = \sum_{(i,j) \in E} q_k \delta_{ijp} \left(c_{ijk} + \pi_{ij} \right) - \sigma_k, \quad \forall k = 1 \ldots K, p \in P(k)$$

In the same way described in Sect. 1.1.2.2, we now formulate the (pricing) subproblem for the column generation method. The subproblem turns out to have a surprisingly nice structure: it is a super-position of *K independent shortest path problems* defined over a graph with the same topology as that of our original graph *G*, but whose arc costs for the *k*th shortest path subproblem are $c_{ijk} + \pi_{ij}$ for each arc (*i,j*) in *E*. The origin of the *k*th shortest path problem is of course the node s_k and the destination is obviously t_k. If the optimal shortest path of the *k*th subproblem is p_k^* with associated cost c_k^* then if for all $k = 1 \ldots K$, $q_k c_k^* - \sigma_k \geq 0$, the LP relaxation of the RMP is optimal for the master problem of the current node. Otherwise, for each commodity type *k* for which $q_k c_k^* - \sigma_k < 0$ the path p_k^* is added as a new column to the RMP, and the LP is re-optimized.

The most successful branching strategy for this problem is a direct extension of the branching rule mentioned earlier when discussing Branch & Price methods for the Pairing Problem. In particular, branching will occur when the variables y_{pk} for at least one commodity type *k* are fractional in the optimal solution of the LP relaxation of the underlying problem at a B&B node. In this case, for some *k*, there will be two or more variables $y_{1,k}$ and $y_{2,k}$ that are non-zero. These variables correspond to two *distinct* paths p_1 and p_2 that both start at the same node s_k and terminate at the same node t_k. Define the *divergence node d* of the two paths p_1 and p_2 as the first node along the route from s_k to t_k where the two paths differ, so that the arcs in paths p_1 and p_2 connecting s_k to *d* appear in common in both paths, but the next arc in path p_1 (d, d_1) is different from the next arc in the path p_2 (d,d_2). Now let *E*(*d*) be the set of arcs emanating from node *d*, $\{e \in E | e = (d,n)\}$. Using any partition of the set $E(d)$ in two disjoint subsets so that $E(d) = E'(d) \cup E''(d) | (d, d_1) \in E'(d), (d, d_2) \in E''(d)$, the branching rule specifies that the left child node of the current node will have to obey the constraint $\sum_{p:p \cap E'(d) \neq \emptyset} y_{pk} = 0$ and the right child node of the current node will have to satisfy $\sum_{p:p \cap E''(d) \neq \emptyset} y_{pk} = 0$. The ingenuity of this branching strategy is that it does not destroy the structure of the shortest path sub-problems as pricing sub-problem for the child nodes of any node in the B&B tree, and that the restrictions require for the child nodes that the k commodity type does not use the arcs in $E'(d)$ or correspondingly in $E''(d)$. But enforcing these constraints in the shortest path sub-problems is almost trivial: simply increase the costs of the arcs in each set to infinity when solving the corresponding sub-problem and the solution will not include the undesired arcs.

Given a fractional multi-commodity flow, yet another decision has to be made on *which* fractional commodity type k to branch on, and which paths to use for the branching rule. A successful strategy is to select from the fractional commodity type k' with the largest flow $q_k{}'$ and from all the paths that k' follows, choose the two paths p_1 and p_2 that carry the greatest fractions of the commodity k'.

This branching strategy has been shown experimentally to divide the problem search space more evenly, a much desired property when searching for optimal solutions to a problem.

For the OD-IMCNF problem a depth-first node selection strategy to grow the B&B tree is usually more beneficial than the best-first search or the hybrid methods mentioned in Sect. 1.2.2.2. Finally, adding cuts is accomplished as follows. Observe that the coupling capacity constraints $\sum_{k=1}^{K} q_k x_{ijk} \leq u_{ij}, \quad \forall (i,j) \in E$ for the OD-IMCNF are essentially the classical 0–1 knapsack constraint of a knapsack with size u_{ij} and K possible items to choose from. Therefore, lifted cover inequalities discussed in Sect. 1.2.2.3 apply in this case. To see how, consider a problem instance whose solution is fractional, so that at least one commodity type k' is assigned to more than one path, and let p' be the shortest *saturated* path that k' is assigned to, where a path is saturated iff there exists at least one arc $e = (i,j)$ in that path with total flow at its capacity limit so $\sum_{k=1}^{K} x_{ijk} = u_{ij}$ and $x_{ijk'} > 0$. Existence of the path p' is guaranteed from the fact that k' is fractional and each sub-problem solution minimizes costs. If k' is the *only* split commodity assigned to e, and C_e is the set of all commodities using e, then C_e is a cover as explained in Sect. 1.2.2.3 and the corresponding cover inequality is violated by the current LP solution. If however there are more than one fractional commodity flows along arc e, then the cover inequality defined in 1.2.2.3 may no longer be valid. Nevertheless, another kind of cutting plane may be in the form of *Lifted Cover Inequalities* which are valid inequalities of the form:

$$\sum_{k \in C} x_{ek} + \sum_{k \in \bar{C}} a_{ek} x_{ek} \leq |C| - 1$$

where $C \subseteq \{1, \ldots K\}$ is a minimal cover, and $\bar{C} = \{1, \ldots, K\} - C$. The coefficients a_{ek} are determined for each arc e by solving $|\bar{C}|$ knapsack problems, one for each commodity type *not* in C. Now, translating a valid lifted cover inequality from the formulation above, into a formulation suitable for the column generation problem is easy, since the arc-flow variables x_{ek} and the path-flow variables y_{pk} are connected by the relationship

$$x_{ijk} = \sum_{p \in P(k)} \delta_{ijp} y_{pk}, \quad \forall (i,j) \in E, k = 1 \ldots K$$

hence a lifted cover inequality can be written in terms of the path-flow variables y_{pk} as:

$$\sum_{k \in C} \sum_{p \in P(k)} \delta_{ep} y_{pk} + \sum_{k \in \bar{C}} a_{ek} \sum_{p \in P(k)} \delta_{ep} y_{pk} \leq |C| - 1$$

At any node in the B&B tree, the algorithm has to solve the LP relaxation of the problem (*OD-IMCNF-CG*) which has been augmented by some branching rule restrictions of the form $\sum_{p: p \cap E'(d) \neq \emptyset} y_{pk} = 0$ using the column generation technique described above. We already saw that the restrictions do not destroy the structure of the sub-problem that has to be solved during column generation, which is still a synthesis of K independent shortest path problems in which however, certain arcs may have infinite costs. Once the LP is solved, if the solution is fractional, the algorithm attempts to identify valid lifted cover inequalities for each of the arcs (i,j) of the network. If any such inequalities are found, they are added to the relaxed problem (row generation) and the resulting LP is reoptimized. Reoptimizing the LP requires solving a new pricing subproblem (possibly many times) which takes into account the valid inequalities added. The sub-problem to be solved turns out again to be a *shortest path problem* on a network with the same topology as G but with different arc costs. In particular, consider a lifted cover inequality for an arc e added to the problem of the form $\sum_{k \in C} \sum_{p \in P(k)} \delta_{ep} y_{pk} + \sum_{k \in \bar{C}} a_{ek} \sum_{p \in P(k)} \delta_{ep} y_{pk} \leq |C| - 1$. (The coefficient a_{ek} will equal 1 if e is in C). The arc cost of arc e for the kth shortest path to be solved as part of the kth pricing sub-problem will have to be modified from its previously computed value according to the equation

$$c'_{ek} = c_{ek} + \pi_e + a_{ek} \gamma_e$$

where π_e is the value of the dual variable associated with the coupling capacity constraint for the arc e, and $-\gamma_e$ is the value of the dual variable associated with the particular lifted cover inequality added to the LP.

Once the new, augmented LP is solved, branching is performed as discussed above according to the rules for fractional commodity selection, and path selection. Computational results have consistently confirmed that Branch, Price & Cut results in fewer nodes of the B&B tree as well as much faster overall execution times, with significantly less memory requirements than other exact methods for such problems. For these reasons, it is a framework that is often used both in the current research and in practical application and systems development today.

1.2.2.5 The Nested Partitions Method for Combinatorial Optimization

A number of randomized meta-heuristics for NP-hard *combinatorial optimization* problems were proposed in the 20 year span between 1970 and 1990, including Simulated Annealing, Evolutionary Algorithms and Genetic Algorithms—already discussed in the context of unconstrained optimization. Later, Tabu Search was proposed as a very effective meta-heuristic method that often comprises an essential component of many successful software codes for the Traveling Salesman Problem and related combinatorial problems (note however that Tabu Search does not heavily rely on randomization of the search as the other methods mentioned above). Later, another randomized method called Nested partitions (Shi and Olafsson 2000) was proposed for global optimization of NP-hard combinatorial problems that is essentially an adaptive sampling method that partitions the search space so as to concentrate the sampling process to promising regions. Similar to Simulated Annealing, the method carries a guarantee of convergence to the global optimum (in a *finite* number of steps for a *finite* search space; but there is no guarantee on the rate of convergence to that point).

The Nested Partitions method (NP) applies to the generic combinatorial optimization problem

$$\min_{x \in S} f(x)$$

where S is a finite set, and $f : S \to \mathbb{R}$ is the objective function of the problem at hand. The NP method partitions the feasible set S into several subsets and samples each subset randomly. Then, the method determines a "*promising index*" for each subset that defines how likely each subset is to contain an optimal solution to the original problem (the higher the promising index, the better the subset). The most promising subset P then is partitioned further among M disjoint subsets, whereas all the other subregions that obtained a lower score than the most promising region are merged into one subset equal to S-P called the *surrounding region*. The method intensifies its search by selecting smaller and smaller subsets of the whole feasible set S that it samples with increasingly higher density.

To describe the algorithm, let $\Sigma \subseteq 2^S$ denote the set of subsets constructed during a run of the algorithm, let $\sigma_k \in \Sigma, \sigma_k \subseteq S$ denote the most promising region in the kth iteration, let $\sigma_{k,i} \subseteq \sigma_k, \quad i = 1...M$ denote the M *disjoint* subsets of the most promising region of the kth iteration that σ_k is partitioned into (so that $\bigcup_{i=1}^{M} \sigma_{k,i} = \sigma_k$) and let $\sigma_{k,M+1} \triangleq S - \sigma_k$. Also, let d_k denote the depth of the nested partitions in the kth iteration and d_* the maximum depth reached by the algorithm, let $\theta_{j,i}$ denote the ith sample point in the jth sub-region (j is in the set $\{1...M\}$), let N denote the total sample size, let z^* denote the best solution found so far, and let $I : \Sigma \to \mathbb{R}$ denote the function that returns the promising index of any subset s of S. A generic description of the NP method is as follows:

Algorithm Generic Nested Partitions

Inputs: Objective function $f : S \rightarrow \mathbb{R}$, the number of partitions for the most promising region in each iteration M & a partitioning strategy $P : 2^S \rightarrow 2^{2^{\wedge}S}$, maximum depth allowed d_*, convergence criteria for stopping the algorithm.
Outputs: A point $x* \in S$ that is a feasible point for $\min_x\{f(x)|x \in S\}$

Begin

/* initialization */

0. Set k=0, d_k=0, σ_k=S, x*=null, z*=+∞.

/* check for convergence */

1. if convergence criteria are satisfied then
   a. return x^*.
2. end-if
   /* partition */
3. if $|\sigma_k| = 1$ then
   a. GOTO 17.
4. else
   a. Set $\Sigma' = P(\sigma_k)$
   b. Set $\sigma_{k,M+1} = S - \sigma_k$.
5. end-if
   /* sampling */
6. for each $\sigma \in \Sigma' \cup \{\sigma_{k,M+1}\}$do
   a. Set $\hat{P}(\sigma) = +\infty$.
   b. for i = 1 to N do
      i. Select at random a point $\theta_{\sigma,i} \in \sigma$.
      ii. if $f(\theta_{\sigma,i}) < \hat{P}(\sigma)$ then
         1. Set $\hat{P}(\sigma) = f(\theta_{\sigma,i})$.
         2. Set $\sigma_k^* = \sigma, x_{[k]}^* = \theta_{\sigma,i}$.
      iii. end-if
   c. end-for
7. end-for
8. Set $z_k = \min_{\sigma \in \Sigma \cup \{\sigma_{k,M+1}\}} \hat{P}(\sigma)$.
9. if z_k<z* then
   a. Set $z^* = z_k, x^* = x_{[k]}^*$

10. end-if
11. if $\sigma_k^* = \sigma_{k,M+1}$ then
    a. GOTO 14.
12. else
    a. Set $\sigma_{k+1} = \sigma_k^*, \quad d_{k+1} = d_k + 1$.
13. end-if
    /* backtracking to surrounding region */
14. Set $\sigma_{k+1} = \sigma_{k,M+1}$, $d_{k+1} = d_k - 1$.
15. Set $k=k+1$.
16. GOTO 1
17. Set $f* = f(x), x \in \sigma_k, |\sigma_k| = 1$.
18. for $i = 1$ to N do
    a. Select a random point $x_i \in S$.
19. end-for
20. Set $x_{\min} = \arg\min\{f(x_i)|i = 1\ldots N\}, f_{\min} = f(x_{\min})$.
21. if $f_{\min} < f^*$ then
    a. Set $\sigma_{k+1} = \sigma_{k,M+1}$, $d_{k+1} = d_k - 1$, $k = k+1$.
    b. if $f_{\min} < z^*$ then
        i. Set $x^* = x_{\min}$, $z^* = f_{\min}$.
    c. end-if
    d. GOTO 1.
22. else GOTO 18

End.

The particular pseudo-code for the NP method essentially implements a function for estimating the promising index $I(\sigma)$ of a region $\sigma \subseteq S$ that randomly samples N points in σ and returns the minimum value the function f takes among these N points. Although this is a very reasonable estimator of the promising index for a region, other options also exist. For example, if a local search method is available that implements *any* algorithm for finding a local minimum of the function starting from a point x_0, then the promising index could return the best result of applying this local search algorithm started using as initial point each of the random samples of σ.

The convergence criteria for the NP method are usually *failure to improve on the best value found so far z^* for a number of iterations*, or *total number of iterations*.

1.2.2.6 The Tabu Search Method for Combinatorial Optimization

Tabu Search (Glover 1989), as mentioned above, is a meta-heuristic method that improves the performance of local search methods for combinatorial optimization problems, and in sharp contrast to all other popular meta-heuristics, it hardly relies on random moves to escape bad quality local optima. In fact, the standard version of Tabu Search does not use any form of randomness at all.

Tabu Search (TS for short), requires the definition of a search space S, in which a neighborhood structure $N(x) \subseteq S$ is defined for every $x \in S$, as well as an objective function $f : S \to \mathbb{R}$. The neighborhood $N(x)$ of any point $x \in S$ should be the set of all points y that result from a given set of "moves" or transformations $T \in \mathcal{T}, T : S \to S$ that have been defined and that can be applied to any point in S. TS attempts to solve the problem $\min_{x \in S} f(x)$ by utilizing the transformations in $\mathcal{T}$ defining a local search in an iterative manner. To avoid the possibility of having the local search return to a previously visited point which then—due to the deterministic nature of the method—would lead to an infinite cycling loop and for other reasons as well, the notion of "tabu" is introduced. As soon as a "move" is selected, that moves the current solution to another point in the search space, some *aspect* of the move is recorded as tabu, so as to avoid "undoing" any benefits gained from the move. The aspect that is recorded is problem-specific, and should be selected so as to guide the search towards better solutions by constraining the "admissible moves" of the next few iterations to a set of moves that will not "undo" the "benefits" of the selected move. TS therefore maintains a "Tabu List" of *rules* that the future moves to be made must obey. The *least* constraining type of rule would be a rule that simply prevents performing a move from any current of near-future solution x to a previously visited solution v. A more constraining type of rule could be that any move applied from the current or future solution x, must not result in a solution in a restricted neighborhood of a previously visited solution v, $N_r(v)$, defined as the set of points that are the result of a particular subset of transformations T applied to v: $N_r(v) = \{y \in S | y = T(v), T \in \mathcal{T}' \subseteq \mathcal{T}\}$. Usually, every "tabu" rule has an associated finite lifetime during which it remains active, meaning that after a particular number of iterations after its creation and insertion in the tabu list, the rule becomes inactive and removed from the list automatically. Also, in most implementations, the tabu list has a finite length, implying that as new tabu rules are created, they are inserted in the list, pushing other older rules out of the list if space is not available.

Finally, another, almost opposite mechanism to the tabu list is usually employed in TS methods: the mechanism of "aspiration criteria" is used to override tabu rules when needed. A rather obvious aspiration criterion mechanism by which tabu rules could be overridden is the "improving-move" criterion: if any move T from a current solution x leads to a new incumbent solution $y = T(x)$ (i.e. better than any solution found so far in terms of objective function value), the move should be allowed, overriding any rules in the tabu list. The combination of tabu lists and aspiration criteria, leads to the concept of the admissible neighborhood $N_a(x), x \in S$ which is defined as the subset of $N(x)$ not disallowed by the tabu list or allowed by the aspiration criteria in the current state of the search process. The generic pseudo-code for TS is shown next.

Algorithm Generic Tabu Search

Inputs: Objective function $f : S \to \mathbb{R}$, neighborhood structure $N(x)$ defined by a set of transformation operators $T{:}S \to S$, a method to create tabu rules, a method to create aspiration criteria, an initial solution x_0, convergence criteria for stopping the algorithm.
Outputs: A point $x^* \in S$ that is a feasible point for $\min_x \{f(x)|x \in S\}$

Begin

/* initialization */

1. Set $x = x_0$, $x^* = x$, $f^* = f(x)$, $Lt = \{\}$, $La = \{\}$.
   /* search */
2. while termination criteria are not satisfied do:
   a. Select x from the set $\arg\min[f(z)|z \in N_a(x)]$.
   b. if $f(x) < f^*$ then
      i. Set $x^* = x$.
      ii. Set $f^* = f(x)$.
   c. end-if
   d. Create and record any tabu rules in the list Lt.
   e. Create and record any aspiration criteria in the list La.
   f. Update the lists Lt and La.
3. end-while
4. return x^*.

End.

Over the last 20 years, TS proved to be a very competitive method for many hard combinatorial optimization problems, such as the Facility Location problem and several of its variants, the Vehicle Routing Problem and its variants (both problems are studied in detail in Chap. 5) including the Traveling Salesman Problem, Graph Partitioning and Graph Coloring problems, Edge Matching problems in graphs, and so on. However, it must be noted that, contrary to the other meta-heuristics discussed so far, TS requires very carefully crafted transformation rules defining the local search itself, and even more carefully designed tabu rules and aspiration criteria that are almost always problem-specific that cannot be generalized to be applicable to other domains as they are usually strongly dependent on the problem search space representation and the local search methods chosen. For these reasons, the Generic Tabu Search algorithm sketched above, can only serve as a very high-level blueprint for algorithm design in a real-world situation.

1.3 Bibliography

As optimization finds applications in almost every aspect of modern life (so that many optimization professors think that "everything is an optimization problem", the quote taken from a recent lecture on advances in convex optimization), the literature on the subject is enormous. Thousands of books covering various aspects of optimization and operations research have been written since the 1950s. Many academic journals are focused *solely* on optimization research or practice, including Mathematical Programming, INFORMS Journal on Computing, SIAM Journal on Optimization, SIAM Journal on Numerical Analysis, Optimization & Mathematical Software, Operations Research, Computers & Operations Research, The European Journal of Operational Research, Computational Optimization & Applications, Operations Research Letters, Optimization Letters, Journal of Global Optimization, Optimization Software and Practice, Interfaces, OR Spectrum, Annals of Operations Research to name just a few. Other journals that often publish optimization research include ACM Transactions on Mathematical Software, IEEE Transactions on Pattern Analysis and Machine Intelligence, IEEE Transactions on Automatic Control, IEEE Transactions on Information Theory, Discrete Applied Mathematics, European Journal on Combinatorics, Journal of Heuristics, etc. There are equally many conferences, with the top conferences being the annual Symposium on Mathematical Programming, the annual INFORMS Conference, and the OR Conference.

With such vast public repository of recorded research available, it is impossible to cover to even a "reasonable" degree the literature on the subject because of space limitations. In the following, we provide only a few important references that are necessarily filtered according to the author's preferences and use in the text.

Optimality conditions for unconstrained optimization, and the "setting the derivative to zero" rule of calculus goes back to Fermat's and Leibnitz's work on extremum problems and the calculus of variations. Excellent treatments of the subject can be found in books such as Apostol (1962), Luenberger (1969), Fletcher (1987) and many others. First-and second-order necessary and/or Sufficient Conditions for Mathematical Programming are beautifully developed in Mangasarian (1994), where theorems of the alternative are extensively used to develop in-depth results. Most of the sufficient conditions for global optimization require some concept of convexity. The best book on convex analysis on $\mathbb{R}^n$ remains Rockafellar (1970), which together with Monotropic Programming has profoundly influenced many network flow algorithms as well.

Historically, algorithms for unconstrained optimization evolved from crude heuristic and brute-force search methods that could not scale with the dimensionality of the problem measured by the number of independent variables—a phenomenon that became known in the 1960s as the "curse of dimensionality"—to theoretically rigorous methods that exploited the mathematical structure of the problem to guide the solution process to a satisfactory solution in a *robust way*. At that time, the term "Artificial Intelligence" was synonymous with advanced tools for optimization, and in particular, Linear Programming (see Chvatal (1983) for a relevant quote from a science fiction movie of that era). Newton and Quasi–Newton methods in conjunction with line-search approaches became the dominant methods for locating saddle points, and are still among the dominant methods for optimization today, even in the absence of analytic derivative formulas.

The BFGS formula derived independently and simultaneously by Broyden (1970), Fletcher (1970), Goldfarb (1970), and Shanno (1970) remains the most widely used formula for updating the inverse Hessian of a function at a point $x_{(k)}$ and not without reason: in a sense it is the optimal approximation update formula (Fletcher 1987).

In the 1980s and 1990s a lot of research efforts were made to improve Conjugate Direction methods because of their potential for high scalability as they avoid memory storage issues that might arise in methods requiring second-order derivative information when the number of variables rises to the order of thousands or tens of thousands. See (Ferris et al. (2007)) for the successful application of such methods embedded in video-game simulations requiring real-time performance. Conjugate-gradient methods for unconstrained and constrained optimization remain the methods of choice for many practical problems in Supply Chain Management today.

Randomized search methods for locating high-quality local optima were developed in the 1980s and the 1990s and research in the performance of such methods continues today. The algorithms described in Sect. 1.1.1.3 describe some of the most successful strategies for locating solutions that are close to the globally optimal value. The reference paper on Simulated Annealing is Kirkpatrick et al. (1983). Evolutionary Algorithms were originally proposed by teams of European researchers, see for example Rechenberg (1973) and Schwefel (1981). Differential Evolution is presented in (Storm and Price 1997). Other very successful methods—some of which are only applicable in the context of discrete optimization—include:

- Tabu Search, for more recent discussion see e.g. Glover and Marti (2006)
- Greedy Randomized Adaptive Search Procedures (GRASP), e.g. Festa and Resende (2009)
- Beam Search, e.g. Lowerre (1976).
- Scatter Search, e.g. Glover (1998)
- Genetic Programming, e.g. Koza and Poli (2005)
- Hybrid exact methods with randomized local search methods
- Hybrid randomized SA and EA methods e.g. Aydin and Fogarty (2004)

Practical aspects of optimization can be found in Gill et al. (1982). Practical algorithms for unconstrained and constrained optimization can be found in Fletcher (2000). The best treatise of linear programming remains Chvatal (1983). Column generation methods for linear programming originate with the Dantzig–Wolfe decomposition principle (Dantzig and Wolfe 1960). Lagrangian methods for nonlinear programming are well covered in Bertsekas (1995). Classical treatise on network flows remains Ford and Fulkerson (1962), whereas modern works on linear network optimization includes Bertsekas (1991), and Ahuja et al. (1993). For a discussion on parallel methods for block-angular problems with emphasis on network flows, see Schultz and Meyer (1989). Coordination issues in the parallel solution of block-angular constrained optimization problems are described in De Leone et al. (1994). De Leone et al. (1999) present a parallel algorithm for separable convex network flow problems based on the ideas of ε-relaxation pioneered by Rockafellar (1970) and Bertsekas (1991), an application of which was detailed in the description of the Auction Algorithm for the Linear Assignment Problem. The book by Bertsekas and Tsitsiklis (1989) describes the state of the art in parallel numerical optimization algorithms up to that time, and remains relevant today, with the exception of algorithms for vector computers which have fallen out of favor in the era of many-core processors.

The classical treatise on Dynamic Programming is of course Bellman (1957). However, see Bertsekas's two-volume set (Bertsekas 2001, 2005) on dynamic programming and optimal control for a more recent introduction to the topic covering deterministic as well as stochastic systems in a unified manner, with many examples from inventory theory and game theory applicable to Supply Chain Management.

The mathematics of discrete and integer optimization are described in Nemhauser and Wolsey (1988), while complexity theory aspects are found in Papadimitriou and Steiglitz (1998), a book however that is recommended mostly for theoretical computer scientists wishing to expand their knowledge of material covered in Garey and Johnson (1979). Reliability branching was proposed in Achterberg et al. (2005), and a recent hybrid branching strategy is described in Achterberg and Berthold (2009).

The Branch-and-Price algorithm for the Crew Pairing Problem was presented in Barnhart et al. (1998). The Branch-Cut-and-Price algorithm for multi-commodity origin–destination network flows was presented in Barnhart et al. (2000).

1.4 Exercises

Hint: For many of the exercises below, the matlab environment offers one of the best tools for prototyping numerical algorithms, with open-source variants of matlab being the mature software codes octave and scilab. However, C/C++ is the preferred language for numerical algorithms development–not necessarily resulting in the fastest compiled code though; fortran implementations often prove to be the fastest codes in many situations. If there is a need to use an interpreted language, Python with the scipy package offers an excellent alternative programming environment favored by many scientific computing developers & programmers. Finally, Java developers can also use great care in the use of data types in their program development and tune their programs to create highly efficient codes. The Colt numerical and scientific computing library for Java offers an excellent API to work with such algorithms.

1. Implement the Conjugate-Gradient algorithm with the Polak–Ribiere update formula of Sect. 1.1.1.1 for unconstrained optimization using any computer programming language and apply the algorithm on the n-dimensional Rosenbrock function $f(x) = \sum_{i=1}^{n-1} \left[100(x_{i+1} - x_i)^2 + (1 - x_i)^2\right]$ for n = 2, 5, 10, and 20, starting from the n different points $[i+1\ i]^T$ for $i = 1 \ldots n$.
2. Implement the Approximate Line-Search method for unconstrained optimization with the Armijo Rule of Sect. 1.1.1.1 and apply the algorithm to the Rastrigin test function $f(x) = 10n + \sum_{i=1}^{n} \left[x_i^2 - 10\cos(2\pi x_i)\right]$ starting from 10 random points chosen uniformly from the hypercube $[-1, 1]^n$. Use n = 2,5, 10.
3. Implement the randomized heuristic algorithms for global unconstrained optimization SA and DE (Sect. 1.1.1.3) and test them on Ackley's n-dimensional test function $f(x) = -ae^{-b\sqrt{\frac{1}{n}\sum_{i=1}^{n} x_i^2}} - e^{\frac{1}{n}\sum_{i=1}^{n} \cos(cx_i)} + a + e$ where $a = 20$, $b = 0.2$, and $c = 2\pi$, for n = 2, 5 and 10. Run the algorithms for 1,000 iterations and observe how the algorithms make any progress.
4. Solve the following LP using the revised simplex method:

$$\max_x .z = x_1$$

$$\text{s.t.} \begin{cases} x_1 - x_2 \geq 0 \\ -x_1 - x_3 \geq -1 \\ -x_2 - 3x_3 \geq 0 \\ x_i \geq 0, \quad i = 1,2,3 \end{cases}$$

   Determine whether your solution is unique or not. Justify your answer.

5. Formulate the dual problem of the LP defined in exercise 4 above. Apply the revised simplex method on the dual problem. Does the optimal solution exist? Is it unique? Justify your answer.
6. For $n > 1$, consider the quadratic function $g : \mathbb{R}^n \to \mathbb{R}$ defined as $g(x) = \frac{1}{2}x^T Ax - b^T x + c$, where the matrix A is symmetric and P.D. Let $x_{(0)}$ be any vector in $\mathbb{R}^n$, and let $p_{(i)}$ $i = 0,\ldots k-1$ be k linearly independent vectors in $\mathbb{R}^n$, and define the set $V_k = \left\{x | x = x_{(0)} + \sum_{i=0}^{k-1} a_i p_{(i)}, a_i \in \mathbb{R}\right\}$. Show that the *unique* minimizer of the restriction of the function g on V_k is given by $x^* = x_{(0)} + P_k\left(P_k^T A P_k\right)^{-1} P_k^T\left(b - Ax_{(0)}\right)$ where the $n \times k$ matrix P_k is given by $P_k = \left[p_{(0)} \quad p_{(1)} \quad \cdots \quad p_{(k-1)}\right]$.

   Hint: Show that V_k is a convex set, which in conjunction with the strong convexity of g implies the uniqueness of the minimizer; to find the minimizer, use the First-Order Necessary Conditions for Mathematical Programming for the problem $\min_{a \in \mathbb{R}^n} f(a) = g(x_{(0)} + P_k a)$.

7. Consider the optimization problem $\min_x \left\{x_1 + x_2 | \varphi(p) x_1^2 \leq x_2\right\}$ where $\varphi : \mathbb{R}^k \to \mathbb{R}$ is a smooth function, with $\varphi(0) = 1/2$.

(a) Show that when $p = 0$, the point $x = y(0) = [-1\ ½]^T$ is the unique minimizer for this problem.
(b) Prove that there exists a smooth function $y : U \to \mathbb{R}^2$ defined over an appropriately small neighborhood U of the origin 0 in $\mathbb{R}^k$ such that $\forall p \in U$, $y(p)$ is the *unique* minimizer for the problem with that value of p.
(c) Compute the derivative of the function y at $p = 0$.

8. An algorithm for solving the all-pairs Shortest Path Problem on a (arc-weighted) graph $G(V,E)$ with edge weights a_{ij} for each arc in E computes the quantities

$$D_{ij}^{1} = \begin{cases} a_{ij}, & \text{if } (i,j) \in E \\ 0, & \text{if } i = j \\ +\infty, & \textit{else} \end{cases}$$

and

$$D_{ij}^{2k} = \begin{cases} \min\limits_{m}\left\{D_{im}^{k} + D_{mj}^{k}\right\}, & \text{if } i \neq j, k = 1,2,\ldots\lfloor\log(N-1)\rfloor \\ 0, & \text{if } i = j, k = 1,2,\ldots\lfloor\log(N-1)\rfloor \end{cases}$$

for all $i,j \in V$ (and $N = |V|$ is the total number of nodes in V). Show that for $i \neq j$, D_{ij}^{k} gives the shortest distance from node i to j using paths with 2^{k-1} arcs or fewer.

9. Consider the following pure Binary Program:

$$\max_{x} z = -x_{n+1}$$
$$\text{s.t.} \begin{cases} 2x_1 + 2x_2 + \ldots + 2x_n + x_{n+1} = n \\ x \in B^{n+1} \end{cases}$$

Show that *any* Branch & Bound algorithm using the LP relaxation to compute upper bounds will require the enumeration of an exponential number of nodes when n is odd.

10. Model as a Mixed Integer Programming Problem the following Many Traveling Salesmen Problem related to vehicle routing problems in distribution management: a company must send all its N representatives to visit each and every city in a network of cities connected by roads (arcs) and have each representative return to the company headquarters in such a way that no city is visited twice by any representative, all representatives start from the same headquarters and return to them, and the total cost of the representatives' tours measured by the sum of the distances they travel is minimized.
11. Prove that the function $c(x,y)$ defined in eq. (1.17) when minimized over all feasible pairs (x,y) will indeed produce an assignment of tasks to workstations such that the total number of stations is minimized and among all assignments

that minimize the required number of work-stations, the assignment produces the least imbalance of workload between any two workstations.

References

Achterberg T (2007) Constraint integer programming. Ph.D. dissertation, Technical University of Berlin, Germany

Achterberg T, Berthold T (2009) Hybrid branching. Lecture Notes in Computer Science. Springer, Heildelberg, 5547:309–311

Achterberg T, Koch T, Martin A (2005) Branching rules revisited. Oper Res Lett 33(1):42–54

Ahuja RK, Magnanti TL, Orlin JB (1993) Network flows: theory, algorithms and applications. Prentice-Hall, Englewood Cliffs

Al-Baali M, Fletcher R (1986) An efficient line-search for nonlinear least squares. J Optim Theory Appl 48:359–377

Anbil R, Gelman E, Patty B, Tanga R (1991) Recent advances in crew-pairing optimization at American airlines. Interfaces 21:62–74

Apostol TM (1962) Calculus. Blaisdel Publishing, NY

Apostol TM (1981) Mathematical analysis, 2nd edn. Addison-Wesley, Reading

Armijo L (1966) Minimization of a function having Lipschitz continuous first partial derivatives. Pac J Math 16(1):1–3

Aydin ME, Fogarty TC (2004) A distributed evolutionary simulated annealing algorithm for combinatorial optimisation problems. J Heuristics 10(3):269–292

Balas E (1965) An additive algorithm for solving linear programs with zero-one variables. Oper Res 13(4):517–546

Barnhart C, Johnson EL, Nemhauser GL, Savelsbergh MWP, Vance PH (1998) Branch-and-price: column generation for solving huge integer programs. Oper Res 46(3):316–329

Barnhart C, Hane CA, Vance PH (2000) Using branch-and-price-and-cut to solve origin-destination integer multi-commodity flow problems. Oper Res 48(2):318–326

Bellman R (1957) Dynamic programming. Princeton University Press, Princeton

Bertsekas DP (1982) Constrained optimization and lagrange multiplier methods. Academic Press, New York

Bertsekas DP (1988) The auction algorithm: a distributed relaxation method for the assignment problem. Ann Oper Res 14(1):105–123

Bertsekas DP (1991) Linear network optimization: algorithms and codes. MIT Press, Cambridge

Bertsekas DP (1995) Nonlinear programming. Athena Scientific, Belmont

Bertsekas DP (2001) Dynamic programming and optimal control, vol 2, 2nd edn. Athena Scientific, Belmont

Bertsekas DP (2005) Dynamic programming and optimal control, vol 1, 3rd edn. Athena Scientific, Belmont

Bertsekas DP, Tsitsiklis JN (1989) Parallel and distributed computation: numerical methods. Prentice-Hall, Englewood Cliffs

Broyden CG (1970) The convergence of a class of double rank minimization algorithms, in two parts. J Inst Math Appl 6:76–90

Cheney W, Kincaid D (1994) Numerical mathematics and computing, 3rd edn. Brooks/Cole Publishing Company, Pacific Grove, CA

Chvatal V (1983) Linear programming. W.H. Freeman and Co, NY

Dantzig GB, Wolfe P (1960) Decomposition principle for linear programs. Oper Res 8(1):101–111

Darwin R (1861) On the origin of species by means of natural selection, or the preservation of favoured races in the struggle for life. J Murray, UK

De Leone R, Meyer RR, Kontogiorgis S, Zakarian A, Zakeri G (1994) Coordination in coarse-grained decomposition. SIAM J Optim 4(4):777–793

De Leone R, Meyer RR, Zakarian A (1999) A partitioned ε-relaxation algorithm for separable convex network flow problems. Comput Optim Appl 12:107–126

Dijkstra EW (1959) A note on two problems in connexion with graphs. Numerische Mathematik 1(1):269–271

Ferris MC, Wathen AJ, Armand P (2007) Limited memory solution of box-constrained convex quadratic problems arising in video games. RAIRO Oper Res 41:19–34

Festa P, Resende MGC (2009) Hybrid GRASP heuristics. In: Abraham A, Hassanien A-E, Siarry P, Engelbrecht A (eds) Foundations of computational intelligence, vol 3: Global Optimization. Springer, Berlin

Fletcher R (1970) A new approach to variable metric algorithms. Comput J 13:317–322

Fletcher R (1987) Practical methods of optimization, 2nd edn. Wiley, Chichester

Fletcher R (2000) Practical methods of optimization, 3rd edn. Wiley, Chichester

Ford LR Jr, Fulkerson DR (1962) Flows in networks. Princeton University Press, Princeton

Garey MR, Johnson DS (1979) Computers and intractability: a guide to the theory of NP-completeness. W.H. Freeman and Co, NY

Gill PE, Murray W, Wright MH (1982) Practical optimization. Emerald, West Yorkshire

Glover F (1989) Tabu search, part I. ORSA J Comput 1:190–206

Glover F (1998) A template for scatter search and path relinking. Lecture Notes in Computer Science. Springer, Heidelberg, 1363:13–54

Glover F, Marti R (2006) Tabu search. In: Meta-heuristic procedures for training neural networks. Springer, Berlin

Goldfarb D (1970) A family of variable metric methods derived by variational means. Math Comput 24:23–26

Goldstein AA (1965) On steepest descent. SIAM J Control 3:147–151

Gomory RE (1958) Outline of an algorithm for integer solutions to linear programs. Bull Am Math Soc 64:275–278

Holland JH (1975) Adaptation in natural and artificial systems. University of Michigan Press, Ann Arbor

Hopp W, Spearman M (2008) Factory physics, 3rd edn. McGraw-Hill/Irwin, NY

Kennington JL (1989) Using KORBX for military airlift applications. In: Proceedings of the 28th IEEE conference on decision and control, Tampa, FL, 13–15 December 1989

Kirkpatrick SC, Gellatt CD, Vecchi MP (1983) Optimization by simulated annealing. Science 220(4598):671–680

Koza J, Poli R (2005) Genetic programming. In: Burke EK, Kendall G (eds) Search methodologies: introductory tutorials in optimization and decision support techniques. Springer, Berlin

Lowerre B (1976) The Harpy speech recognition system. Ph.D. dissertation, Carnegie-Mellon University

Luenberger DG (1969) Optimization by vector space methods. Wiley, NY

Mangasarian OL (1994) Nonlinear programming. SIAM, Philadelphia

Megiddo N, Supowit KJ (1984) On the complexity of some common geometric location problems. SIAM J Comput 13(1):182–196

Meyer RR (1992) Lecture notes on integer programming. Dept Comput Sci, University of Wisconsin-Madison

Michalewicz Z (1994) Genetic algorithms + data structures = evolution programs, 2nd edn. Springer, Berlin

Nemhauser GL, Wolsey LA (1988) Integer and combinatorial optimization. Wiley, NY

Papadimitriou CH, Steiglitz K (1998) Combinatorial optimization: algorithms and complexity. Dover, Mineola

Rechenberg I (1973) Evolutionsstrategie: optimierung technisquer systeme nach prinzipien der biologischen evolution. Frommann-Holzboog, Stuttgart, Germany (in German)

Rockafellar RT (1970) Convex analysis. Princeton University Press, Princeton

Ryan DM, Foster BA (1981) An integer programming approach to scheduling. In: Wren A (ed) Computer scheduling of public transport urban passenger vehicle and crew scheduling. North-Holland, Amsterdam

Schultz GL, Meyer RR (1989) A three-phase algorithm for block-structured optimization. In: Proceedings of the 4th SIAM conference on parallel processing for scientific computing, pp 186–191, Philadelphia, PA

Schwefel HP (1981) Numerical optimization of computer models. Wiley, NY

Shanno DF (1970) Conditioning of quasi-Newton methods for function minimization. Math Comput 24:647–656

Shi L, Olafsson S (2000) Nested partitions method for global optimization. Oper Res 48(3):390–407

Storm R, Price K (1997) Differential evolution: a simple and efficient heuristic for global optimization over continuous spaces. J Glob Optim 11(4):341–359

Vanden Berghen F (2004) CONDOR: A constrained nonlinear derivative-free parallel optimizer for continuous high-computing load, noisy objective functions. Ph.D. dissertation, Universite Libre de Bruxelles, Belgium

Chapter 2
Forecasting

Since the beginning of civilizations, the ability to predict future events has been one of the most important abilities and capacities of the human mind, greatly assisting in its survival. The ability to foretell the future has always been a major source of power. On the other hand, the example of Cassandra, the ancient princess who could clearly see and prophesize catastrophic near-future events but was dismissed as insane by her people, underscores the importance of the fact that the forecaster must not only be able to make accurate forecasts, but also convince others of the accuracy of her/his forecasts. In today's world, the ability to accurately forecast near-term as well as medium-term events such as demand for existing or new products is among the most crucial capacities of an enterprise. In general, the use of forecasts falls under one of three major types: (a) economic forecasts, which attempt to measure and predict macro-economic quantities such as business cycles, inflation rates, money supply and currency exchange rates, (b) technological forecasts, whose main purpose is to predict imminent and upcoming technological break-through and innovation, and to a lesser degree market penetration of completely new products and (c) demand forecasts, whose main purpose is to predict short-and medium term sales of existing products, whose sales' history exists and is accurately recorded.

Regardless of the shift of emphasis on pull-based production models, agile manufacturing, or demand information sharing among supply chain partners, forecasting remains an invaluable tool for planning the activities of any organization—manufacturing, financial, or otherwise. In this chapter, we examine the most successful forecasting techniques available today. These include classical statistical quantitative methods such as time-series analysis that attempt to decompose a signal into a number of components such as trend, seasonality, cycle, and random noise and determine each component as accurately as possible, causal methods such as regression, as well as new techniques including Artificial Neural Networks (ANN) and Prediction Markets (PMs). All of them are statistical tools at heart.

Before discussing the above-mentioned methods in detail, it is worth noting that regardless of the forecasting method used, a few things are always true:

I. T. Christou, *Quantitative Methods in Supply Chain Management*,
DOI: 10.1007/978-0-85729-766-2_2,

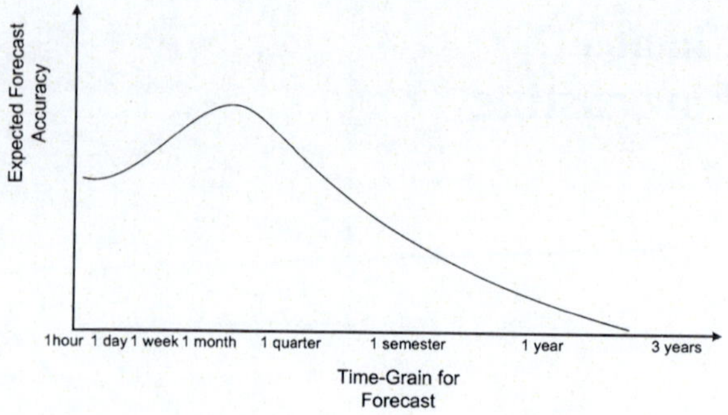

Fig. 2.1 Forecasting accuracy of an aggregate variable of time as a function of the time-grain used

(1) any forecast is by necessity only an estimate of the future, and therefore will always be wrong! The only question worth asking and answering is: "*how wrong?*"
(2) any aggregate forecast (e.g., total demand of a product sold in different packages through different distribution channels) will be more accurate than a forecast for an individual item in the aggregation. This is intuitively easy to understand, as random fluctuations of the values of individual items would tend to "cancel out" each other, making the variability of the aggregation less than the variability of its constituents. Indeed, this intuition is statistically correct: assume n items $x_1, \ldots, x_n$ make up together an aggregate y whose value we wish to estimate; further assume each item's value is a random variable with the *same* expected value μ_x and standard deviation σ_x. Then, the random variable $y = x_1 + \cdots + x_n$ has expected value $\mu_y = n\mu_x$ and standard deviation $\sigma_y = \sqrt{n\mu_x}$ and therefore its coefficient of variation (c.v.) is $1/\sqrt{n}$ times the c.v. of the individual items (Halikias 2003), which means that as n gets larger, the variable y is much more predictable than the individual variables x_i and the percentage error of the forecast tends to zero (its relative dispersion around the mean is much less than that of the constituent variables). This also holds true about aggregations in the time dimension: the forecast of *total* demand for a particular item during the next month is likely to be more accurate than the best estimate for tomorrow's demand, *assuming of course that each day in the month is statistically the same as any other.*
(3) any short-term forecast is in general more accurate than forecasts for events in the far future. This is also intuitively easy to understand as the farther into the future one attempts to "see", the more uncertainty about the course of future events enters into the picture to make the variability of more distant events much greater than that of near-future events. The forecast of a day's demand of a product 1 year from now, is likely to be much less accurate than the forecast of tomorrow's demand of the same product.

It is interesting to notice that the combined effect of the last two observations implies that the accuracy of an aggregate forecast as a function of the time window of the aggregation must have the shape shown in Fig. 2.1: there exists an optimal length of time for which we can make accurate forecasts; attempting to aggregate further into the future will decrease the accuracy of the forecast because the distribution of the random variables representing the far-future events will be much wider than the distribution of the variables representing the near future events. In other words, while a forecast of next quarter's sales may be more accurate than a forecast of tomorrow's sales, a forecast of the next 3 years sales will likely be much less accurate than a forecast of next quarter's sales.

The most fundamental premise, upon which all forecasting is based on, is that the future will somehow resemble the past. Therefore, as mentioned earlier, all forecasting activities will always be wrong. However, this does not mean that any forecast is useless; on the contrary, a forecast that contains a small error can be extremely useful for Supply-Chain Management purposes. Before describing the main methods of analysis of historical data for a given quantity such as monthly demand for a product, *we describe the quantities that are used to measure the accuracy of a forecast.* Suppose we have a time-series describing a particular quantity in past times, $d_i|i = 1, 2, \ldots, n$ for which we are interested in predicting its value in time $n + 1$. Let F_{n+1} be the value we forecast for the quantity d_{n+1} at time n.

The quantity

$$e_t = d_t - F_t, \quad t > 1$$

defines the (one-period or instant) forecasting error. For $t > 1$, we define the following "*forecasting accuracy measures*" for any method that provides the forecasts F_{t+1} at time $t = 1,2,\ldots$:

(1) Mean deviation $\mathrm{MD}_t = \frac{\sum_{i=2}^{t} e_i}{t-1}$

(2) Mean percentage deviation $\mathrm{MPD}_t = 100 \times \frac{\sum_{i=2}^{t} \frac{e_i}{d_i}}{t-1}\%$

(3) Mean absolute deviation $\mathrm{MAD}_t = \frac{\sum_{i=2}^{t} |e_i|}{t-1}$

(4) Mean absolute percentage deviation $\mathrm{MAPD}_t = 100 \times \frac{\sum_{i=2}^{t} \frac{|e_i|}{d_i}}{t-1}\%$

(5) Mean square error $\mathrm{MSE}_t = \frac{\sum_{i=2}^{t} e_i^2}{t-1}$

(6) Root mean square error $\mathrm{RMSE}_t = \sqrt{\frac{\sum_{i=2}^{t} e_i^2}{t-1}}$

(7) Tracking signal $S_t = \frac{E_t}{MAD_t} = \frac{\sum_{i=2}^{t} e_i}{\frac{\sum_{i=2}^{t} |e_i|}{t-1}}$

(8) Directional symmetry $\mathrm{DS}_t = \frac{\sum_{i=3}^{t} [\mathrm{sgn}((d_i - d_{i-1})(F_i - F_{i-1}))]_+}{t-2}$

(9) U-Statistics $U_1 = \frac{\sqrt{\frac{1}{t-1}\sum_{i=2}^{t} e_i^2}}{\sqrt{\frac{\sum_{i=2}^{t} d_i^2}{t-1}} + \sqrt{\frac{\sum_{i=2}^{t} F_i^2}{t-1}}}, \quad U_2 = \sqrt{\sum_{i=2}^{t-1} \frac{(F_{i+1} - d_{i+1})^2}{(d_{t+1} - d_t)^2}}.$

The sgn(x) function takes the value 1 if x is non-negative and -1 if the argument is negative. The function x_+ is defined as the $\max\{x, 0\}$. When analyzing a time-series to produce a forecast, all of the above measures are significant and should be monitored so as to get an understanding of the accuracy of the forecast.

There is no established rule to indicate what accuracy is optimal or near-optimal, but in many practical situations, a value of MAPD_t that is less than 10% is often considered excellent—although in some situations it is possible to obtain values of MAPD_t that are well below 1% in relatively stable market environments. MAPD_t values between 10 and 20% are generally considered good while values between 20 and 30% are considered moderate. Forecasts with MAPD_t scores worse than 30% are in general poor and should be discarded.

The *Directional Symmetry* metric provides an indication of the *direction* of prediction. Its value is always in the range [0, 1] and a value close to 1 indicates that the forecasting procedure produces forecasts in a direction that most of the time agrees with the direction (upward or downward) of the actual time-series, and that if the time-series is about to increase in the next period, the forecast will also be higher for the next period than the forecast produced for the current period. The *tracking signal* can be useful for identifying systematic biases in the forecasts made: if it is (significantly) greater than zero, the forecast systematically underestimates the time-series d_i, whereas if it is (significantly) less than zero, the forecast overestimates the time-series. The *Root Mean Square Error* quantity (RMSE_t) is very useful as an estimate of the standard deviation of the forecast errors, which can be derived from it using the formula

$$s_t^e = \sqrt{\frac{t-1}{t-2}}\mathrm{RMSE}_t.$$

Under the hypothesis that the errors are symmetrically distributed around zero and unbiased, the effectiveness of the forecast procedure can be affirmed by checking whether each error $e_i | i = 2, \ldots, t$ is in the interval $\left(-3s_t^e, +3s_t^e\right)$ to which it should lie within with a probability 99.8%. If this test fails, the forecast process needs to be revised as the errors are most likely *not* due to random noise alone.

The (Theil's) U-statistics have the characteristic that the more accurate the forecast, the lower their value. The U_1-statistic is bounded in the interval [0,1], whereas the value of the U_2-statistic provides a measure of how much better is the forecast method to the naïve method of forecasting (discussed immediately below). Values of this statistic less than 1 indicate better accuracy than the naïve method. Both U-statistics represent a compromise between absolute and relative measures of forecast error.

2.1 Smoothing Methods for Time-Series Analysis

2.1.1 Naïve Forecast Method

Obviously, the easiest—and most naïve—way to forecast the value of a time-series is to think that the immediate future will be the same as the immediate past, and to assume therefore that the next value of a time-series will be the same as the last one available, i.e.

$$F_{t+1} = d_t$$

This forecast, would be optimal in terms of accuracy if the values in the time-series were coming from a stochastic process that generated values according to the formula $d_{t+1} = d_t + R_t$ where R_t are independent, identically distributed (i.i.d) random variables with zero mean and constant variance σ^2. Such a series would have a constant expected value $E[d_t] = \mu$ and a variance $\text{Var}[d_t] = t\sigma^2$ that increases linearly in time. Such a process is called a *Random Walk*, and often arises in financial-related time-series such as stock market prices, and other financial indices. More general stochastic processes where the $R_t = d_{t+1} - d_t$ variables are not necessarily independent nor have a constant variance are called *martingales* and are of great importance in financial applications as well. However, in Supply-Chain Management, such processes are quite rare, and therefore the naïve forecast method is *rarely useful* in practical situations in this domain.

2.1.2 Cumulative Mean Method

Assume that demand for a particular product is statistically constant over the time window of interest, or more generally, assume that we are interested in predicting the value of a time-series that is an instantiation of a stochastic process that generates values around a constant mean, so that the process is described by the equation $d_t = D + R_t$ where D is an unknown constant, and R_t are independent identically distributed random variables with zero mean. Under these—rather restrictive— assumptions, assuming we know the values $d_1,\ldots, d_n$ the cumulative mean method computes the *best* estimate for the value d_{n+1} by the formula

$$F_{n+1} = \frac{\sum_{i=1}^{n} d_i}{n}$$

If at time n, any estimates of the more distant future $n + 2$, $n + 3\ldots$ are absolutely required, one must necessarily set $F_{n+i} = F_{n+1}$ for all $i > 1$ as well (but in general, one should not attempt to make long-range predictions using time-series methods as the error associated with them increases very fast, independent of the method). It should be rather easy for the reader to verify that the average of the observed values is the best estimator possible for the constant D and for the next value in the time-series. The graph in Fig. 2.2 shows how close the cumulative mean estimate is to a time-series that comes from the above-mentioned stochastic process, but also how close it comes to estimating the values of a time-series that is not generated from a stationary stochastic process.

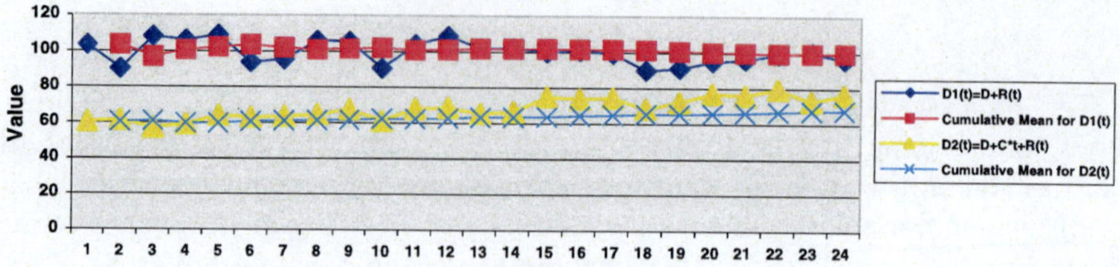

Fig. 2.2 Forecasting using the cumulative mean method: the forecast is optimal for time-series D1(t), but not nearly as good when the generating stochastic process does not have a constant mean

2.1.3 Moving Average Method

One of the most well-known and established methods for demand forecasting among supply chain managers is the moving average method. The method initially looks as a computational optimization of the Cumulative Mean method, in that to compute the forecast of the next period we compute the average of a small number of the immediate past values in the time-series, thus avoiding the computation of the summation of the entire time-series. However, using only the most recent values in the time-series rather than the entire history has significant advantages when the generating stochastic process does not generate values that hover around a constant mean. If the time-series is generated from a stochastic process with a mean value that somehow drifts in time, or if the time-series is a realization of a stair-shaped stochastic process, i.e. a process that is governed by equation such as

$$d_{t+1} = D + E_{k_t} + R_t\,, \quad t \geq k_t$$

where D, and E_k are unknown constants, k_t is a sequence of numbers, and R_t are independent identically distributed random variables with zero mean, then the Moving Average method (with an appropriate choice of the parameter M) is a better estimator for the next value in the time-series. The Moving Average method with parameter M computes the next value in a time-series as the average of the last M observed values:

$$F_{t+1} = \frac{\sum_{i=1}^{M} d_{t-i+1}}{M}$$

with $M < t+1$ of course. If estimates are needed for more distant future times, the value $F_{t+m} = F_{t+1}$ (for $m > 1$) is used as well. The method is attractive because it is very easy to understand, very easy to implement, and requires only maintaining a history of the previous M periods for each data-item to forecast. For this reason, in many surveys (Sanders and Manrodt 1994), it was shown that it ranked first among professional supply-chain managers for use in short-term and medium-term forecasts.

The graph in Fig. 2.3 shows how close the forecasts generated by the Moving Average method come to the actual values of a time-series that is generated from the stochastic process formulated above, for different values of the parameter M. However, notice in Fig. 2.4, how all the forecasts of the Moving Averages

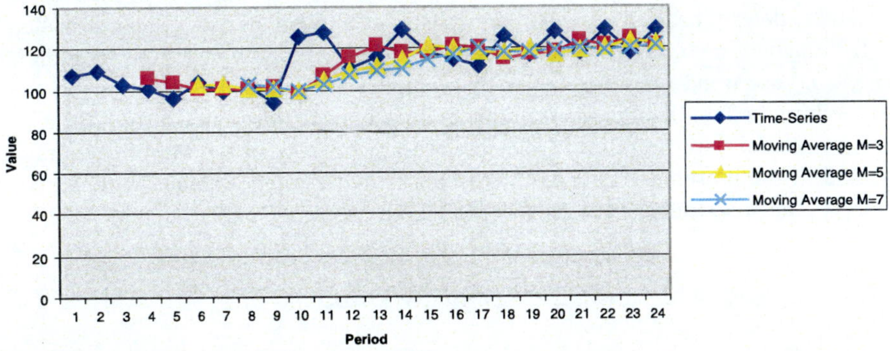

Fig. 2.3 Forecasting using the moving averages method for 3 different values of the parameter M: the generating process is a single stair-step whose values hover around a mean that changes from the value 100 to the value 120 in period 10

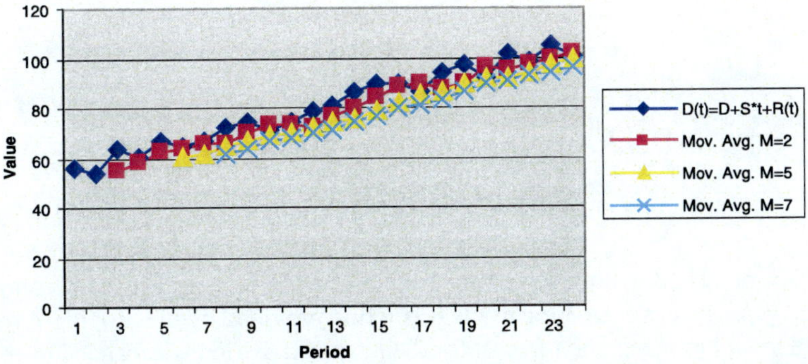

Fig. 2.4 Forecasting a time-series with a non-zero trend using the moving averages method leads to systemic errors

method consistently under-estimate a time-series that is inherently increasing (has a *trend* component). This is a major disadvantage of the method, in that by averaging the previous history, its predictions will always be in the range $[v_{min}, v_{max}]$ where v_{min} and v_{max}, are respectively the minimum and maximum value attained in the time-series in the previous M periods.

Even if we extend the Moving Averages Method to produce a forecast as the weighted average of the last M observations, using M *non-negative weights* $w_1,\ldots,w_M$ as follows:

$$F_{t+1} = \sum_{i=1}^{M} w_i d_{t-i+1} \Big/ \sum_{i=1}^{M} w_i$$

this disadvantage still persists.

Nevertheless, it is obvious from the graph that the Moving Average depicts the increasing trend in the data, despite its inability to "catch up" with the trend in the actual forecasts. The smoothing provided by the Moving Average allows the analyst to visually determine in an instant if there exists any trend in the data even in so-called high-frequency data, i.e. time-series that fluctuates widely with a high frequency around a base trend line. The larger the value of the parameter M, the larger the "smoothing" effect that eliminates high-frequency oscillations in the time-series.

Two similar ways to improve the forecasting accuracy of the Moving Average method in the presence of trends in the data is the Moving Average with Trends method, and the Double Moving Average, both presented next.

2.1.4 Moving Average with Trends Method

To avoid the systemic problem of underestimating an inherently increasing time-series (or vice versa, over-estimating an inherently decreasing one), the following set of recursive equations is used to predict at time t, the value of a time-series at *any* time $t + h$ for any integer $h > 0$ given the values d_i $i = 1,\ldots, t$:

$$m_t = \frac{\sum_{i=1}^{M} d_{t-i+1}}{M}, \quad t > M$$

$$F'_t = \begin{cases} d_t, & t \le M \\ F'_{t-1} + \frac{6}{M(M^2-1)}((M-1)d_t + (M+1)d_{t-M} - 2Mm_{t-1}), & t > M \end{cases}$$

$$F_{t+h} = m_t + \left(h + \frac{M-1}{2}\right)F'_t, \quad h \ge 1$$

The set of equations above (easily implementable in a spreadsheet computer program) represents a correction to the prediction provided by the Moving Average (denoted by m_t) in which the forecast for the next time-period is enhanced by a term that is analogous to the forecast for the current period which in turn is the forecast of the previous time-period plus a weighted average of the extreme points of the window considered in the time series and the average of that window. This weighted average is the best estimator of the slope of the series in a least-squares optimization sense, in which it provides the best estimate of the slope of a time-series that is linearly increasing according to the equation $d_t = D + Ct + R_t$ with R_t being independent normal random variables with zero mean. In Fig. 2.5 we show the results of the application of this method on a trended time-series, as the one in Fig. 2.4. The method fits the data better than the Moving Average method can.

2.1.5 Double Moving Average Method

Similar to the Moving Average with Trends method, the Double Moving Average method is an improvement over the classical Moving Average method that attempts to fit data generated from a linear model obeying the equation

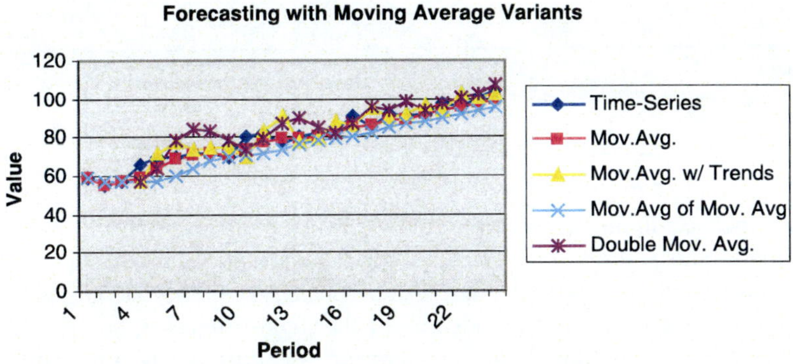

Fig. 2.5 Forecasting a time-series with a non-zero trend using the moving average, moving averages with trend method and double moving average, all with $M = 3$. the moving average of moving average assumes the time-series values for $t = 1, 2, 3$ to initialize the double moving average method. the moving average with trends method avoids systemic over- or under- estimations

$d_t = D + Ct + R_t$. The Double Moving Average computational scheme therefore evolves an estimate of the current level of the time-series and the current slope of the time-series according to the equations

$$F_{t+m} = a_t + mb_t, \quad m \geq 1$$
$$a_t = 2\gamma_t - \eta_t$$
$$b_t = \frac{2}{r-1}(\gamma_t - \eta_t)$$
$$\gamma_t = \begin{cases} \frac{1}{r}\sum_{i=0}^{r-1} d_{t-i}, & t \geq r \\ d_t, & t < r \end{cases}$$
$$\eta_t = \frac{1}{r}\sum_{i=0}^{r-1} \gamma_{t-i}, \quad t \geq r$$

F_{t+m} is the forecast for the value of the time-series d_i at points $t + m$ for any $m > 0$ given the values d_i $i = 1,\ldots, t$. Clearly, γ_t is a Moving Average with parameter r, whereas η_t is the moving average of the moving average time-series. A graphical representation of the time-series γ_t, η_t, and F_t offers some initial insight into the workings of the Double Moving Average method. The illustration in Fig. 2.5 shows how the method compares to the Moving Average with Trends method. As it turns out, the Moving Average with Trends method is superior to the Double Moving Average which tends to oscillate more than the former method.

In the following table we compare the Moving Average method with the two variants of the Moving Average that directly attempt to handle trends in the data. The Moving Average method clearly introduces systemic errors since the Tracking Signal value for period 24 is 16.92 (larger than the values produced by both variants of the method). The Moving Average with Trends method has an acceptable Tracking Signal value of -5.28, obtains the best Mean Deviation value, and has a very good $MAPD_{24}$ score of less than 6%.

Method ($M = 3$)	MD_{24}	$MAPD_{24}(\%)$	$RMSE_{24}$	S_{24}
Mov. Avg.	2.20	3.3	3.31	16.92
Mov. Avg. w/Trends	−1.17	5.8	5.83	−5.28
Double Mov. Avg.	−3.33	7.6	7	−11.7

As we shall see in the next sections, more advanced methods (Holt's method for trended data, and Holt-Winters' method for trended and seasonal data) directly compute any trend—or periodic component—present in the series, and then synthesize the time-series as the composition of their constituents. It is interesting to notice however a common theme that appears in most of the methods to be discussed below: components, or the whole signal itself are computed as a composition of two basic estimates, an estimate of the signal itself, plus an adjustment for the previous error in the estimate, which leads to a convex combination of the estimate and the previous value of the data-series.

2.1.6 Single Exponential Smoothing Method

The easiest way to incorporate a feedback loop into the prediction system is to add to the last forecast made, a small fraction of the last forecast *error* made. The formula for Single (sometimes referred to as Simple) Exponential Smoothing Forecast (SES) is therefore the following

$$F_{t+1} = F_t + ae_t, \quad a \in (0, 1)$$

Feedback loops are extremely important in most scientific and engineering processes as they allow a system to be controlled and stabilized by making continuous adjustments to its state based on its previous performance. Fundamentally, a system that does not take into account its past performance, cannot know what adjustments to make to its processes so as to improve in the future. The Single Exponential Smoothing Method (and all Exponential Smoothing Methods to be discussed below) applies a small correction to the previous forecast if it was good and large corrections if the forecast was bad, in the direction of minimizing the error. Taking into account the definition of the error e_t, we get

$$F_{t+1} = F_t + a(d_t - F_t) = ad_t + (1 - a)F_t$$

which tells us that the next forecast is a convex combination of the last observed value in the data-series and the last forecast for the data-series. The name Exponential Smoothing derives from the fact that if we expand the formula in its convex combination form, it becomes

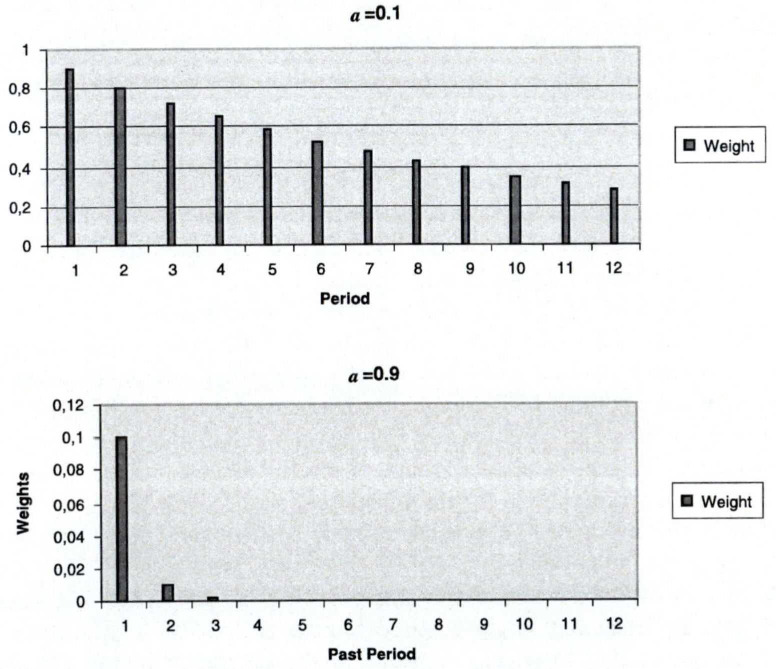

Fig. 2.6 Impact of the parameter a on the past values of the time-series in computing the SES forecast

$$\begin{aligned} F_{t+1} &= ad_t + (1-a)(ad_{t-1} + (1-a)F_{t-1}) = \cdots \\ &= a\sum_{i=0}^{t-1}(1-a)^i d_{t-i} + (1-a)^t F_1 \end{aligned}$$

which shows that the Single Exponential Smoothing Method is similar to a weighted average method applying *exponentially decreasing weights* (the $(1-a)^i$ terms) to the past values of the time-series. (The sum of the exponential weights does *not* sum to one though as in weighted average methods, as the reader can easily verify.) Higher values of the parameter a imply more rapid depreciation of the past, where values of a near zero make the forecasting process behave more like the Cumulative Mean Method. Initialization of the SES method requires an initial forecast value F_1. Usually, the choice $F_1 = d_1$ is made (other options, such as averaging demand prior to the first period and defining it as initial forecast value F_1 are also possible). In Fig. 2.6 we show how much a discounts the past values of the data-series as a function of the past.

SES methods were devised as general tools for forecasting, but work best for time-series that do *not* have inherent trends or seasonal (or business cycle) fluctuations. This is because, assuming the time-series is stationary and is generated from a process of the form $d_t = D + R_t$, it is not hard to show that the SES method computes forecasts in such a way so that the following sum of *discounted* residual squares is minimized:

$$S' = \sum_{j=0}^{\infty} (1-a)^{j+1} (d_{t-j} - F_t)^2$$

This fact provides the theoretical justification for the method *when the time-series is generated from a (wide-sense) stationary stochastic process*. In the graphs in Fig. 2.7, we show how the value of the parameter a affects the quality of the forecast for a (trended) real-world time-series.

It is easy to see in Fig. 2.7 that higher values of parameter a allow the forecast to "catch-up" to sudden increases or decreases in the time-series much faster. To select an "optimal" value for the parameter a, one must first define a criterion which they wish to optimize. Minimizing the Mean Square Error of forecast (MSE) is often utilized as it can be expressed as a smooth polynomial of the parameter a (with degree $2t$), and local search algorithms for nonlinear optimization will easily locate a local minimizer for the criterion.

The parameter a can also be dynamically and automatically adjusted to reflect changing patterns in the data.

The *Adaptive Response Rate Single Exponential Smoothing (ARRSES) method* extends SES by assigning larger a values during periods of highly-fluctuating time-series values, and lowering the value of the parameter during periods of steady values.

The set of equations describing the application of ARRSES method are as follows

$$\begin{aligned}
F_{t+1} &= a_t d_t + (1 - a_t) F_t \\
a_t &= \left| \frac{A_t}{M_t} \right| \\
A_t &= \beta e_t + (1 - \beta) A_{t-1}, \quad t > 0 \\
M_t &= \beta |e_t| + (1 - \beta) M_{t-1}, \quad t > 0 \\
A_0 &= 0, \quad M_0 = 0, \quad F_1 = 0, \quad \beta \in (0, 1)
\end{aligned}$$

The method *extends* SES in that the basic SES formula is modified to include a changing a_t parameter, which is the absolute value of the ratio of a feedback loop estimate of the error and the same estimate for the absolute forecast error. The method still requires selecting a value for the single parameter β. If the error $e_i = d_i - F_i$ is consistently large in absolute value, then a_i will tend to the value 1, thus making the method more responsive to changes in the time-series whereas if the forecast errors are consistently small, the parameter values a_i will tend to zero, making the method "smooth out" random variations in the signal. In Fig. 2.8, we show how the ARRSES method forecasts the time series used in Fig. 2.7.

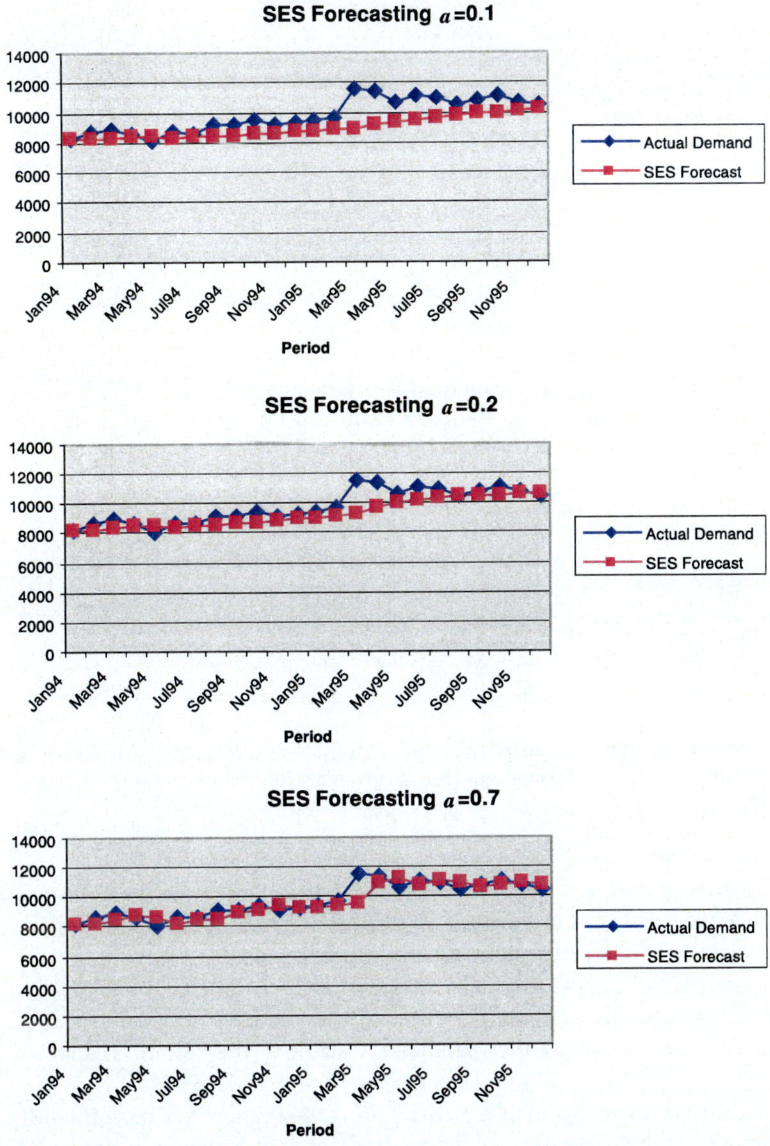

Fig. 2.7 Forecasting using the single exponential smoothing method with a = 0.1, 0.2, 0.7. The time-series represents the monthly avicultural meat exports (in hundreds of kg) from Italy to Germany during 1994–1995 (Source: Ghiani et al. 2004)

At this point we repeat that the Single Exponential Smoothing method, independent of how the parameter a is chosen or adapted, *being a special case of a weighted average method with non-negative weights*, has the same fundamental problem of weighted averaging methods: it produces forecasts that are

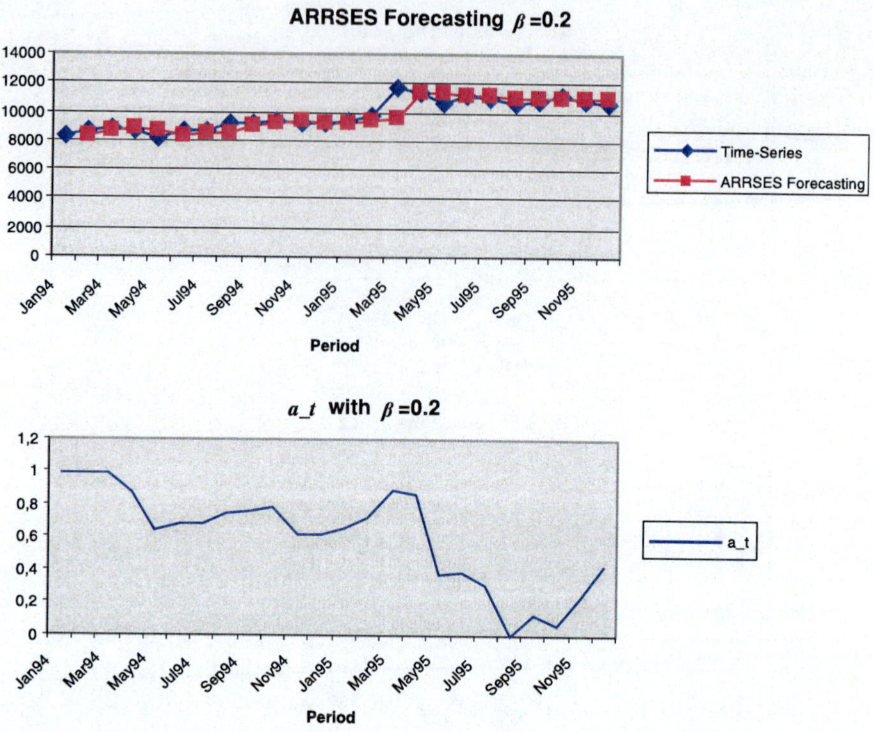

Fig. 2.8 Forecasting using the adaptive response rate single exponential smoothing method (ARRSES) with $\beta = 0.2$ and the corresponding adaptation of the values a_t. The time-series is the same as the one in Fig. 2.7

systematically under-estimating the time-series when there is an inherent positive trend, and vice versa, the forecast errors are systematically over-estimating demand when the time-series is inherently declining.

The experiment shows that the parameters can fluctuate wildly as the time-series changes its "pattern" even in the short term. In Fig. 2.9, we show that smaller values of the parameter β dampen the rate of change of the parameters a_i.

Despite the intuitive appeal of the ARRSES method, there has been a significant body of evidence that suggests that the method does not outperform simple SES in the long run, and that, on the contrary, careless use of the ARRSES method in practice can lead to significant forecast errors, due to long-term instabilities that the method sometimes incurs.

For this reason, many researchers and practitioners recommend great care to be taken if ARRSES is to be applied in practice (for example, by often resetting the method, or by using it within a forecasting ensemble where other methods are used to forecast the time-series as well and a fusion scheme computes the final forecast value).

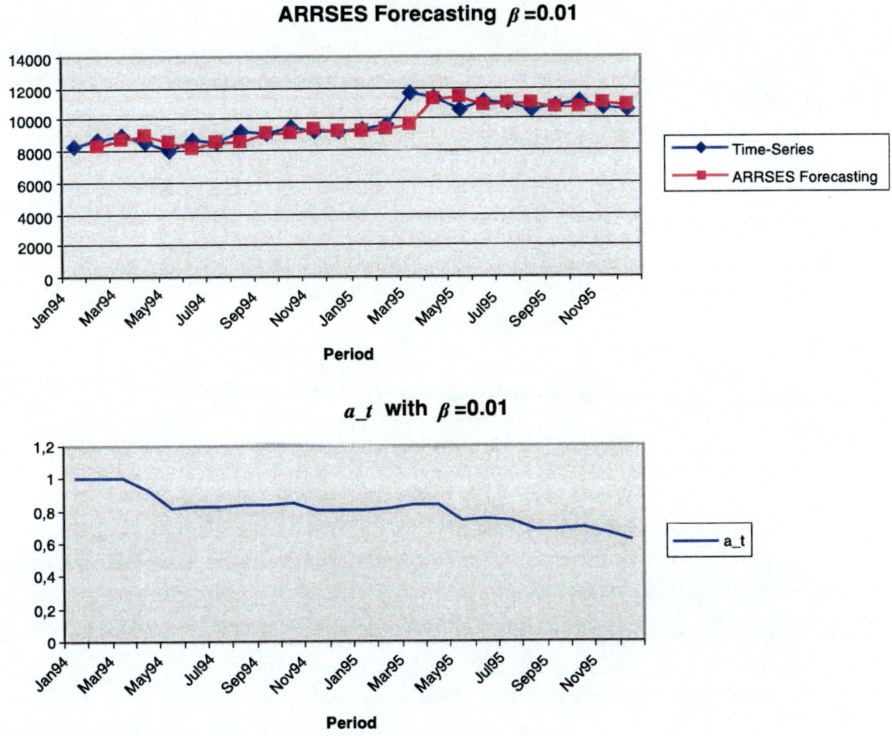

Fig. 2.9 Forecasting using the adaptive response rate single exponential smoothing method (ARRSES) with $\beta = 0.01$ and the corresponding adaptation of the values a_t. The time-series is the same as the one in Fig. 2.7

2.1.7 Multiple Exponential Smoothing Methods

Double and Triple Exponential Smoothing Methods are direct extensions of SES. The idea is to apply the same method twice or thrice on the initial time-series, in the hope that the SES method applied to a smoothed version of the initial time-series may provide better results than a single application of the method.

It can also be thought of as applying a control feedback on an already controlled forecasting system. The set of recursive equations describing the Double Exponential Smoothing Method (DES) are as follows:

$$
\begin{aligned}
F_{t+1} &= aF'_{t+1} + (1-a)F_t \\
F'_{t+1} &= ad_t + (1-a)F'_t \\
F_0 &= F'_0 = d_1,\ a \in (0,1)
\end{aligned}
$$

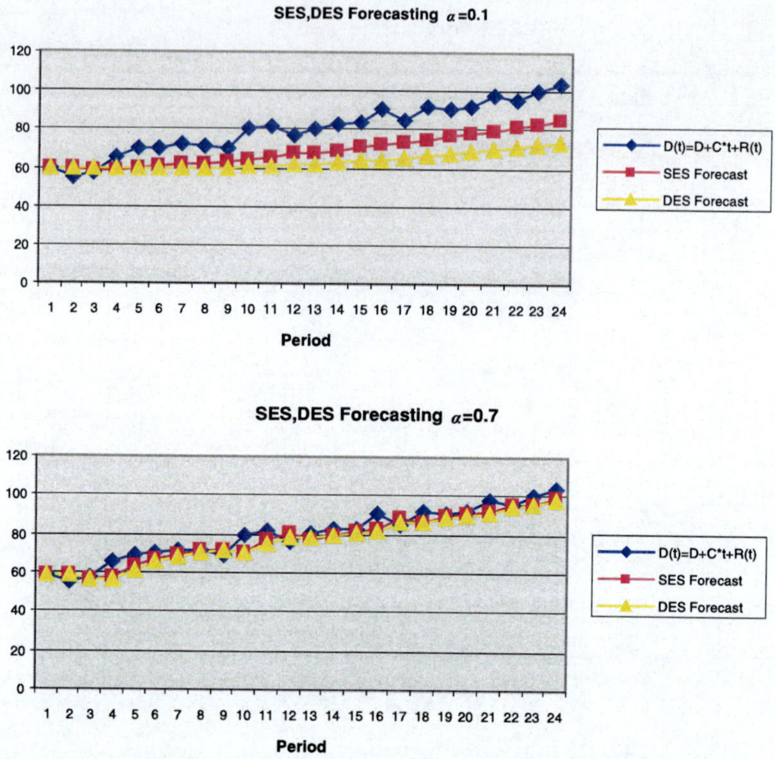

Fig. 2.10 Forecasting the next period using the double exponential smoothing method (DES) with $a = 0.1, 0.7$. The original time-series the same as the one in Figs. 2.4 and 2.5

Applying this method has the same complexity as the SES method (although it requires twice as many computations as SES), and the results are still strongly affected by the choice of the parameter a, as we can see in Fig. 2.10. We check the forecasting accuracy of DES against a time-series that grows according to the equation $d_t = D + Ct + R_t$ as in the example time-series of Figs. 2.4 and 2.5. As can it be seen, DES forecasts are smoothed-out versions of the smoothed-out time-series, and for this reason *lag significantly behind the original time-series when there are trends in the data*. The effect is of course less pronounced for larger values of a.

For data with a quadratic trend of the form $d_t = D + Ct + Et^2 + R_t$, Triple Exponential Smoothing can be applied. Triple Exponential Smoothing (TES) is computed via the following recursive equations:

$$F_{t+m} = 3L_t - 3L'_t + L''_t + b_t m + \frac{1}{2}c_t m^2, \quad m \geq 1$$

$$L_t = ad_t + (1-a)L_{t-1}$$

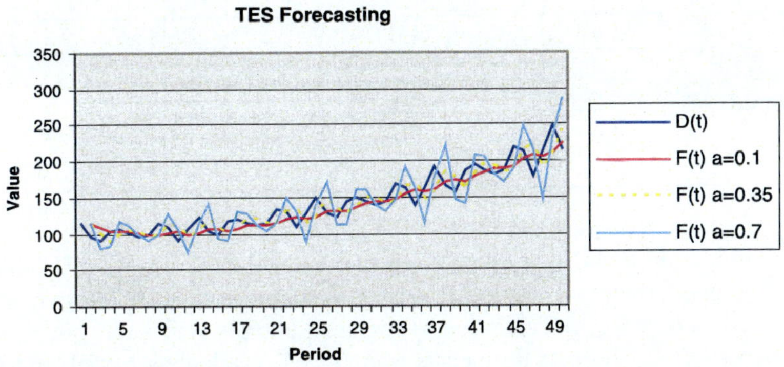

Fig. 2.11 Forecasting the next period using the triple exponential smoothing method for three different a values. the forecast with a up to 0.35 in general follows the increasing trends of the time-series, which exhibits a slow quadratic increasc in time accompanied with sinusoidal oscillations. the one-period lead-time forecast using triple exponential smoothing for quadratic trends exhibits a "jerky" behavior for higher values (0.7) of the smoothing constant a

$$
\begin{aligned}
&L'_t = aL_t + (1-a)L'_{t-1} \\
&L''_t = aL'_t + (1-a)L''_{t-1} \\
&b_t = \frac{a^2}{2(1-a)^2}\left[(6-5a)L_t - (10-8a)L'_t + (4-3a)L''_t\right] \\
&c_t = \frac{a^2}{(1-a)^2}\left[L_t - 2L'_t + L''_t\right] \\
&a \in (0,1), \quad t \geq 1 \\
&L_0 = L'_0 = L''_0 = d_0
\end{aligned}
$$

As mentioned already, one should not attempt to make forecasts for values F_{t+m} for values of m significantly greater than 1, as the error increases rapidly to unacceptable levels.

In Supply-Chain Management and many other real-world domains, time-series exhibiting quadratic trends are rather rare and therefore forecasting a time-series using Triple Exponential Smoothing for Quadratic Trends should never be used without extreme care, as it is likely to eventually give forecasts that are highly over-estimating the time-series. In the graph of Fig. 2.11 we show the effect of TES on a high-frequency oscillating time-series with a slow quadratic trend.

It is easy to see the "dampening" effect that the repeated applications of Exponential Smoothing have on the forecasts, but it is not easy to judge by the graphs of Fig. 2.10 alone the accuracy of the methods, mainly due to the high contribution of the random chance in the time-series. Some forecast accuracy scores for each method are shown in the table below

Method	MD_{24}	$MAPD_{24}$(%)	$RMSE_{24}$	S_{24}
SES(a = 0.1)	6.685	12.2	19.129	9.2
DES(a = 0.1)	8.171	13.3	22.055	10.03
SES(a = 0.7)	0.805	9.6	16.691	1.4
DES(a = 0.7)	1.580	9.5	16.388	2.8

From this table, it is very easy to see that even though the $MAPD_t$ score is reasonably good (below 15% in all cases), systemic errors are present in the forecasts when $a = 0.1$, as evidenced by the unacceptably very large values of the Tracking Signal S_t. This bias disappears when $a = 0.7$, allowing models to follow the data more closely.

2.1.8 *Double Exponential Smoothing with Linear Trends Method*

In the same fashion that was used to extend the Moving Average method to handle linear trends in the data, the Double Exponential Smoothing with Linear Trends method extends the DES method to take into account any linear trends in the data. The method estimates the two sequences that together implement the DES method, but provides a final forecast value that is a weighted average of the last values of the two sequences, with both positive and negative weights.

$$\begin{aligned}
F_{t+1} &= 2F'_{t+1} - F''_{t+1} + \frac{a}{1-a}\left(F'_{t+1} - F''_{t+1}\right) \\
F''_{t+1} &= aF'_{t+1} + (1-a)F''_t \\
F'_{t+1} &= ad_t + (1-a)F'_t \\
F''_0 &= F'_0 = d_1,\ a \in (0,1)
\end{aligned}$$

If required to predict further data points in time, the method would obviously use the only available option $F_{t+m} = F_{t+1}$ for any $m > 1$. The method's results when forecasting the next period are graphically shown in Fig. 2.12 for different values of a, corresponding to high—or low—dampening effects. While for $a = 0.1$ the method consistently underestimates the actual signal, for a value equal to 0.7 the method responds more quickly to changes but as the random component causes high fluctuations in the data, the method consistently overshoots or undershoots the signal. Indeed, the error measures indicate that for $a = 0.7$, the MD_{24} value is –1.779, $MAPD_{24} = 12.5\%$ and $RMSE_{24} = 20.91$. On the other hand, for $a = 0.1$, the method follows the data with $MD_{24} = 5.034$, $MAPD_{24} = 11\%$, and $RMSE_{24} = 17.35$. For $a = 0.25$, the error measures are even better, exhibiting $MD_{24} = 0.63$, $MAPD_{24} = 9.3\%$, $RMSE_{24} = 16.8$, and a tracking signal value $S_{24} = 1.17$, well within the limits of control for the forecasting process.

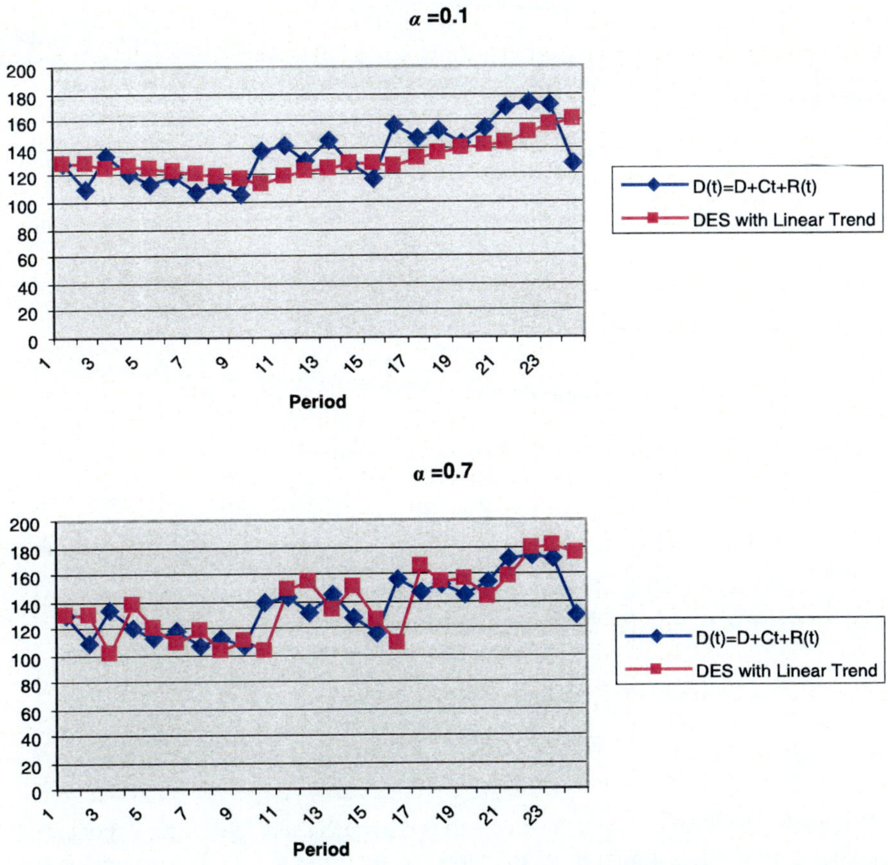

Fig. 2.12 Forecasting using the double exponential smoothing with linear trend method with $a = 0.1$ and 0.7

2.1.9 The Holt Method

The DES with Linear Trends method that was discussed above is an example of an adaptation of the general idea of exponential smoothing to explicitly account for linear trends in the time-series. A better method is known as Holt's method for Forecasting; the underlying model assumes data that are generated from a stochastic process of the form $D_t = D + Ct + R_t$ so they exhibit a linear relationship with time, and attempts to estimate both the level of the data-series at each time as well as the slope of the series at that point, using a feedback loop in the spirit of exponential smoothing.

The forecast is then computed as the sum of the estimated level of the time-series at the current time L_t, plus the estimated slope b_t of the series at the current point. Estimates of the time-series at further points in time are computed as $F_{t+m} = L_t + mb_t$

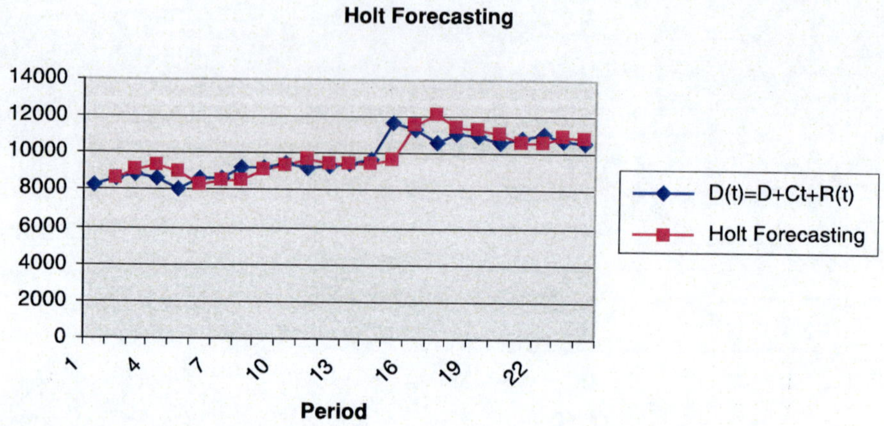

Fig. 2.13 Forecasting using the holt method with $a = 0.61$ and $\beta = 0.5$. the values for the parameters were chosen via grid search to optimize the MAPD criterion

The Holt forecasting method is given by the following set of recursive equations:

$$
\begin{aligned}
L_t &= ad_t + (1-a)F_{t-1} \\
b_t &= \beta(L_t - L_{t-1}) + (1-\beta)b_{t-1} \\
F_{t+1} &= L_t + b_t \\
L_1 &= d_1,\ b_1 = d_2 - d_1 \\
a,\ \beta &\in (0,1)
\end{aligned}
$$

Estimating the level at which the time-series is done via standard single exponential smoothing, as is the estimation of the slope of the time-series. The values of the parameters a and β are usually chosen after some kind of search so as to optimize a specific criterion, which is more often than not the MSE. In Fig. 2.13 we show the results of the application of Holt's method in the time-series used in Fig. 2.12. The application of the method gives an optimal MAPD_{24} score of 11.1% which is considered good, but is nevertheless inferior to that obtained by SES or DES with linear trends methods with optimized parameters. The RMSE_{24} score for the Holt method has a value 20.61. The reason is that the particular time-series, even though generated from a stochastic process that has a linear trend, also includes a very high-noise component (the random component), which prevents the method from operating optimally.

The power of Holt's method is more evident when forecasting trended data that do not suffer from *significant* inherent random fluctuation. If we apply Holt's method to the (real-world) time-series used in Fig. 2.7, setting $a = 0.7$ and $\beta = 0.2$ (found via grid search) the results are much better: the MAPD_{24} score becomes 4.02% which indicates very good accuracy, and the RMSE_{24} score is 595.3. This score is almost exclusively due to the presence of three single large forecast errors for periods 5, 15, and 17. The errors for periods 15 and 17 are in turn caused by a large spike of the time-series value at period 15 which however was then returned to normal, while Holt's method expected the time-series to lower their values more slowly.

The equation to estimate the slope of the data-series does not use the actual last two values of the time-series but rather the estimates of the level for the last two periods. An obvious question could be, why not use the actual values and set

$$b_t = \beta(d_t - d_{t-1}) + (1 - \beta)b_{t-1}$$

The answer is that the actual values have in them the random component which is essentially "factored out" when applying the Holt method. If one used the actual time-series values to forecast the slope at each point, setting $a = 0.5$ and $\beta = 0.55$, the ME_{24} score would be 1.82, the $MAPD_{24}$ score would be 11.2%, and the $RMSE_{24}$ metric would be 20.4. These values are only marginally worse than the Holt method.

2.1.10 The Holt–Winters Method

In Supply-Chain Management, the time-series to forecast often exhibit *cyclical fluctuations*. For example, demand for air-conditioning units typically increases significantly during summer months and decreases as the fall and then winter sets in. Such repeated cycles in demand that are directly related to the seasons of the Earth are known as the seasonal component of a time-series. A time-series can of course exhibit cyclic variations of much lower frequency, known as business cycles, but such low-frequency oscillations are useful in long-term predictions with horizons spanning several years. In any case, the models for time-series forecasting that were discussed so far are inadequate when applied to data that have inherent periodic oscillations.

The Holt–Winters' method is a direct extension of Holt's method for trended data that models the seasonal variation in a time-series as either a multiplicative component, or an additive component. *Both models assume the length of the "season" is known* and provided as input parameter s. In the *multiplicative seasonality model*, there are four equations used to obtain the forecast and they are as follows:

$$\begin{aligned}
L_t &= a\frac{d_t}{S_{t-s}} + (1 - a)(L_{t-1} + b_{t-1}) \\
b_t &= \beta(L_t - L_{t-1}) + (1 - \beta)b_{t-1} \\
S_t &= \gamma\frac{d_t}{L_t} + (1 - \gamma)S_{t-s} \\
F_{t+m} &= (L_t + b_t m)S_{t-s+m}, \quad t \geq s, m \geq 1 \\
&a, \beta, \gamma \in (0, 1)
\end{aligned}$$

$$b_{1 \leq i \leq s} = d_{i+1} - d_i,\ L_{1 \leq i \leq s} = d_i,\ S_{1 \leq i \leq s} = d_i s \Big/ \sum_{j=1}^{s} d_j$$

The first equation computes a smoothed estimate of the level of the time-series at time t, in the same way as Holt's method does except that the time-series value d_t at time t is "seasonally adjusted" by dividing it with the (multiplicative) seasonal index S_{t-s}. Therefore, the series L_t estimates the "seasonally adjusted" level of the original time-series, taking into account the trend b_t in the data, which is estimated exactly as in Holt's method by the second equation. The third equation computes the estimated seasonality index S_t of the time-series, again in the spirit of exponential smoothing methods. A rough estimate of the (multiplicative) seasonality index would be of course the value d_t/L_t so the third equation smoothes these estimates taking into account the season length. Finally, the forecast value for next period is the Holt-based estimate of the time-series multiplied by the most recent estimate available for the seasonality index of that period (S_{t-s+m}). The initialization of the procedure requires an estimate for the seasonality index for each period of the first seasonal cycle. This estimate is usually computed as simply the demand of the corresponding period divided by the mean demand throughout the first season cycle. The initial level and slope estimates are by default initialized as in the Holt method.

A special-case of the Holt–Winters' method that can be used to *forecast seasonal data that exhibit* ***no*** *trend* is easily derived by simply removing the second equation from the Holt–Winters' model and the variables b_t from the model, to obtain:

$$
\begin{aligned}
&L_t = a\frac{d_t}{S_{t-s}} + (1-a)L_{t-1}\\
&S_t = \gamma\frac{d_t}{L_t} + (1-\gamma)S_{t-s}\\
&F_{t+m} = L_t S_{t-s+m}, \quad t \geq s, \ m \geq 1\\
&a, \gamma \in (0,1),\\
&L_{1\leq i\leq s} = d_i, S_{1\leq i\leq s} = d_i s \Big/ \sum_{j=1}^{s} d_j
\end{aligned}
$$

To test the multiplicative Holt–Winters method, we first apply the method to a time-series generated by a stochastic process described by the equation $D_t = (D + Ct)(2 + \sin(\omega t))(1 - R_t)$ where $R_t \sim N(0,\sigma^2)$ is a random normal variable with zero mean. With $\omega = \pi/2$, the season length equals 4. The results of applying the Holt–Winters' method to this time-series are shown in Fig. 2.14.

For this synthetic time-series example, the Holt–Winters method achieves a Mean Error $MD_{36} = -1.4995$, a $MAPD_{36}$ score of 7.82%, and an $RMSE_{36} = 12.56$. Notice the particularly good fit of the forecast after period 21, also witnessed in the reduction of the Tracking Signal S_t. These numbers are much better than Holt's method for trended data could achieve.

When the time-series oscillates in more than one frequency, the Holt–Winters method gives results that are, as expected, less favorable. Consider for example another synthetic time-series generated by an equation of the form $D_t = (D + Ct)(2 + c_1 \sin(\omega_1 t) + c_2 \sin(\omega_2 t))(1 - R_t)$ where R_t is, as before, small random noise. This time-series clearly oscillates in more than one frequency.

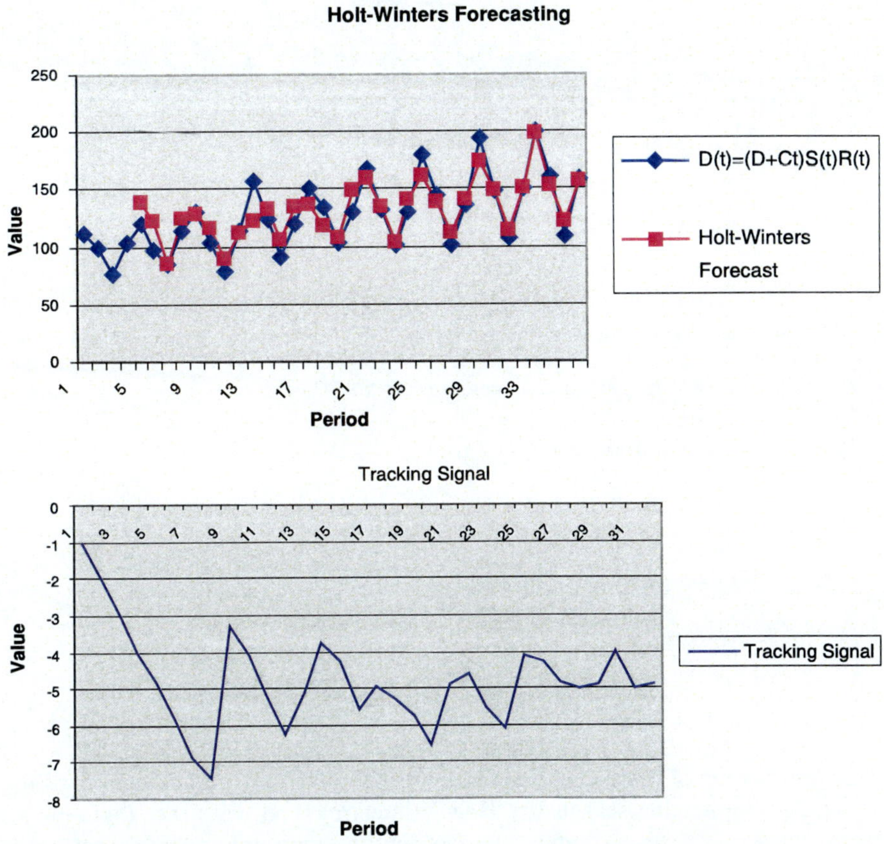

Fig. 2.14 Forecasting using the multiplicative Holt–Winters method with $s = 4$, $a = 0.79$, $\beta = 0.25$, $\gamma = 0.999$. The values for the parameters were chosen via grid search to optimize the MAPD criterion while maintaining a tracking signal within a reasonable range. The tracking signal is also plotted in the second plot

Setting $\omega_1 = \pi/2$, $\omega_2 = 2\pi/3$, and a total season length equal to 6, and carrying out a grid-search on the parameters α, β, γ to minimize the MAPD score, the parameter values are set to $a = 0.13$, $\beta = 0.3$, and $\gamma = 0.1$.

The best MAPD_{50} value attained is 17.2% which is not useless, but is less than ideal. In order to compare the Holt–Winters' method with the Holt method, we apply Holt's method to this time-series; the best MAPD_{50} score found by grid search when setting the parameters $a = 0.4$ and $\beta = 0.45$, is approximately 29.7% which is on the border of being useless for practical purposes, and far worse than the forecast provided by the Holt–Winters' method that directly takes into account periodic fluctuations in the data. Figure 2.15 illustrates how the method follows the data.

The ***additive seasonality Holt–Winters method*** assumes that the time-series fluctuates according to a model of the form $D_t = D + Ct + S_t + R_t$

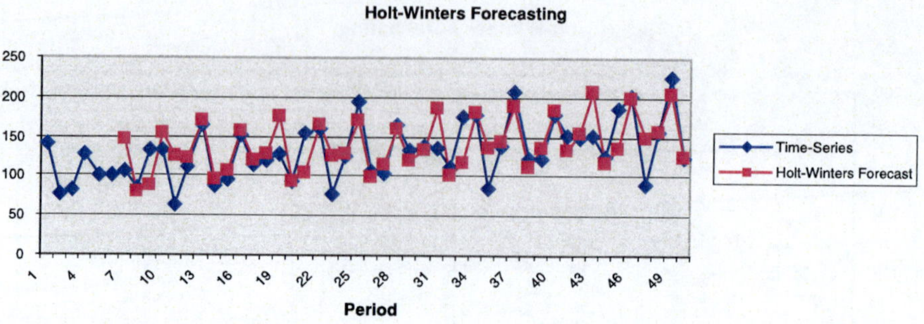

Fig. 2.15 Forecasting an oscillating time-series with more than one frequency using the multiplicative Holt–Winters method with $s = 6$, $a = 0.13$, $\beta = 0.3$, $\gamma = 0.1$. The values for the parameters were chosen via grid search to optimize the MAPD criterion

The forecasting equations are as follows:

$$
\begin{aligned}
&L_t = a(d_t - S_{t-s}) + (1-a)(L_{t-1} + b_{t-1}), \quad t > s \\
&b_t = \beta(L_t - L_{t-1}) + (1-\beta)b_{t-1}, \quad t > 1 \\
&S_t = \gamma(d_t - L_t) + (1-\gamma)S_{t-s}, \quad t > s \\
&F_{t+m} = L_t + mb_t + S_{t-s+m}, \quad m \geq 1, t \geq s \\
&L_{1 \leq i \leq s} = d_i, b_1 = d_2 - d_1, S_{1 \leq i \leq s} = sd_i \Big/ \sum_{j=1}^{s} d_j \\
&a, \beta, \gamma \in (0, 1)
\end{aligned}
$$

The justification of these equations follows the same rationale that was used to justify the multiplicative seasonality Holt–Winters method, with the single modifications to account for and reflect the fact that the seasonality component does not multiply the sum of level and trend component of the time-series but is rather added to it. The additive seasonality Holt–Winters method is not used very often in practice for the simple reason that the time-series that arise in real-world Supply-Chain Management problems are best modeled via a multiplicative seasonality component.

2.2 Time-Series Decomposition

Besides smoothing methods, a widely used methodology for time-series analysis and prediction is based on a decomposition of the time-series into its constituent parts, namely, (*T*) trend, (*S*) seasonality, (*C*) cycles, and (*R*) random variations. As mentioned in Sect. 2.1, the two major models for time-series analysis assume that the time-series is either the sum of the 4 parts (additive model), or alternatively, the product of the four parts (multiplicative). In the additive model, expressed by the equation

$$d_t = T_t + S_t + C_t + R_t$$

all constituent time-series are expressed in the same measurement unit. For example, if the time-series represents the mass of bananas imported to UK in a month in metric tons, each of the series *T, S, C,* and *R* express values measured in metric tons. On the other hand, in the multiplicative model, expressed by the equation

$$d_t = T_t\, C_t\, S_t\, R_t$$

the trend component T_i is the only time-series that is expressed in the same measurement unit as the original time-series d_t. The components *S, C,* and *R* are pure indices, i.e. simple numbers that do not express any quantity in some measurement unit. In the multiplicative model of time-series decomposition therefore, the influence of the *S, C,* and *R* components is measured as percentages and not as absolute numbers.

The idea behind time-series decomposition is to estimate and forecast as accurately as possible each of the contributing components of a time-series and then obtain a final forecast of the time-series by composing the components together again (by adding them together in an additive model, or multiplying them in case of the multiplicative model). Before discussing the technique in detail, it is important to make two observations:

(1) the distinction between trend and cyclic components in the time-series is somewhat artificial, and for this reason, most decomposition techniques do not separate the two, but instead directly estimate a single trend-cycle component.
(2) the time-series decomposition method, while intuitively appealing, has significant theoretical weaknesses. Despite this fact, practitioners have applied the method with significant success in a wide range of business-oriented forecasting problems, and this fact alone more than makes up for its theoretical issues.

2.2.1 Additive Model for Time-Series Decomposition

In the additive model for time-series decomposition, the trend-cycle component is first estimated. The most widely-used method for trend-cycle estimation in this model is the *centered* Moving Average method with parameter *k*

$$A_t = \begin{cases} \dfrac{\sum_{i=-\lfloor k/2 \rfloor}^{\lfloor k/2 \rfloor} d_{t+i}}{k} & t \geq \lfloor k/2 \rfloor, k \text{ odd} \\ \dfrac{d_{t-k/2}/2 + \sum_{i=-k/2+1}^{k/2-1} d_{t+i} + d_{t+k/2}/2}{k} & t \geq k/2, k \text{ even} \end{cases}$$

Note that, in contrast to Sect. 2.1.3, the value corresponding at period *t* is obtained by considering *k/2* past values and *k/2* future values of the time-series d_t.

At this point, the usual method is to consider a linear regression of the series A_t (to be discussed in detail in the next section) to compute the line which is the best mean square error estimator of the time-series. This line $T_t = a + bt$ forms the trend, and the difference $C_t = A_t - T_t$ would form the business cycle component. The slope b of the line and the intercept a are given by the formulae

$$b = \frac{N' \sum_{t=1}^{N'} tA_t - \sum_{t=1}^{N} t \sum_{t=1}^{N'} A_t}{N' \sum_{t=1}^{N} t^2 - \left(\sum_{t=1}^{N'} t\right)^2}, a = \frac{\sum_{t=1}^{N'} A_t - b\sum_{t=1}^{N'} t}{N'}$$

where

$$N' = N - \left\lfloor \frac{k}{2} \right\rfloor.$$

and N is the total length of the time-series where its values are known.

Next, the estimate of the seasonal components S_t are easily computed from the de-trended time-series $SR_t = d_t - A_t$ which must represent the seasonal component plus the random variations. Assuming that the seasonal length s is known, the seasonal components are estimated as the average of all the *homologous* values of the de-trended series SR_t at a point $t = 1,\ldots,s$ in the period. The formula for computing the seasonality component is therefore as follows:

$$S'_t = \frac{\sum_{i=0}^{\lfloor N/s \rfloor} SR_{t\%s+is+1}}{\lfloor \frac{N}{s} \rfloor + 1}, \quad t = 1,\ldots$$

where $t\%s$ denotes the remainder of the *integer* division of t by s. Because the seasonality indices are assumed to complete a cycle in exactly s periods, their sum over this cycle must equal zero. However, the random white noise will almost always prevent this sum from equaling zero, and for this reason, the indices are further adjusted to sum to zero using the following formula to provide the final estimates S_t:

$$S_t = S'_t - \frac{\sum_{i=1}^{s} S'_i}{s}$$

Forecasting the time-series then reduces to adding back together the future estimates for the trend component T_t and the seasonal component S_t at a future time t:

$$F_t = T_t + C_{N'} + S_t, \quad t \geq N$$

Note that the estimate does not include the random variation which of course cannot be predicted, as it is assumed to be essentially white noise. However, also note that the cyclic component C_t is assumed to be constant. This is because it is very hard to predict the fluctuations of business cycles; however, by their nature business cycles oscillate in very low frequencies and therefore, in the short range, the estimate $C_t \simeq C_{N'}$ is valid for values of t close to N.

2.2.2 Multiplicative Model for Time-Series Decomposition

In an analogous fashion to the additive model for time-series decomposition, in the multiplicative model, it is assumed that $d_t = T_t\, C_t\, S_t\, R_t$. The procedure is almost identical to the procedure used for the additive model, with the only difference being divisions being made instead of subtractions, and thus the name "*ratio-to-moving-averages*" often used for this method.

The estimate for the trend-cycle component is identical to the additive model (Sect. 2.2.1). The formula

$$A_t = \begin{cases} \dfrac{\sum_{i=-\lfloor k/2\rfloor}^{\lfloor k/2\rfloor} d_{t+i}}{k}, & t \geq \lfloor k/2\rfloor, k \text{ odd} \\ \dfrac{d_{t-k/2}/2 + \sum_{i=-k/2+1}^{k/2-1} d_{t+i} + d_{t+k/2}/2}{k} & t \geq k/2, k \text{ even} \end{cases}$$

provides the estimate of the trend-cycle component parameterized by the value k, and a linear regression provides the optimal estimate for the trend as $T_t = a + bt$, with the parameters a,b taking on the values

$$b = \frac{N' \sum_{t=1}^{N'} tA_t - \sum_{t=1}^{N'} t \sum_{t=1}^{N'} A_t}{N' \sum_{t=1}^{N'} t^2 - \left(\sum_{t=1}^{N'} t\right)^2}, \quad a = \frac{\sum_{t=1}^{N'} A_t - b \sum_{t=1}^{N'} t}{N'},$$

where $N' = N - \left\lfloor \dfrac{k}{2} \right\rfloor$

An estimate of the cyclic component is then $C_t = A_t/T_t$ for values of t up to N'.

The de-trended time-series $SR_t = d_t/A_t$ now represents the *ratio of actual to moving-averages*, and expresses the seasonal components multiplied by the random variations R_t. Assuming the season length s is known, the initial seasonal indices are computed in the same way as for the additive model, according to the formula:

$$S'_t = \frac{\sum_{i=0}^{\lfloor N/s \rfloor} SR_{t\%s+is+1}}{\lfloor \frac{N}{s} \rfloor + 1}, \quad t = 1, \ldots$$

However, since in the multiplicative model, the seasonal indices are interpreted as percentages rather than absolute numbers, their sum should equal 1 (100%). Therefore the final seasonal indices are adjusted according to:

$$S_t = S'_t \Big/ \sum_{i=1}^{s} S'_i$$

The final forecast for the original time-series is then given as

$$F_t = T_t C_{N'} S_t, \quad t > N$$

The rationale behind setting the cyclic component's value to the computed value for time N' is the same as in the previous section.

As a final note on time-series decomposition, sometimes it may be beneficial to replace the global homologous de-trended series values averaging by a moving average, i.e. to compute the seasonal components S_t according to an equation of the form

$$S_t = \frac{\sum_{i=\lfloor N/s \rfloor - m}^{\lfloor N/s \rfloor} SR_{t\%s+is+1}}{m+1}, \quad t = 1, \ldots$$

where m is a small integer representing the number of previous seasons over which the homologous time-periods will be averaged. This can be advantageous in cases where the seasonal component is not stable but changes significantly within the length of the time-series [1,…,N].

2.2.3 Case Study

The following table is the monthly demand for electricity throughout Greece between 2002 and 2004 in MWh. We shall use this real-world time-series to draw some conclusions about the demand for energy in Greece in this time-period.

Year	Month	Extracte energy (MW h)
2002	1	3876335.96
I	2	3288467.95
	3	3485829.77
	4	3350508.41
II	5	3282427.63
	6	3737809.103
	7	4152938.09
III	8	3764802.17
	9	3336105.56
	10	3448071.245
IV	11	3218141.1
	12	3487888.82
2003	1	3369915.94
I	2	3343150
	3	3652744.9
	4	3486522.03
II	5	3665734.19
	6	3802478.11
	7	4226681.92
III	8	4038016.82
	9	3610959.24

(continued)

(continued)

Year	Month	Extracte energy (MW h)
	10	3538465.09
IV	11	3440257.73
	12	3875609.42
2004	1	4028306.21
I	2	3656400.53
	3	3640526.75
	4	3432973.48
II	5	3363492.61
	6	3713392.58
	7	4428404.18
III	8	4165200.75
	9	3813872.16
	10	3699573.23
IV	11	3590192.74
	12	

The data in the above table are plotted in Fig. 2.16 .

The data as given are in monthly granularity. Since it is well established that electrical energy demand has a strong seasonal component that is directly related to the Earth's seasons, we shall first aggregate the data in a time-decomposition fashion in quarters. The quarterly data are presented in the following table

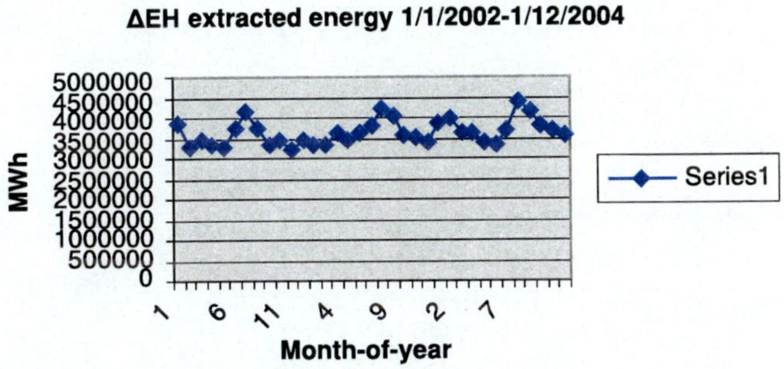

Fig. 2.16 Total electrical energy demand in Greece

Quarter	Quarterly energy
2002	
I	10650633.68
II	10370745.14
III	11253845.82
IV	10154101.17
2003	
I	10365810.84
II	10954734.33
III	11875657.98
IV	10854332.24
2004	
I	11325233.49
II	10509858.67
III	12407477.09
IV	

Next, in the aggregate, quarterly data time-series, we compute the Moving Average with period four, thus covering an entire year and all the seasons in the year, resulting in a series that has smoothed-out any seasonal variations. In the table below, we also show the centered-moving-average as the mean of two consecutive aggregate quarterly time-series values.

Quarter	Quarterly energy	Moving average of quarterly energy with M = 4	Centered moving average of 4 quarters
2002			
I	10650634		
II	10370745		
		10607331.45	
			10571728.6
III	11253846		
		10536125.74	
			10609124.39
IV	10154101		
		10682123.04	
2003			10759849.56
I	10365811		
		10837576.08	
			10925104.96
II	10954734		
		11012633.85	
			11132561.68
III	11875658		

(continued)

(continued)

Quarter	Quarterly energy	Moving average of quarterly energy with M = 4	Centered moving average of 4 quarters
		11252489.51	
			11196880.05
IV	10854332		
		11141270.6	
2004			11207747.98
I	11325233		
		11274225.37	
			5637112.686
II	10509859		
III	12407477		

The deviations of the actual quarterly demands from the centered 12-month Moving Average are shown in the table below.

Quarters	Centered moving average of 4 quarters	Deviations from moving average	Quarterly energy	Seasonally adjusted data
2002				
I			10650633.68	10885754.96
II			10370745.14	10255755.77
	10571729			
III		682117.2231	11253845.82	10638083.73
	10609124			
IV		–455023.2253	10154101.17	10649731.35
2003	10759850			
I		–394038.7188	10365810.84	10600932.12
	10925105			
II		29629.36688	10954734.33	10839744.96
	11132562			
III		743096.3013	11875657.98	11259895.89
	11196880			
IV		–342547.8125	10854332.24	11349962.43
2004	11207748			
I		117485.5063	11325233.49	11560354.77
	5637113			
II		4872745.984	10509858.67	10394869.3
III			12407477.09	11791715

The seasonal adjustments were made using the formulae in Sect. 2.2.1, and the operations are shown in tabulated form in the following table:

	Seasonal index estimation			
	I	II	III	IV
2002			682117.2	–455023
2003	–394039	29629.37	743096.3	–342548
2004	117485.5	394038.7		
Average	–138277	211834	712606.8	–398786
Seasonal indices adjusted	–235121	114989.4	615762.1	–495630

A plot of the seasonally adjusted quarterly electricity demand data is shown in Fig. 2.17.

The plot clearly shows an upward trend in demand during the period 2002–2004, with a spike in the third quarter of 2004 (the last point in the plot), which is partly due to the Olympic Games that took place at the time.

Finally, in Fig. 2.18, we plot the best Holt–Winters forecast on the monthly energy demand.

2.3 Regression

Consider a set of ordered pairs of observations $\{(t_1, d_1), (t_2, d_2), \ldots, (t_n, d_n)\}$, where $t_i < t_j$ whenever $i < j$. If the observations $d_i = d(t_i)$ are the result of a process $d(.)$ applied to the points t_i that is expected to be linear in its argument but there is the possibility of some noise to interfere with the measurement of the observations, one valid question is how to obtain the "best" line describing the data. If the noise interfering with the measurements is "white noise", i.e. follows a Gaussian distribution with zero mean, then the line that optimally describes the observations is the line that minimizes the ℓ_2 norm of the errors, or the square errors of the observation points from that line. The optimal line that best describes the observations is the line $y = ax + b$ where the real coefficients a and b minimize the error function $\varphi(a,b) = \|at + be-d\|^2$ where $t = [t_1\ t_2\ \ldots\ t_n]^T$, $d = [d_1\ d_2\ \ldots\ d_n]^T$ and $e = [1\ 1\ \ldots\ 1]^T$. The method is known as the *least squares method.* Using Fermat's theorem, taking the partial derivatives of the function $\varphi(a,b)$ with respect to both a and b and setting them to zero in order to locate the unique (and thus global) minimum of the function, we obtain the following

$$\frac{\partial \varphi}{\partial a} = \sum_{k=1}^{n} 2(at_k + b - d_k)t_k = 0$$

$$\frac{\partial \varphi}{\partial b} = \sum_{k=1}^{n} 2(at_k + b - d_k) = 0$$

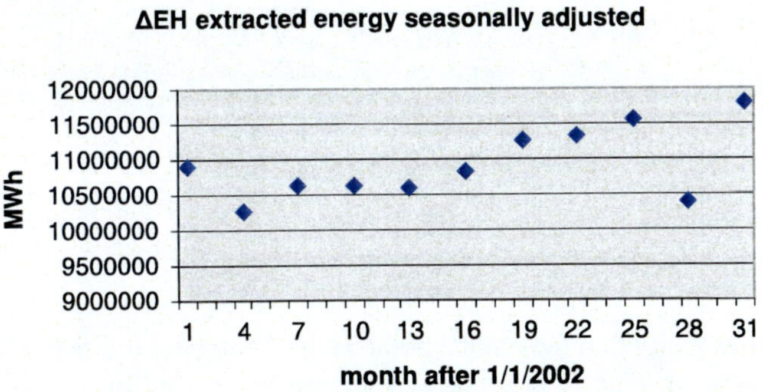

Fig. 2.17 Plot of quarterly demand for electrical energy in Greece, seasonally adjusted

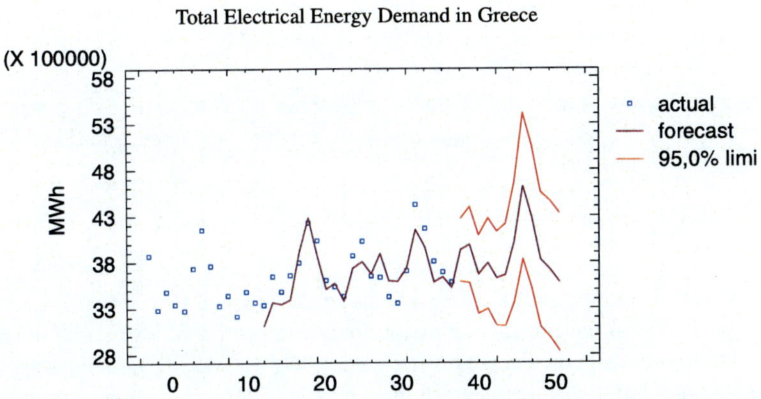

Fig. 2.18 Best Holt–Winters method forecast of monthly demand for electrical energy in Greece between January 2002 and December 2004. The figure is produced from the statistical software package StatGraphics

and rearranging terms, we obtain a system of two linear equations in the two unknowns a and b:

$$\left(\sum_{k=1}^{n} t_k^2\right) a + \left(\sum_{k=1}^{n} t_k\right) b = \sum_{k=1}^{n} t_k d_k$$
$$\left(\sum_{k=1}^{n} t_k\right) a + nb = \sum_{k=1}^{n} d_k$$

Solving this 2×2 linear system yields the values for the parameters of the best line interpolating the data in the ℓ_2 norm:

$$a = \frac{1}{D}\left(n\sum_{k=1}^{n} t_k d_k - \sum_{k=1}^{n} t_k \sum_{k=1}^{n} d_k\right)$$

$$b = \frac{1}{D}\left(\sum_{k=1}^{n} t_k^2 \sum_{k=1}^{n} d_k - \sum_{k=1}^{n} t_k \sum_{k=1}^{n} t_k d_k\right)$$

$$D = n\sum_{k=1}^{n} t_k^2 - \left(\sum_{k=1}^{n} t_k\right)^2$$

It is trivial to verify that the unique point where $\nabla\varphi(a, b) = 0$ is a minimum for the function $\varphi(\cdot,\cdot)$ and so it is left as an exercise for the reader.

The technique of finding the best line that fits a set of observations is known as "linear regression", and one possible use in forecasting should be immediately clear. Assume the observations form a time-series $\{(1, d_1), (2, d_2),\ldots, (n, d_n)\}$. If the data seem to agree to a large degree with the best line found by linear regression, or in other words if the minimum value of the function φ is small enough, then a reasonable estimate of the value of the time series at any time $n + m$ $m > 0$ is given as $F_{n+m} = a(n + m) + b$.

A somewhat more subtle use of linear regression for forecasting purposes is as follows. Define the *correlation coefficient* between the variables t_i and d_i as

$$r = \frac{\sum_{i=1}^{n}(t_i - \bar{t})(d_i - \bar{d})}{\sqrt{\sum_{i=1}^{n}(t_i - \bar{t})^2 \sum_{i=1}^{n}(d_i - \bar{d})^2}}, \quad \bar{t} = \frac{\sum_{i=1}^{n} t_i}{n}, \quad \bar{d} = \frac{\sum_{i=1}^{n} d_i}{n}.$$

The correlation coefficient takes on values in the interval $[-1, 1]$ and is 1 when there is a perfect positive linear relationship between the variables t and d meaning the two are increasing or decreasing together, is -1 when there is a perfect negative linear relationship between them (i.e. whenever t increases, d decreases) and is zero if statistically there is no relationship between the two variables. Assume that

(1) the correlation coefficient r between t and d is close to 1.
(2) the value of the variables t_{n+i}, $i > 0$ can be accurately predicted.

Then, the value of the quantity d_{n+i} for $i > 0$ can be predicted according to the formula $F_{n+i} = at_{n+i} + b$.

Even when the correlation coefficient between two variables is high, it is not necessarily true that the two variables are truly related. Spurious correlations arise when the correlation coefficient is high, but is not statistically significant—but instead, presumably it may be high because of sampling errors. To double check the validity of the hypothesis of the strong relationship between the two variables, which is essential in order to have any confidence on the forecasting results based on the regression analysis of the two variables, the following test statistic is often used:

$$t_{n-2} = \frac{r}{\sqrt{\frac{1-r^2}{n-2}}}$$

If the value $\|t_{n-2}\| > \|t_{n-2,a/2}\|$ then the hypothesis that the correlation coefficient is different from zero and is accepted at confidence level $1 - a$ (Halikias 2003). The value $\|t_{n-2,a/2}\|$ is simply the t-student criterion, and its value can be found in standard statistical tables for most values of interest for the parameter a; it is also available in every statistical software package or spreadsheet software as well.

2.3.1 Generalized Regression

The method of least squares is also applicable when one wishes to compute the linear combination of a set of given functions that optimally—in the sense of least squares—matches a set of observations, i.e. a set of points in $\mathbb{R}^2$ ordered in the first dimension. If we are given a set of *basis functions* $g_j\ (x)\ j = 0,\ldots,m$ then the following optimization problem provides the least-squares approximation of any linear combination of the basis functions to the data $\{(t_1, d_1), (t_2, d_2), \ldots, (t_n, d_n)\}$:

$$\min_{c_0,\ldots,c_m} \varphi(c_0, c_1, \ldots, c_m) = \sum_{k=1}^{n} \left[\sum_{j=0}^{m} c_j g_j(t_k) - d_k \right]^2$$

As before, applying Fermat's theorem to determine the minimizing point of the error function $\varphi(c)$, where c is the $m + 1$ dimensional column vector collecting the c_j parameters, we obtain

$$\frac{\partial \varphi}{\partial c_i} = \sum_{k=1}^{n} 2 \left[\sum_{j=0}^{m} c_j g_j(t_k) - d_k \right] g_i(t_k) = 0, \quad i = 0, \ldots, m$$

which can be re-written as a system $Ac = b$ of $m + 1$ linear equations in $m + 1$ unknown coefficients $c_0, \ldots, c_m$:

$$\sum_{j=0}^{m} \left[\sum_{k=1}^{n} g_i(t_k) g_j(t_k) \right] c_j = \sum_{k=1}^{n} d_k g_i(t_k), \quad i = 0, \ldots, m$$

Solving the above linear system yields the optimal approximation in a least squares sense of the data using the basis functions.

2.3.2 Non-linear Least Squares Regression

It is also possible to apply the idea of least squares to optimally fit observations to any model function with parameters $\beta = [\beta_1\ \beta_2\ \ldots\ \beta_k]^T$. Suppose we want to compute the parameters β so that the observations $\{(x_i, y_i)\ i = 0, \ldots, n\}$ fit the known function $g(x,\beta)$ as best as possible in a least squares sense. The objective is therefore to solve the following unconstrained optimization problem:

$$\min_{\beta} f(\beta) = \sum_{i=0}^{n} (y_i - g(x_i,\beta))^2$$

According to the First Order Necessary Conditions, the optimal β^* will satisfy the following conditions:

$$\nabla f(\beta^*) = 0 \Leftrightarrow \frac{\partial f}{\partial \beta_j}(\beta^*) = -\sum_{i=0}^{n} 2[y_i - g(x_i,\beta^*)]\frac{\partial g}{\partial \beta_j}(x_i,\beta^*) = 0, \quad j = 1\ldots k$$

The above is a system of k nonlinear equations in k unknowns and can be written as

$$\sum_{i=0}^{n} g(x_i,\beta^*)\nabla_{\beta} g(x_i,\beta^*) = \sum_{i=0}^{n} y_i \nabla_{\beta} g(x_i,\beta^*)$$

Obtaining a solution to the above system may be of the same complexity as the original least squares optimization problem; also, as we have seen in Chap. 1, a solution to the above problem is not guaranteed to be the global minimum to the original problem. In fact, depending on convexity properties of the function $g(x,\beta)$, the solution of the nonlinear system of equations may not even be a local minimum, and therefore checking the Second Order Sufficient Conditions is required to eliminate the possibility of having located a local maximum or a saddle point of the original objective function $f(\beta)$. As discussed in the first chapter, there is in general no guarantee that a local minimizer β^* of the objective function f is the actual global minimizer, and for this reason, usually an algorithm is chosen for nonlinear optimization and is repeatedly run with a number of starting points β_0 to increase confidence in the belief that the best minimizer found is actually the global optimum. The algorithms discussed in Sect. 1.1.1.3 are particularly useful in this context.

2.3.3 Exponential Model Regression

The method of least squares regression can also be applied when the data are generated from an *exponential* curve. Such data are sometimes found in economics and business-related time-series and especially when studying macroeconomic data. A highly successful model in such cases is the curve $y(x) = c_1 c_2^x$. To find the optimal fit of data on such a curve, one works with the equivalent equation $\ln y = \ln c_1 + x \ln c_2$ which is linear in the unknowns $c_1' = \ln c_1, c_2' = \ln c_2$, and therefore is amenable to standard linear regression. The optimal values for the parameters are given by the equations:

$$c_1 = \exp\left(\frac{1}{D}\left(\sum_{k=1}^{n} t_k^2 \sum_{k=1}^{n} \ln d_k - \sum_{k=1}^{n} t_k \sum_{k=1}^{n} t_k \ln d_k\right)\right)$$

$$c_2 = \exp\left(\frac{1}{D}\left(n \sum_{k=1}^{n} t_k \ln d_k - \sum_{k=1}^{n} t_k \sum_{k=1}^{n} \ln d_k\right)\right)$$

$$D = n \sum_{k=1}^{n} t_k^2 - \left(\sum_{k=1}^{n} t_k\right)^2$$

To illustrate the point, let us fit the Greek GDP in million (1980-adjusted) Euro values between the years 1948 and 1998 to the curve $y(x) = c_1 c_2^x$. The data (Halikias 2003) are tabulated in the following table

Year	GDP	Year	GDP	Year	GDP
1948	678	1965	2243	1982	5056
1949	803	1966	2380	1983	5093
1950	846	1967	2509	1984	5246
1951	920	1968	2675	1985	5424
1952	926	1969	2942	1986	5511
1953	1054	1970	3176	1987	5482
1954	1086	1971	3397	1988	5701
1955	1168	1972	3698	1989	5869
1956	1266	1973	3965	1990	5660
1957	1348	1974	3823	1991	5851
1958	1412	1975	4061	1992	5897
1959	1464	1976	4317	1993	5869
1960	1526	1977	4459	1994	6030
1961	1695	1978	4763	1995	6161
1962	1720	1979	4937	1996	6277
1963	1894	1980	5021	1997	6467
1964	2050	1981	5028	1998	6690

The plot in Fig. 2.19 shows the actual GDP data versus the best fit curve of the form $y(x) = c_1 c_2^x$.

Applying the same procedure to the data for the time-period 1948–1980, we obtain the optimal values for $c_1 = \exp(-116.1)$ and $c_2 = \exp(0.06) \approx 1.06503$ and the plot in Fig. 2.20 gives a visualization of the fit of the data to the exponential model for that period. The exponential model regression is a much better fit for the data of this period. This analysis shows clearly that Greek GDP time-series essentially "changed" its pattern after 1980.

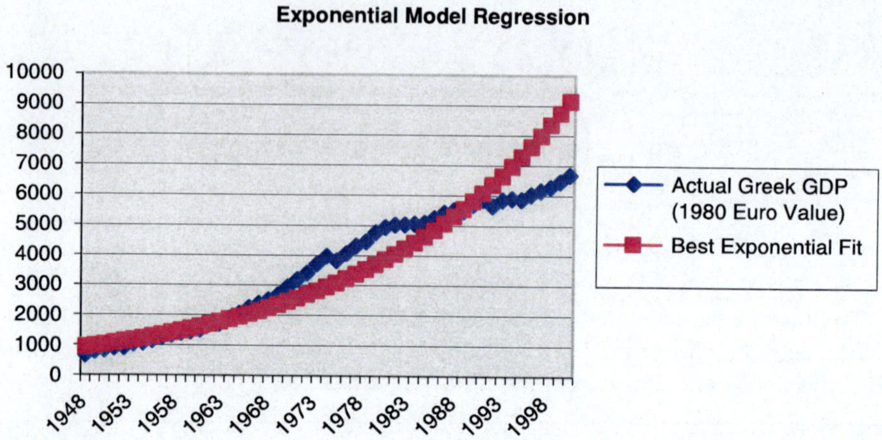

Fig. 2.19 The optimal exponential regression model to fit the Greek GDP data. Although it has a very high-correlation coefficient ($r = 0.965$) with time, it shows a clear deviation from the real data after 1968

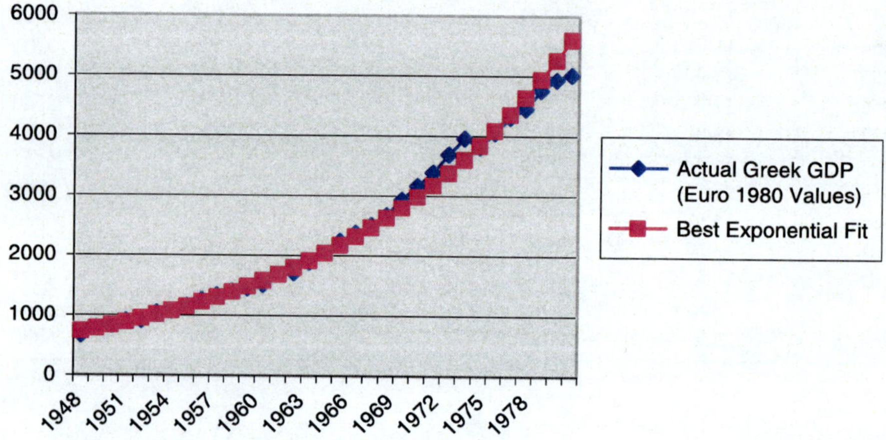

Fig. 2.20 The optimal exponential regression model to fit the Greek GDP data for the time-period 1948–1980

2.4 Auto-Regression-Based Forecasting

The regression method presented in Sect. 2.3 computes the line $y = at + b$ that minimizes the sum of the square differences of the data-points in a given time-series from the points of the line at each particular time. Exponential smoothing methods on the other hand use a number of previous data-points to forecast the immediately next values. Combining the two ideas with this ultimate goal of forecasting the time-series in mind, a new method can be derived as the answer to the question "how to optimally, in a least squares sense, combine the last p points in the time-series so as to predict the next value". In this Auto-Regressive (AR) model of a time-series therefore, the current value of the series d_n is expressed as a linear combination of a constant number of the values from the immediate past plus an error term (that is assumed to have zero mean and constant variance). The time-series in such a case is obviously wide-sense stationary. If the time-series is not stationary, i.e. its mean drifts in time, then an AR model cannot be expected to provide good forecasts; in such cases *integrating* the time-series by applying an appropriate differencing operator to produce a new time series $\Delta d_t = d_t - d_{t-1}$, perhaps multiple times, as in $\Delta^m d_t = \Delta(\Delta^{m-1})d_t$ may help. If the time-series $\Delta^m d_t$ for some value of $m > 0$ is stationary, the AR model can be applied on this integrated series, and then forecasts for the original time-series can be easily calculated.

To formalize the idea of AR modeling, let $d(n) = [d_n\, d_{n-1} \ldots d_{n-p}]^T$ denote the column vector of the last $p + 1$ observations of the time-series $d = \{d_0, d_1, \ldots, d_{N-1}\}$ at time n. We can always write $d_n = -(a_1 d_{n-1} + a_2 d_{n-2} + \cdots a_p d_{n-p}) + e_n$ where e_n is an error term that we hope to render almost always as small as possible via the optimal selection of the parameters a_i, $i = 1,\ldots,p$. Therefore, we would like to compute a vector of p coefficients $a = [a_1 \ldots a_p]^T$ so that the inner product $[1\ a^T]d(n)$ is minimized in a least squares sense. Therefore, the following *least squares error problem* must be solved:

$$\min_a \sum_{i=p}^{N-1} \left([1 \quad a^T]d(i)\right)^2 = \min_a [1 \quad a^T] \left[\sum_{i=p}^{N-1} d(i)d(i)^T\right] \begin{bmatrix} 1 \\ a \end{bmatrix}$$

The quantity $d(i)d(i)^T$ is a $(p + 1) \times (p + 1)$ matrix whose (r, c) element is the quantity $d_{i-r+1}d_{i-c+1}$ so that the matrix $R_{p+1} = \sum_{i=p}^{N-1} d(i)d(i)^T$ is also a $(p + 1) \times (p + 1)$ matrix and has at its (k, l) position the quantity $\sum_{i=p}^{N-1} d_{i-k+1}d_{i-l+1}$. If $p \ll N$ then the elements of the matrix along *any* diagonal (i.e. the elements whose indices have a constant difference $k - l =$ const) should be essentially the same. This is because the difference

$$\sum_{i=p}^{N-1} d_{i-k+1}d_{i-k+c+1} - \sum_{i=p}^{N-1} d_{i-k+2}d_{i-k+c+2} = d_{p-k+1}d_{p-k+c+1} - d_{N-k+1}d_{N-k+c+1}$$

will be much smaller than the value $\sum_{i=p}^{N-1} d_{i-k+1}d_{i-k+c+1}$ when $p \ll N$. Notice that the value $\sum_{i=p}^{N-1} d_{i-k+1}d_{i-l+1} = r_{|k-l|}$ and is independent of p or N when the time-series d_n is generated from a wide-sense stationary process.

Since the elements along the diagonals of the matrix R_{p+1} are the same, by definition the matrix is a Toeplitz matrix. From the arguments above, the matrix is also symmetric. Therefore denote the elements of the matrix R_{p+1} as follows:

$$R_{p+1} = \begin{bmatrix} r_0 & r_1 & \cdots & r_p \\ r_1 & r_0 & \cdots & r_{p-1} \\ \vdots & \cdots & \cdots & \vdots \\ r_p & r_{p-1} & \cdots & r_0 \end{bmatrix}$$

where $r_i = \sum_{j=0}^{N-i-1} d_j d_{j+i}$. The optimization problem becomes the unconstrained minimization of the function $f(a) = [1\ a^T]R_{p+1}[1\ a^T]^T$ which is a smooth function. So, we can rewrite the least-squares optimization problem as follows:

$$\min_{a=[a_1 \ldots a_p]^T} \begin{bmatrix} 1 & a^T \end{bmatrix} \begin{bmatrix} r_0 & \tilde{r}_p^T \\ \tilde{r}_p & R_p \end{bmatrix} \begin{bmatrix} 1 \\ a \end{bmatrix} = r_0 + 2a^T \tilde{r}_p + a^T R_p a$$

where $a = [a_1 \ldots a_p]$, $\tilde{r}_p = [r_1 \quad \ldots \quad r_p]^T$ and

$$R_p = \begin{bmatrix} r_0 & r_1 & \cdots & r_{p-1} \\ r_1 & r_0 & \cdots & r_{p-2} \\ \vdots & \vdots & \cdots & \vdots \\ r_{p-1} & r_{p-2} & \cdots & r_0 \end{bmatrix}$$

From Fermat's theorem, setting the gradient to zero, we obtain $\nabla f(a) = 2\tilde{r}_p + 2R_p a = 0$ or, equivalently,

$$R_p a = -\tilde{r}_p \tag{2.1}$$

The linear system (2.1) of p linear equations in p unknowns, namely the values of the coefficients $a_1, \ldots, a_p$, is known as the *Yule-Walker equations,* also known as the *Wiener–Hopf equations* in the theory of linear prediction. Solving this system yields the optimal in the least squares sense predictor of the time-series d_n using a linear combination of the p last values of the signal. The minimum total error is then $E_{\min} = r_0 + a^T \tilde{r}_p$. The optimal prediction in the least-squares sense for the new value d_N as a linear combination of the last p values is given as

$$F_N = -\left(a_1 d_{N-1} + a_2 d_{N-2} + \cdots + a_p d_{N-p}\right).$$

Even though it is possible to solve the system $R_p a = -\tilde{r}_p$ using any standard algorithm for linear systems of equations, the special structure of this particular linear system of equations allows its solution via a very elegant and fast *order-recursive algorithm*, known as the *Levinson-Durbin algorithm*. Time-recursive methods also exist in the form of extensions of the ideas to be presented next. Let $\tilde{a}_m$ be the solution of the mth order system $R_m \tilde{a}_m = -\tilde{r}_m$ for any m less than or equal to p. We will develop a recursive set of equations that compute the solution of the system $R_{m+1}\tilde{a}_{m+1} = -\tilde{r}_{m+1}$ given the solution $\tilde{a}_m$. First, consider the matrix

$$J_m = \begin{bmatrix} 0 & \dots & 0 & 1 \\ \vdots & \ddots & 1 & 0 \\ 0 & 1 & \ddots & \vdots \\ 1 & 0 & \dots & 0 \end{bmatrix}$$

of dimensions $m \times m$ that when it multiplies any $m \times 1$ vector, it *reverses the order of the vector's elements*. Notice that $J_m^T = J_m$, $J_m J_m = I_m$ and that $J_m R_m = R_m J_m$ where I_m is the unit $m \times m$ matrix. Further, let $L_m = [I_m|0]$. The matrix L_m has dimensions $m \times (m+1)$ and has the property that when it multiplies an $(m+1) \times 1$ vector it results in an $m \times 1$ vector that is identical to the vector that was multiplied except that the last element of the original vector is dropped. The matrix R_{m+1} can be written as

$$R_{m+1} = \begin{bmatrix} R_m & | & J_m \tilde{r}_m \\ - & | & - \\ \tilde{r}_m^T J_m & | & r_0 \end{bmatrix}$$

and the system $R_{m+1}\tilde{a}_{m+1} = -\tilde{r}_{m+1}$ can now be written in a decomposed form as

$$\begin{bmatrix} R_m & J_m \tilde{r}_m \\ \tilde{r}_m^T J_m & r_0 \end{bmatrix} \begin{bmatrix} L_m \tilde{a}_{m+1} \\ a_{m+1} \end{bmatrix} = - \begin{bmatrix} \tilde{r}_m \\ r_{m+1} \end{bmatrix}$$

where a_{m+1} is the last element of the vector $\tilde{a}_{m+1}$. The above system can now be written as

$$\begin{aligned} R_m L_m \tilde{a}_{m+1} + a_{m+1} J_m \tilde{r}_m &= -\tilde{r}_m \\ \tilde{r}_m^T J_m L_m \tilde{a}_{m+1} + r_0 a_{m+1} &= -r_{m+1} \end{aligned}$$

Multiplying the first equation above by J_m keeping in mind the properties of the matrix, that R_m is invertible, and the fact that $\tilde{r}_m = -R_m \tilde{a}_m$ we obtain

$$R_m(J_m L_m \tilde{a}_{m+1}) = R_m(J_m \tilde{a}_m + a_{m+1}\tilde{a}_m) \Leftrightarrow L_m \tilde{a}_{m+1} = \tilde{a}_m + a_{m+1} J_m \tilde{a}_m$$

The last equation expresses the first m components of the vector $\tilde{a}_{m+1}$ as a product of the vector $\tilde{a}_m$ and a coefficient that is linear in the value a_{m+1}. Substituting this quantity to the last equation in our system of equations we then obtain:

$$\tilde{r}_m^T(J_m \tilde{a}_m + a_{m+1}\tilde{a}_m) + r_0 a_{m+1} = -r_{m+1} \Leftrightarrow a_{m+1} = -\frac{r_{m+1} + \tilde{r}_m^T J_m \tilde{a}_m}{r_0 + \tilde{r}_m^T \tilde{a}_m}$$

At this point the $m+1$ dimensional column vector $\tilde{a}_{m+1}$ can be expressed as a linear combination of the $m \times 1$ vector $\tilde{a}_m$. The recursive equation that obtains the solution of the $(m+1)$st order system of equations from the mth order solution is given by:

$$\tilde{a}_{m+1} = \begin{bmatrix} \tilde{a}_m \\ 0 \end{bmatrix} + k_{m+1} \begin{bmatrix} J_m \tilde{a}_m \\ 1 \end{bmatrix}$$

$$k_{m+1} = -\frac{r_{m+1} + \tilde{r}_m^T J_m \tilde{a}_m}{r_0 + \tilde{r}_m^T \tilde{a}_m} = -\frac{b_{m+1}}{\alpha'_m}, \quad b_{m+1} = r_{m+1} + \tilde{r}_m^T J_m \tilde{a}_m, \quad \alpha'_m = r_0 + \tilde{r}_m^T \tilde{a}_m \tag{2.2}$$

It is not hard to verify that the quantities α'_m can be computed recursively as well as they obey the equation

$$\alpha'_m = \alpha'_{m-1} + k_m b_m = \alpha'_{m-1}(1 - k_m^2)$$

The above equations lead to a very fast order-recursive algorithm for the computation of the parameters of the optimal forward linear predictor of the time-series d_n. The following is a formal description of the Levinson-Durbin algorithm.

Algorithm Levinson-Durbin

Input: A time-series $d = \{d_0, \ldots d_{N-1}\}$ and the order p of the optimal least-squares linear predictor.
Output: Optimal Coefficients vector $a = [a_1 \ldots a_p]^T$ of dimension $p \times 1$ to be used in forward time-series prediction $F_N = -(a_1\, d_{N-1} + \cdots + a_p\, d_{N-p})$

Begin
/*Initialization*/

1. Set $r_0 = \sum_{i=0}^{N-1} d_i^2$, $\alpha_0 = r_0$, $b_1 = r_1 = \sum_{i=0}^{N-2} d_i d_{i+1}$, $k_1 = -r_1/r_0$, $a_1^{[1]} = k_1$, $m = 1$.
   /* Loop over the orders */
2. while $m < p$ do:
   a. Set $a^{[m]} = \left[a_1^{[m]} \ldots a_m^{[m]}\right]^T$.
   b. Set $r_{m+1} = \sum_{i=0}^{N-m} d_i d_{i+m+1}$, $\tilde{r}_m = [r_1 \ldots r_m]^T$
   c. Set $\alpha_m = \alpha_{m-1} + b_m k_m$.
   d. If $\alpha_m \le 0$ then ERROR ('R_{m+1} matrix is not symmetric Toeplitz').
   e. Set $b_{n+1} = a^{[m]T} J_m \tilde{r}_m + r_{m+1}$.
   f. Set $k_{m+1} = -b_{m+1}/\alpha_m$.
   g. Set $a^{[m+1]} = \begin{bmatrix} a^{[m]} \\ 0 \end{bmatrix} + k_{m+1} \begin{bmatrix} J_m a^{[m]} \\ 1 \end{bmatrix}$.
   h. Set m=m+1.
3. end-while
4. return $a^{[p]}$.

End

The algorithm runs in $O(p)$ time, as can be easily verified from the description of the algorithm's loop.

It is also possible to efficiently update the optimal auto-regression coefficients vector a as new points in the time-series become available. The update is to be done continuously, as soon as each new point of the time-series becomes known. The algorithm minimizes an exponential smoothing-based variant of the squares of forecasting errors. In particular, the algorithm minimizes the following error function

$$\min E_m(t) = \sum_{j=M}^{t} \lambda^{t-j} \left(\varepsilon_m^f(j)\right)^2$$

where

$$\begin{aligned}
\varepsilon_m^f(j) &= d_j + \tilde{a}_m(j)^T \tilde{d}_m(j-1) \\
\tilde{d}_m(j) &= [d_j d_{j-1}, \ldots, d_{j-m+1}]^T \\
\tilde{a}_m(j) &= [a_1(j), \ldots, a_m(j)]^T
\end{aligned}$$

and λ in (0,1] plays a role analogous to the α factor in SES. Following the same steps in the analysis of the equations resulting from the First Order Necessary Conditions for the optimization of the new objective function, we obtain the following linear system of "normal equations":

$$\begin{aligned}
R_m(t-1)\tilde{a}_m(t) &= -\tilde{r}_m(t) \\
\tilde{r}_m(t) &= \sum_{j=M}^{t} \lambda^{t-j} d_j \tilde{d}_m(j-1) \\
R_m(t) &= \sum_{j=M}^{t} \lambda^{t-j} \tilde{d}_m(j-1) \tilde{d}_m(j-1)^T
\end{aligned}$$

The optimum total error will then be $E_m^f(t) = r_{om}^f(t) + \tilde{a}_m(t)^T \tilde{r}_m(t)$ with the first term $r_{om}^f(t) = \sum_{j=M}^{t} \lambda^{t-j} d_j^2$. By decomposing the $R_m(t)$ matrix and the vectors $\tilde{d}_{m+1}(t) = \left[\tilde{d}_m(t)^T \;\; d_{t-m}\right]^T = \left[d_t \;\; \tilde{d}_m(t-1)^T\right]^T$ and $\tilde{r}_m(t), \tilde{a}_m(t)$ we finally obtain the following discrete time-recursions:

$$\begin{aligned}
\tilde{a}_m(t+1) &= \tilde{a}_m(t) + e_m^f(t+1)\tilde{w}_m^*(t) = \tilde{a}_m(t) + \varepsilon_m^f(t+1)\tilde{w}_m(t) \\
e_m^f(t+1) &= d_{t+1} + \tilde{a}_m(t)^T \tilde{d}_m(t) \\
\varepsilon_m^f(t+1) &= d_{t+1} + \tilde{a}_m(t+1)^T d_m(t) \\
R_m(t)\tilde{w}_m^*(t) &= -\tilde{d}_m(t) \\
\lambda R_m(t)\tilde{w}_m(t+1) &= -\tilde{d}_m(t+1)
\end{aligned}$$

The quantities $e_m^f(t)$ are known as the a priori errors whereas the quantities $\varepsilon_m^f(t)$ are known as the a posteriori errors of the prediction process. The vectors $\tilde{w}_m^*(t)$ are known as the *Kalman gain vectors*. Now, the Sherman-Morrison equality from linear algebra states that *if* x is a vector of n components, the matrix xx^T is of rank 1 and the matrix R is invertible, *then* for every non-zero λ it holds that

$$(\lambda R + xx^T)^{-1} = \frac{1}{\lambda}R^{-1} - \frac{\lambda^{-2}R^{-1}xx^TR^{-1}}{1+\lambda^{-1}x^TR^{-1}x}$$

Using this fact, we obtain a formula to compute recursively the inverse of the matrix $R_m(t)$ as follows:

$$P_m(t+1) = R_m(t+1)^{-1} = \lambda^{-1}R_m(t)^{-1} - \tilde{w}_m^*(t+1)\tilde{w}_m(t+1)^T$$

From this, the classical time-recursive optimal linear prediction algorithm follows.

Algorithm Time-Recursive Optimal Linear Predictor

Input: A time-series $d = \{d_0, \ldots d_t\}$, the order p of the optimal least-squares linear predictor, the smoothing factor λ, the quantities $P_p(t)$, $\tilde{w}_p^*(t)$, $\tilde{a}_p(t)$ from the previous iteration, and the latest time-series point d_{t+1} .
Output: Updated Optimal Coefficients vector $\tilde{a}_p(t+1)$ of dimension $p \times 1$ to be used in forward time-series prediction $F_{t+2} = -(a_1(t+1)\,d_{t+1} + \cdots + a_p(t+1)\,d_{t+2-p})$

Begin

0. If t=0 then Set $P_p(0) = \sigma^{-2}I$, $0<\sigma<<1$ /* initialization */
1. Set $\tilde{w}_p(t+1) = \lambda^{-1}P_p(t)\tilde{d}_p(t+1)$
2. Set $\alpha_p(t+1) = 1 - \tilde{d}_p(t+1)^T\tilde{w}_p(t+1)$
3. Set $\tilde{w}_p^*(t+1) = \frac{1}{\alpha_p(t+1)}\tilde{w}_p(t+1)$
4. Set $P_p(t+1) = \lambda^{-1}P_p(t) - \tilde{w}_p^*(t+1)\tilde{w}_p(t+1)^T$
5. Set $e_p^f(t+1) = d_{t+1} + \tilde{a}_p(t)^T\tilde{d}_p(t)$
6. Set $\tilde{a}_p(t+1) = \tilde{a}_p(t) + e_p^f(t+1)\tilde{w}_p^*(t)$
7. return $\tilde{a}_p(t+1)$.

End

This algorithm runs in $O(p^2)$ complexity since Step 4 is of this complexity. A faster time-recursive algorithm that runs in $O(p)$ iterations was first developed in 1978, and is known as the *Fast Kalman Algorithm*. Details of this algorithmic scheme can be found in (Karagiannis 1988).

Auto-regressive models are special cases of the more general case where a time-series d_n is well approximated by a model of the form

$$d_n = -\sum_{k=1}^{p} a_k d_{n-k} + \sum_{k=0}^{q} b_k u_{n-k}. \tag{2.3}$$

where the input driving sequence u_n is white noise process that is inherent in the model (and cannot be attributed to some "observation error"). This model is known as an *Auto-regressive Moving Average model* (ARMA (p, q) process) with parameters p and q. Clearly, the AR model of order p is a special case of the ARMA (p, 0) model with $q = 0$ and $b_0 = 1$. As with AR models, an ARMA model can be applied only to wide-sense stationary processes generating the sequence dn so if the time-series is not wide-sense stationary, the differencing operators should be applied to the time-series until the resulting time-series appears to be stationary in the wide sense. If application of the differencing operator is necessary, then the model is called an ARIMA (p, d, q) model, where the parameter d refers to the number of times the difference operator had to be applied to turn the time-series into a wide-sense stationary time-series. In time-series where noise plays a significant role in the signal values, application of an ARMA model for prediction may yield much superior results than AR-based models, but the downside of the application of an ARMA model is the fact that it requires the optimization of a highly nonlinear objective function; this in turn implies that standard gradient-descent-based methods of nonlinear optimization can only guarantee convergence of the model parameters a_k and b_k to a saddle point (or at most a local minimum, see Chap. 1).

Another special case worth noting concerns the development of a forecasting algorithm assuming an MA (Moving Average) process of order q. Assuming that the time-series is generated from a process of the form

$$d_n = \sum_{k=0}^{q} b_k u_{n-k}$$

where u_n is white noise, it is also possible to model the time-series as an infinite-order Auto-Regressive-based time-series according to the model AR(∞):

$$d_n = \sum_{k=1}^{\infty} a_k d_{n-k} + u_n$$

(assuming the time series exists or can be extended to negative infinite time). Durbin (1960) showed that by considering a sufficiently large order, an AR(p) model optimized via the Levinson-Durbin algorithm can approximate very well a time-series arising from a MA(q) process. Having obtained the vector a of length p that must satisfy $q \ll p \ll N$, the length of the time-series, the optimal estimation for the parameters $b_1,\ldots, b_q$ are then obtained as the solution to the linear system

$$\begin{aligned} Rb &= -\tilde{r} \\ R_{i,j} &= \frac{1}{p+1}\sum_{n=0}^{p-|i-j|} a_n a_{n+|i-j|}\,, \quad i,j = 1,\ldots,q \\ \tilde{r}_i &= \frac{1}{p+1}\sum_{n=0}^{p-i} a_n a_{n+i}\,, \quad i = 1,\ldots,q \end{aligned}$$

It must be noted that the b vector thus obtained contains the optimal parameters in a Maximum Likelihood Estimation sense.

2.5 Artificial Intelligence-Based Forecasting

Among the many tools that Artificial Intelligence researchers developed during the past half-century, Artificial Neural Networks, ANNs for short, by far have been the most popular tools for forecasting financial and other time-series. Many different factors contributed to the ANNs' popularity; the ability to predict continuous-valued outputs (or classes) for a given input vector ranks certainly high among those factors. Because of the enormous popularity of ANNs as forecasting tools in stock-market data and other business and financial data, a short description of the ANN architecture as well as the most popular algorithm for training ANNs is given below.

The discussion on ANNs will follow a fairly standard approach where an ANN will be considered as a system that can be trained to fit a finite data set $X = \{x^{[1]},\ldots,x^{[N]}\} \subset \mathbb{R}^n$ to an accompanying value set $V = \{v_1,\ldots,v_N\}$ of real numbers by constructing a function $g(x)$ such that the sum of squared errors $\sum_{i=1}^{N} \left(v_i - g(x^{[i]})\right)^2$ is minimized.

We shall restrict attention to *feed-forward, multi-layer perceptron models* (MLP). A MLP ANN is a network of individual nodes, called perceptrons organized in a series of layers as shown in Fig. 2.21. Each node (perceptron) in this network has some inputs and outputs as shown in more detail in Fig. 2.22, and represents a processing element that implements a so-called activation function. This activation function is a function of one real variable that is the weighted sum of the values of the node's inputs, so that

$$v = \varphi\left(\sum_{i=0}^{k} w_k u_k\right).$$

The most often used function for the hidden and output layers nodes is the sigmoid function:

$$\varphi(x) = (1 + e^{-x})^{-1}$$

which has the property that is differentiable everywhere, and its derivative is $\varphi'(x) = \varphi(x)(1 - \varphi(x))$, plus near the origin the function is almost linear. For the input layers, the activation function most often used is the identity function, $\varphi(x) = x$. Finally, a less often used activation function is the threshold function:

$$\varphi(x) = \mathrm{sgn}(x_+)\,, \quad \mathrm{sgn}(0) = 0$$

where the sign function sgn(x) is defined in the beginning of this chapter. The interest in MLP networks stems from a fundamental theorem due to Kolmogorov stating that a MLP with just one hidden layer and threshold nodes can approximate any function with any specified precision, and from an algorithmic technique developed in the 1980s that became known as the "BackPropagation Algorithm" that could train an MLP network to classify patterns given a input training set.

The (Error) BackPropagation algorithm (BP) is a form of gradient descent-based optimization algorithm that attempts to adjust the weights of each edge

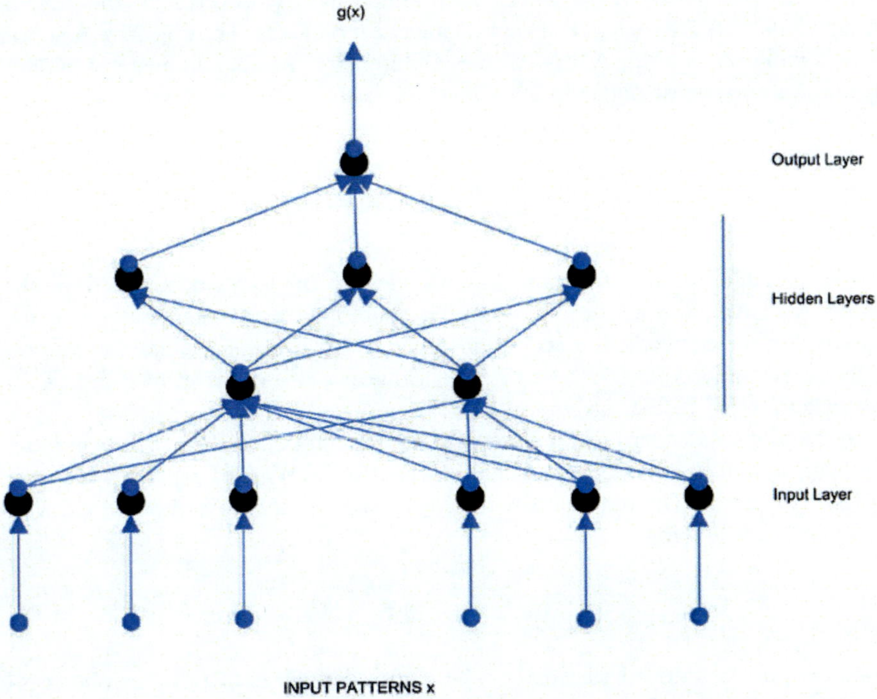

Fig. 2.21 A feed-forward multi-layer perceptron artificial neural network. notice how each node accepts inputs only from the immediate layer of nodes beneath it, and transmits the same value to the nodes in the layer immediately above it. Edges connecting two nodes have associated weights to them that multiply the value outputted by the node at the lower end of the edge

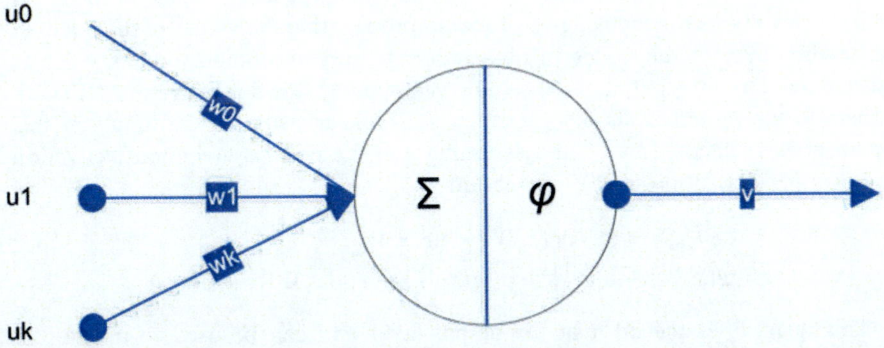

Fig. 2.22 Schematic representation of an individual perceptron. The node has associated weights $w_0,\ldots, w_k$ that multiply each of its input values $u_0,\ldots, u_k$ and an activation function $\varphi()$. Its output is the value $v = \varphi(w_0\, u_0 + w_1\, u_1 + \ldots ,w_k\, u_k)$. By default, w_0 represents the bias of the node and the corresponding variable value u_0 is constantly set to -1

connecting two nodes in the MLP to minimize the square error of the ANN's predicted outputs for each vector input pattern x it receives. The objective function to be *minimized* by the ANN—whose architecture, namely number of hidden layers and number of nodes in each layer is assumed to have been fixed somehow—is therefore the following:

$$E(W) = \frac{1}{2}\sum_{j=1}^{N} \left(g(x^{[j]}, W) - v_j\right)^2$$

The variables in this function are the weights of each node that form the matrix W. Given the above discussion, it should be now clear that given the topology of the network, the activation function for each node, and the weights W that essentially define the network, the value $g(x, W)$ is trivial to compute for any pattern x fed to the ANN.

BP employs a standard gradient descent-based method (see Sect. 1.1.1) in order to obtain a *saddle point* of the objective function $E(W)$ (i.e. a point where the derivative is zero). The rule that BP employs to update the ith node's weight $w_{ij}$is therefore the following

$$w_{ij} \leftarrow w_{ij} - \eta \frac{\partial E}{\partial w_{ij}} \tag{2.4}$$

where η is a user-defined parameter. The partial derivative $\partial E/ \partial w_{ij}$ can be computed using the chain rule of differential calculus. Let the node's input sum be denoted by ξ, having accepted input values $u_{i0}, \ldots, u_{ik}$ with weights for each input $w_{i0}, \ldots, w_{ik}$. The partial derivative of $E(W)$ with respect to w_{ij} by the chain rule is

$$\frac{\partial E}{\partial w_{ij}} = \frac{\partial E}{\partial \xi}\frac{\partial \xi}{\partial w_{ij}} = \frac{\partial E}{\partial \xi}u_{ij}$$

Now, the partial derivative $\partial E/ \partial \xi = \delta$ is called *the error*. Given an input pattern x, and its associated value v, we can calculate the output value of the ANN with given weights for each edge by forward propagating the values of the nodes at the lowest layer to the nodes in higher layers, until the output node's value is determined. Once the output node's value is computed, the derivative of $E(W)$ with respect to the *output* is simply $g_o(x,W) - v$. If the net sum of the inputs of this output node is denoted by ξ^o and assuming that the node is a sigmoid activation node so that $g(x, W) = \varphi(\xi^o)$, we obtain

$$\begin{aligned}\delta^o &= \partial E/\partial \xi^o = \partial E/\partial g(x, W) \times \partial g(x, W)/\partial \xi^o \\ &= [g(x, W) - v]\varphi'(\xi^o) = [g(x, W) - v]g(x, W)[1 - g(x, W)]\end{aligned}$$

This quantity is the error at the output layer and can be used to update the weights at the *most upper hidden layer* using the BP rule:

$$w_{ko} \leftarrow w_{ko} - \eta\delta^o v_k$$

where w_{ko} are the weights of the edges connecting the kth most upper hidden layer node and the output node, and v_k is the output of that kth node. Now, to update the weights of the edges between the previous hidden layer and the upper-most hidden layer, the chain rule can be applied again. As before, let ξ^k denote the net input sum of the kth upper-most hidden layer node, so that $v_k = \varphi(\xi^k)$, having assumed sigmoid activation hidden nodes (except at the input layer nodes where the activation function is assumed to be the identity function). Now, $\delta^k = \partial E/\partial \xi^k = \partial E/\partial v_k \times \partial v_k/\partial \xi^k = \partial E/\partial v_k \varphi'(\xi^k)$. The partial derivative of E with respect to v_k can be easily computed via the chain rule again as $\partial E/\partial v_k = \partial E/\partial \xi^o \times \partial \xi^o/\partial v_k = \delta^o w_{ko}$. Substituting this expression in the previous equation for the error δ^k we obtain the following Back-Propagation of errors equation:

$$\delta^k = \delta^o w_{ko} \varphi'(\xi^k) = \delta^o w_{ko} v_k [1 - v_k] \tag{2.5}$$

Applying the BP rule, we see that the weights of the edges connecting the upper-most hidden layer node i and node k on the layer immediately below it are updated according to

$$w_{ki} \leftarrow w_{ki} - \eta \delta^i v_k$$

The detailed BP algorithm for MLP ANNs can now be stated.

Algorithm Online BackPropagation MLP Training

Input: A dataset $S = \{x^{[1]}, \ldots, x^{[N]}\}$ of vectors in $\mathbb{R}^n$, together with associated real values $r_1, \ldots, r_N$, number of hidden layers L, number of nodes at each layer M, learning rate η, maximal number of epochs T and maximum acceptable error tolerance ε (optionally an activation function for the hidden and output layers other than the sigmoid).
Output: matrix of ANN adjusted weights.

Begin

0. Set the weights of each edge in the network to a small random value, Set $E = +\infty, t = 1, j = 1$ /* initialization */
1. while ($E > \varepsilon$ AND $t \leq T$) do
   a. Pass $x^{[j]}$ as input, and compute the output v of every node in the ANN. /* Forward Propagation */
   b. Set $\delta^o = [g(x^{[j]}, W) - r_j] g(x^{[j]}, W) [1 - g(x^{[j]}, W)]$ /* output layer error */
   c. Set $\delta^k = \delta^o w_{ko} v_k [1 - v_k]$for each node k at the upper-most hidden layer. /* upper-most hidden layer error */
   d. for each layer l below the upper-most hidden layer do /* BackPropagation */
      i. for each edge (m,p) connecting layer l with upper layer l+1 do
         1. Set $\delta^m = \delta^p w_{mp} v_m [1 - v_m]$.
      ii.end for

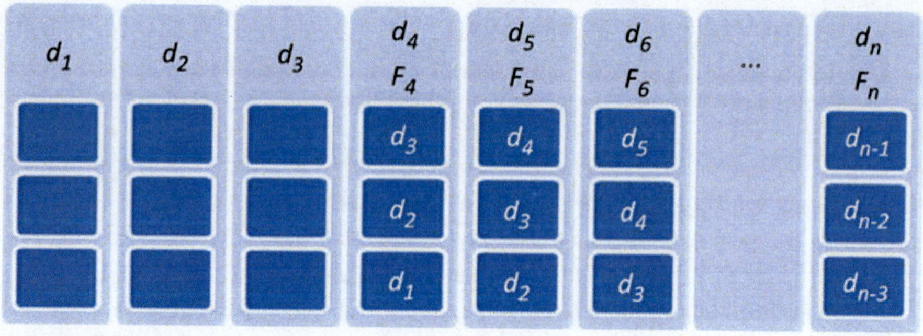

Fig. 2.23 Example break-down of time-series for use with ANN

e. end-for.
f. for each edge (*m,o*) connecting upper-most hidden layer node *m* to output node *o* do

   i. Set $w_{mo} = w_{mo} - \eta\delta^o v_m$.

g. end-for.
h. for each edge (*k,i*) connecting node *k* at hidden layer *l* to node *i* at hidden layer *l*+1 do

i. Set $w_{ki} \leftarrow w_{ki} - \eta\delta^i v_k$.

   i. end-for

   j. Set $E = \frac{1}{2}\sum_{j=1}^{N} \left(g(x^{[j]}, W) - v_j\right)^2$using current weights *W*.

k. if $j = N$ then Set $t = t + 1$, Set $j = 0$ else Set $j = j + 1$. /* epoch update */

3. end-while
4. return *W*.

End.

The Algorithm is known as the *online* version of BackPropagation because weights are updated as soon as each pattern $x^{[j]}$ is fed to the algorithm. In the *batch* version of BackPropagation, weights are not updated until a full epoch has passed, meaning all instances have been fed to the ANN. In such a case, as patterns are fed to the system, the system maintains and stores the changes that should be made to each weight, but does not actually modify the edge weights and applies the cumulative changes as soon as all patterns have been fed to the system.

To apply MLP ANNs in forecasting, a trivial procedure may have to be applied first. Assume the time-series d_i $i = 1,\ldots, N$ is available. Then, one can obtain the data-set of patterns *S* by first choosing a time-window *w* that will serve as the dimensionality of the input pattern data-set, and define the *N*–*w* patterns $x_i = [d_i \ldots d_{i+w-1}]^T$ in R^w with associated values $v_i = d_{i+w}$. Figure 2.23 illustrates this procedure for $w = 3$.

Several points should be made regarding MLP ANNs. The most important point has to do with the generalization capacity of an ANN. It is true that the more hidden nodes one selects as input network architecture the more likely it is for the system to be able to minimize the training error function $E(W)$. The same holds true for the maximum number of epochs T the system is allowed to run for. However, minimization of the training error E does not necessarily guarantee good performance. Indeed, ANNs are often plagued by a phenomenon known as *over-fitting*, whereby the ANN simply "memorizes" the training input patterns and so even though its performance on the training set is very good, its performance on previously unseen instance patterns is unacceptably bad. For this reason, ANNs are usually trained using a (reasonably large) subset S' of the training set S, and at the end of each epoch, the performance of the ANN is evaluated on $S - S'$ and when this (test-set) performance stops improving, the algorithm stops.

A last point to be made about ANNs is their inherent instability. By instability we mean that small differences in the inputs to the algorithm can somctimes result in significantly differently trained ANNs. This observation does not have only bad consequences. A good consequence is that combining different ANNs (that are obtained starting from different random initializations of the network's weights and applying the same BP algorithm) in a classifier ensemble can lead to better performance due to the "diversity" of the base classifiers involved in the scheme, caused precisely by the ANN's inherent instability.

2.5.1 Case Study

We illustrate the use of ANN-based forecasting in the case of stock-market forecasting. We try to forecast the daily closing price of a common stock that trades in the Athens Stock Exchange. The time-series contains data for approximately 15 months (290 data points) between January 2009 and March 2010. These data points were broken down in patterns of 10 points, where we hypothesized that the closing prices of two weeks should provide a reasonable indicator of the price of the stock for the next day. The ANN architecture was set to have a single hidden layer with 20 nodes, and one output node. Using the MATLAB Neural Network Toolbox, we trained the system using cross-validation as the criterion to be optimized. The historical performance of the training of the system is shown in Fig. 2.24.

The results of the training are very good as the trained ANN can forecast the next day closing price of the stock with reasonable accuracy, certainly much better than the accuracy obtainable from exponential smoothing methods. Figure 2.25 shows how closely the ANN can match the historical data.

The accuracy of the forecast provided is more evident in the following regression plot (Fig. 2.26)—output by the matlab toolbox—that plots the forecast values versus the target values for the time-series in our case-study.

The performance of the trained ANN is also shown in Fig. 2.27, where the *Percentage* Deviations of the forecasts from the actual values are shown. Notice

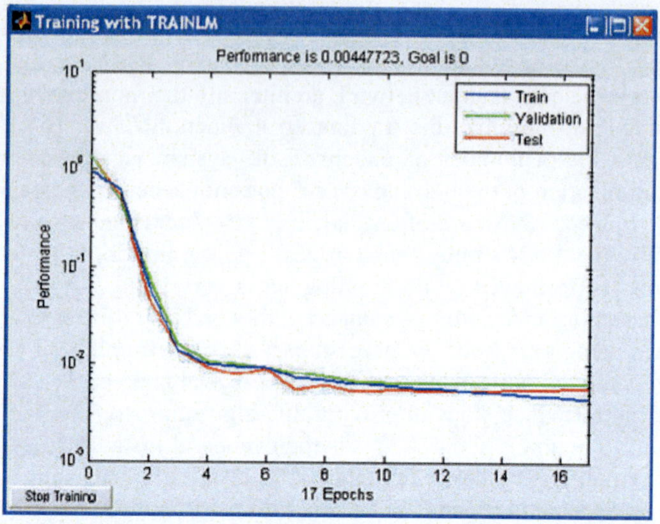

Fig. 2.24 ANN training using matlab NN-toolbox

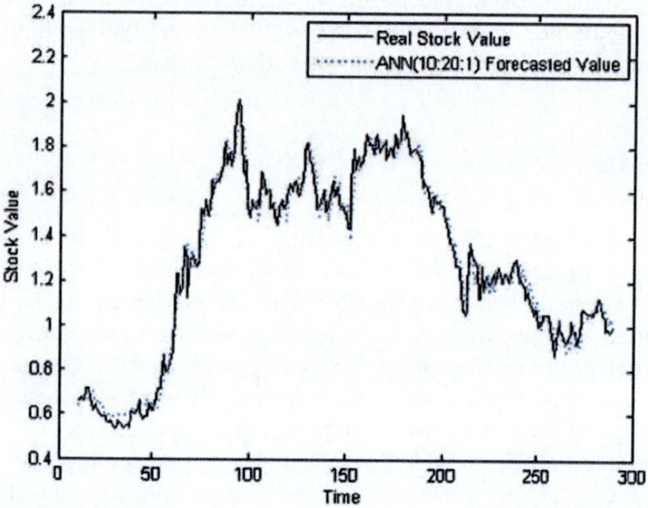

Fig. 2.25 Forecasting performance of the trained ANN on closing price of common stock

that the daily percentage fluctuations of a common stock are usually in a comparable range to the fluctuations observed in the figure.

The Mean Square Error of the Neural Network's forecasts is $MSE_{289} = 0.0032$. This value is actually better (but *not* significantly better) than the Mean Square Error of the forecasts obtained by the Naïve Forecast Method forecasting $F_{t+1} = d_t$, which obtains an MSE_{289} value of 0.0038. In Fig. 2.28 we show a plot of the Percentage Forecast Deviations obtained by applying the Naïve Forecast method.

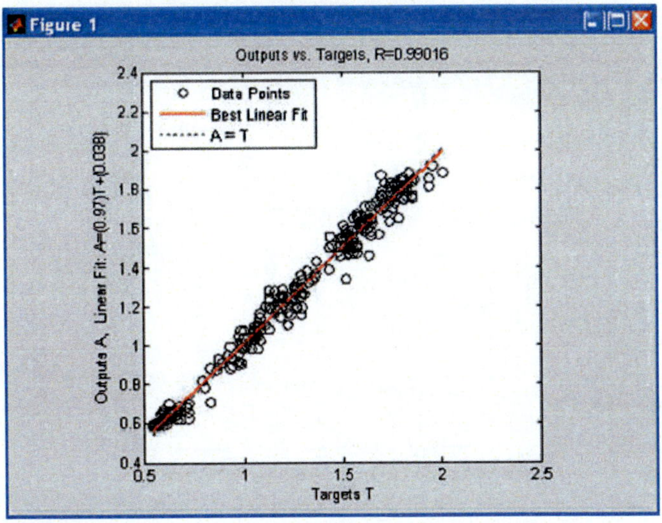

Fig. 2.26 Forecasting performance of the trained ANN as regression line between actual and forecasted value pairs

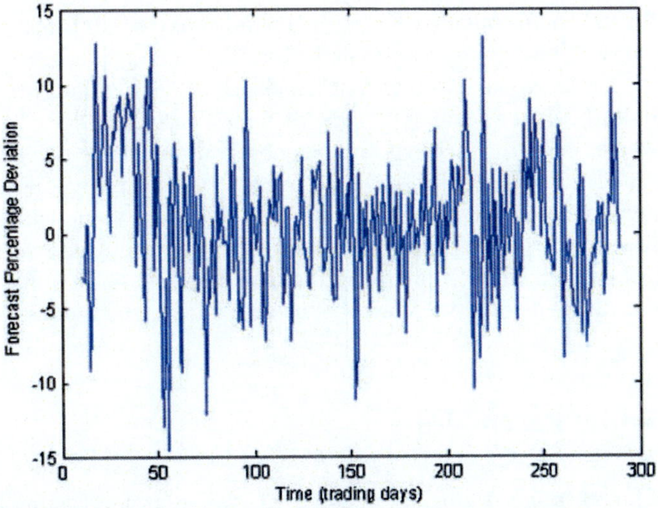

Fig. 2.27 Forecast percentage deviation of the trained ANN on the closing value of a common stock time-series

The predictions made by the trained ANN are reasonable as one can easily check by looking at Figs. 2.25 and 2.26. However, this does *not* mean that one can make money by speculating on the stock's price as predicted by the ANN. Indeed, suppose we use the ANN's predictions to predict whether buying that stock on any particular day is likely to be a good investment in that within the next 5 days

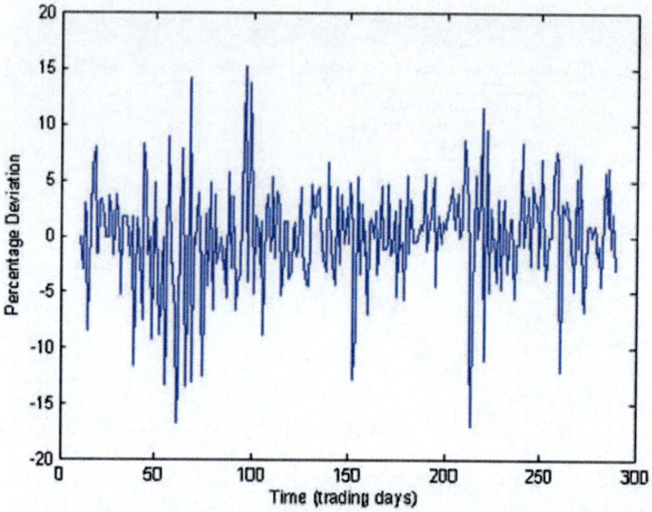

Fig. 2.28 Forecasting percentage deviation of the Naïve forecasting method on the closing value of a common stock time-series. This figure essentially represents the daily percentage fluctuations of the closing price of the stock

the stock will increase its value (speculative buying behavior). It turns out that the particular ANN will be wrong more than 50% of the time in its trend prediction. Nevertheless, this does not imply that stock short-term trend prediction is impossible. Indeed, when a Support Vector Machine (Vapnik 1995) is trained with input data a vector comprising 10 continuous observations of the twice integrated (i.e. differentiated) closing stock price time-series labeled, each vector labeled as "positive" if the twice-differentiated time-series increases its value within the next 5 observations, and "negative" otherwise, the results show a testing accuracy (on unseen data) of more than 70%, meaning that it is possible to a certain extent to predict short-term trends in stock market time-series for certain stocks at least.

2.6 Forecasting Ensembles

Ensembles of forecasting systems have been used in weather forecasting for a long time with great success. In weather forecasting, a system of highly non-linear differential equations is solved many times, each time with a slightly perturbed initial condition. Because of the nonlinearity of the system dynamics involved, even slight perturbations to the initial conditions of the system can lead to widely-differing solutions within a short amount of time. Because the initial conditions are not exactly known, the system is solved many times to see in a statistical sense how it will likely evolve. The results of the differently-initialized system are then combined to produce a final prediction about the weather. The predictions of an

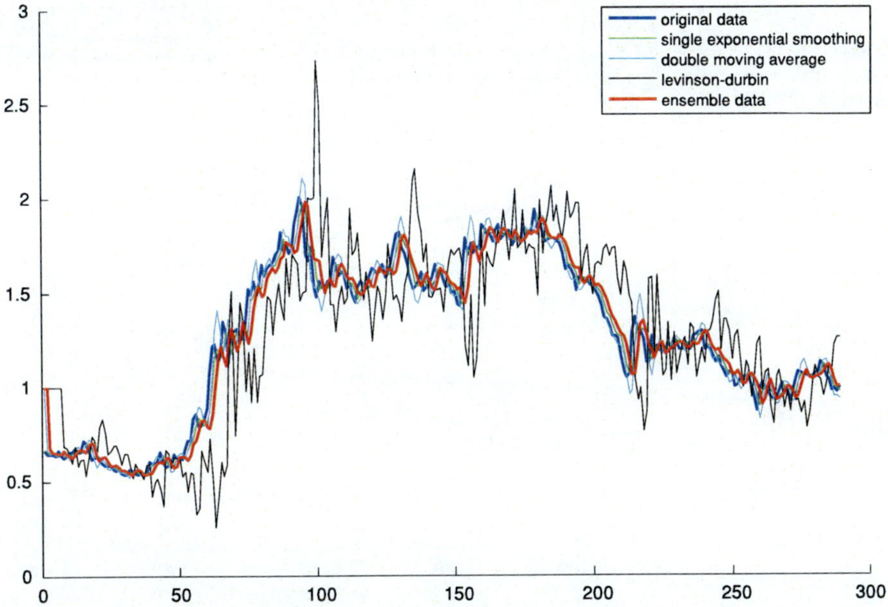

Fig. 2.29 Forecasting performance of an ensemble of three different methods on closing price of common stock. The x-axis represents time (in trading days)

event are given a probability that is often equal to the frequency of appearance of the event in the solutions of the system in the ensemble.

Similarly to forecasting ensembles for weather forecasting, general time-series forecasting ensembles (sometimes also known as committee forecasting) comprise of a collection of forecasting models and algorithms that operate on the same time-series to produce a forecast for the next period. The predictions are then combined in a *fusion scheme* to produce a final forecast. The most obvious fusion method is to average the predictions of the methods to obtain the final forecast. A more intelligent but still straightforward method to combine the ensembles' forecasts would be to compute a weighted average of the forecasters' predictions, where the weight of each forecaster would be proportional to its accuracy as measured by one of the metrics discussed at the beginning of this chapter. More formally, let $\Im = \{P_1, \ldots, P_L\}$ denote a set of L forecasting algorithms. Assume that the forecasts F^i_{n+1}, $i = 1\ldots, L$ for the $(n + 1)$st element of the time-series d_n have associated MAPD$_i$ values $w_1, \ldots w_L$. The combined forecast according to the *weighted-average fusion scheme* is then given as

$$F_{n+1} = \frac{\sum_{i=1}^{L} w_i^{-1} F^i_{n+1}}{\sum_{i=1}^{L} w_i^{-1}}$$

For example, combining the predictions of the Single Exponential Smoothing model, the Double Moving Average model, and the Auto-Regressive model in one

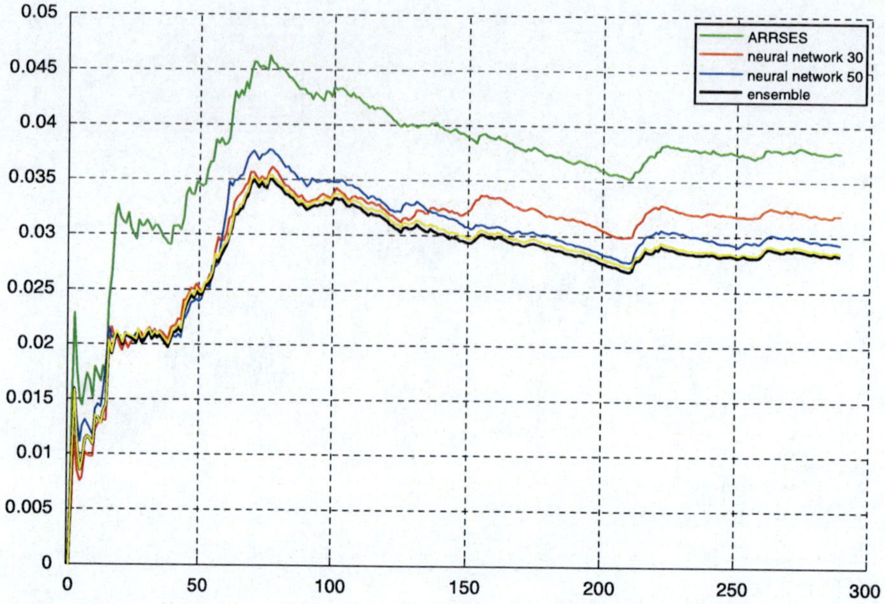

Fig. 2.30 Forecasting performance of an ensemble of three different methods on closing price of common stock. The *x*-axis represents time (in trading days). The *y*-axis represents the MAPD metric for each forecasting method

ensemble and applying it to the same (undifferentiated) time-series as that of Fig. 2.25, we get the forecasts shown in Fig. 2.29.

If we combine the predictions of ARRSES, an ANN with 30 hidden nodes, and an ANN with 50 hidden nodes forecasting the same common stock time-series as before, we get the $MAPD_k$ measure plotted in Fig. 2.30. As can be seen, combining the forecasts of the three methods actually leads to a consistently better forecast in terms of MAPD error metric. Such results are *not* always typical of forecasting ensembles however.

Testing the ensemble idea on a much more “predictable” data-set, that of CO_2 concentrations measured at the Mauna-Loa Observatory provided by the NIST/SEMATECH e-Handbook of Statistical Methods (http://www.itl.nist.gov/div898/handbook), we get the graphs shown in Fig. 2.31. The ensemble consists of a SES forecaster, a DES forecaster, and an AR-based forecaster. Notice that *none* of the forecasting methods comprising the ensemble of Fig. 2.31 have any notion of seasonality (it is obvious by looking at the graph that seasonality and a trend component is inherent in the data).

Experiments with forecasting ensembles of ANNs on stock-market data have often produced poor results (the ensemble prediction being often worse than the best individual forecasting method in the ensemble.) However, when we test the performance of an ensemble of 9 ANNs, each with a different number of hidden nodes, ranging from 10, 20,…,90 on three stock-market time-series data sets, the

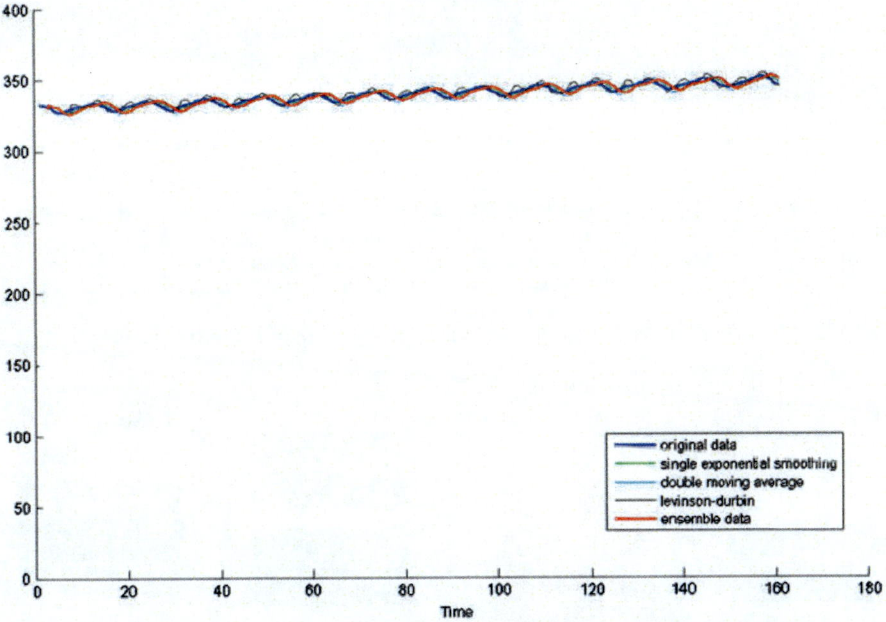

Fig. 2.31 Forecasting performance of an ensemble of three different methods on CO_2 concentrations measurements at the Mauna-Loa observatory

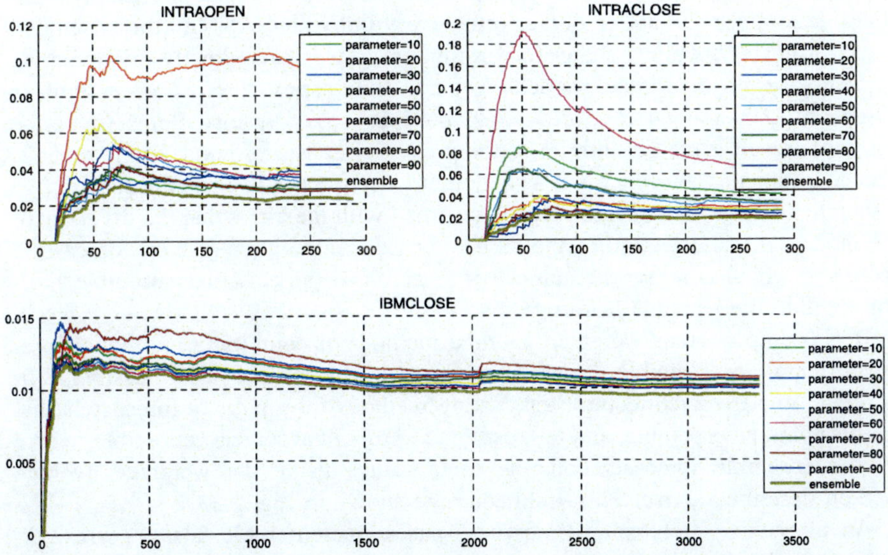

Fig. 2.32 Forecasting performance of an ensemble of 9 ANNs on three different stock market data sets. The *x*-axis represents trading days. The *y*-axis represents MAPD error. The ensemble outperforms each individual base ANN

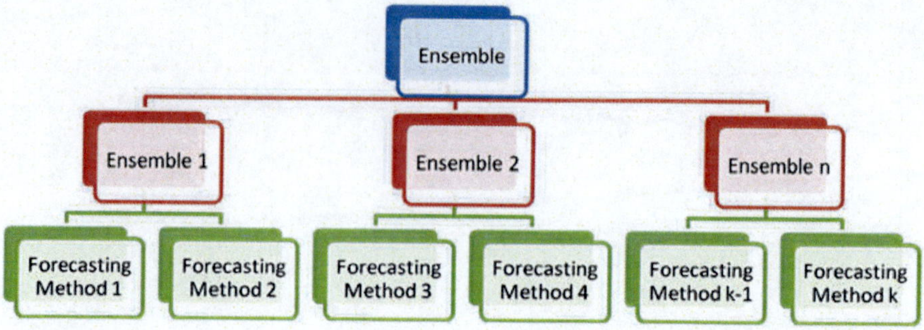

Fig. 2.33 Structure of a tree forecasting ensemble. obviously, $n = k/2$ in the figure

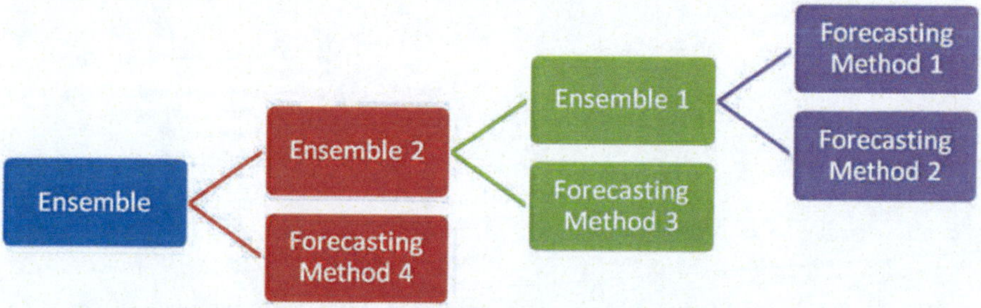

Fig. 2.34 Structure of a cascading forecasting ensemble

results shown in Fig. 2.32 show that the ensemble of ANNs combined with the simple rule discussed above outperforms every individual ANN in the ensemble.

Unfortunately, while weather forecasting ensembles have greatly improved the performance of individual models, general time-series forecasting ensembles based on ANN trained for forecasting, produce mixed results. Some times the ensemble result is clearly inferior to the predictive power of good, single models. The conditions under which such phenomena occur are still the subject of active research. Another area of research is concerned with the effect that the diversity of the models in the selection of the base forecasting algorithms has on forecasting accuracy, which is a favorite subject of research in the classifier ensemble literature within the machine learning community.

Besides the obvious weighted average method for combining individual forecasters into an ensemble forecast, other architectures are also possible. For example, in a tree architecture, forecasting methods form pairs of forecasters that are fused in one ensemble, and the result is fed to a final forecast ensemble, where all the ensemble forecasts compute their values using the weighted average method described above. This architecture is shown in Fig. 2.33.

An alternative architecture known as cascading ensemble, also "borrowed" from the classifier research community is shown in Fig. 2.34. Preliminary experiments in (Luo 2010) show that a tree ensemble classifier may perform slightly better than a standard ensemble or a cascading ensemble.

2.7 Prediction Markets

When historical data about the demand for a product exist, one can use any or all of the methods discussed in the previous sections regarding time-series analysis to predict the future demand within a short *lead-time*. However, when new products are developed, the more innovative the products are, the less clues one has as to what the demand for such products might be. In the absence of any quantitative data to work with, marketing research tools including *market surveys* could be launched to "probe" the market so as to gauge the "acceptance" of the product in the market before it is launched. Such market surveys are usually conducted using questionnaires that are filled in by participants in the survey, and are then processed using segmentation and clustering tools from classical pattern recognition that can eventually lead to some kind of forecast demand. Unfortunately, the sample population that must fill out such questionnaires so that the survey results are statistically meaningful is usually of very big size, making such tools and methods very expensive. An alternative approach, based on ideas borrowed from the mechanisms of stock-markets is known as Prediction Markets (PMs) and has been proved to be successful enough that it is currently in use by most major corporations world-wide, at least when other quantifiable data are not available for the prediction problem at hand. PMs can be thought of as a (virtual) stock market for events: player/traders place "bids" to buy/sell options on a future event: each option will pay a number of (virtual) dollars if a particular future event materializes, and pay nothing otherwise. Usually, derivative options are also allowed .

Wolfers and Zitzewitz (2007) restricted their attention to simple, binary option PMs, where traders buy and sell an all-or-nothing contract that will pay $1 if a specific event occurs, and nothing otherwise (the $1 may be real or "virtual" money used only inside the particular PMs). The specific event could be the re-election of the US President to office for a second term, or whether the latest smart-phone from Apple will exceed the sales of its predecessor within 3 months. The market players (traders) have in general different subjective beliefs about the outcome of the specific event, and the belief of trader j on the occurrence of the event is considered a random variable, denoted by q_j, drawn from a distribution $F(x)$. Assuming further that traders are price-takers—so there are no oligopoly dynamics—who maximize their expected utility defined to be a log function that is often assumed in economics theory, the optimization problem each trader j faces is the following:

$$\max U(x) = q_j \log(y + x(1 - \pi)) + (1 - q_j)\log(y - x\pi)$$

where π is the price for the contract, y is the trader's wealth level, and x is the decision variable representing the number of contracts trader j should buy to maximize their expected subjective utility. Setting $dU(x)/dx = 0$, we obtain the optimal buying quantity for trader j at price π to be

$$x_j = y(q_j - \pi) \,/\, [\pi(1 - \pi)]$$

From this, one immediately sees that individual demand will be zero when the price π equals the belief of the trader q_j –as expected. Also, trader's j demand will increase linearly with their belief, and is decreasing in risk (represented by a price close to 0.5).

Now, the PMs will be in equilibrium when supply will equal demand, which can be written down mathematically as

$$\int_{-\infty}^{\pi}\int y\frac{q-\pi}{\pi(1-\pi)}\mathrm{d}G(y)\mathrm{d}F(q) = \int_{\pi}^{+\infty}\int y\frac{\pi-q}{\pi(1-\pi)}\mathrm{d}G(y)\mathrm{d}F(q)$$

where $G(y)$ is the Cumulative Distribution Function (cdf) of the wealth levels of all traders in the market. Denoting the Probability Density Function (pdf) of the traders' beliefs by $f(q)$, and assuming that wealth and beliefs are uncorrelated so that $E[q, y] = 0$, the above equation implies that

$$\frac{y}{\pi(1-\pi)}\int_{-\infty}^{\pi}(q-\pi)f(q)\mathrm{d}q = \frac{y}{\pi(1-\pi)}\int_{\pi}^{+\infty}(\pi-q)f(q)\mathrm{d}q \Leftrightarrow$$

$$\pi = \int_{-\infty}^{+\infty} qf(q)\mathrm{d}q = E[q]$$

The latter directly shows that under this simple but elegant and easily expandable model, *market equilibrium is achieved when the market price equals the mean belief of the population about the probability of the outcome of the event.* As Wolfers and Zitzewitz argue, the monotonicity of demand in expected returns implies that this is the only price which results in equilibrium of the market (resulting in zero aggregate demand).

Interestingly, even when the model is generalized to account for various degrees of correlation of traders' wealth levels and beliefs (which should be expected to exist), or other utility functions, it remains true that the deviation of the equilibrium market price from the mean beliefs of the traders is very small—and usually negligible. Therefore, PMs should be expected to work very efficiently as (approximate) information aggregators. Indeed, this has been verified in several experiments as well.

2.8 Bibliography

Forecasting, as mentioned in the beginning of this chapter is as old as civilization itself. However, rigorous methods for time-series analysis are not that old. While statistical methods such as regression go back to the work of mathematicians including Euler, Gauss and Sir Francis Galton, work on Auto-Regressive models goes only back to Yule (1927) and Walker (1931), while traces of the basic ideas can be found in Schuster (1906). The definitive text on Auto-Regressive models and their expanded form, ARIMA models remains Box and Jenkins (1976); the current 4th edition of this book by Box et al. (2008) contains very useful material on outlier points detection in time-series, quality control etc. Fast algorithms for solving the Yule-Walker equations are usually the subject of signal processing courses in Electrical Engineering curricula. The Levinson-Durbin algorithm was developed in the 1960s, see Durbin (1960), and at around the same time, Kalman filters were introduced in the signal processing literature as well, e.g. Kalman (1960). Exponential smoothing and its variants (SES, DES, TES, Holt–Winters method etc.) were invented in the 1950s within the context of Operations Research and Statistics research. See for example Holt (1957), Winters (1960), and the book by Brown (1962). The combination of forecasting methods has been a subject of research since the 1980s, e.g. Winkler and Makridakis (1983).

See the book by Makridakis et al. (1998) for a detailed discussion of time series decomposition. For more details on the regression and numerical computing aspects of the method see Cheney and Kincaid (1994).

The literature on ANNs is as impressive as that on forecasting methods. Early analysis of single layer ANNs (perceptrons) was carried out in Minsky and Papert (1969). A classical treatment on ANNs remains the two-volume set (Rumelhart et al. 1987). Computational Intelligence techniques involving ensembles of ANNs used for forecasting time-series are discussed in some detail in Shi and Liu (1993), Yong (2000), and more recently, Palit and Popovic (2005). For a case-study of the application of Computational Intelligence techniques in electric load forecasting see Tzafestas and Tzafestas (2001).

Tree forecasting ensembles and cascading forecasting ensembles are described in detail in Yonming Luo's Master Thesis (Luo 2010).

2.9 Exercises

1 Show that the forecasts produced by the Single Exponential Smoothing method minimize the discounted cost criterion $S' = \sum_{j=0}^{\infty} (1-a)^{j+1}(d_{t-j} - F_t)^2$.

2 Show that for the two-variable function φ defined as $\varphi(a, b) = \|at + be - d\|^2$ where $t = [t_1\ t_2\ \ldots\ t_n]^T$, $d = [d_1\ d_2\ \ldots\ d_n]^T$, and $e = [1\ 1\ \ldots\ 1]^T$ are given n-dimensional column vectors, the unique point (a^*, b^*) where $\nabla\varphi(a^*, b^*) = 0$ is its global minimizer.

3 Two methods for predicting weekly sales for a product gave the following results

Period	Method 1	Method 2
1	80	80
2	83	82
3	87	84
4	90	86
5	81	86
6	82	83
7	82	82
8	85	82

The actual demand observed was the following:

Period	Actual demand
1	83
2	86
3	85
4	89
5	85
6	84
7	84
8	83

Determine the $MAPD_i$ and MSE_i value of the two methods for $i = 1,\ldots,8$, as well as the Tracking Signal S_i of the two methods. Based on this information, determine whether it is possible to safely use one method or the other.

4 Implement the SES and DES formulae on a spreadsheet program, and use it to determine the optimal parameter a in the SES method giving the best $MAPD_8$ error on the time-series of the previous exercise.

5 Implement the Levinson-Durbin algorithm.

(a) Test the implementation with the following values of $p = 2,3,5$ on the following time-series:

Period i	d_i
1	20
2	18
3	17
4	17
5	19
6	22
7	20
8	20
9	19
10	18
11	19
12	22
13	20
14	21
15	22
16	24
17	23
18	25
19	27
20	26

(b) Test the same implementation on the differentiated time-series $\Delta d_i = d_i – d_{i-1}$

6 For the time-series d_i of Exercise 5, compute the best estimate for the value d_{21} using the *additive* model for time-series decomposition of Sect. 2.2.1. To estimate trend in the data, use the centered-moving-average method with parameter $k = 6$. Assume that the seasonality length $s = 6$.

7 Assume demand for some good is a *martingale* process where $d_{t+1} = d_t + R_t$ where the R_t are independent random variables normally distributed, with zero mean, and variance $\sigma_t = \sqrt{t}$. Which of the forecast methods discussed so far would give—when optimized in its parameters– the best results in the mean square error sense?

References

Box GEP, Jenkins GM (1976) Time series analysis: forecasting and control. Holden Day, San Francisco CA

Box GEP, Jenkins GM, Reinsel GC (2008) Time series analysis: forecasting and control., 4th edn. Wiley, Hoboken, NJ

Brown RG (1962) Smoothing, forecasting and prediction of discrete time series. Prentice-Hall, Englewood Cliffs, NJ

Cheney W, Kincaid D (1994) Numerical mathematics and computing, 3rd edn. Brooks/Cole Publishing Company, Pacific Grove, CA

Durbin J (1960) The fitting of time-series models. Rev Int Stat Inst 28:233–244

Ghiani G, Laporte G, Musmanno R (2004) Introduction to logistics systems planning and control. Wiley, Chichester, UK

Halikias I (2003) Statistics: analytic methods for business decisions, 2nd edn. Rosili, Athens, Greece (in Greek)

Holt CC (1957) Forecasting trends and seasonals by exponentially weighted moving averages. O.N.R. Memorandum 52, Carnegie Institute of Technology, Pittsburgh

Kalman RE (1960) A new approach to linear filtering and prediction problems. J Basic Eng 82:35–45

Karagiannis G (1988) Digital signal processing. National Technical University of Athens, Athens, Greece (in Greek)

Luo Y (2010) Time series forecasting using forecasting ensembles. M.Sc. thesis, Information Networking Institute, Carnegie-Mellon University

Makridakis S, Wheelwright SC, Hyndman RJ (1998) Forecasting: methods and applications, 3rd edn. Wiley, Hoboken, NJ

Minsky M, Papert S (1969) Perceptrons. MIT Press, Cambridge, MA

Palit AK, Popovic D (2005) Computational intelligence in time series forecasting: theory and applications. Springer, Berlin, Germany

Rumelhart DE, McClelland JL et al (1987) Parallel distributed processing: explorations in the micro-structure of cognition, volume I: foundations. MIT Press, Cambridge, MA

Sanders NR, Manrodt KB (1994) Forecasting practices in US corporations: survey results. Interfaces 24(2):92–100

Schuster A (1906) On the periodicities of sunspots. Philos Trans R Soc A206:69

Shi S, Liu B (1993) Nonlinear combination of forecasts with neural networks. In: Proceedings of the international joint conference on neural networks, Nagoya, Japan

Tzafestas S, Tzafestas E (2001) Computational intelligence techniques for short-term electric load forecasting. J Intell Robot Syst 31(1–3):7–68

Vapnik VN (1995) The nature of statistical learning theory. Springer, New York

Walker G (1931) On periodicity in series of related terms. Proc R Soc A131:195–215

Winkler R, Makridakis S (1983) The combination of forecasts. J R Stat Soc Ser A 137:131–165

Winters PR (1960) Forecasting sales by exponentially weighted moving averages. Manag Sci 6(3):324–342

Wolfers J, Zitzewitz E (2007) Interpreting prediction markets as probabilities. NBER working paper #12200, The Wharton school of business, University of Pennsylvania

Yong Y (2000) Combining different procedures for adaptive regression. J Multivar Anal 74:135–161

Yule GU (1927) On a method of investigating periodicities in disturbed series with special reference to Wolfer's sunspot numbers. Philos Trans R Soc A226:267–298

Chapter 3
Planning and Scheduling

Operational planning and scheduling rank among the most important activities an industrial organization has to carry out, as they lie at the heart of the operations of the enterprise. Production planning, together with personnel scheduling comprise the decision making procedures regarding *what* will be *produced*, *when*, *how*, and *by whom*. Of course, planning and scheduling is not confined to the manufacturing sector alone. In the airline and transportation industry for example, planning relates to long-term decisions regarding routes to fly, fleets to travel the decided routes, and frequency of flights. Scheduling then refers mainly to two problems: (1) crew-pairing, or the problem of matching consecutive flights (known as legs) to form round-trips from a given base (known as pairings) so that all legs are covered in pairings with minimal cost; and (2) crew-assignment, where the problem is to match appropriately qualified personnel to pairings so that no constraints are violated, and an appropriately formulated total cost function is minimized.

Planning and scheduling research has progressed very significantly since the early 1950s, when the operations research community produced the first results in the field, and often received significant boost from computational complexity theorists and algorithm developers that devised important algorithms for large classes of problems in this area. On the other hand, it is important to realize that production planning models and algorithms for these models are not always depicting the exact reality of a plant or an organization. For example, almost all of the hard constraints we shall encounter in job-shop scheduling and due-date management, in reality are not that "hard" but are soft constraints in that often, violating one of them by a small slack does not violate any physical laws nor does it hurt company profitability in the long run.

3.1 Aggregate Production Planning

The production planning and control literature is full of models and algorithms for finding efficient plans for most types of industries. In this chapter, we present some of the most successful approaches to the general problem.

I. T. Christou, *Quantitative Methods in Supply Chain Management*,
DOI: 10.1007/978-0-85729-766-2_3, © Springer-Verlag London Limited 2012

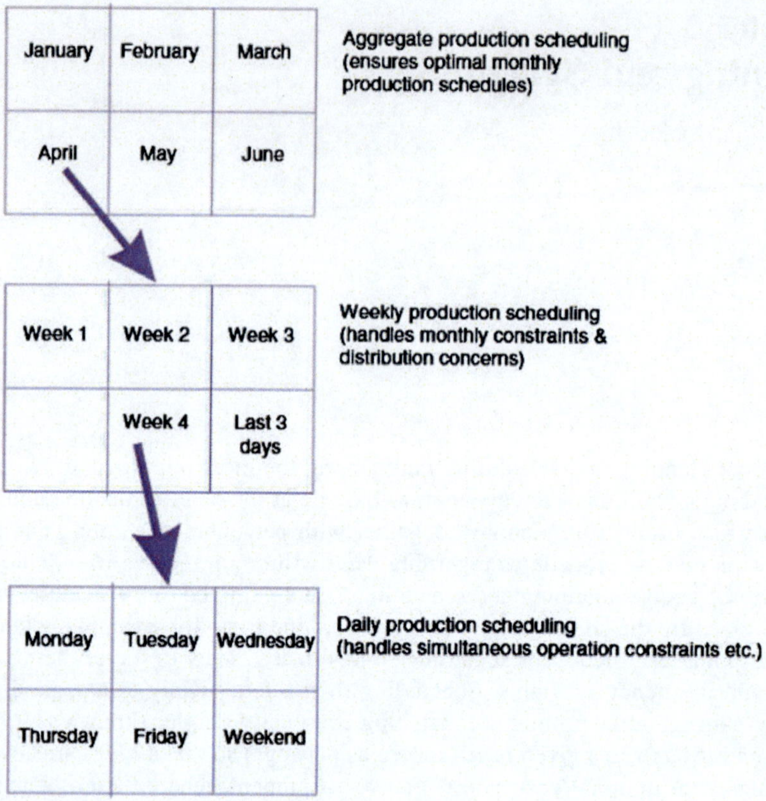

Fig. 3.1 Hierarchical decomposition for production planning and scheduling

One particularly effective technique for obtaining robust and efficient production plans is based on the principle of decomposition, also known as "divide-and-conquer" (see Fig. 3.1).

To find the optimal plan over a time-horizon, time itself is decomposed in two or more levels of granularity, forming a hierarchy: in the aggregate level, a time period may comprise one month or one quarter, and the *aggregate production planning problem* is to decide what to produce for a number of upcoming periods, given an aggregate forecast as well as the expected capacity of the production lines during these periods. These production plans are particularly useful for staffing decisions, including decisions regarding possible overtime that the company should use in a given aggregate period, as well as decisions regarding new hires/layoffs or flex working schedules. They are also useful in determining if an expansion or, rarely, contraction of available capacity is needed. Then, once an aggregate production plan is determined, the finer-level production planning problem becomes the problem of computing the optimal plan for each finer-level time period contained within the current and the next aggregate periods. In this fine-level granularity of time, a time period can be for example one week, or one day, depending on the number of levels in the time-hierarchy. Additional (and widely different) constraints may be imposed at this fine-level of time granularity, indicating for example personnel and union constraints, required down-times of the machines, etc.

We begin our discussion on aggregate production planning with the simplest possible example. Consider a monthly demand forecast $d_i\, i = 1, \ldots, N$ for N periods, for an imaginary company that produces a single product, and consider the problem of building inventories of finished goods during each period so that the monthly demand can be met, while maintaining minimal levels of inventories, thus avoiding inventory holding costs accruing from opportunity costs, risk of inventory obsolescence, risk of inventory damage due to natural or other disasters, etc. If no capacity constraints are taken into account, then the optimal policy is to produce everything in-time, in a just-in-time (JIT) fashion, since there is enough capacity to meet any level of demand at any period. However, when capacity is not enough to meet demand during peak seasons—as is usually the case—inventories have to be built ahead of time, obviously as late as possible, to avoid accumulation of inventory costs.

The following linear program (LP) determines the optimal solution to our first production planning problem.

$$
\min_{x,h} \sum_{i=1}^{N} h_i
$$

$$
\text{s.t.} \begin{cases} h_i = h_{i-1} + x_i - d_i, & i = 1, \ldots, N \\ 0 \le x_i \le c, & i = 1, \ldots, N \\ 0 \le h_i, & i = 1, \ldots, N \\ h_0 = I_0 & \end{cases}
$$

The decision variables x represent the production for each period, while the variables h represent the inventory at the end of each period. At the beginning, we assume an existing inventory build-up of I_0 units. This LP can be solved in a fraction of a second on any workstation even when the horizon is very large (which makes sense only as a feasibility exercise, as any plans longer than a few years are not very likely to be valid after a few months into their execution). Notice however that in case that demand cannot be met because of the capacity constraints, the problem becomes infeasible, and no solution will be provided. Interestingly, there is a linear-time algorithm that solves the above problem, and that has the additional advantage of providing a solution that meets demand as best as possible even when meeting demand completely is infeasible as well. The algorithm is based on the dynamic programming principle (see Sect. 1.1.3). In particular, notice that the above problem can be modeled by the following dynamic program:

$$z_i = \min\{d_i + e_{i+1}, c_i\} \quad i = N-1, \ldots, 1$$
$$e_i = \max\{0, d_i + e_{i+1} - c_i\} \quad i = N-1, \ldots, 1$$
$$e_N = 0, z_N = \min\{d_N, c_N\}$$

where c_i $i = 1,\ldots, N$ is the (time-varying) plant capacity, and z_i is the quantity to be produced in each period i. This dynamic program is directly implemented in the algorithm below. The algorithm works backwards in time, and guarantees the optimal solution, but only for the model given above.

Algorithm SimplePlanner

Inputs: array d_i i=1,…, N of forecast demands for periods 1,…, N, array of plant capacities cap_i i=1,…, N.
Outputs: array x_i i=1,…, N of optimal production levels for the N periods 1,…, N.

Begin

1. Set $t = N$.
2. Set *excess* = 0.
3. Create new array $x[1,\ldots, N]$.
4. while $t > 0$ do:
   a. if d_t + *excess* $\leq cap_t$ then
   – Set $x_t = d_t$ + *excess*.
   – Set *excess* = 0.

   b. else
   – Set $x_t = cap_t$.
   – Set *excess* = d_t + *excess* – x_t.

   c. end-if
   d. Set $t = t-1$.
5. end-while
6. return x.

End.

It is not very hard to verify the correctness of the algorithm under the assumptions made in the model. The algorithm works in a JIT fashion, attempting to produce material as late as possible so as to meet demand of future periods. Its run-time complexity is $O(N)$ making it optimal since any planning algorithm has to go through each planning period and make a decision on how much to produce. Unfortunately, the algorithm (and the associated dynamic programming model) cannot be extended to handle the case of any more complicated situations. For example, if there is more than one product to be produced, and there is more than one production line available, the algorithm cannot be modified to compute in linear time the optimal production quantities of each product for each period. The usefulness of the SimplePlanner algorithm is limited to that of an introduction to the subject. Parenthetically, we note that unfortunately, variants of this algorithm are used in practice often with very bad results (in many cases, practitioners aggregate the demand of all products in one aggregate demand forecast, and using the aggregate capacity of the plant as the sum of the capacities of each production line apply a variant of the algorithm to compute an approximate quantity of total monthly production, which is then used for staffing computations).

Yet another model for production planning that is based on dynamic programming, is the Wagner–Whitin dynamic lot-sizing procedure (Wagner and Whitin 1958). This approach, despite its simplicity—and strong assumptions that include an *infinite capacity assumption*—played a key role in the development of information systems for manufacturing resource management; indeed, material requirements planning and manufacturing resources planning (MRP and MRP II, respectively) systems are based on the key concepts of the Wagner–Whitin model, a version of which we describe next.

Consider the case of known but time-varying demand for a product $D_i|i = 1, \ldots, M$ for M periods. The cost of unit-production during each period is $c_i|i = 1, \ldots, M$, and can be varying in time (a property that often is true in the real-world), and there is also a set-up cost for initiating production during period $i = 1,\ldots, M$, denoted by A_i. This set-up cost is a so-called *fixed-charge cost*, meaning that it is incurred only in order to initiate production activities during a period, and is otherwise independent of the total amount of production of that period. Assuming a holding cost h_i to carry a unit of inventory from period i to period $i + 1$, the *dynamic lot-sizing problem* is the problem of computing the optimal production plan that produces exactly the quantities of product demanded during each period, on or before that period.

Assuming infinite capacity during each period, we can immediately see that the following property must hold true: in an optimal plan, either the demand for a period j must be fully produced during that period (JIT fashion) or else all of the demand for period j must be produced in earlier periods. The reason is that if in the optimal plan, we have to produce anything during a period j (and thus incur the set-up cost A_j), then it must be more economical to produce all of the demand for that period JIT, instead of producing any part of that demand in a previous period. Now, the cost Z_M^* of the optimal production plan covering periods $\{1,\ldots, M\}$ for the problem can be determined from the following dynamic programming recursive equation:

$$Z_k^* = \min_{r=j_{k-1}^*,\ldots,k-1} \cdot \left\{ Z_r^* + A_{r+1} + \sum_{i=r+2}^{k} D_i \sum_{j=1}^{i-1} h_j + c_{k+1} \sum_{i=r+1}^{k} D_i \right\}, \quad k=1,\ldots,M, \; Z_0^*=0$$

$$j_k^* = \operatorname*{argmin}_{r=j_{k-1}^*,\ldots,k-1} \cdot \left\{ Z_r^* + A_{r+1} + \sum_{i=r+2}^{k} D_i \sum_{j=1}^{i-1} h_j + c_{k+1} \sum_{i=r+1}^{k} D_i \right\}, \quad k=2,\ldots,M, \; j_1^*=0$$

The Wagner–Whitin procedure is recursive going forward in time. At the beginning, we compute the optimal cost when the planning horizon is just one period long, which is of course the cost of producing D_1 units JIT. In general, in the end of iteration $k-1$, we have computed the optimal production plan that covers periods 1,..., $k-1$ for any positive k less than the total number of periods in the overall planning horizon, and we have determined the last period in the interval $\{1, \ldots, k-1\}$ during which production must occur when solving for only that period interval of length $k-1$, which we denote by j^*_{k-1}. Now, in iteration k, we can compute the optimal plan for the interval $\{1,\ldots, k\}$ by picking the minimum cost plan from the $k - j^*_{k-1}$ alternatives of producing so as to cover the whole k periods by producing only up to period j^*_{k-1}, or by producing only up to period $j^*_{k-1} + 1$, or ... or by producing up to period k. For each alternative choice, we evaluate the optimal cost of such a plan by adding up production, fixed-charge and holding costs incurred by the choice, and we pick the best one.

The model has a strong theoretical elegance and, as mentioned before, it has been at the basis of the algorithms that were developed within the context of MRP and MRP II systems. However, the assumption of infinite capacity and known demand throughout the planning horizon are very strong assumptions whose violation in real-world settings should be considered inevitable. How much the computed plans will deviate from optimality, is a question that can best be answered by planners doing sensitivity analysis of their results (with regard to the demand variability) to get an understanding of how flat the landscape of the optimal solutions is in each case. Finally, notice that under the more realistic assumption of finite period capacities $C_i | i = 1, \ldots, M$ the Wagner–Whitin property by which, in an optimal plan there is no production until the inventory level falls to zero, no longer holds. In this case, the following mixed-integer program (MIP) provides a model for the optimal solution to the dynamic lot-sizing problem, but is no longer solvable via dynamic programming techniques in polynomial time.

$$\min_{x,y,I} \sum_{i=1}^{M} \{c_i x_i + A_i y_i + h_i I_i\}$$

$$\text{s.t.} \begin{cases} I_i = I_{i-1} + x_i - D_i & i = 1, \ldots, M, \quad I_0 = 0 \\ x_i \leq C_i y_i & i = 1, \ldots, M \\ 0 \leq x_i & i = 1, \ldots, M \\ 0 \leq I_i & i = 1, \ldots, M \\ y_i \in \{0, 1\} & i = 1, \ldots, M \end{cases}$$

The MIP above has M binary variables $y_i | i = 1, \ldots, M$ each of them representing the decision to produce in the respective period or not. If any quantity is produced during the ith period, the value of the variable y_i will equal 1, because this is the only way to satisfy the second set of constraints $x_i \leq C_i y_i$, and vice versa, if in the optimal plan no production occurs in that period, then the value of the variable y_i will go to zero in order to have a minimum cost solution. The introduction of the binary variables is a usual trick in mathematical programming (MP) to represent logical implications (logical implications are logical expressions of the form "IF x holds THEN y must also hold") as discussed in Chap. 1. Whenever logical implications must be expressed, binary or discrete variables must necessarily be introduced to express this relationship. Nevertheless, lately, optimization software packages have advanced solver features that allow one to merge traditional mixed linear integer programming with constraint programming (CP), so that logical implication constraints can be expressed directly into the problem to be solved without the need to convert it into a new MIP using the technique mentioned above. The open-source software package SCIP (http://scip.zib.de) features such a state-of-the-art solver. Most advanced commercial optimization codes (cplex®, gurobi®, etc.) also have similar features.

The first constraint of the above MIP model is the material-flow balance constraint that we have seen in our earlier models. The problem complexity which is dealt with by the introduction of the binary variables y, arises from the fixed-charge set-up costs assumed in the dynamic lot-sizing problem. Again, if demand cannot be met because of the capacity constraints of the problem, the MIP stated above, as its predecessor, is infeasible.

In this case (when the demand during the planning horizon is such that it cannot be met using the production capacity available), both the mathematical programs discussed above become infeasible, and no solution is provided. Therefore, returning to the first problem discussed, to solve the problem of minimizing inventory costs while meeting demand as best as possible we re-formulate the problem as follows:

$$\max_{x,s,h} . \sum_{i=1}^{N} (s_i - h_i)$$

$$\text{s.t.} \begin{cases} h_i = h_{i-1} + x_i - s_i & i = 1, \ldots, N \\ \min\{c_i, d_i\} \leq s_i \leq d_i & i = 1, \ldots, N \\ 0 \leq x_i \leq c_i & i = 1, \ldots, N \\ 0 \leq h_i & i = 1, \ldots, N \quad h_0 = I_0 \geq 0 \end{cases}$$

The above LP introduces the variables $s_i | i = 1, \ldots, N$ which represent the actual amount of sales of each period. This problem is always feasible regardless of any peaks in the forecasted demand, and its size remains very reasonable: there are $3N$ variables, N equality constraints and $3N$ boxing constraints, which makes the problem trivial to solve using any available implementation of the simplex method even for large values of N (usually N will not be more than 100).

In the following, we make a number of assumptions about the business processes that determine what constitutes optimality of a production plan. We are concerned with perishable products which pose the extra constraint that they have limited life-times. Alternatively, we could add depreciation functions that discount the value of inventory as time goes by, and maximize a global profit function.

First, we discuss a number of generic constraints that are present at the aggregate decomposition level, for companies possessing multiple (n) production lines $L = \{\ell_1, \ldots, \ell_n\}$, each of them capable of producing a finite set of products $\Pi(\ell)$, at a rate of production $r_i | i = 1, \ldots, n$. We denote the set of lines along which a product π can be produced by $L(\pi)$. These constraints, in a general form can be stated as follows:

c1 Capacity constraints incurred by the finite rates of line production, and by maintenance constraints.
c2 Budgetary or flexibility concerns limiting the number of shifts each line may operate.
c3 Product expiration date constraints.
c4 Distribution concerns influencing the distribution of production to the various plants.
c5 Line coupling constraints (lines tied together).
c6 Maturation constraints forcing certain products to be scheduled for production at least one week before their scheduled distribution to the markets.
c7 Operational practice considerations favoring the scheduling of certain products to start on or no later than a certain day of the week.

At the aggregate production planning level, only constraints c[1–4] are taken into consideration, as constraints c[5–7] are by their nature tied to the finer levels of planning and scheduling.

There is a time horizon of M aggregate time periods (usually, a period is one month long) for which we will specify a production plan for each line. A production plan for a line l consists of specifying quantities $x_{i,j}^{\pi,l}$ for each product π that the line may produce throughout the periods $i = 1, \ldots, M$ of the time horizon; these quantities are to be forwarded to the distribution centers in periods $j = i, \ldots, i + T(\pi)$ where $T(\pi)$ is the life-time of product π so as to meet forecast demands $d_i^{\pi} \geq 0$ for each product throughout the periods of the time horizon. This (long-term) production plan also includes the number of shifts needed to operate each line for each time period. The shifts a_i^l allocated to each line during each period i are the minimum necessary to produce the quantities described in the production schedule and should be less than a certain pre-specified number h_i^l dictated by the budgetary concerns constraint. However, coming from budgetary concerns, this last requirement does not represent a hard constraint whose violation would render a production schedule physically impossible. Rather, it forms a strong guideline that (heavily) influences the schedules produced. In fact, these budgetary concerns on the number of shifts are so strong that it is always better to produce quantities of products ahead of time (no matter the storage costs associated) than introduce more shifts than the maximum desired number of shifts indicated by the company for a given period. An added benefit is that some lines may be left with unused planned capacity during some periods to accommodate a possible sudden increase in the product demands of subsequent periods, thus allowing the company more flexibility for its operations throughout the time horizon. Still, the number of shifts cannot exceed a certain maximum $H_i^l = D_i S$ dictated by the total number of working calendar days D_i in a period and the maximum number S of shifts in a day. This upper bound imposes a true hard constraint on the maximum number of shifts (for any line) for a given period, and places a true upper bound on the maximum productivity of any line for any given period of time that is $u_i^l = r_l H_i^l t$ where t is the number of hours in any shift (constant). The objective of the multi-commodity aggregate planning problem (MCAP) problem then, is to find a plan that determines production of quantities of each product in each line throughout the time horizon so as to:

- Meet demands of each period for each product as best as possible.
- Minimize the total number of extra shifts needed to meet the demand, and
- Minimize storage and inventory costs while maximizing product quality in terms of freshness when production of commodities ahead of time is needed.

The above goals are clearly in conflict with each other, and therefore a trade-off has to be made somehow. This is usually achieved by setting priorities on each objective. We will discuss the resulting mathematical program that arises when the hierarchy of objectives is as follows:

1. The produced plan must provide the maximum level of customer service possible, i.e. meet demand as best as possible (even if it means that extra shifts must be utilized).
2. Among all plans that meet objective 1, find the plan that minimizes extra shifts needed, and finally.
3. Among all plans that are optimal with respect to the previous objective, choose the one that minimizes inventory costs (and simultaneously maximizes product quality).

The procedure that we present below applies equally well when the ordering of the objectives changes in any arbitrary way, even though the resulting mathematical model in its objective function changes as well. Notice also, that constraints arising from geographical considerations, namely that different sets of lines in different plants widely separated geographically should service the demand of nearby distribution centers, can easily be handled within this generic model. The constraint is well justified, because a major source of costs for the company is the transportation costs. In this case, a series of MCAP problems can be solved, where in each problem only the lines of a given production plant are taken into account, and only the demands of the nearest markets are met. Then, in a final coupling stage, an MCAP problem is solved where all the production lines are asked to use the remainder of their capacities to meet demands of distant markets.

The above objectives give rise to a hierarchical view of the MCAP problem itself, where various (conflicting) goals must be met in a well defined order (specified above). The conflicts between the goals are resolved using their relative priorities. It is now possible to formulate a total cost function that must be minimized subject to a certain set of constraints. These constraints consist of a large number of flow balance (network) constraints, and integrality constraints regarding the maximum number of shifts to be used each period in each production line.

3.1.1 Formulation of the Multi-Commodity Aggregate Planning Problem

In order to formulate MCAP as a MIP, we introduce a cost K associated with the introduction of any extra shift above the desired level h_i^l but below the maximum number of shifts in the line H_i^l. This cost K is big enough so that in the optimal solution, extra shifts should never be used when a feasible production plan exists (covers demands of all products throughout the time horizon) that does not require extra shifts. In order to always produce a schedule that covers the products' demands of all periods as much as possible—even when complete coverage is not possible—we allow the production $s_i^\pi | i = 1, \ldots, M, \pi \in \Pi$ of an unlimited number of the commodities. However, we incur a cost J associated with such commodities. This cost is extremely high since such commodities cannot physically be produced (they exceed the capabilities of the lines). In particular, this cost has to be so high, that in an optimal solution the variables s_i^π are always zero when there exists a way to meet the products' demands in each period without exceeding the capacities of any line. Finally, in order to differentiate between preferences among products that face possible stock-out (e.g. it might be better to not be able to meet demand for a product of low demand that few people buy than face stock-outs for popular products) we introduce weights $w_i^\pi \geq 1$ that multiply the quantities s_i^π to form the total cost of stock-outs.

At this point, the inputs and parameters of the problem are as follows:

$$\text{inputs}\begin{cases} M, \text{time horizon} \\ L, \text{number of lines} \\ \Pi, \text{total set of products with cardinality P} \\ S, \text{number of shifts per day} \\ t, \text{number of hours in a shift} \\ T(\pi) \in \mathbb{N}, \pi \in \Pi, \text{the life - times of the products} \\ D_i \in \mathbb{N}, i = 1, \ldots, M, \text{the calendar days of each period} \\ r_l, l = 1, \ldots, L, \text{the line rates} \\ h_i^l \in \mathbb{N}, i = 1, \ldots, M, l = 1, \ldots, L, \text{the maximum desired number of shifts} \\ d_i^\pi \in \mathbb{N}, i = 1, \ldots, M, \pi \in \Pi, \text{the monthly product demands} \\ \Pi(l), l = 1, \ldots, L, \text{the } \mathit{set} \text{ of products each line produces} \end{cases}$$

$$\text{parameters}\begin{cases} w_i^\pi \geq 1, i = 1, \ldots, M, \pi \in \Pi, \text{relative importance of the products} \\ K, \text{cost coefficient associated with more shifts than desired} \\ J, \text{cost coefficient associated with product stock} - \text{outs} \end{cases}$$

We can now formulate MCAP as follows:

$$\text{(MCAP)} \quad \min_{x,y,a,s} \sum_{i\in M}\sum_{\ell\in L}\sum_{\pi\in\Pi(\ell)}\sum_{j=i}^{\min[M,i+T(\pi)]} 2^{j-i}x_{i,j}^{\pi,\ell} + K\sum_{i\in M}\sum_{\ell\in L} y_i^\ell + J\sum_{i\in M}\sum_{\pi\in\Pi} w_i^\pi s_i^\pi$$

subject to:

$$\sum_{\ell\in L(\pi)}\sum_{j=\max(1,i-T(\pi))}^{i} x_{j,i}^{\pi,\ell} + s_i^\pi = d_i^\pi \quad \forall i\in M, \pi\in\Pi$$

$$\sum_{\pi\in\Pi(\ell)}\sum_{j=i}^{\min(i+T(\pi),M)} x_{i,j}^{\pi,\ell} \le a_i^\ell r_\ell t \quad \forall \ell\in L, i\in M$$

$$y_i^\ell \ge a_i^\ell - h_i^\ell \quad \forall \ell\in L, i\in M$$

$$x_{i,j}^{\pi,\ell} \in \mathbb{N} \quad \forall \pi\in\Pi, \ell\in L(\pi), i\in M, \quad j=i,\ldots,\min(i+T(\pi),M)$$

$$y_i^\ell \in \mathbb{N} \quad \forall i\in M, \ell\in L$$

$$s_i^\pi \in \mathbb{N} \quad \forall i\in M, \pi\in\Pi$$

$$a_i^\ell \le D_i^\ell S, a_i^\ell \in \mathbb{N} \quad \forall i\in M, \ell\in L$$

Note the introduction of the exponential costs of storage 2^{j-i} in the objective function. They ensure that between two alternative solutions that meet the demands of various months with the same number of shifts the one that sends products to be consumed faster—and therefore maintains a higher level of freshness for the oldest product—is preferred. Cost coefficients that are linear in storage time do not share this property (exponentially increasing cost-coefficients will also be needed in the preferential bidding problem and will be introduced in the section discussing personnel scheduling.) This problem always has an optimal solution that indeed satisfies the hierarchy of goals set forth previously.

Theorem 3.1 *The (MCAP) problem with* $K = K^* = 2^M \sum_{i\in M}\sum_{\pi\in\Pi} d_i^\pi$ *and* $J = J^* = K^*\left(\sum_{i\in M}\sum_{l\in L}(D_i^l S - h_i^l) + 1\right)$ *has optimal solutions (x*, y*, a*, s*) that among all points in the feasible set:*

1. *Minimize* $\sum_{i\in M}\sum_{\pi\in\Pi} w_i^\pi s_i^\pi$
2. *Minimize the number of extra shifts required to produce the quantities* $d_i^\pi - (s^*)_i^\pi$
3. *Minimize inventory costs among all feasible points that minimize the last two terms of the objective function.*

Proof Notice that (MCAP) is a linear mixed-integer programming problem and has at least one feasible solution, i.e. $(x, y, a, s) = (0, 0, 0, d)$. The objective function is also bounded from below by zero. Therefore the problem has an optimal solution.

Now observe that K^* is an upper bound on the value of the first term in the objective function (when the variables are subject to the constraints), and that J^* is an upper bound on the value of the first two terms of the objective function for all points in the feasible set. Let (x^*, y^*, a^*, s^*) be an optimal solution, with cost z^*. Then, the value $p^* = \sum_{i\in M}\sum_{\pi\in\Pi} w_i^\pi s_i^{*\pi}$ is minimum among all points in the feasible set. This is because, if there was a point in the feasible set with $\sum_{i\in M}\sum_{\pi\in\Pi} w_i^\pi s_i^\pi = p < p^*$, then, this point's cost z would be at most $K^* + K^*\left(\sum_{i\in M}\sum_{\ell\in L}\left(D_i S - h_i^\ell\right)\right) + J^* p = J^*(p+1) \leq J^* p^* < z^*$ where we have used the fact that the weights are greater than or equal to one, and that the variables $s_i^\pi \in \mathbb{N}$, and also the fact that under the assumption of the existence of a solution with $p < p^*$ the production variables x of the optimal solution cannot all be zero. But then, this feasible solution has a value strictly lower than the optimum value, a contradiction.

Further, for any feasible point (x, y, a, s), define $q = \sum_{i\in M}\sum_{\ell\in L} y_i^\ell$. Among all feasible points that minimize the last term of the objective function, the optimal solution (x^*, y^*, a^*, s^*) minimizes the number of extra shifts needed to produce the quantities x^*. Otherwise, there would exist a feasible point (x, y, a, s) with minimal sum of weighted unmet demands $s_i^\pi \in \mathbb{N}$ that would require fewer total extra shifts than $\sum_{i\in M}\sum_{\ell\in L} y_i^{*l}$ therefore, its cost would be at most $K^*(q+1) + J^* p^* \leq K^* q^* + J^* p^* < z^*$, again, a contradiction.

Finally, among all feasible points that minimize the last two terms of the objective function of (MCAP), the optimal solution also minimizes inventory (storage) costs, due to the exponential storage costs coefficients which not only minimize total storage time, but also maximize "product quality" in terms of "freshness" as they force products to be scheduled for consumption as soon as possible QED.

Now the constraints of the (MCAP) model can be divided in two categories. The first category contains the flow balance constraints coupled with capacity constraints implied from the production rates of each line and number of shifts each line operates with during each period, while the second category consists of the constraints determining the number of shifts (and extra shifts) used to produce the quantities x. Once a decision has been made for the values of the variables a_i^l (Shift allocation or SA part), solving the resulting problem (Production schedule or PS) only requires solving a linear network flow problem (see Fig. 3.2).

In Fig. 3.2, the demand of the leftmost node (labeled 'Sink') is equal to the sum of capacities of all the lines over all periods given the number of shifts a_i^l that each line should operate. The next column of nodes has supplies equal to the summation of the corresponding line's capacity over all periods. The nodes in the next column have zero demand, and the last column of nodes has demands equal to the forecast product demands d_i^π. The rightmost node (labeled 'Infeas') has supply equal to the sum of all products' demands over all periods. The arcs connecting the first column of nodes to the second have zero costs and capacities equal to the line's capacity for a given period. The arcs connecting the second column of nodes to the third have unlimited capacities but costs that are exponential in the period

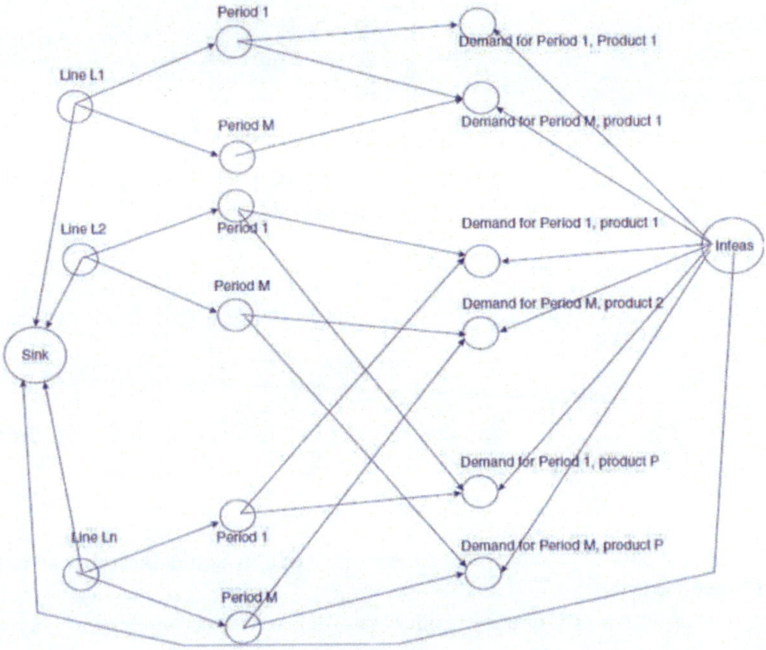

Fig. 3.2 Network flow structure of PS sub-problem of MCAP

difference (2^{j-i}). (Optionally, such arc costs between nodes representing the same production and distribution periods i, j for the same product can be decreasing in line rate so as to prefer faster lines.) And the arcs connecting the rightmost node to the last column of nodes have unlimited capacities and costs equal to $w_i^{\pi} J^*$. All other arcs have zero cost and unlimited capacity.

3.1.2 Solving Multi-Commodity Aggregate Planning Problem

Even though it is possible to solve the (MCAP) as an instance of a MIP problem, it is also possible to use a two-stage decomposition approach that exploits the network flow structure of the problem when the shift variables are set (see Fig. 3.3).

In particular, we split the problem into two parts, namely SA and PS, with a final post-processing local optimization phase, as described in the following. First, we fix $a_i^l = h_i^l$ for all periods and products and solve the corresponding network flow problem (using the network simplex method). If the optimal solution of this problem does not set any s_i^{π} to a value greater than zero, the algorithm ends. We have found an optimal solution to the problem. Otherwise, there exist some periods during which some products face stock-outs. For each such product, we perform the following steps (in order of importance of each product):

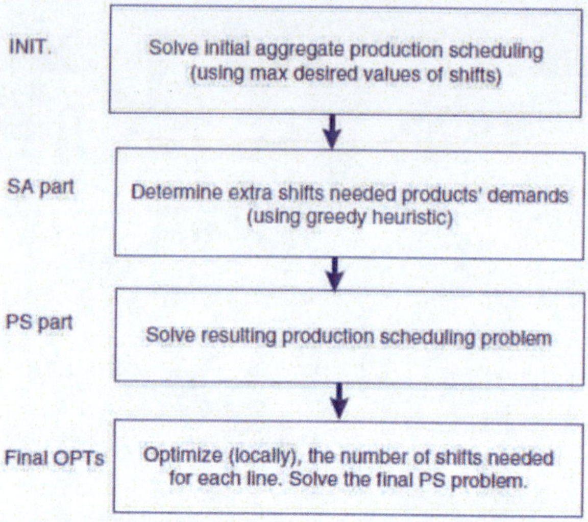

Fig. 3.3 Two-stage decomposition approach to MCAP

0. while there exists a product that faces stock-out in some period that has not been examined do.
1. Sort the lines which produce the product in order of decreasing rates into a list PL.
2. For each period m of unmet demand do:
3. Increase the number of shifts for each line l in PL that has not yet been assigned the maximum number of shifts (starting with periods closest to the problematic period) until the demand can be met (or the lines reach their maximum capacity).
4. Solve the corresponding network flow problem using the network simplex method.
5. end for
6. end while
7. Set $a_i^l = \left\lceil \frac{\sum_{j,\pi} x_{i,j}^{\pi,l}}{r_l t} \right\rceil$

End

3.1.3 Multi-Commodity Aggregate Planning Problem as Multi-Criterion Optimization Problem

From the beginning of the discussion of the MCAP problem, it becomes clear that several conflicting objectives have to be optimized simultaneously. This is in contrast to standard optimization, where a single function has to be optimized subject to many constraints. Multi-objective optimization then, attempts to determine the set of so-called pareto-optimal solutions, all of which share the property that there exists no feasible point that has objective values for each of the objective criteria that is better than the corresponding values of the pareto-optimal solution. In other words, a pareto-optimal solution x_P for a multi-criterion objective optimization problem may not yield optimal solution for any of its many objectives, but there cannot exist any feasible point that is better than the solution x_P in *all* the objectives.

If an order of importance is given for the objectives, the so-called lexicographic method can be used to determine the pareto-optimal solution. Consider the problem MMP defined as follows.

$$(\text{MMP}) \min_{x \in S} f(x) = \begin{bmatrix} f_1(x) \\ \vdots \\ f_m(x) \end{bmatrix}$$

$$\text{s.t.} \begin{cases} c_i(x) \geq 0, & \forall i \in I \\ c_i(x) = 0, & \forall i \subset E \end{cases}$$

where I is a possibly empty index set of inequality constraints, E is another possibly empty index set of equality constraints, $f : \mathbb{R}^n \supseteq S \to \mathbb{R}^m$ is a vector function with domain of definition the set $S \subseteq \mathbb{R}^n$ which is also the domain of definition of each of the functions $c_i(x)$. Now, assume further that the objective functions f_i $i = 1, \ldots, m$ are ordered in order of importance, so that in the optimal solution, the fist objective function must be at its optimal value subject to the constraints, the second objective function value must be at its minimum among all solutions that optimize the first objective subject to the problem constraints, and so forth. In that case, the lexicographic method consists of solving a series of m single-objective MP problems. The first problem of this series is the following

$$\begin{aligned} &\min_{x \in S} f_1(x) \\ &\text{s.t.} \begin{cases} c_i(x) \geq 0, & \forall i \in I \\ c_i(x) = 0, & \forall i \in E \end{cases} \end{aligned} \tag{3.1}$$

Let $x_{(1)}$ denote the solution of (3.1), and let $f^{[1]} = f_1\left(x_{(1)}\right)$. The ith mathematical programming problem in the series ($i = 2,\ldots, m$), denoted by $(\text{MP})_i$ now becomes

$$\begin{aligned} &(\text{MP})_i \quad \min_{x \in S} \quad f_i(x) \\ &\text{s.t.} \begin{cases} c_i(x) \geq 0, & i \in I \\ c_i(x) = 0 & i \in E \\ f_j(x) = f^{[j]}, & j = 1, \ldots, i-1 \end{cases} \end{aligned} \tag{3.2}$$

where the values $f^{[j]}$ are defined as the optimal values of problem $(\text{MP})_j$ and $x_{(j)}$ is the minimizer of the corresponding problem. The final solution $x^* = x_{(m)}$ is clearly a feasible point that among all points in the feasible set optimizes the first objective. Further, the point $x^*$is in the set $S_i = \left\{x \in \mathbb{R}^n | x \in \arg\min(MP)_i\right\}$ for all $i = 1,\ldots, m$ and therefore it happens to optimize the second objective among all feasible points that optimize the first objective, and so on.

Some times—though rarely—one may add a small slack into the formulation of the problems $(MP)_i$ by modifying the last set of constraints in Eq. 3.1 to be $f_j(x) \leq f^{[j]} + a_j \quad j = 1, \ldots, i-1$. In this case, the values a_j indicate how much we are willing to worsen a higher-priority objective in order to obtain a better lower-priority objective. In the original formulation of the problem such an action of course does not make sense. But if the prioritization of the objectives is not completely clear and imposed for very good reasons, then it is conceivable that one may want to run some sensitivity analysis scenarios by solving the problems with various values of the slack parameters a_j to see if it is possible to worsen a bit one objective in exchange of a serious improvement in another. Nevertheless, one must not confuse this procedure for a search of pareto-optimal solutions. Indeed, when introducing these slack variables, there is no guarantee that the resulting solutions will be in the efficient frontier (i.e. be pareto optimal).

We can now easily formulate MCAP as a series of three standard MP problems. The first problem in the series is the following $(\text{MCAP})_1$

$$(\text{MCAP})_1 \quad \min_{x,y,a,s} \sum_{i \in M} \sum_{\pi \in \Pi} w_i^\pi s_i^\pi$$

subject to:

$$\sum_{\ell \in L(\pi)} \sum_{j=\max(1,i-T(\pi))}^{i} x_{j,i}^{\pi,\ell} + s_i^\pi = d_i^\pi \quad \forall i \in M, \pi \in \Pi$$

$$\sum_{\pi \in \Pi(\ell)} \sum_{j=i}^{\min(i+T(\pi),M)} x_{i,j}^{\pi,\ell} \leq a_i^\ell r_\ell t \quad \forall \ell \in L, i \in M$$

$$y_i^\ell \geq a_i^\ell - h_i^\ell \quad \forall \ell \in L, i \in M$$

$$x_{i,j}^{\pi,\ell} \in \mathbb{N} \quad \forall \pi \in \Pi, \ell \in L(\pi), i \in M, \quad j = i, \ldots, \min(i + T(\pi), M)$$

$$y_i^\ell \in \mathbb{N} \quad \forall i \in M, \ell \in L$$

$$s_i^\pi \in \mathbb{N} \quad \forall i \in M, \pi \in \Pi$$

$$a_i^\ell \leq D_i^\ell S, a_i^\ell \in \mathbb{N} \quad \forall i \in M, \ell \in L$$

Let $f^{[1]}$ denote the optimal value of $(\text{MCAP})_1$. Now, the second problem $(\text{MCAP})_2$ is as follows:

$$(\text{MCAP})_2 \quad \min_{x,y,a,s} \sum_{i \in M} \sum_{\ell \in L} y_i^\ell$$

subject to:

$$\sum_{\ell \in L(\pi)} \sum_{j=\max(1,i-T(\pi))}^{i} x_{j,i}^{\pi,\ell} + s_i^\pi = d_i^\pi \quad \forall i \in M, \pi \in \Pi$$

$$\sum_{\pi \in \Pi(\ell)} \sum_{j=i}^{\min(i+T(\pi),M)} x_{i,j}^{\pi,\ell} \leq a_i^\ell r_\ell t \quad \forall \ell \in L, i \in M$$

$$y_i^\ell \geq a_i^\ell - h_i^\ell \quad \forall \ell \in L, i \in M$$

$$x_{i,j}^{\pi,\ell} \in \mathbb{N} \quad \forall \pi \in \Pi, \quad \ell \in L(\pi), \quad i \in M, \quad j = i, \ldots, \min(i + T(\pi), M)$$

$$y_i^\ell \in \mathbb{N} \quad \forall i \in M, \ell \in L$$

$$s_i^\pi \in \mathbb{N} \quad \forall i \in M, \pi \in \Pi$$

$$a_i^\ell \leq D_i^\ell S, a_i^\ell \in \mathbb{N} \quad \forall i \in M, \ell \in L$$

$$\sum_{i \in M} \sum_{\pi \in \Pi} w_i^\pi s_i^\pi = f^{[1]}$$

Now, let the solution of $(\text{MCAP})_2$ be $f^{[2]}$. The third and final mathematical program that optimally solves the original (MCAP) problem can be written down as follows:

$$(\text{MCAP})_3 \quad \min_{x,y,a,s} \sum_{i \in M} \sum_{\ell \in L} \sum_{\pi \in \Pi(\ell)} \sum_{j=i}^{\min(M, i+T(\pi))} 2^{j-i} x_{i,j}^{\pi,\ell}$$

subject to:

$$\sum_{\ell \in L(\pi)} \sum_{j=\max(1, i-T(\pi))}^{i} x_{j,i}^{\pi,\ell} + s_i^\pi = d_i^\pi \quad \forall i \in M, \pi \in \Pi$$

$$\sum_{\pi \in \Pi(\ell)} \sum_{j=i}^{\min(i+T(\pi), M)} x_{i,j}^{\pi,\ell} \leq a_i^\ell r_\ell t \quad \forall \ell \in L, i \in M$$

$$y_i^\ell \geq a_i^\ell - h_i^\ell \quad \forall \ell \in L, i \in M$$

$$x_{i,j}^{\pi,\ell} \in \mathbb{N} \quad \forall \pi \in \Pi, \ell \in L(\pi), i \in M, j = i, \ldots, \min(i + T(\pi), M)$$

$$y_i^\ell \in \mathbb{N} \quad \forall i \in M, \ell \in L$$

$$s_i^\pi \in \mathbb{N} \quad \forall i \in M, \pi \in \Pi$$

$$a_i^\ell \leq D_i^\ell S, a_i^\ell \in \mathbb{N} \quad \forall i \in M, \ell \in L$$

$$\sum_{i \in M} \sum_{\pi \in \Pi} w_i^\pi s_i^\pi = f^{[1]}$$

$$\sum_{i \in M} \sum_{\ell \in L} y_i^\ell = f^{[2]}$$

Even though the series of problems $(\text{MCAP})_{1,\ldots,3}$ optimally solves our original problem, computationally it is much more efficient to use the two-stage decomposition algorithm of Sect. 3.1.2 than solving these three problems. The theoretical reason why is left as an exercise to the reader, as are the implementation details of an algorithm based on this approach.

3.2 Production Scheduling

At the bottom of the hierarchical decomposition of the planning and scheduling problems lies the production scheduling problem. It is concerned with the daily or weekly operations of an organization. In case of manufacturing organizations, production scheduling is sometimes synonymous to job-shop scheduling, discussed in the next section. However, production scheduling also may refer to the short-range planning of operations (regardless of whether or not a job-shop problem will have to be solved at an even more fine-grained level of time decomposition).

Production scheduling in the multi-commodity production planning context discussed in the previous section, is the problem where line coupling constraints as well as transportation costs have to be taken into account when deciding for the next week, how much to produce, of what product, and where to produce it, given that production lines are dispersed in widely different geographic locations and that there is a forecast of the local demand for products in certain known geographic locations.

Line coupling constraints, if present, can be easily handled using a simple trick. If two lines l_1, l_2 must be coupled together within an aggregate time period, consider them as one new line $l_{1,2}$with rate $r_{1,2} = r_1 + r_2$ that can produce the product line $\Pi(l_{1,2}) = \Pi(l_1) \cap \Pi(l_2)$.

The multi-commodity production scheduling problem then (MCPS) takes into account fine-grain demand information $d_i^{\pi,p} | i = 1, \ldots, T', \pi \in \Pi, p \in U$ where T' represents the number of fine-grain periods remaining in the current aggregate period T (each fine-grain period usually representing one week), as well as unit transportation costs $t_\pi^{l,p}$ to transport a unit of product π from the plant where line l is located to location place $p \in U$. Assuming that the objectives at this level can be ordered in order of importance as follows:

1. Maximize service level.
2. Minimize extra shifts costs and transportation costs.
3. Minimize inventory holding costs,the MCPS can be solved by solving a series of three Mathematical Programs, in the spirit of the multi-objective optimization discussed in the previous section. Notice that at this level of time granularity, costs of extra shifts incurred are accurately known, so the second objective can be modeled well. Assume that the cost of an extra shift for line l in period $i \in M' = \{1, \ldots, T'\}$ is b_i^l.

The first problem, $(\text{MCPS})_1$is defined as follows:

$$(\text{MCPS})_1 \qquad \min_{x,y,a,s} \sum_{i \in M'} \sum_{\pi \in \Pi} w_i^\pi s_i^\pi$$

subject to:

$$\sum_{\ell \in L(\pi)} \sum_{p \in U} \sum_{j=\max(1,i-T(\pi))}^{i} x_{j,i}^{\pi,\ell,p} + s_i^\pi = \sum_{p \in U} d_i^{\pi,p} \quad \forall i \in M', \pi \in \Pi$$

$$\sum_{\pi \in \Pi(\ell)} \sum_{p \in U} \sum_{j=i}^{\min(i+T(\pi),M')} x_{i,j}^{\pi,\ell,p} \leq a_i^\ell r_\ell t \quad \forall \ell \in L, i \in M'$$

$$y_i^\ell \geq a_i^\ell - h_i^\ell \quad \forall \ell \in L, i \in M'$$

$$x_{i,j}^{\pi,\ell,p} \in \mathbb{N} \quad \forall \pi \in \Pi, \ell \in L(\pi), i \in M', p \in U, j = i, \ldots, \min(i + T(\pi), M')$$

$$y_i^\ell \in \mathbb{N} \quad \forall i \in M', \ell \in L$$

$$s_i^\pi \in \mathbb{N} \quad \forall i \in M', \pi \in \Pi$$

$$a_i^\ell \leq D_i^\ell S, a_i^\ell \in \mathbb{N} \quad \forall i \in M', \ell \in L$$

Let $f^{[1]}$ denote the optimal value of $(\mathrm{MCPS})_1$. The $(\mathrm{MCPS})_2$ which minimizes extra shifts plus transportation costs subject to the constraint that service level is maximum attainable, is then defined as follows

$$(\mathrm{MCPS})_2 \quad \min_{x,y,a,s} \sum_{i \in M'} \sum_{l \in L} b_i^l y_i^l + \sum_{p \in U} \sum_{\pi \in \Pi} \sum_{l \in L(\pi)} \sum_{i \in M'} \sum_{j=i}^{\max(T',i+T(\pi))} t_\pi^{l,p} x_{i,j}^{\pi,l,p}$$

subject to:

$$\sum_{\ell \in L(\pi)} \sum_{p \in U} \sum_{j=\max(1,i-T(\pi))}^{i} x_{j,i}^{\pi,\ell,p} + s_i^\pi = \sum_{p \in U} d_i^{\pi,p} \quad \forall i \in M', \pi \in \Pi$$

$$\sum_{\pi \in \Pi(\ell)} \sum_{p \in U} \sum_{j=i}^{\min(i+T(\pi),M')} x_{i,j}^{\pi,\ell,p} \leq a_i^\ell r_\ell t \quad \forall \ell \in L, i \in M'$$

$$y_i^\ell \geq a_i^\ell - h_i^\ell \quad \forall \ell \in L, i \in M'$$

$$x_{i,j}^{\pi,\ell,p} \in \mathbb{N} \quad \forall \pi \in \Pi, \ell \in L(\pi), i \in M', p \in U, j = i, \ldots, \min(i + T(\pi), M')$$

$$y_i^\ell \in \mathbb{N} \quad \forall i \in M', \ell \in L$$

$$s_i^\pi \in \mathbb{N} \quad \forall i \in M', \pi \in \Pi$$

$$a_i^\ell \leq D_i^\ell S, a_i^\ell \in \mathbb{N} \quad \forall i \in M', \ell \in L$$

$$\sum_{i \in M', \pi \in \Pi} w_i^\pi s_i^\pi = f^{[1]}$$

Assume the optimal value of the problem $(\mathrm{MCPS})_2$ is $f^{[2]}$. The final problem to be solved then is the following $(\mathrm{MCPS})_3$

$$(MCPS)_3 \quad \min_{x,y,a,s} \sum_{i \in M'} \sum_{\ell \in L} \sum_{\pi \in \Pi(\ell)} \sum_{p \in U} \sum_{j=i}^{\min(M',i+T(\pi))} 2^{j-i} x_{i,j}^{\pi,\ell,p}$$

subject to:

$$\sum_{\ell\in L(\pi)}\sum_{p\in U}\sum_{j=\max(1,i-T(\pi))}^{i} x_{j,i}^{\pi,\ell,p} + s_i^{\pi} = \sum_{p\in U} d_i^{\pi,p} \quad \forall i\in M', \pi\in\Pi$$

$$\sum_{\pi\in\Pi(\ell)}\sum_{p\in U}\sum_{j=i}^{\min(i+T(\pi),M')} x_{i,j}^{\pi,\ell,p} \le a_i^{\ell} r_{\ell} t \quad \forall \ell\in L, i\in M'$$

$$y_i^{\ell} \ge a_i^{\ell} - h_i^{\ell} \quad \forall \ell\in L, i\in M'$$

$$x_{i,j}^{\pi,\ell,p}\in\mathbb{N} \quad \forall \pi\in\Pi, \ell\in L(\pi), i\in M', p\in U, j=i,\ \ldots,\ \min(i+T(\pi),M')$$

$$y_i^{\ell}\in\mathbb{N} \quad \forall i\in M', \ell\in L$$

$$s_i^{\pi}\in\mathbb{N} \quad \forall i\in M', \pi\in\Pi$$

$$a_i^{\ell}\le D_i^{\ell}S, a_i^{\ell}\in\mathbb{N} \quad \forall i\in M', \ell\in L$$

$$\sum_{i\in M',\pi\in\Pi} w_i^{\pi}s_i^{\pi} = f^{[1]}$$

$$\sum_{i\in M'}\sum_{l\in L} b_i^l y_i^l + \sum_{p\in U}\sum_{\pi\in\Pi}\sum_{l\in L(\pi)}\sum_{i\in M'}\sum_{j=i}^{\max(T',i+T(\pi))} t_{\pi}^{l,p} x_{i,j}^{\pi,l,p} = f^{[2]}$$

Notice that solving this series of problems can be considerably harder than the multi-commodity aggregate production planning problem, due to the decomposition of the variables $x_{i,j}^{\pi,l}$ to sums of finer-level variables $x_{i,j}^{\pi,l,p}$ that determine the level of quantity of product π that must be produced in period i in line l to be distributed in period j to market p. Clearly, the equation $\sum_{p\in U} d_i^{\pi,\bar{p}} = d_i^{\pi}$ holds for all periods i and products π.

The issue of determining the sequence of operations in the production lines on a given day is the objective of Job-Shop Scheduling, and it will be discussed next.

3.3 Job-Shop Scheduling

Short-range production scheduling and job-shop scheduling applies at the level of individual work-stations in the factory floor (Silver et al. 1998), where operators and supervisors have to know which job to process next as well as when to start processing; in addition, some times they will have to make a decision as to where to route a job in progress (in case multiple routings are permissible), and on which of a set of identical parallel machines to assign the job to (if there are such multi-processor machines.) It is important to realize that in the context of supply chain management, such problems appear in factories that are configured as job-shops, or that employ MRP, but do not apply for example in factories that are set-up for continuous flow of production, or in any other setting that makes the entire factory look as if it is a single machine.

Job-shop scheduling, is the central part of the more general shop floor control (SFC) process (Hopp and Spearman 2008). Shop floor control has been rightly defined as the module where "planning meets process" and it has been the source of a multitude of research problems that arose from the efforts of more than five decades to optimize production processes in their daily basis. From a standpoint of practical operations management, such efforts to optimize the daily schedules may not contribute significantly to company-level goals such as firm profitability or productivity, as it has been shown by numerous simulations and case-studies that decisions regarding the shape of the production environment are far more important than decisions affecting material flow on the shop floor. The fact that currently there exist heuristics that find rather satisfactory solutions to large job-shop scheduling problems indicates that investing effort in trying to optimize an already reasonable schedule can be a waste in financial terms. On the other hand, from a theoretical stand-point, Job Shop Scheduling has been the source of inspiration for countless heuristic and exact methods for combinatorial optimization, and has been the arena for testing the effectiveness of many widely successful general-purpose search meta-heuristics.

The deterministic job shop scheduling problem (JSSP) is widely considered one of the most stubborn scheduling problems (Lawler 1982); a refractoriness which originates from the JSSP's NP-hard status even for small problem instances for which the machine number (m) is larger than three (Graham et al. 1979; Lenstra and Rinnooy Kan 1979; Brizuela and Sannomiya 2000). Due to its complex nature it is also widely regarded as a platform for testing new algorithmic concepts and mathematical techniques (Zobolas et al. 2009).

In the following we shall describe JSSP from a theoretical standpoint which enables one to see how it has applications in supply chain management and in particular in SFC, but also in many areas of computer science and engineering; for example, multi-processor task scheduling in parallel and distributed computing, or operating systems process and memory management for single or multi-processor environments.

A formal (not the first) description of the JSSP was given by French (1982): a $n \times m$ deterministic JSSP problem consists of finite set J of n job orders, with$\{J_i\}_{1 \leq i \leq n}$, that have to be processed on m machines, with$\{M_k\}_{1 \leq i \leq m}$, which constitute the finite set M. Every single job order must be processed on all the machines and is comprised of a set of operations $\{O_{ij}\}_{1 \leq i \leq n, 1 \leq j \leq m}$ that have to be scheduled in a predefined manner. This manner differs amongst jobs and forms the precedence constraints set. The fact that each machine can only execute a single job at any time forms the resource constraints set. Semantically, operation O_{ik} is the operation of job order J_i that has to be processed on machine M_k for an uninterrupted period of time τ_{ik}. The aforementioned problem form uses discrete time intervals, meaning that all the processing and setup times are integers.

Considering all the above, the completion time of operation O_{ij} on machine M_j is C_{ij}. The duration in which all operations are completed is $C_{\max}$ and is often referred to as the makespan. The makespan, more often than not, is utilized as the optimization criteria for the deterministic JSSP, with other criteria employed being minimizing average cycle time on a single machine, minimizing maximum lateness on a single machine, or minimizing average tardiness on a single or multiple machines, tardiness defined as zero if a job is completed on or before its due-date, and as the difference between actual job completion time and due date otherwise. When makespan is used as the criterion of choice, JSSP is the minimization of the following function:

$$C^*_{\max} = \min(C_{\max}) = \min_{\text{feasible-schedules}} (\max(t_{ik} + \tau_{ik}) : \forall J_i \in J, m_k \in M)$$

In the above equation $t_{ik} \geq 0$ is the starting time of operation O_{ik}. Consequently, the objective is to determine the starting times for each operation, $t_{ik} \geq 0$, in order to minimize the makespan without violation of the precedence and capacity constraints.

The solution space for any JSSP problem contains up to $(n!)^m$ solutions, where each schedule can be regarded as the aggregation of the operation sequences per machine. Since each operation sequence can be permuted individually from the sequences of the other machines, the $n!$ combinations are raised to the power of the machine number. For example a small 5×5 instance has 24.883.200.000 possible solutions. For this very reason, even small square problem instances that have an equal number of jobs to machines are very hard to solve and the complexity grows exponentially as the number of machines grows.

Pardalos and Shylo (2006) define the JSSP as a problem consisting of scheduling the set of jobs J on a set of machines M with the objective of minimizing the makespan, subject to two sets of constraints: the machines can only process one job at a time (resource constraint set) and each job has a predefined processing order through all the machines (precedence constraints set). Once a machine commences with the processing of job it cannot be interrupted until it finishes, i.e. there is no preemption. Other JSSP characteristics worth noting in order to complete its definition are as follows:

- The processing times of the operations are known in advance and are problem independent.
- Each job order has to be processed on every machine.
- There are no parallel machines. Each machine is unique and every operation has to be processed only to its corresponding machine.
- Nothing unforeseen ever happens, e.g. there are no machine breakdowns, rush orders, delays, transportation times between machines, etc.

3.3.1 Scheduling for a Single Machine

In order to understand the problem better, we will start the discussion of the JSSP with the special case of $|M| = 1$, i.e. with the case of a single machine. In this case, the JSSP objective function will normally be the average or maximum *tardiness* of the jobs, since the makespan will be the same regardless of how we sequence the jobs in the single machine we have available. Indeed, when there is only one machine, JSSP becomes simply the problem of sequencing the available jobs on this single machine, but even then, the problem can be far from easy in terms of determining an optimal schedule. This special-case is still of particular interest, because many successful algorithms for the JSSP rely on good (reliable and fast) heuristics for sequencing jobs on a single machine. The following sequencing rules each constitute a heuristic for computing a schedule for the single-machine case, and, besides having being thoroughly investigated in theory, are often used in practice—probably due to the ease of understanding and implementation.

- *FCFS—first come first served*. This is an extremely well known rule in Queuing theory and systems, which dictates that the first job that arrives in the floor is the first to be served–as soon as the machine becomes free and available for processing. The rule has the property of being "fair" in many contexts, but can often be sub-optimal regarding many objectives in many contexts as well. The resulting "algorithm" has $O(1)$ complexity as there is nothing to be done. The initial input sequence of jobs is the final schedule.
- *SPT—shortest processing time first*. This rule has the property that minimizes the average waiting time that jobs will experience, and is implemented in many time-sharing computer operating systems (where each user provides an estimate of how long each of their jobs will take.) The resulting algorithm simply sorts the jobs according to increasing order of their processing time requirements, and the sorted sequence is the schedule for the machine. Obviously, the complexity of the algorithm is $O(nlogn)$ for n jobs to be sequenced.
- *EDD—earliest due-date first*. This rule recognizes that jobs ought to be completed before their due-date requirements, and therefore sorts the jobs in increasing order of their due-date requirements, in hope that most or all jobs will be finished before their due-date. It is a rule aiming to minimize average lateness (defined as the time the job completes minus its due-date). Its time complexity is again, $O(nlogn)$ for n jobs to be sequenced.
- *MSF—minimum slack first*. This rule sorts jobs according to their current slack, and selects as the next job to dispatch, the job with the least slack. The slack of job j at time t, is defined as $\max\{d_j - p_j - t, 0\} = (d_j - p_j - t)_+$ where d_j is the due-date of job j and p_j is its requested processing time.

- *ATCF—apparent tardiness cost first*. This is yet another example of a dynamic rule heuristic. The rule intends to minimize total (weighted) tardiness. Let d_j be the due-date of job j and p_j its requested processing time, and let w_j denote the relative weight (importance) of job j. If the machine is freed at time t, the rule sequences next the job with index j that maximizes the quantity $I_j(t) = \frac{w_j}{p_j}\exp\left(-\frac{(d_j-p_j-t)_+}{K\bar{p}}\right)$ where K is a scaling parameter, and $\bar{p}$ is the mean of the processing times of the jobs not yet processed. Sequencing the jobs using this rule can be accomplished in $O(n^2)$ time using the following algorithm:

Algorithm ATCF

Inputs: sequence of n jobs, $j_1,\ldots, j_n$, with due-dates d_i, $i=1,\ldots, n$, processing times p_i, $i=1,\ldots, n$, and weights w_i, $i=1,\ldots, n$, scaling constant K.

Outputs: sequence seq_i $i=1,\ldots, n$ of indices indicating order of jobs to be processed in the machine.

function getNext(…)
Inputs: array marked$_i$ $i=1,\ldots, n$ of booleans, double t, array p_i $i=1,\ldots, n$ of doubles, array d_i $i=1,\ldots, n$ doubles, array w_i $i=1,\ldots, n$ doubles.
Outputs: integer.

Begin

1. Set $bi = -1$.
2. Set $bv = +\infty$.
3. Set $pbar = 0$.
4. Set $nact = 0$.
5. for $i=1,\ldots, n$ do:
5.1. if ($marked_i$=false) then
5.1.1. Set $pbar = pbar+p_i$, Set $nact = nact+1$.
5.2. end-if
6. end-for
7. Set $pbar = pbar/nact$.
8. for $i=1,\ldots, n$ do
8.1. if ($marked_i$=true) continue.
8.2. Set $v = w_i$*exp($-$max(d_i-p_i-t,0))/(K*$pbar$)).
8.3. if ($bv > v$) then
8.3.1. Set $bi = i$, Set $bv = v$.
8.4. end-if
9. end-for
10. return bi.

End.

Begin /* Main Routine */

1. Set $t = 0$.
2. for $i=1,\ldots, n$ do
2.1. Set $marked_i$ = false.

3. end-for
4. for j=1,…, n do
4.1. Set m = getNext($marked$, t, p, d, w).
4.2. Set $t = t+p_m$.
4.3. Set $seq_j = m$.
4.4. Set $marked_m$ = true.
5. end-for
6. return seq.

End.

- *CR*—critical ratio first. This rule states that the jobs to be sequenced, have to be split in two sets: the sets of already late jobs (past their due-date) and the rest. If the set of late jobs is non-empty, choose among the late jobs according to the SPT rule. Else, choose the job with the highest ratio of estimated processing time remaining until completion over (due-date minus current–time.) Intuitively, the algorithm attempts to minimize average lateness, as its strategy is to pick the job that is running the most risk of becoming late at any stage, but if there are jobs that are already late, it sorts them according to the SPT rule so that the average lateness of the late jobs is minimized. The following algorithm implements this rule.

Algorithm CR

Inputs: sequence of n jobs, $j_1,\ldots, j_n$, with due-dates d_i, i=1,…, n and processing times p_i, i=1,…, n.
Outputs: sequence seq_i i=1,…, n of indices indicating order of jobs to be processed in the machine.

function getNext(…)
Inputs: array $marked_i$ i=1,…, n of booleans, double t, array p_i i=1,…, n of doubles, array d_i i=1,…, n of doubles.
Outputs: integer.

Begin

1. Set $crfi = -1$.
2. Set $crfv = 0$.
3. Set $sptfi = -1$.
4. Set $sptfv = +\infty$.
5. for i=1,…, n do
5.1. if ($marked_i$ =true) continue.
5.2. Set $cri = p_i / (d_i - t)$.
5.3. if $cri > crfv$ then
5.3.1. Set $crfi = i$.
5.3.2. Set $crfv = cri$.
5.4. end-if

5.5. if ($p_i < sptfv$ AND $cri < 0$) then
5.5.1. Set $sptfv = p_i$, Set $sptfi = i$.
5.6. end-if
6. end-for
7. if $sptfi > 0$ return *sptfi* else return *crfi*.

End.

Begin /* Main Routine */

1. Set $t = 0$.
2. for $i = 1,\ldots, n$ do
2.1. Set $marked_i$ = false.
3. end-for
4. for $j = 1,\ldots, n$ do
4.1. Set m = getNext(*marked*, *t*, *p*, *d*).
4.2. Set $t = t+p_m$.
4.3. Set $seq_j = m$.
4.4. Set $marked_m$ = true.
5. end-for
6. return *seq*.

End.

- Again, the complexity of the algorithm above is $O(n^2)$ for n jobs to be sequenced, as it is not enough to simply sort the jobs according to the SPT or CR criterion. As it can be seen, when we are in the ith step of the main algorithm, n iterations must be performed in the function getNext(), which results in the quadratic complexity of this particular algorithm that implements the critical ratio rule.

3.3.2 Scheduling for Parallel Machines

In the case of a single work-center with P identical parallel machines, the objective of minimizing makespan of a set of n jobs with processing times $p_1,\ldots p_n$, defined as the total time required to process all jobs is no longer trivial, as the makespan is no longer the same for all schedules. In this case, and since all parallel machines are assumed identical, minimizing makespan is equivalent to the load-balancing problem (arising frequently in parallel processing applications in computer science and engineering): partition the jobs among the P machines so that the deviation of the load of each machine, defined as the sum of the processing times of all jobs assigned to the machine from the mean is minimized. The problem is also known as the number partitioning problem, and it is known to be NP-hard. Because of its NP-nature, polynomial-time algorithms to solve the problem to optimality most likely do not exist. Nevertheless, when the processing times of the various jobs do not differ by many orders of magnitude, a variety of algorithms can find the optimal solution (zero

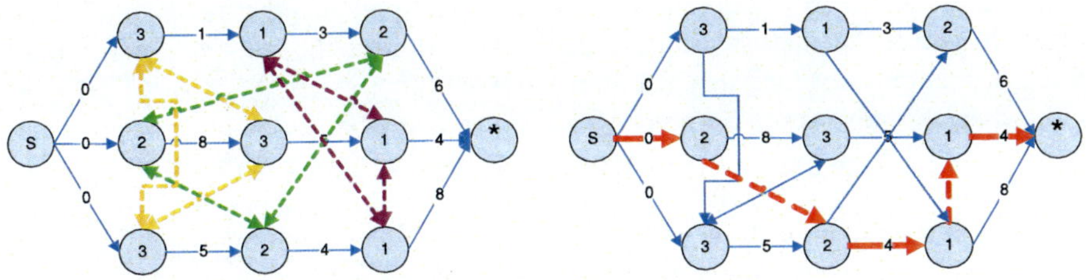

Fig. 3.4 Disjunctive graph representation of a 3 × 3 JSSP

deviation from the mean load for each machine) easily. One such algorithm that performs very robustly for large data sets, as long as the gap between smallest and largest number is not many orders of magnitude is recursive in nature and exact, meaning it will always return with the optimal solution. The algorithm is recursive in nature as recursion in this case captures the essence of the logic behind the computation in the best possible way. The logical argument is that in order to minimize the total time required to process a number of jobs with given processing times on any machine, one should try to balance the (identical) machines optimally, and therefore one should attempt to load each of the P machines with a load as close as possible to the value

$$L = \frac{\sum_{i=1}^{n} p_i}{P}$$

Considering each machine m sequentially, if we have k_m jobs available to select from, we would like to select from them a subset of jobs so that the sum of their processing times equals L. If the jobs to be allocated to the machines are sorted according to decreasing order of the processing time required, we may add the first item in the set with value p_1, and see if we can reach with the rest of the items the value $L - p_1$. If we cannot reach this value, we do not include the first item in the set of jobs that machine m will have to process, and we attempt to reach with the rest of the items the value L. If we cannot do that either, then a perfect balancing of the jobs to the machines is impossible.

3.3.3 *Shifting Bottleneck Heuristic for the General Job-Shop Scheduling Problem*

For the general case of a work-shop with many different machines and jobs that have to be processed in each (or some) of these machines, it will be useful to reformulate more formally the JSSP. Notice first that the JSSP can be represented with a disjunctive graph as in Fig. 3.4.

The disjunctive graph $G = (N,(A,B))$ consists of a set of nodes N, corresponding to all operations O_{ij} that must be performed plus two dummy nodes representing the "start" node and the "sink" node, and two distinct sets of arcs, namely A, and B. The first set of arcs, the set A, comprises the conjunctive arcs that represent the routes of the jobs; Therefore, if a conjunctive arc $((i,j) \to (h,j)) \in A$ exists, then the job j must be processed on machine i before being processed on machine h. Conjunctive arcs are represented in the figure with solid directed arrows. Now, any two operations belonging to different jobs that have to be processed at the same machine will be connected to one another by two disjunctive arcs going in opposite directions. Such disjunctive arcs (shown with dashed doubly-directed arrows in the figure) form of course a clique for each machine in the problem and collectively form the set B. Clearly then, all operations in the same clique have to be processed on the same machine. A feasible schedule is obtained when for each pair of opposite directed disjunctive arcs, one is removed and the resulting directed graph is acyclic. We can now formulate the JSSP.

In the Disjunctive Mathematical Programming formulation of the problem, p_{ij} represents the processing time of job j on machine i (i.e. the time required for j on machine i, and $C_{\max}$ representing the total makespan of the operation o_{ij}). The decision variables are y_{ij}, representing the start time of jobschedule, i.e. the total time needed to finish all jobs. The JSSP is then formulated as follows:

$$\min_{y, C_{\max}} C_{\max}$$

$$\text{s.t.} \begin{cases} y_{k,j} - y_{i,j} \geq p_{i,j} & \forall((i,j) \to (k,j)) \in A \\ C_{\max} - y_{i,j} \geq p_{i,j} & \forall(i,j) \in O \\ y_{i,j} - y_{i,l} \geq p_{i,l} \, OR \, y_{i,l} - y_{i,j} \geq p_{i,j} & \forall((i,l) \rightleftarrows (i,j)) \in B \\ y_{i,j} \geq 0 & \forall(i,j) \in O \end{cases}$$

The above problem has no discrete variables, but of course this does not reduce its complexity, as the complexity lies in the 3rd set of constraints, the disjunctive constraints, which essentially state that for every pair of jobs that must be executed on a machine, either the first, or the second operation of the pair has to be first, and that while an operation executes, it cannot be pre-empted. A pure MIP formulation of the above program is given below.

$$\min_{y, C_{\max}, z} C_{\max}$$

$$\text{s.t.} \begin{cases} y_{k,j} - y_{i,j} \geq p_{i,j} & \forall((i,j) \to (k,j)) \in A \\ C_{\max} - y_{i,j} \geq p_{i,j} & \forall(i,j) \in O \\ y_{i,j} - y_{i,l} \geq p_{i,l} - (1 - z_{i,j,l}) \sum\limits_{(i,j) \in O} p_{i,j} & \forall(i,j), (i,l) \in A \\ y_{i,j} \geq 0 & \forall(i,j) \in O \\ z_{i,j,l} + z_{i,l,j} = 1 & \forall(i,j), (i,l) \in O \\ z_{i,j,l} \in B = \{0, 1\} & \forall(i,j), (i,l) \in O \end{cases}$$

Notice the introduction of the binary variables z: z_{ijl} is set to 1 if job l precedes job j on machine i, and is zero otherwise. Given that the start-time of any operation cannot be after the sum of the processing times of all operations in all machines, the 3rd set of constraints actively force the start time of an operation O_{ij} to be after the operation O_{il} if and only if job l precedes job j on machine i, but will be an inactive constraint otherwise. The last two constraints ensure that for each pair of operations that must be performed on the same machine, one must be first, and the other second.

We now proceed to describe one of the most successful heuristic algorithms for minimizing makespan in the JSSP, the Shifting Bottleneck Heuristic. As its name implies, the algorithm iteratively determines which machine seems to be the "bottleneck" in the current job-shop schedule and optimizes its sequencing using some variants of the rules we discussed in the single-machine problem. It then re-optimizes the sequences determined for the machines previously considered, and begins a new iteration until all machines have been sequenced. The description follows that of Pinedo (2008).

The operating assumptions of the algorithm is that each job is to be processed by a number of machines in the job-shop in a given order, there is no pre-emption, and the same job cannot be executed on the same machine more than once.

Algorithm Shifting-Bottleneck

Inputs: sequence of n jobs, $j_1,\ldots, j_n$, with processing times of job k on machine i p_{ik}, i=1,…, m, k=1,…, n.
Outputs: start-times y_{ij} of each operation O_{ij} with the objective of minimizing makespan $C_{\max}$.

Begin
/* initialization */

1. Set $M_0 = \{\}$.
2. Create the disjunctive graph G for the problem.
3. Remove all disjunctive arcs from G.
4. Set $C_{\max}(M_0)$ = longest-path-makespan(G).

   /* Analysis of non-scheduled machines */

5. for each m in set $M-M_0$ do:
   a. Set SMP_m = setup-single-machine-problem(m,G,p).
   b. Solve min. $L_{\max}(SMP_m)$ using any algorithm discussed in 3.3.1.
6. end-for

   /* Bottleneck selection and scheduling */

7. Set $h = \arg\max_{i \text{ in } M-M0} L_{\max}(SMP_i)$.
8. Sequence machine h according to the results of optimization step 5.b.
9. Set G = add-disjunctive-arcs(G,h).
10. Set $M_o = M \text{ U } \{h\}$.

   /* Re-optimize previously scheduled machines */

11. for each λ in M_0–$\{h\}$ do:
    a. Set G = remove-disjunctive-arcs(G,λ).
    b. Set r = release-dates(longest-path(G));
    c. Set d = due-dates(longest-path(G));
    d. Set SMP_λ = setup-single-machine-problem(λ,G,p,r,d).
    e. Solve min. $L_{max}(SMP_m)$ using any of the algorithms in 3.3.1.
12. end-for
13. Set $h = \arg\max_{i \text{ in } M0-\{h\}} L_{max}(SMP_i)$.
14. Sequence machine h according to the results of optimization step 11.e.
15. Set G = add-disjunctive-arcs(G,h).
16. if $M_0 = M$ end.
17. GOTO 5.

End.

3.4 Personnel Scheduling

In manufacturing, personnel scheduling [also known as man-power shift planning (MSP)], is the problem of optimally allocating personnel to the shifts in a period. Schedule horizon can range from short-term (daily schedules) to medium term (weekly schedules) to long-term schedules (monthly or quarterly schedules). Usually, the schedule horizon is determined by the rules governing schedule feasibility. These rules in turn, are often negotiated by management and worker unions, but there are almost always rules that are dictated by higher-level associations or organizations; for example, when constructing monthly schedules for the crew of an airline, a number of complex rules set forth by the Federal Aviation Administration (FAA) must be strictly obeyed as they represent flight safety regulations.

In the following, we will present two increasingly complex models and corresponding systems for personnel scheduling. The first model is often used in the manufacturing sector and considers only a single constraint when building personnel schedules. The second model is a real-world case drawn from the airline industry.

3.4.1 Scheduling Two Consecutive Days-Off

The problem of scheduling personnel to work weeks with two consecutive days off in any week must be dealt with by any manager who has in their payroll hourly waged workers (this is a fairly ubiquitous requirement in personnel scheduling set forth by the Fair Labor Standards Act; in airline crew scheduling, the requirement is even more strict, and requires that within any seven day sliding window, each crew must have at least two consecutive days off to rest).

A simple algorithm to obtain a feasible schedule then works as follows: the requirements for shifts for each day of the week to be scheduled are given. Then workers are added as rows in a matrix A whose (i,j) cell denotes the requirements for shifts on day j before the ith worker is added to the schedule. Once a worker row has been added, the so-called "lowest pair" of numbers in the row is marked as the two consecutive days off for that worker, where the lowest pair in the row (Nanda and Browne 1992) is defined as the pair of consecutive numbers in the row—with wrapping allowed—such that the highest number in the pair is lower or equal to the highest number in any other pair in that row. Ties are broken by choosing the pair with the lowest requirements on any adjacent day. In case ties still exist, they are broken arbitrarily. Workers are added as rows in the weekly schedule until all days' requirements for shifts have been met. The pseudo-code for this algorithm is given below.

Algorithm 5-2WeekScheduler

Inputs: array r_i, i=1,..., 7 of shifts requirements for the week.
Outputs: 2-D Matrix $A_{n\times 7}$ of worker weekly schedules, obeying the constraint of having two consecutive days off.

Function getLowestPair1stIndex()
Inputs: array *row* of 7 integers
Outputs: integer indicating the position of the first number in the lowest pair in the row

Begin

1. Set $ind = -1$, $best = +\infty$, $ladj = +\infty$.
2. for i = 1,..., 7 do
   a. Set $h = \max\{row_i\,,\ row_{i\%7+1}\}$.
   b. if $h <$ best then
      i. Set $best = h$, $ind = i$, $ladj = +\infty$.
   c. else if $h = best$

      Set $adj = \min\left\{row_{\max\{(i+6)\%7,[7-(i+6)\%7]_-\}}, row_{i\%7+2}\right\}$.

   d. if $adj < ladj$ then
      i. Set $ladj = adj$, $ind = i$.
   e. end if
3. end-for
4. return *ind*.

End

```
Begin /* Main routine */
/* initialize first row */
1.      for j = 1,…, 7 do
1.1.    Set A_{1,j} = r_j.
2.      end-for
3.      Set i = 1.
4.      Set stop = false.
5.      while stop = false do
5.1.    Set array a = A_{i,1,…, 7}.
5.2.    Set ind = getLowestPair1stIndex(a).
5.3.    Set stop = true.
5.4.    for j = 1,…, 7 do
5.4.1.  Set A_{i+1,j} = A_{i,j}−1.
5.4.2.  if j = ind or j = ind%7+1 then Set A_{i+1,j} = A_{i,j+1}.
5.4.3.  if A_{i+1,j}>1 then stop = false.
5.5.    end-for
5.6.    if stop=false then Set i = i+1.
6.      end-while

End.
```

The reader should realize that the algorithm, although it guarantees a feasible schedule, does not carry any "certificates of optimality", in the sense that there is no guarantee that the number of workers that are returned as the result of running the algorithm will be the minimum number needed. The algorithm is essentially "greedy" in nature, as at each step, it attempts to maximize the gain of inserting a worker by having them work during the days of highest requirements in shifts.

3.4.2 Air-Line Crew Assignment

Crew assignment in the air-line industry refers to the problem of assigning actual crew (pilots and flight attendants) to the trips (also known as pairings) scheduled for the next period for a particular fleet operating from a given base. There are two broad categories of crew assignment systems, namely preferential seniority bidding systems, and bid-line generation systems. In preferential bidding, each crew member declares their preferences for his/her schedule for the next period. Preferences can specify particular trips the crew member would like to fly or not to fly, particular days-off they would like to have, interval length of days-off between trips and so on. The objective of the preferential bidding system is then to create schedules for each crew-member that satisfy all rules and constraints set forth by the Federal Aviation Administration as well as by union negotiated rules, and to maximize the satisfied preferences of the crew personnel, with the understanding that more senior personnel's preferences are infinitely more important than less senior personnel preferences. This essentially implies that among all feasible schedules for all personnel, the optimal schedule is one that satisfies the most senior crew member's preferences the most, and among all feasible schedules that satisfy the most senior crew member's preferences the most, the optimal schedule is one that satisfies the second most senior crew member's preferences the most, and so on.

Every crew member $c = 1,\ldots, N$ can assign preference values $p_1^c > p_2^c, \ldots, > p_{n_c}^c$ to each of their preferences, from which—conceptually at least—one can build all feasible schedules (from here on called *lines*) that can be assigned to this crew member and rank them according to their preferences. Let L_m denote the matrix whose columns are the lines comprising all feasible schedules for crew-member m, sorted in ascending order of cost, so that $c([L_m]_i) \leq c([L_m]_{i+1})\ \forall i = 1, \ldots, |L_m| - 1$. Each line is the indicator vector of trips that the line contains. The Preferential Bidding Problem for Crew Assignment can be formulated as follows:

$$\min_{x^{[1]},\ldots,x^{[N]}} c^T \begin{bmatrix} x^{[1]} \\ \vdots \\ x^{[N]} \end{bmatrix}$$

$$\text{s.t.} \begin{cases} [L_1, \ldots, L_N] \begin{bmatrix} x^{[1]} \\ \vdots \\ x^{[N]} \end{bmatrix} = e \\ e^T x^{[i]} = 1 \quad \forall i = 1, \ldots, N \\ x_j^{[i]} \in \{0, 1\} \quad \forall i = 1, \ldots, N, j = 1, \ldots, |L_i| \end{cases}$$

The cost vector c elements are determined by $c_j = -2^{\prod_{i=1}^{N} |L_i| - j}$ $j = 1, \ldots, N$. The introduction of the exponentially decreasing cost-coefficients ensures that as long as there is a feasible schedule for which the most senior crew-member is assigned their most favorable line, this line will be assigned to that crew-member regardless of how the schedules of the less senior crew-members are constructed. The same idea is present in the formulation of the multi-commodity aggregate production planning problems in Sect. 3.1.1. The model is an instance of a set partitioning problem, but unfortunately, even for small-size problems cannot be solved exactly due to its enormous size. It is also of interest to realize that the formulation suffers significantly from the introduction of the exponential coefficients: the number of columns $|L_i|$ of each matrix L_i is an astronomical number, so the product $\Pi_{i=1}^{N} |L_i|$ is an incomprehensibly large number, and raising two to this product would require more bits than the memory of any computer can store today. To approximately solve such models, heuristic methods are unavoidable, although they are some times used in conjunction with column generation techniques to solve the LP relaxation of a much smaller master problem.

Among the most efficient heuristics for solving the preferential bidding problem is a backtracking iterative roster building method employing best-first search of an appropriate search tree with look-ahead [Rehwinkel (1996) Private communication]. The algorithm exploits the fact that most senior personnel must get their preferences before junior personnel do unless satisfying a senior crew member's preference would make it infeasible to build the schedules of the more junior personnel. Therefore, the algorithm builds the schedules of the crew in order of seniority satisfying their preferences in order unless:

- Satisfying a preference of a crew member conflicts with their already satisfied preferences and the constraints of the system;
- or, satisfying a preference of a crew member would make it impossible to build the rest of the crew-members' schedules.

To see how satisfying a crew member's preference can render infeasible the rest of the schedules consider the so-called Christmas problem: during Christmas, almost all personnel wants that day-off. Clearly, satisfying this request for all crew-members is impossible. If the airline has m trips scheduled to be flying on that day, then at least the m most junior personnel that is not scheduled for any other activity (e.g. training) during that day will have to have their request for a day-off on Christmas rejected. However, while the algorithm builds the schedule of the some junior crew member, can possibly award this crew member their preference, only to find out several steps below that all schedules cannot be built (since trips during Christmas remain unassigned), in which case it has to backtrack, start undoing previously built schedules and try new schedules until it rolls back to the mth most junior personnel and after trying all other alternatives, it rejects the crew-member's request for that day-off.

The Look-ahead functionality builds a global "stacking picture" maintaining a look-up table of the number of unassigned trips for each day of the schedule. Together with the information of how many crew-members are available to fly on each particular day of the scheduling period, the system can decide early on to reject such requests for "days-off" due to overloaded "off-periods".

Building an individual crew-member's schedule makes use of Best-First Search techniques combining the strengths of Depth-First and Breadth-First techniques. While building a line, the ability to determine whether a line that is partially filled can be completed is of paramount importance. *Completability* of a partially filled line must be guaranteed before placing any trip in it else the line cannot form a valid schedule for the crew member. A path-construction mechanism is used to determine the best path (the best combination of trips that will allow the line to become valid). It is this path construction mechanism that creates a tree whose nodes are trips to be assigned. By traversing the tree from any of its nodes towards its root along the edges of the tree, valid combinations of trips are formed that can be placed on the line. Among all paths that allow the line to reach the goal (i.e. to become a valid, complete line), the best one according to the crew member's preferences is then selected. If no path in the tree allows the line to reach goal, the line is non-completable and intelligent backtracking has to occur. In Fig. 3.5, node

Fig. 3.5 Line building tree

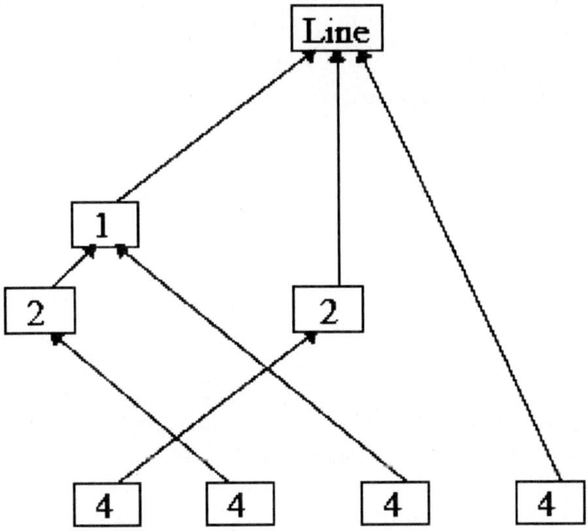

numbers represent trips and nodes represent *trip instances*, which are the combination of a trip starting on a particular day. Time flows left to right in the figure, so that the left-most node labeled "4" represents a trip with id 4, starting much earlier than the trip 4 represented in the right-most node in the figure labeled "4".

Since this tree can grow very large, memory and speed considerations force a limit on the maximum number of paths to be created in the tree (usually set to 50,000 paths, or nodes). The root node contains the line as it currently is (empty). Then, the preferences of the crew member are scanned, from most important to least important one: requests for trips defined that can be placed in the line without violating any rules are examined (one at a time) for assignment right after each of the nodes of the tree have been built so far. Whenever a trip can fit in the line after all the trips in the path beginning from the node that is being checked have been assigned, a new node representing this trip is created that points to the node being examined. Negative requests for trips are handled by temporarily removing the indicated trip from the pool of trips available to complete the line and checking if the line can still be completed. Requests for days-off are handled by removing all trips that intersect the particular day-off period request from the pool of available trips to complete the line, checking using look-ahead for stacking picture violations, and ensuring the resulting pool of trips can still complete the current line.

Despite the advantages (for crew-members) of preferential bidding systems, most North American airlines still use bid-line generation systems to create their monthly or quarterly period schedules. Bid-line Generation is the problem of assigning trips to schedules for the crew members of an airline regardless of each crew member's individual preferences. The pilot and flight-attendants' unions negotiate with airline management what "collectively" constitutes good schedules, and it is the job of the planners to determine as many "good" schedules as possible without violating any rules or regulations set-forth. The nature of these rules makes the problem at least NP-hard: every trip has a property called pay or value (that roughly corresponds to the total flying time of the trip); the total pay rule states that each produced line must have a total pay that is within a predetermined pay window. Therefore, solving the bid-line generation problem involves solving at least the *k*-way number partitioning problem, a generalization of the number partitioning problem (Garey and Johnson 1979) which is NP-complete.

Crew members submit bids for the generated lines based on their seniority. The objective of the assignment is to maximize the quality of the produced schedules as well as maximize the average value (pay) of the lines. The latter objective aims at improving pay opportunities for crew and also improving the efficiency of the airline by reducing staffing needs.

The problem, therefore, is a complex assignment problem which can be formulated as an integer non linear multi commodity network flow problem. Each trip represents a commodity to be transferred to a sink node through exactly one intermediate node (line or open-time). The costs along each arc (from a trip instance to a line or from a line to the sink node) are complex, non-linear functions that become infinitely large when the flow of trips into a line makes the line illegal.

Formulating bid-line generation (BLP) as a set partitioning problem (as was done in the case of the preferential bidding system) is also possible. In order to give an exact mathematical definition of the BLP, let $L = [l_1|l_2, \ldots, |l_n]$ be the matrix whose columns represent all the valid lines of time l_i that can be built using any valid combination of trips in the current category. We represent these lines as p-dimensional vectors in $\{0,1\}^p$ where p is the number of trip instances in our category. Assuming we have a cost function $q(l)$ that assigns to every legal bid-line a cost representing the quality of the line, we can formulate the BLP as follows:

$$\max_{x,s} \sum_{i=1}^{n} q(l_i)x_i$$

$$\text{s.t.} \begin{cases} [L|I]\begin{bmatrix} x \\ s \end{bmatrix} = e \\ [0|c_1, \ldots, \quad c_p]\begin{bmatrix} x \\ s \end{bmatrix} \le C_l \\ x \in \{0,1\}^n \\ s \in \{0,1\}^p \end{cases}$$

where c_i is the credit of the ith trip, and C_l represents the threshold of the so-called coverage rule stating that the lines must cover all the trips except for a small number of trips whose total credit will be less than this threshold. The problem as formulated is therefore a set partitioning problem with a *side-constraint* that allows for a few trips to remain unassigned.

A particularly successful approach to solving the BLP that is capable of solving some of the largest real-world problems involves a two-phase approach.

The system, in its first phase, constructs many high-quality lines taking fully into account the concept of purity, a term used to define and measure the quality of the produced schedules. In the second phase, a Genetic Algorithm (see Sect. 1.1.1.3) is utilized in order to arrange the remaining open trips into valid lines of time. In this sense, the GA solves a *feasibility* problem that is already highly constrained, and is therefore different in many respects from the traditional use of GA's that are used as first-order optimizers that locate the neighborhood of an optimal or almost optimal solution.

In the 1st phase of the system, the high-quality line construction phase, the same path-building mechanism described above for the case of preferential bidding systems (see Fig. 3.5) is used to construct as many *pure* lines as possible. Now, *purity* is a broad term in the airline industry, used to describe the quality of a line. There are two types of purity, namely *trip purity* and *day purity*. A line is *trip-pure* when all the trips in it are essentially the same trip; for example a line that consists of the trip 3415 (ATL(departs:0800) → EWR(departs:1300) → YYZ (layover) (departs:0930) → CVG(departs:1400) → EWR (layover) (departs:0800) → ATL) departing on the 1st, the 8th, the 15th, and the 24th is *trip-pure*. A line that consists of trips that depart on the same day of the week (e.g. every Tuesday) is called *day-pure*; for example, a line that consists of the trip 3415 departing on the 1st, the 8th, the 15th, and the 22nd of the month is day-pure (and trip-pure as well, making it a *perfectly pure line*.) However, a line can be trip-pure even if the trip identifications in the line are not all the same: as long as two trips have the same duty periods (every duty period begins and ends within a few minutes of the corresponding duty period of the other trip, and they have the same layover cities) they are "essentially" the same, and so a line consisting of such trips is still trip-pure. A *family of trips* refers to a set of trips that are essentially the same for purity purposes.

The following algorithm implements the 1st phase of the system:

Algorithm PureLineBuilder

1. Select a family of essentially the same pairings (according to some input criterion, usually maximization of total pay).
2. Compute the estimated number of lines, N, to be built from the family, and break them into groups of seven lines (so that the resulting lines are highly day-pure).
3. Place the ith trip of the family (in chronological order) into the ($i\%N$)th line, subject to the constraint that no rule is violated and that the line remains completable after this assignment.
4. Complete lines that were left incomplete in the previous step using a single filler trip that if possible maintains the day purity of the line, as well as the trip purity of the line. If there are more than one such trips, choose the one with the highest total pay.
5. Complete any lines that were left incomplete in the previous step, using more than one filler, by choosing a combination that minimizes the number of

Fig. 3.6 Pure bid-lines

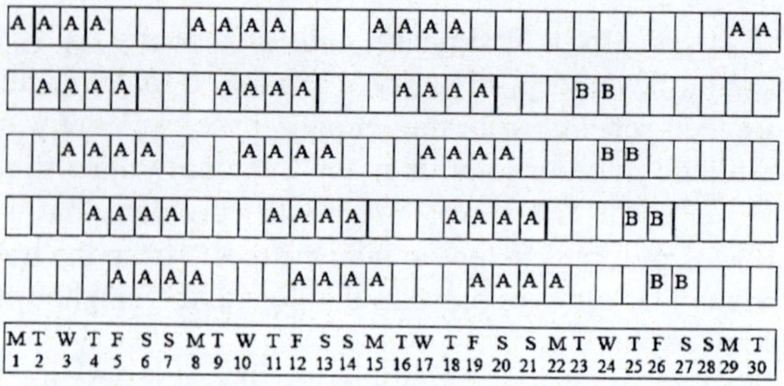

different trips in the line. If more than a certain threshold (usually three) of different trips are needed to complete the line, the line is reset and the assigned trips are freed.

6. Do a stack test to ensure that no period of time requires more crew members to cover the open trips that operate during this interval than the current estimate of the total number of lines. If the stack test fails, undo lines built from the family, one by one, until the stack test succeeds again.
7. If one or more lines are completed, GOTO 1.

Steps 1 through 5 of the algorithm PureLineBuilder create lines as in Fig. 3.6. Step 6 solves a *semi-assignment problem*. Semi-assignment is a generalization of the linear assignment problem (discussed in Sect. 1.1.2.2) where a task requires more than one person to be assigned to it (as a consequence, in semi-assignment problems, the number of persons is larger than the number of tasks). This step detects infeasibility of the BLP after the assignment of a combination of trips into a line. Note that it cannot guarantee the feasibility of the assignment, but serves very well as an indicator that a period of time is left with too many open trips during the purity phase. The purity phase ends either when all trips have been assigned into complete lines of time, or when a family yields no lines. In the latter case, the second phase (the GA phase) is initiated to complete the assignment, i.e. to place the remaining open trips to valid lines.

The 2nd phase of the algorithms then is a Genetic Algorithm that attempts to complete the assignments of the unassigned trips into lines of high-total pay. The number of lines remaining to be built is known because of the Pay-Window rules that specify how much credit each line must have. To create feasible lines, a Genetic Algorithm encodes the assignment of a trip into one of the remaining lines as a chromosome of length equal to the number of unassigned trips left-over from phase 1. The representation, population breeding and evolution of the GA is schematically shown in Fig. 3.7.

The representation chosen is the following: every position (allele) in the individual's string represents an open trip, so that the length of the genetic string is the number of open trips. The letter in the allele is a number in the range $-1,\ldots, L-1$ where L is the total number of lines to be built. The number -1 indicates that the trip must not be assigned to any line at this time. Any other number in the range $0,\ldots, L-1$ represents a guideline for placing the trip in a line. In particular, if the

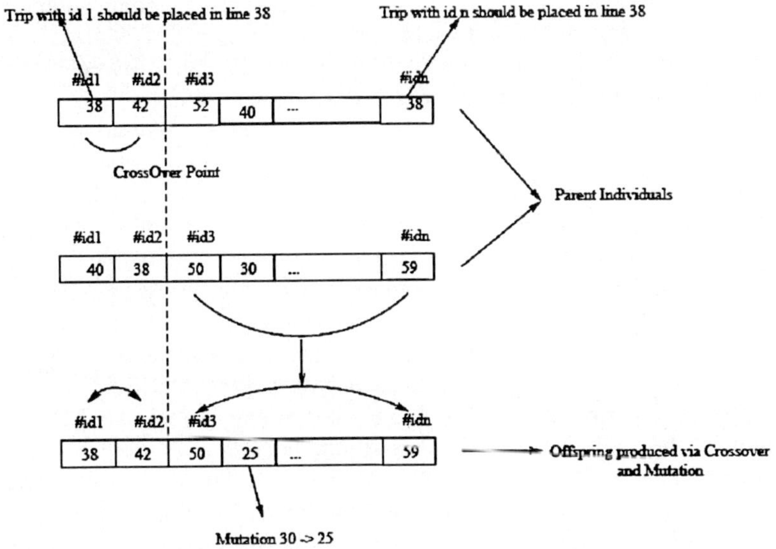

Fig. 3.7 Individual representation and breeding

trip can be assigned in the indicated line without violating any rules and without rendering the line incompletable, the assignment is made; else, the trip is assigned to the closest line (in number order) to the one indicated by the chromosome that will remain completable after the assignment (which should violate no constraint). Note that it is quite possible that a trip won't fit into any line. After all the open trips represented in the individual have been checked for assignment, local improvement methods are executed in an attempt to further improve upon the proposed solution, thus acting as semi-repair methods.

The evaluation function returns the total open time of the unassigned trips after every line left incomplete has been cleared from any assignments in it. This way, the objective function value is an exact metric of the objective of the problem, which is the placement of the open trips into valid, complete lines of time leaving a total number of trips open that are no more than the lower end of the total pay window worth. Once an individual gets an objective value below this threshold (the total open time below the lower limit of the total pay window) the GA stops; a feasible solution has been found and the resulting assignments are written in the database as the proposed solution.

To speed up the search, as already mentioned, after the individual's string has been interpreted and various trips placed into lines a local improvement heuristic based on swaps begins to further improve upon the current solution. In particular, the evaluation function executes a swapping procedure that checks every line that is not complete yet, for completion: if, after the line gets cleared from all assignments, it can be completed from the open trips, the highest total pay combination of trips is assigned to it; else we check it against any other line that was built in the GA phase for swaps that will allow both lines to become completed by rearranging their assignments and using open trips.

Finally, if there is only one line left incomplete (which, by means of the previous step, cannot be completed using any swaps with any other line built in the GA phase) yet another swapping heuristic is used; the evaluation function now executes a procedure to check if the line can be completed by swapping trips with lines that were built in the purity phase (these are lines that normally one would not undo as they are high quality lines). If this fails to complete the last line and bring the total open time below the lower limit of the pay window, yet another procedure attempts to finish off the assignment by checking every line built in the GA phase first, then every line built in the purity phase, for any possible increase in the total pay of the lines by undoing some previously made assignments and using other trips from the ones left open that have higher total pay. This attempt stops as soon as the trips left open add up their total pay to a number less than the lower limit of the total pay window. This simple heuristic has very often enabled the search to find a feasible solution rather early (within less than five generations) and thus cut a lot the computational costs, without significantly sacrificing the overall quality of the assignments.

As a note, we mention that the reason that the heuristic for increasing the credit of individual bid-lines helps very often to complete the BLP, is that in the purity phase, many lines are built very pure, but can improve their total pay by significant amounts, thus reducing the open time. When the open time is already close to the feasible region, a valid solution can be easily found this way. If, after a certain number of generations, a feasible solution has not been found, a fixed number of lines (with least total pay) that were built in the purity phase are undone; the Genetic Algorithm starts again, using this expanded set of open trips and lines, trying to rearrange them into a feasible solution. Note that the more pure lines are undone, the easier it is for the GA to find a feasible solution as the problem becomes less constrained, but the longer it takes to perform a fitness function evaluation as more trips and lines have to be assigned to each other.

3.5 Due-Date Management, Available To Promise Logic and Decoupling Point Coordination

Due-date management is the operational/tactical level process of deciding lead-time quotations as well as shop-floor control for meeting the quoted lead-times to customers. Traditionally, due-dates quotation involved little more than quoting a standard constant lead-time for each product using past observations about the average time it took to produce a certain product. More recent approaches (Keskinocak and Tayur 2004) have advocated the use of customized lead-times taking into account customer order importance as well as shop-floor status and constraints, thereby involving production managers into the due-date management process. Even more recent approaches (Wu et al. 2010), building on work done on available-to-promise (ATP) and related issues, propose full *order admission control* schemes that determine the optimal decision whether to accept or not an order request so as to maximize the expected profitability within a finite planning horizon for a make-to-order (MTO) company in the B2B industry subject to resource utilization constraint, which results in a dynamic stochastic knapsack problem that they solve using dynamic programming.

There are many benefits from employing such procedures since they have the potential to drastically reduce due-date quotations in many cases, as well as reduce the delayed orders and increase both customer service levels and profitability. Optimized due-date management has the potential to alleviate many shop-floor control issues as well, as was already mentioned in Sect. 3.3. The problems of aggregate and short-term production planning, personnel scheduling, and due-date management are highly inter-dependent. Their inter-play is very well manifested in the decision-making process known as ATP and the related areas of capable-to-promise (CTP).

3.5.1 Introduction

The concepts of ATP originate from a set of business practices that were eventually captured in the association of operations management (APICS) dictionary as the method whereby a firm examines its available finished goods inventory, in order to agree on the quantity and promise a due date against a customer order request (Blackstone and Cox 2004). As supplier reliability became a prime concern in supply chain management and customer relationship management as well, best practices emerged for the optimal set of policies upon which the company should rely when making promises.

A few decades ago, the concepts of ATP were focusing on the efficient search among the company's warehouses and depots for available inventory to promise to a customer. As such, ATP was clearly a sales and operations activity, where operations were directly involved only in the distribution of the products. ATP was not a concept that had any linkage to the planning processes or the day-to-day shop-floor operations.

3.5.2 The Push–Pull Interface

Ball et al. (2004) were the first to (re-)define ATP as a set of business controls that operate on the *interface* of push and pull mechanisms of a company. Its objective is to match in the most profitable way possible the manufacturing resources and production capabilities of the company with the market demand. Push mechanisms include the necessary planning and scheduling processes that a company has to execute in order to fulfill its operational requirements as effectively as possible. The core characteristic of the (traditional) push mechanisms is the forecasting process, by which the marketing and planning functions of the organization predict as accurately as possible future market needs that the organization should cover. Since the days of the Oracle at Delphi a number of statistical tools and techniques has been invented and refined during the years for the unbiased and accurate estimation of marketplace demand (see Chap. 2). Recently (Chen-Ritzo 2006), it has been suggested that the demand forecasts should be viewed as a sales target rather than as an estimate for the Production Planning process of a make-to-stock business (MTS).

Once the company has finalized its demand forecasts for the next planning horizon, the production planning process calculates a schedule that will efficiently manufacture the right quantities for each product at the right time. In doing so, the production schedule takes into account the bill-of materials (BOM) of each product, the lead times of each sub-component of the final product, the time and quantities when raw material should become available and so on, using plain MRP or MRPII logic.

At one extreme of the range of possible business operational environments, a MTS practice is a push mechanism where the organization predicts the future market demand (or defines sales targets that should be met) and produces quantities of products that it "pushes" to the market. The organization is anticipating future demand and builds its operations around that estimate.

At the other extreme, MTO business practices (a radically different approach to inventory control based on the principle of producing nothing until it is needed) are a pull-based approach to manufacturing: actual orders initiate production. Inventory of work-in-progress (WIP) or finished goods is kept at minimal levels, which minimizes the risks associated with unsold inventory, obsolescence of products, cost of capital tied up in inventory, etc. The essence of such business controls is the principle of *reacting to demand instead of anticipating it*. As work is carried out only on confirmed orders, it is the market that pulls the products from the factory, instead of the factory pushing products to consumers.

Unfortunately, neither of the above practices is without risks. MTO and related JIT practices aim at the reduction of inventory to the minimal possible levels as mentioned in the previous chapters. This practice can only be successfully applied to environments of relatively steady demand and steady supply. Any sudden demand or supply fluctuations leave the organization unable in the short to medium term to cope with demand. There are many examples of spectacular failures of companies to keep up with the competition because of singular events of failures in their supply chain. It has been successfully argued that without sufficient inventory buffers, the supply chain becomes extremely vulnerable in the presence of turbulent markets (Christopher 2005). But push-based controls also run serious risks, especially in the face of turbulent markets. In such situations, push-based inventory controls have no advantage over pull-based controls, other than the increased probability of having somewhat increased inventory levels of raw materials to finished products because of the planning horizons that are covered. In other words, if the planning process has dictated early production of goods to be distributed several periods later, then in the event of a short-term shortage of raw materials, current demand can be met by inventory that was meant to be distributed in later periods, and new plans can be made for the upcoming periods later on. In such cases, inventory acts as the buffer that prevents the serious disruption of the supply chain. Therefore, a competitive and proactive organization should make every possible effort to combine the advantages of pull-based controls (reacting to customer orders) with those of push-based controls (planning early anticipating demand). *The optimal interplay of those two controls can be achieved via appropriate Available-To-Promise logic mechanisms*.

Regarding the Customer Order Decoupling Point, it is frequently observed that this point is too far down the pipeline and that, secondly, real demand is hidden from view and all that is visible are orders (Christopher 2005). Another, equivalent definition of the demand penetration point is that it occurs at the point in the logistics chain where real demand meets the plan. Upstream from this point everything is driven by a forecast and/or a plan. Downstream we can respond to customer demand. Clearly in an *ideal world* we would like everything to be demand-driven so that nothing is purchased, manufactured or shipped unless there is known requirement (the main goal and practice of JIT as well). A key concern of logistics management should be to seek to identify ways in which the demand penetration point can be pushed as far as possible upstream. This might be achieved by the use of information so the manufacturing and purchasing get to hear of what is happening in the marketplace faster than they currently do.

Perhaps the greatest opportunity for extending the customer's order cycle is by gaining earlier notice of their requirements, which can lead under fairly general conditions to strongly stable supply chains, as we shall see in the next chapter. But in so many cases the supplying company receives no indication of the customer's actual usage until an order arrives. If the supplier could receive "feed-forward" on what was being consumed they would anticipate the customer's requirement and better schedule their own logistics activities. We shall exploit this idea (and its natural extension, that of commitment-based ordering policies) in greater detail in the next chapter in a sub-section on the stability of supply chains and the bullwhip effect.

3.5.3 Business Requirements from Available-To-Promise

From a business point of view, ATP should be the set of processes that allow the company to decide in the best possible way, whether to accept or decline a customer order request, and to (optionally) best negotiate the request fulfillment's due date. These processes should be fast enough so as to allow sales personnel to respond to such requests in time-frames that are deemed acceptable by the customer. Such processes of course, need to be properly aligned with the business model the company implements. Further, they should not violate other established business practices or other hard constraints (such as production capacity, or product life-times constraints, etc.).

Of course, the above "definition" does not define what is meant by "optimal decision", even though, in business, optimal is usually "most profitable". The definition also stays short of explaining what would be the decision variables of the problem, what are the constraints and so on. In the following we will explain in some more detail the above definition. For many companies in the foods and beverages sectors, belonging to the general category of perishable consumer goods products, the business rules dictate that customer service level should be first priority in the long-term production planning process (see discussion in Sects. 3.1 and 3.1.1 in particular). In other words, the planning process should aim to meet forecasted demand as best as possible. If the shifts required to meet the forecasted demand exceed the desired number of shifts the company sets as target, then such shift violations are acceptable, but only if no other feasible schedule exists. And of course, product freshness should be maximized, which also has a direct positive correlation with minimization of finished goods inventory holding costs. Also, if stock-outs are unavoidable (there exists no feasible way to meet all forecasted demand with the given production capacity of the company), then priorities should be set, so as to favor certain products over other products in different time periods.

3.5.4 Problem Inputs

Below, we list a number of inputs that a business could ask to be taken into account when considering whether or not to accept a customer order:

1. Customer order request data including product item code, description, quantity, and delivery date requested.
2. Customer importance. Companies usually classify their customers in ABC-analysis (Christopher 2005) on characteristics such as their profitability or the sheer size of their account. Key customers have associated key accounts and are treated specially. A usual practice in ATP is to divide customers among "demand-classes" (Ball et al. 2004; Kilger and Schneeweiss 2005) and build a hierarchy of such classes to be used in "Allocated ATP" as shown in Fig. 3.8, explained further below.
3. A long-term aggregate planning horizon and the decomposition of each aggregate period into fine-grain periods (see Fig. 3.9).
4. Existing current product demand forecasts for the planning horizon.
5. Existing production and distribution plans and schedules.
6. Existing inventory levels for each warehouse and depot of the company and associated geographic considerations and rules.
7. Existing promised orders and order details.
8. Factory capacity and personnel work-schedules (shifts per period, union rules, etc.).

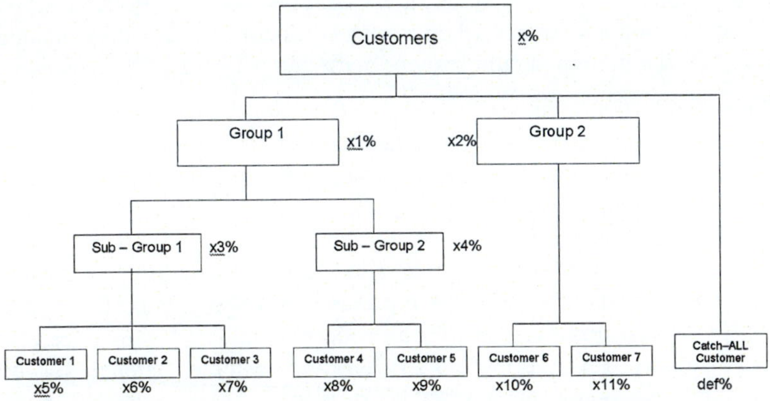

Fig. 3.8 Planning horizon hierarchical decomposition

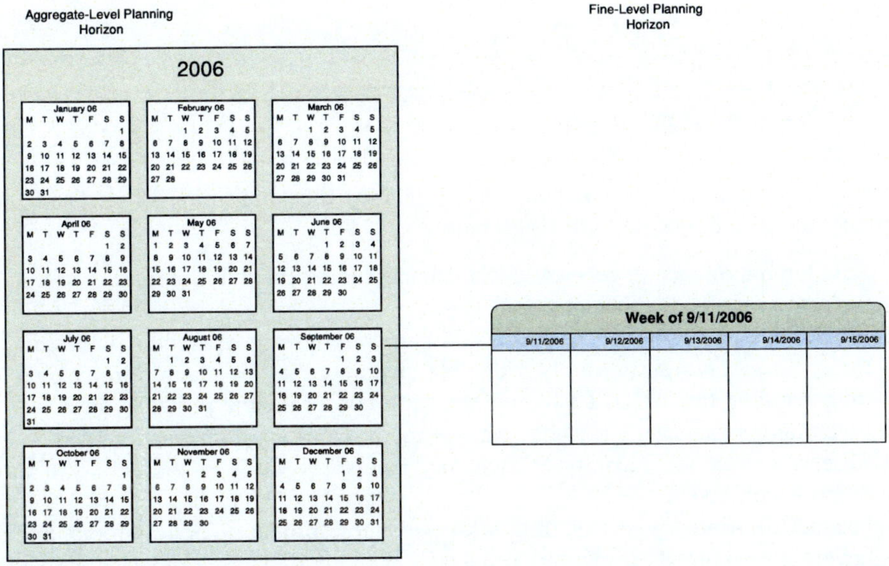

Fig. 3.9 Customers hierarchy. Each box represents a demand class

9. Raw materials and semi-finished goods inventories, together with Bill-Of-Material for each product.
10. Procurement schedules (raw material availability plans).
11. Master production schedules including scheduled down-times according to maintenance policy.
12. Product profitability details.

13. Company business rules relating to service levels for each product. A business rule could also indicate that when a request cannot be fulfilled by the due date requested, the system should respond with another proposed later date.

In the example figure above, customers are grouped in a hierarchy of groups, with actual customer accounts being the leafs of the customer hierarchy tree. Companies that service customers with no accounts associated, create a so-called "Catch-ALL" customer account.

There is a different tree for each product the company sells. Each node in the hierarchy tree is given a percentage of allocated ATP quantity for the product (shown in the figure as x#% where # represents an index number). When a customer order request arrives, a search procedure is initiated which searches for sufficient inventory from the leaf node where the customer belongs towards the root of the tree. As soon as enough inventory is found so as to satisfy the request, the search stops with an indication that the order can be fulfilled. Otherwise, the order cannot be fulfilled even though the company may still have inventory of this product. In the latter case, this inventory is allocated ATP for other customers. The percentages of the figure above indicate how much of the initial inventory can be allocated to each customer class.

3.5.5 Problem Parameters

Similarly, a number of parameters that influence the application of ATP in a production business setting are listed below:

1. Relative importance of service levels versus product profitability in the form of weighted factors or other means.
2. Importance of short-term cash-flows versus long-term relationships with customers; e.g. a currently highly profitable customer may be given priority over long-standing traditional customers ordering less profitable products, in a setting where short-term cash-flow matters more than long-term relationships.
3. Reserve production capacity or inventory levels.
4. Safety stock levels.
5. Product life-times (when dealing with perishable products such as foods and beverages having short expiration dates).
6. Cost (or profit) function to optimize or exact description of the business priorities for answering an ATP case instance.

3.5.6 Problem Outputs

Clearly, solving the ATP problem requires deciding whether to accept or deny a customer order request. Besides that, in the event of a positive answer, the solution to the problem has to provide details about which inventory is to be used for servicing the request, and which changes to production and distribution scheduling have to be made to service both the request as well as all other previously committed orders.

3.5.7 Modeling Available-To-Promise as an Optimization Problem

We argue that we can formulate ATP as a combination of deterministic optimization problems (Christou and Ponis 2009). These problems operate on the interface of the push and pull control practices as described earlier, and align properly the long-term production and sales plans with the day-to-day sales and production operations, dealing with market volatility by appropriate reservation and commitment mechanisms.

We view ATP as the problem of deciding whether to accept a customer order request given the available inventory and planned production plus the remaining production capacity and the business rules concerning covering demand from certain customer demand classes, for given products and for a given time window. Whenever there is sufficient inventory allocated to a given customer for a certain product for a given time-period, the ATP problem becomes a simple search problem in the company warehouses and depots for the appropriate amounts of product requested. Actually, this is what most, if not all, current Supply Chain and Planning software packages implement. However, when the allocated—existing or planned—inventory is not sufficient to cover an order request, there is still the possibility of modifying the production schedule (by utilizing "reserved" capacity and resources) to cover the extra demand. In fact, very often, long-term aggregate planning builds plans that reserve extra capacity for periods of high demand exactly because the company realizes that there may be significantly higher demand for its products during such periods and wants to have the agility to respond quickly to such surges (Christou et al. 2007).

We formulate three models for production planning and allocating demand and production capacity that together with some straight-forward search algorithms decide whether to accept a customer order, and if so, how to select inventory and possibly modify the production schedule to satisfy all committed customer orders so far, without deviating from the original aggregate long-term and medium-term plans of the company.

The models we formulate operate on two different time-granularities on inventory and production controls as depicted in Fig. 3.10:

At the push-control level, an extension to the multi-commodity aggregate production planning (eMCAP) model of Sect. 3.1.1 provides production plans to meet aggregate product demand for each aggregate period. The demand allocation ATP (DAATP) model provides a way to ration the aggregate production of a period among fine-grain level periods and among customer demand classes in a way to maximize company profits. And finally, the multi-commodity fine-grain

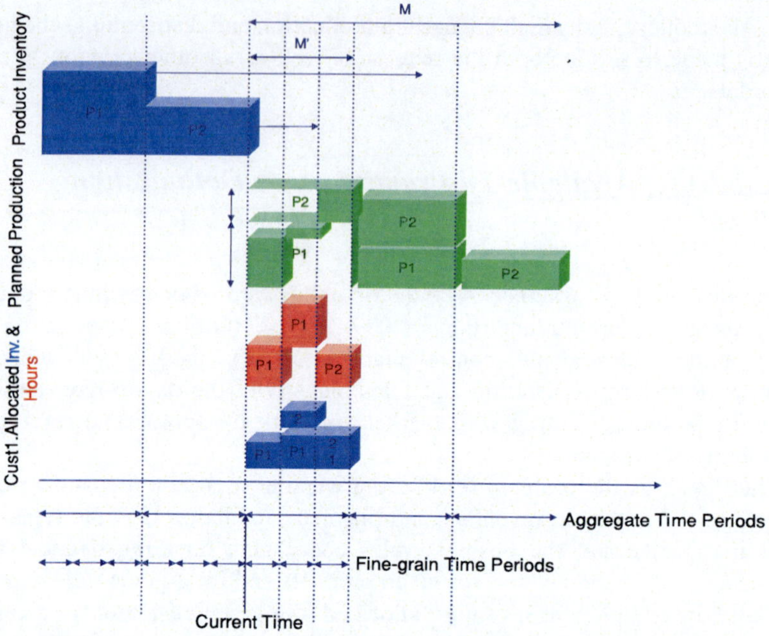

Fig. 3.10 Production and inventory controls in varying grain time periods

production planning (MCFP) problem together with traditional allocated ATP search procedures operates at the pull-control level of actual sales as an order admission control mechanism.

Regarding aggregate production planning, we assume as in Sect. 3.1.1 a multi-factory setting, where each factory has multiple production lines. Each line ℓ can produce a set of products denoted by $\Pi(\ell)$. Each line ℓ, can produce a product $\pi \in \Pi(\ell)$ at a rate that is r_ℓ^π measured in product units per hour where a product unit could be a package of 6, or 1 kg of finished material, etc. Each product has a life-time $T(\pi)$ that starts the moment it is produced during which it may reach the downstream customer. There is a forecasting horizon of $M = \{1, \ldots, \mathrm{M}\}$ periods, and for each period $i \in M$ there is a demand forecast d_i^π for each product π. Finally, each product has a relative weight w_i^π for period i that signifies the importance of no stock-outs for this product relative to the other products in the range of products the company manufactures in period i. As is the practice in many companies in the foods and beverages industries, budgetary and planning concerns dictate a desired number of shifts h_i^ℓ for line ℓ to be used in period i. This soft constraint can be viewed as an attempt to *reserve* line capacity during or immediately before periods of high demand (holiday seasons, periods that will follow promotional activities such as advertising campaigns, penetration into new markets, etc.). For each period i, line ℓ can operate a total of D_i^ℓ calendar days. Each day has a number S of shifts (usually three), and each shift is t hours long (usually eight). The above numbers do not include dates during which a line is scheduled to be down for maintenance or any other reasons. We denote the number of hours a line ℓ will work on a product $\pi \in \Pi(\ell)$ in period i as $o_i^{\ell,\pi}$. From the BoM we also have the quantity $b_{r,\pi}$ of raw material $r \in R$ that is required to build a unit of product π. And from Material Request Planning we have the amount $q_{r,i}$ of raw material r that will become available in period i.

The extended Multi-Commodity Aggregate Production Planning problem (eMCAP) determines the optimal quantities for production $x_{i,j}^{\pi,\ell}$ for line ℓ, of product $\pi \in \Pi(\ell)$ in period i to be sold in later period j, along with the number of total shifts a_i^ℓ to be used in period i in line ℓ. The number of extra shifts that will be required in period i on line ℓ will be denoted by y_i^ℓ. And the excess demand for a product π in a period i that cannot be physically produced by the company will be denoted by s_i^π. We formulate the eMCAP problem as follows:

$$\min_{x,y,a,o,s} \sum_{i\in M} \sum_{\ell\in L} \sum_{\pi\in\Pi(\ell)} \sum_{j=i}^{\min(M,i+T(\pi))} 2^{j-i} x_{i,j}^{\pi,\ell} + K^* \sum_{i\in M} \sum_{\ell\in L} y_i^\ell + J^* \sum_{i\in M} \sum_{\pi\in\Pi} w_i^\pi s_i^\pi$$

subject to:

$$\sum_{\ell\in L(\pi)} \sum_{j=\max(1,i-T(\pi))}^{i} x_{j,i}^{\pi,\ell} + s_i^\pi = d_i^\pi \quad \forall i \in M, \pi \in \Pi$$

$$\sum_{j=i}^{\min(i+T(\pi),M)} x_{i,j}^{\pi,\ell} = o_i^{\ell,\pi} r_\ell^\pi \quad \forall \ell \in L, i \in M, \pi \in \Pi(\ell)$$

$$\sum_{\pi\in\Pi(\ell)} o_i^{\ell,\pi} \leq a_i^\ell t \quad \forall i \in M, \ell \in L$$

$$y_i^\ell \geq a_i^\ell - h_i^\ell \quad \forall \ell \in L, i \in M$$

$$x_{i,j}^{\pi,\ell} \geq 0 \quad \forall \pi \in \Pi, \ell \in L(\pi), i \in M, j = i, \ldots, \min(i + T(\pi), M)$$

$$y_i^\ell \geq 0 \quad \forall i \in M, \ell \in L$$

$$s_i^\pi \geq 0 \quad \forall i \in M, \pi \in \Pi$$

$$a_i^\ell \leq D_i^\ell S, a_i^\ell \in \mathbb{N} \quad \forall i \in M, \ell \in L$$

$$o_i^{\ell,\pi} \geq 0 \quad \forall i \in M, \ell \in L, \pi \in \Pi$$

The quantities $K^* = 2^M \sum_{\pi\in\Pi, i\in M} d_i^\pi$, $J^* = K^* \left(\sum_{i\in M, \ell\in L} (D_i^\ell S - h_i^\ell) + 1 \right)$ guarantee that the eMCAP problem has an optimal solution $(x^*, y^*, a^*, o^*, s^*)$ that among all points in the feasible set:

1. Minimize the quantity $\sum_{i\in M, \pi\in\Pi} w_i^\pi s_i^\pi$,
2. Minimize the number of shifts above the desired shifts needed to produce the quantities $d_i^\pi - (s^*)_i^\pi$,

3. Minimize inventory holding time and costs among all feasible points that minimize the last two terms of the objective function.

The proof of the above statement follows the same line of arguments made in the proof of Theorem 3.1 in Sect. 3.1.1. The eMCAP problem as it is defined above takes into account line production rates that vary depending on the product being produced, different number of calendar days in each period that each line can be operational, and finally, it determines the optimal number of hours $(o^*)_i^{\ell,\pi}$ each line will have to work every period to produce product $\pi \in \Pi(\ell)$. Because of these details, it is no longer effective to decompose the problem as we did before into two parts, namely Shift-Allocation and then solve a Production-Scheduling resulting part as a linear minimum cost network flow problem, because, even if we could somehow determine the shifts to allocate to each line during each period, the remaining problem does not have a network flow structure to exploit.

As a final comment, we note that it is of course possible to also solve a series of three MIP problems each of which has as its objective function only the corresponding trade-off, in the lexicographic order of importance fashion, just as was detailed in the end of Sect. 3.1.1.

Regarding allocation of products inventory to demand (customer) classes, we use a model inspired from the push-based ATP model presented in Ball et al. (2004). The model we present takes explicitly into account product life-time constraints and the particularities of a non-homogeneous multi-line manufacturing setting.

The model rations available raw materials and production capacity among a set of demand classes K. The model operates over a finer-level time horizon than the aggregate time horizon of MCAP. This finer-level horizon analyzes each aggregate period i into sub-periods (usually with a grain of one week) $M'(i)$ that for brevity, when not ambiguous, will be denoted simply by M'. As in the MCAP model, the set of products is denoted by Π. There is a set of raw materials R. The results of the eMCAP problem above provide the aggregate quantities

$$\hat{X}_i^\pi = \sum_{\ell \in L(\pi)} \sum_{j=i}^{\min(M, i+T(\pi))} x_{i,j}^{\pi,\ell}$$

of each product π that should be produced in aggregate period i to meet the demands of the planning horizon, obeying as best as possible the soft constraints on the number of shifts to be used in the production plan.

The input data of the DAATP problem then are as follows:

- $\bar{d}_i^{\pi,k}$—a forecasted upper bound on the demand for product π from demand class k in period i.
- d_i^π—forecasted total demand for product π in period i.
- $v^{\pi,k}$—per unit net revenue for demand for product π from demand class k.
- $b_{r,\pi}$—raw material $r \in R$ that is required to build a unit of product π.
- $q_{r,i}$—amount of raw material r that will become available in period i.

- c_i^{π}—cost of producing a unit product π in period i.
- h_i^{π}—cost of holding inventory of a unit product π in period i.
- $h_i^{\prime r}$—cost of holding inventory of a unit raw material r in period i.
- $\hat{X}_i^{\pi}$—quantity of product π to be produced in aggregate period i.
- M'—a fine-level planning horizon covering the aggregate period i.

The decision variables are as follows:

- $Y_{j,i}^{\pi,k}$—the quantity of product π produced in period j allocated to class k in period $i \geq j$.
- $Y_i^{\pi,k}$—the total quantity of product π allocated to class k in period i.
- $I_{j,i}^{\pi}$—inventory of product π produced in period j held in period $i \geq j$.
- I_i^{π}—total inventory of product π held in period i.
- J_i^r—inventory of raw material r held in period i
- X_i^{π}—quantity of product π to be produced in fine-grain period i.

The push-based DAATP problem can now be stated as the following LP:

$$\max_{Y,I,J,X} \sum_{i \in M'} \sum_{k \in K} \sum_{\pi \in \Pi} v^{\pi,k} Y_i^{\pi,k} - \sum_{i \in M', \pi \in \Pi} h_i^{\pi} I_i^{\pi} - \sum_{i \in M', r \in R} h_i^{\prime r} J_i^r - \sum_{i \in M', \pi \in \Pi} c_i^{\pi} X_i^{\pi}$$

subject to:

demand and availability limitations

$$\sum_{k \in K} Y_i^{\pi,k} \leq I_i^{\pi} + X_i^{\pi} \quad \forall i \in M', k \in K, \pi \in \Pi$$

$$\sum_{k \in K} Y_i^{\pi,k} \leq d_i^{\pi} \quad \forall i \in M', k \in K, \pi \in \Pi$$

$$Y_i^{\pi,k} \leq \bar{d}_i^{\pi,k} \quad \forall i \in M', k \in K, \pi \in \Pi$$

product inventory balance subject to life-time constraints

$$I_i^{\pi} = \sum_{j=\max(1,i-T(\pi))}^{i-1} I_{j,i}^{\pi} \quad \forall i \in M', \pi \in \Pi$$

$$Y_i^{\pi,k} = \sum_{j=\max(1,i-T(\pi))}^{i} Y_{j,i}^{\pi,k} \quad \forall i \in M', \pi \in \Pi, k \in K$$

$$I_{j,i}^{\pi} = \begin{Bmatrix} X_j^{\pi} - \sum_{k \in K} \sum_{j'=j}^{i-1} Y_{j,j'}^{\pi,k} & j \leq i-1 \\ 0 & \text{else} \end{Bmatrix} \quad \forall i,j \in M', \pi \in \Pi$$

material inventory balance

$$J_{i-1}^{r} + q_{r,i} = J_i^r + \sum_{\pi \in \Pi} b_{r,\pi} X_i^{\pi} \quad \forall i \in M', r \in R$$

aggregate production requirements

$$\sum_{i' \in M'(i)} X_{i'}^{\pi} = \hat{X}_i^{\pi} \quad \forall i \in M, \pi \in \Pi$$

Initialization and non-negativity

$$I_{i,j}^{\pi} \geq 0, I_{j,0}^{\pi} = p_j^{\pi}, J_0^r = q_{r,0}, J_i^r \geq 0, X_i^{\pi} \geq 0, Y_{i,j}^{\pi,k} \geq 0$$

Finally, the third optimization problem, the MCFP operates at the pull-control level. In the case when customer classes do not form a tree hierarchy but instead form a flat partitioning of the total customer accounts, it accepts as input the allocated inventory $Y_i^{\pi,k}$ for each customer class per period per product, as computed by the solution of the DAATP model. It also accepts as second input a set of pending customer orders; each customer order is a set of quadruples $(i, \pi, k, d_i^{\pi,k})$ where k is the customer class where the customer belongs. This set of open customer orders is denoted by *CustOrder*. The third input is the remaining unutilized possible number of shifts $u_i^{\ell} = D_i^{\ell} S - a_i^{\ell} \quad \forall i \in M, \ell \in L$ for the current *aggregate* period i, for each line, as determined by the eMCAP problem solution. The fourth input is a decomposition $M'(i)$ of the current aggregate period—i.e. the same time horizon used in the DAATP problem—and a rationing of the total number of currently unused hours $o_i^{\pi,k} \forall \pi \in \Pi, i \in M', k \in K$among customer demand classes $k \in K$ per product per period. This rationing is such so that the percentage of unused production hours per customer per product per period is the same as the allocated ATP determined by the solution of the DAATP problem. The MCFP problem determines whether a feasible schedule exists that will produce within the current aggregate period, all the product quantities identified in the customer order request within the fine-grain time period specified.

The decision variables of MCFP are:

- $x_{i,j}^{\pi,\ell,k}$—quantity of product π to be produced in line $\ell \in L(\pi)$ during fine-grain period i to be delivered in fine-grain period j to customer class $k \in K$.
- $e_i^{\ell,\pi,k}$—the hours that line ℓ must operate in fine-grain period i, on product π for customer $k \in K$.

The pull-based MCFP problem becomes the following LP:

$$\min_{x,e} \sum_{i \in M'} \sum_{\ell \in L} \sum_{\pi \in \Pi(\ell)} \sum_{k \in K} \left(\sum_{j=i}^{\min(i+T(\pi),M')} 2^{j-i} x_{i,j}^{\pi,\ell,k} \right)$$

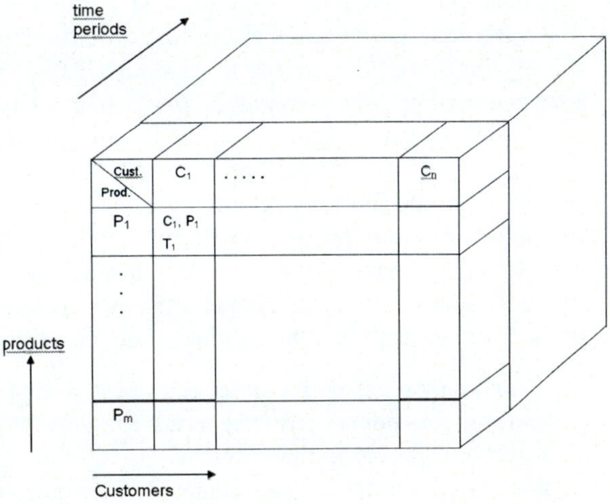

Fig. 3.11 Customer/product/period allocated ATP inventory cube

subject to:

$$Y_i^{\pi,k} + \sum_{\ell \in L(\pi)} \sum_{j=\max(1,i-T(\pi))}^{i} x_{j,i}^{\pi,\ell,k} \geq \sum_{(i,\pi,k) \in \text{CustOrder}} d_i^{\pi,k} \quad \forall (i,\pi,k) \in \text{CustOrder}$$

$$\sum_{j=i}^{\min(i+T(\pi),M')} x_{i,j}^{\pi,\ell,k} = e_i^{\ell,\pi,k} r_\ell^\pi \quad \forall \ell \in L, i \in M', \pi \in \Pi(\ell), k \in K$$

$$\sum_{j \in M'(i)} \sum_{\pi \in \Pi} \sum_{k \in K} e_j^{\ell,\pi,k} \leq u_i^\ell t \quad \forall i \in M, \ell \in L$$

$$\sum_{\ell \in L(\pi)} e_i^{\ell,\pi,k} \leq o_i^{\pi,k} \quad \forall i \in M', (\pi,k) \in \text{CustOrder}$$

$$x_{i,j}^{\pi,\ell,k} \geq 0, e_i^{\ell,\pi,k} \geq 0$$

Notice that the above problem—as opposed to the eMCAP problem which is always feasible—may well be infeasible. This would be the case if there are not enough extra hours allocated for production of products from a customer in fine-grain periods $i \in M'$.

The whole system workflow consists of a number of steps:

1. *eMCAP*: solve the eMCAP problem to determine next periods' aggregate production requirements based on the latest updates of market forecasts for the company's products.
2. *DAATP*: solve the DAATP problem to determine how to allocate current and planned product inventory among the current aggregate period's finer level time intervals and customer classes.
3. *Inventory cube*: using the solution of the DAATP problem in step 2, compute a customer/product/period cube that contains the allocated quantities of each product to each customer class in each fine-grain level period. See Fig. 3.11.

4. *Capacity cube*: using the same portion of the allocated products per customer class per period, allocate the total extra hours o_i^{π} of unused capacity available during this aggregate period, to customer classes per period per product.
5. *Customer order acceptance decision*: when a customer order request arrives, first check via simple search among the customer classes to which the customer belongs in a bottom-up fashion for available inventory in the periods up to the period requested. For any remaining product quantities that cannot be found in the inventory cube, proceed to step 6.
6. *MCFP*: solve the MCFP problem to determine whether a production plan exists that will satisfy all current constraints, and will produce the required remaining product quantities until the customer requested due-date.
   - If such a plan exists, the order request is accepted and the appropriate book-keeping procedures are triggered to modify the production schedule to accommodate for the new order.
   - Else, the order is rejected or countered by the quantities that can be found in step 5.

This workflow is schematically shown in Fig. 3.12.

The following UML Swim-Lane Activity Diagram shows the responsibilities of different functions within the same company and the required coordination in order to implement the above workflow (Fig. 3.13).

To summarize, the eMCAP problem determines an aggregate production plan that has to be followed to meet market demand and maintain the highest possible service level, and market share (push-based controls). The DAATP problem rations the planned production and by proportion the reserved production capacity as well, among customer demand classes hierarchies in the most profitable way for the company. Then, in real-time, when a new order arrives, the allocated ATP plus the MCFP model is used to determine whether to accept an actual order request based on whether the order can be feasibly produced using the customer's allocated inventory and production capacity (pull-based controls). For this to work of course, the solution to the MCFP problem must be found in real-time since this model will be solved each time allocated inventory is not sufficient to cover a customer order request.

The problems that are solved to compute feasibility or not of a new customer order request allow solving again for the upcoming periods with input data all pending customer orders, together with their promised due dates to see if a re-arrangement of production schedules can be made so as to satisfy all previously committed customer order requests plus the latest one.

3.5.8 A Simplified Example

For the shake of better understanding and absorption of the method and models used, an almost trivial example of the whole algorithm for solving an ATP instance is presented. Assume a case company that manufactures only one product with a

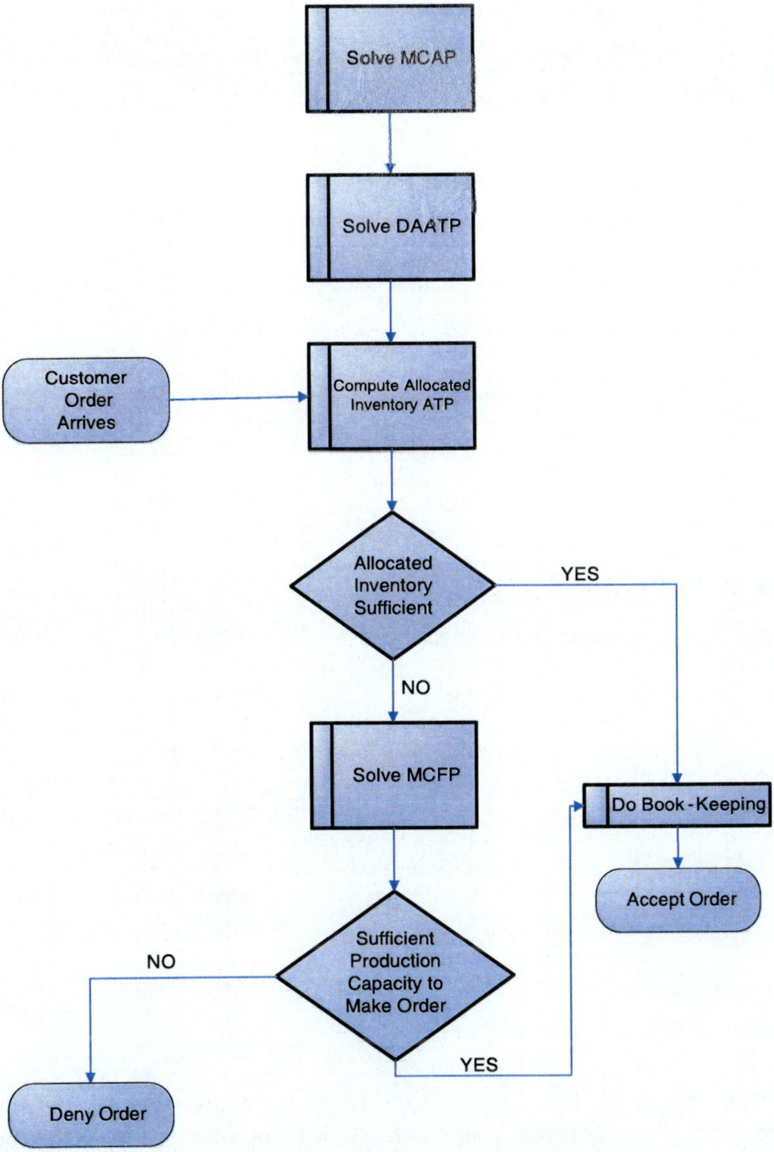

Fig. 3.12 ATP workflow operating on the push/pull control interface

life-time of four weeks, A product unit is produced in 0.5 h. An aggregate period is one month (4 weeks) long. The company has two customer classes. Class A consists of a single, very profitable customer; all other orders are categorized as belonging to a second all-encompassing bucket class B. There is only one production line. The eMCAP problem determined that for the current aggregate period the total production should reach 40 units, and have reserve capacity of

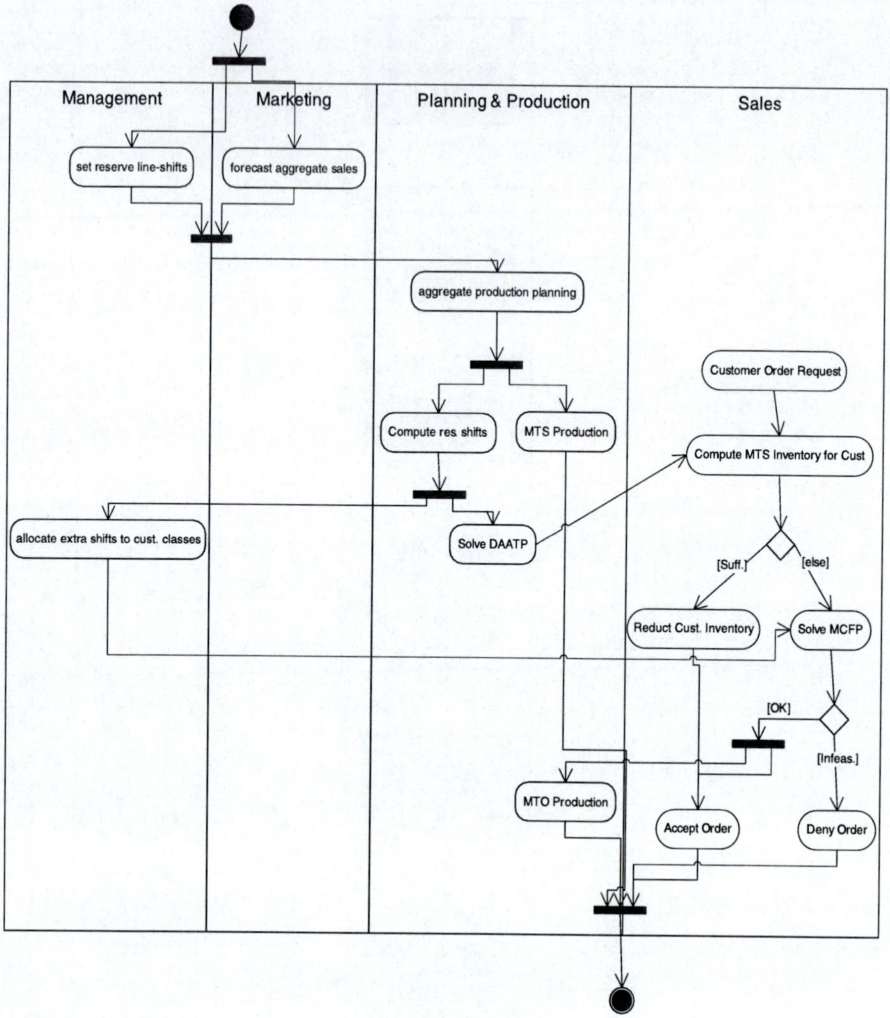

Fig. 3.13 Coordination of activities for ATP among various functions

1 week only during the last period. At the beginning of the period there was no inventory. The allocated production via the DAATP problem, taking into account the weekly demand forecasts of each customer is as follows:

Production	Week 1	Week 2	Week 3	Week 4
Customer A	15	15	5	0
Customer B	5	0	0	0

Now, the allocated extra shift-hours available for each customer become:

Extra-hours	Week 1	Week 2	Week 3	Week 4
Customer A	0	0	0	$35 \times 5 \times 8/40 = 35$ h
Customer B	0	0	0	$5 \times 5 \times 8/40 = 5$ h

Now, assume the following orders arrive:

1. customer A posts a customer order request of 30 units to be delivered in end of period 3. This request is immediately accepted as there will be an inventory of 35 units at that time.
2. one type-B customer posts a request of six units to be delivered in the end of week2. After running MCFP, the request is rejected as no inventory + allocated production capacity can suffice to meet the order demands.
3. another type-B customer posts a request of four units to be delivered in the end of week2. The order request is accepted as there is sufficient inventory.
4. another type-B customer posts a request of eight products to be delivered in the end of week 4. After running MCFP, the request is accepted as the combination of available inventory (1 left) and extra production capacity will suffice to meet the order request. The plan now becomes to produce seven products in week 4.
5. yet another type-B customer posts a request of 1 item to be delivered in the end of week 3. By running MCFP, we see that it is still possible to satisfy this and all previously accepted customer orders by committing the customer B-class allocated remaining inventory to this last arrived customer order, and change the production plan to produce eight products in week 4 (to be delivered all as the order #4 requested).

3.5.9 Implementation

We first compute the size of each of the three models comprising the proposed ATP system:

3.5.9.1 Extended Multi-Commodity Aggregate Production Planning Problem Model

Let the size of the time horizon be M, the number of different products the company produces Π, and let there be L lines. The eMCAP problem is a MIP problem with

$$N = \frac{M(M+1)}{2}\Pi L$$

variables for the $x_{i,j}^{\pi,\ell}$ variables of production quantities, ML variables for the y_i^ℓ variables of extra shifts above the desired shifts that might be used in a line on a given month, $M\Pi$ variables for the s_i^π variables denoting stocked-out quantities of a product on a given month, ML variables denoting the actual shifts a_i^ℓ that will be used in a line on a given month and $ML\Pi$ variables for the hours $o_i^{\ell,\pi}$ that each line will be used each month for a particular product. The total number of variables in this model is therefore

$$C = \frac{M(M+1)}{2} + M(2L + \Pi(L+1))$$

of which ML are integer variables (with corresponding integrality constraints).

The number of constraints is $M(\Pi(L+1)+L)$ plus $M(3L+\Pi(L+1+\frac{L(M+1)}{2}))$ box constraints of which $2ML$ are variable upper bound constraints and the rest are non-negativity constraints on the variables. The non-trivial constraints of eMCAP therefore are $M(\Pi(L+1)+L)$.

3.5.9.2 Demand Allocation Available-To-Promise Model

Let M', Π, K, R be respectively the number of fine-grain periods in the short-term planning horizon, the number of different products being planned, the total number of customer classes, and the total number of different raw materials. The DAATP model consists of $\frac{M'(M'+1)}{2}\Pi(K+1) + M'(\Pi + R)$ variables and $3M'K\Pi + \frac{M'(M'+1)}{2}\Pi + M'R + \Pi$ non-trivial constraints and $\frac{M'(M'+1)}{2}\Pi(K+1) + 2M'\Pi + R(M'+1)$ variable lower and upper bound constraints. These numbers are derived after substituting the quantities $I_i^\pi, Y_i^{\pi,k}$ in the model by the sums from which they are computed.

3.5.9.3 Multi-Commodity Fine-Grain Production Planning Model

Finally, for the MCFP model let M', Π, K, L, O be respectively the number of fine-grain periods in the short-term planning horizon, the number of different products being planned, the total number of customer classes, the total number of production lines in the organization, and the total number of orders to be received within the current planning horizon. The MCFP model is a Linear Program with $(\frac{M'(M'+1)}{2} + M')\Pi LK$variables and $O(M'+1) + LM'\Pi K$ non-trivial constraints and $M'\Pi LK(\frac{M'+1}{2}+1)$ variable non-negativity constraints.

Experiments presented in (Christou and Ponis 2008, 2009) show that the eMCAP problem even without resorting to heuristics can be solved within seconds of computing time in a modern server, despite its combinatorial nature. This is mostly due to the few integer variables appearing in the model, and its underlying

Table 3.1 eMCAP formulation in GAMS running times on NEOS servers

Name	#Lines	#Products	#Periods	eMCAP time (s)
Ex5	5	8	12	1
Ex6	5	8	18	1.1
3E	14	8	12	5.5
CF	3	15	12	4

structure, which, even though in the case of variable line capacities for different products is not network flow, is still sparse enough to allow for very fast computation times.

Solving the DAATP problem presents no difficulties either, as it is a standard Linear Program with an inventory control structure suitable for dynamic programming techniques. In any case, there are no essential response time constraints for this problem as it has to be solved off-line.

Finally, solving the MCFP problem can be accelerated if we take into account that all the orders entered in the system can be represented with at most $M'\Pi K$ constraints. Indeed, every order to be considered is a set of quadruples of the form $(i, \pi, k, d_i^{\pi,k})$. Different orders for the same period, same product coming from the same customer class can be concatenated into one quadruple containing the sum of their individual demands.

Formulating the resulting LP in GAMS format and solving it on the NEOS servers shows that the problem can be solved in less than one tenth of a second for a time horizon of four periods, for a company with five lines, eight different products and three customer classes, independent of the total number of different orders received.

3.5.10 Computational Results

Solving instances of each of the three optimization problems on the NEOS servers of course depends on the size of each problem, and the speed of the server. As mentioned before, the first two push problems (MCAP and DAATP) are off-line processes and as such do not impose strict requirements on execution time of the model. Nevertheless, for instances whose size corresponds to the real size of two different SME European manufacturers in the food and beverages industry, the solution time is always in the order of seconds. In particular, the following tables show response times for a number of different problems tested on the NEOS server at Argonne National Laboratory in USA. The response times are very reasonable and well within any response time constraint any company would be likely to set for its operations (Table 3.1).

The DAATP problem can also be solved very fast as indicated in Table 3.2.

And finally, the MCFP problem is solvable in real-time even on a commodity laptop computer running GAMS as shown in Table 3.3.

Table 3.2 DAATP formulation in GAMS running times on NEOS servers

Name	#Products	#Raw-materials	#Periods	#Customers	DAATP time (s)
Ex51	8	3	2	5	15
Ex61	8	3	4	5	18
CF	10	5	4	3	20

Table 3.3 MCFP GAMS model running times on a 1.8 GHz pentium M laptop

Name	#Lines	#Products	#Periods	#Customers	MCFP time (s)
Ex7	2	2	4	2	0.8
CF1	3	5	4	5	1.5
CF2	3	10	4	10	5
CF3	3	20	4	20	9
3E3	14	8	4	20	8

So it can be seen that an order can be accepted or denied in real-time even when the company produces 20 different products and monitors 20 key accounts. In reality, one of the two companies we are targeting has many more product codes than the 20 tested here (more than 100); however, most codes are C-category items in an ABC analysis and therefore are not worthy of reserving capacity or monitoring. The scheme we have developed can be used to guide decision making about any subset of a company's product family and customer accounts. Normally, it would offer most value if applied to the most profitable (or most demanded) products and customers in a company's product line. And indeed, the common practice in both target companies is to monitor only a few key accounts and key products on a weekly basis, and use a standard MTS approach for all other products.

3.6 Bibliography

Regarding job-shop scheduling, a plethora of exact and approximate algorithms have been proposed for the JSSP over the years. The successful implementation of exact algorithms, such as Branch and Bound (Brucker et al. 1994), is limited to small problem instances due to their sizeable computational overhead. For problems with more than 15 jobs and machines such techniques are rendered impractical since they may run on days on end on modern hardware and still not find the optimum solution. Literature has showed that the larger problem instances are best tackled by efficient approximation meta-heuristic algorithms such as tabu search (Nowicki and Smutnicki 2005; Watson et al. 2003; Taillard 1994), simulated annealing (Van Laarhoven et al. 1992; Kolonko 1999; Aydin and Fogarty 2004), greedy adaptive randomized search procedure (Binato et al. 2001; Aiex et al. 2003; Fernandes and Lourenco 2007), global equilibrium search (Pardalos and Shylo 2006), threshold accepting (Lee et al. 2004; Tarantilis and Kiranoudis 2002), variable neighborhood search—a discrete optimization method we shall briefly discuss in Chap. 5 on location theory—(Sevkli and Aydin 2006) and genetic algorithms (Mattfeld 1996; Dorndorf and Pesch 1995; Vásquez and Whitley 2000). Other approaches based on ant colony optimization and particle swarm optimization have also been proposed more recently as well. Regarding crew assignment, the GA-based approach presented in the text is based on Christou et al. (1999), and Christou and Zakarian (2000). More recently, commercial MIP solvers from ILOG and other companies have been able to attack large scale crew assignment problems using exact methods based on the Branch-Cut-and-Price scheme (Sect. 1.2.2.4).

Ball et al. (2004) provide an excellent overview and review of the research in the area of ATP scheduling. First of all, they make the deep observation that the purpose of ATP is to operate on the boundary of Push-based control and Pull-based control. They point out that the conventional ATP systems that were associated with traditional MTS supply chains are being updated to accommodate the make/manufacture-to-order (MTO) supply chains prevalent today. So, ATP problems are classified as being either ‘push based’ or ‘pull based’. Push based ATP models allocate resources to products or demand classes prior to receiving orders, while pull based ATP models perform the allocation in response to incoming orders. The primary advantages of push based scheduling over pull based scheduling is that order promising decisions can incorporate long term objectives and can be provided to customers immediately.

Pull based ATP scheduling has an advantage over push based scheduling in that it can be responsive to disparities between actual and forecasted demand. Since pull based ATP scheduling makes resource allocations after demand is realized, there is the need to repeatedly determine allocations. The more frequently the allocation problem is solved, the more myopic it becomes. On the other hand, push-based ATP, as it is more closely linked with advanced planning, offers the possibility to more efficiently and profitably schedule resources and capabilities, and to utilize any excess capacity that is available but unused at any given moment.

The notions of allocated ATP are more explicitly detailed in Kilger and Schneeweiss (2005), where they draw heavily from concepts used in yield management in the airline industry. Allocated ATP is clearly inspired if not directly related to yield management (Smith et al. 1992). Yield management problems however, tend to emphasize the use of pricing as a method of controlling the allocation of fixed, perishable resources. In ATP, resources are typically not perishable and pricing cannot typically be treated as a decision variable. Kilger and Schneeweiss (2005) describe ATP as working along three major dimensions: *customer, product and time*. They define hierarchies along each of these dimensions and use a simple search procedure to find available-to-promise quantities. No optimization model is employed in their work.

Ervolina and Dietrich (2001) studied deterministic push based ATP scheduling models for CTO systems with multiple products, components and time periods. Their work is based on the resource allocation software engines developed at IBM Research, and provides an important part of the foundation for the research developed in this dissertation. They sketch two different heuristic approaches for determining the ATP schedule, but do not provide any computational results. Since Ervolina and Dietrich (2001) acknowledge that product configurations are uncertain, they use an 'average box demand' to represent them.

In the deterministic pull-based ATP scheduling realm, Chen et al. (2002) consider a rolling horizon ATP problem for a configure-to-order (CTO) product. In their model, orders are batched over some pre-specified period of time, after which order commitment dates and the production schedule are obtained by solving a multi-period mixed-integer program. In addition to specifying a due date, customers are allowed to specify a range of acceptable delivery quantities and a set of substitutable suppliers for a given component. They find that while profits initially increase with the length of the batching period due to increased information about demand, profits eventually drop as more orders with shorter due dates are lost. In an earlier work, Chen et al. (2001) consider a similar problem where customer due dates are flexible and charge a penalty for allowing component inventory levels to drop below a pre-specified reserve level at the end of each batching interval. In any given run, they use the same reserve level for all resources. An experimental study shows the use of such a reserve level can increase profits by anticipating the arrival of more profitable orders in future batching intervals.

A deep study of ATP using deterministic optimization models is presented in Zhao et al. (2005) where the authors present a MIP model tailored to the specific requirements of ATP for an electronic product at Toshiba, taking into account due date violations rules, manufacturing orders, production capability and capacity. A rather complicated model involving millions of variables and constraints is decomposed using aggregation into weekly and daily problems each involving a few thousands of variables and constraints that is then solved using commercial state-of-the-art LP/MIP software (CPLEX). The results were very promising in that they were able to simultaneously optimize both due date violations and inventory holding costs. Bilgen and Gunther (2009) study planning and ATP in fast moving consumer goods industries.

Regarding software implementations, in Friedrich and Speyerer (2002), the authors present an Information Systems open architecture based on XML document exchanges for implementing ATP in the standard context of MRP/MRPII logics.

Finally, logistics models for product class importance depreciation were developed as part of Apostolopoulos's Master of Science thesis (Apostolopoulos 2008).

3.7 Exercises

1. Formulate the MCAP problem discussed in Sect. 3.1.1 when the priorities for the company are as follows:

    1. Never use any extra shifts (but outsourcing is allowed to meet demand),
    2. Among all plans that obey the above constraint, choose a plan that fully meets demand,
    3. Among all plans that are optimal with respect to the previous objective, choose the one that minimizes inventory costs (and simultaneously maximizes product quality).

    Does the problem have any special structure that can be exploited?
2. Modify the MCAP model in Sect. 3.1.1 under the assumption that all products are non-perishable, i.e. they have infinite life-times. Is the model easier to solve under this assumption, and if so, how?
3. Implement the shortest processing time (SPT) algorithm and critical ratio first (cr) algorithm for sequencing jobs on a single machine, and experiment with a set of 1000 randomly generated jobs, with processing times normally distributed around $\mu = 100$ min and with standard deviation $\sigma = 30$. The due-date for each job $i = 1, \ldots, 1000$ should be a random variable $D_i = \sqrt{p_i}(\sqrt{p_i} + R)$ where p_i is the processing time of job i, and R is a random variable following the *uniform distribution* in [0, 100]. Which algorithm yields better results in terms of average tardiness? What about average lateness?

References

Aiex RM, Binato S, Resende MGC (2003) Parallel GRASP with path-relinking for job shop scheduling. Parallel Comput 29:393–430

Apostolopoulos P (2008) A decision making system for orders for the pull-based part of the available-to-promise strategy. M.Sc. thesis, Athens Information Technology

Aydin ME, Fogarty TC (2004) A distributed evolutionary simulated annealing algorithm for combinatorial optimisation problems. J Heuristics 10:269–292

Ball MO, Chen C-Y, Zhao Z-Y (2004) Available to promise. In: Simchi-Levi D, Wu SD, Shen ZM (eds) Handbook of quantitative supply chain analysis: modeling in the e-business era. Springer, NY

Bilgen B, Gunther H-O (2009) Integrated production and distribution planning in the fast moving consumer goods industry: a block planning application. OR Spectrum, 18 June 2009

Binato S, Hery W, Loewenstern D, Resende MGC (2001) GRASP for job shop scheduling. In: Essays and surveys on meta-heuristics. Kluwer, Amsterdam

Blackstone JH, Cox JF (2004) APICS Dictionary, 11th edn. McGraw Hill, Falls Church

Brizuela CA, Sannomiya N (2000) A selection scheme in genetic algorithms for a complex scheduling problem. In: Proceedings of the GECCO 2000 genetic and evolutionary computation conference, Las Vegas, NV

Brucker P, Jurisch B, Sievers B (1994) A branch and bound algorithm for the job-shop scheduling problem. Discret Appl Math 49:107–127

Chen C-Y, Zhao Z-Y, Ball MO (2001) Quantity and due-date quoting available to promise. Inform Syst Frontiers 3(4):477–488
Chen C-Y, Zhao Z-Y, Ball MO (2002) A model for batch advanced available to promise. Prod Oper Manag 11(4):424–440
Chen-Ritzo C-H (2006) Availability management for configure-to-order supply chain systems. Ph.D. dissertation, College of Business Administration, Pennsylvania state University
Christopher M (2005) Logistics and supply chain management: creating value-adding networks, 3rd edn. Prentice-Hall, Harlow
Christou IT, Ponis S (2008) Enhancing traditional ATP functionality in open source ERP systems: a case-study from the food and beverages industry. Int J Enterp Inf Syst 4(1):18–33
Christou IT, Ponis S (2009) A hierarchical system for efficient coordination of available-to-promise logic mechanisms. Int J Prod Res 47(11):3063–3078
Christou IT, Zakarian A (2000) Domain knowledge and representation issues in genetic algorithms for scheduling problems. In: Proceedings of the GECCO 2000 genetics and evolutionary computation conference, Las Vegas, NV
Christou IT, Zakarian A, Liu J-M, Carter H (1999) A two phase genetic algorithm for solving large scale bid-line generation problems at Delta Air Lines. Interfaces 29(5):51–65
Christou IT, Lagodimos AG, Lycopoulou D (2007) Hierarchical production planning for multi-product lines in the beverage industry. J Prod Plan Control 18(5):367–376
Dorndorf U, Pesch E (1995) Evolution based learning in a job-shop scheduling environment. Comput Oper Res 22:25–40
Ervolina T, Dietrich B (2001) Moving toward dynamic available to promise. In: Gass PI, Jones AT (eds) Supply chain management practice and research: status and future directions. Manufacturing Engineering Laboratory, RH School of Business, University of Maryland
Fernandes S, Lourenco HR (2007) A GRASP and branch-and-bound meta-heuristic for job shop scheduling. Lecture notes in computer science, vol 4446, pp 60–71
French S (1982) Sequencing and scheduling: an introduction to the mathematics of the job shop. E. Horwood, Chichester
Friedrich J-M, Speyerer J (2002) XML-based available-to-promise logic for small and medium enterprises. In: Proceedings of the 35th international conference on system sciences, Hawaii, HW
Garey MR, Johnson DS (1979) Computers and intractability: a guide to the theory of NP-completeness. WH Freeman, NY
Graham RL, Lawler EL, Lenstra JK, Rinnooy Kan AHG (1979) Optimization and approximation in deterministic sequencing and scheduling: a survey. Ann Discret Math 5:287–326
Hopp W, Spearman M (2008) Factory physics, 3rd edn. McGraw-Hill/Irwin, NY
Keskinocak P, Tayur S (2004) Due date management policies. In: Simchi-Levi D, Wu SD, Shen ZM (eds) Handbook of quantitative supply chain analysis: modeling in the e-business era. Springer, NY
Kilger C, Schneeweiss L (2005) Demand Fulfillment and ATP. In: Stadtler H, Kilger C (eds) Supply chain management and advanced planning: concepts, models, software and case studies, 3rd edn. Springer, Berlin
Kolonko M (1999) Some new results on simulated annealing applied to the job shop scheduling problems. Eur J Oper Res 113:123–136
Lawler EL (1982) Preemptive scheduling of precedence-constrained jobs on parallel machines. In: Dempster MAH, Lenstra JK, Rinnooy Kan AHG (eds) Deterministic and Stochastic Scheduling. D Reidel Publishing Company, Dordrecht
Lee DS, Vassiliadis VS, Park JM (2004) A novel threshold accepting meta-heuristic for the job-shop scheduling problem. Comput Oper Res 31:2199–2213
Lenstra JK, Rinnooy Kan AHG (1979) Computational complexity of discrete optimization problems. Ann Discret Math 4:121–140
Mattfeld DC (1996) Evolutionary search and the job shop: investigations on genetic algorithms for production scheduling. Physica-Verlag, Heidelberg
Nanda R, Browne J (1992) Introduction to employee scheduling. Van Nostrand-Reinhold, NY

Nowicki E, Smutnicki C (2005) An advanced tabu search algorithm for the job shop problem. J Sched 8:145–159
Pardalos PM, Shylo O (2006) An algorithm for the job shop scheduling problem based on global equilibrium search techniques. Comput Manag Sci 3(4):331–348
Pinedo M (2008) Scheduling: theory, algorithms and systems. Springer, NY
Sevkli M, Aydin ME (2006) A variable neighborhood search algorithm for job shop scheduling problems. Lecture notes in computer science, vol 3906, pp 261–271
Silver EA, Pyke DF, Peterson R (1998) Inventory management and production planning and scheduling, 3rd edn. Wiley, Hoboken
Smith B, Leimkuhler J, Darrow R, Samuels J (1992) Yield management at American Air-Lines. Interfaces 22:8–31
Taillard ED (1994) Parallel taboo search techniques for the job shop scheduling problem. ORSA J Comput 6:108–117
Tarantilis CD, Kiranoudis CT (2002) A list-based threshold accepting method for the job-shop scheduling problems. Int J Prod Econ 77:159–171
Van Laarhoven PJM, Aarts EHL, Lenstra JK (1992) Job shop scheduling by simulated annealing. Oper Res 40:113–125
Vásquez M, Whitley L (2000) A comparison of genetic algorithms for the static job shop scheduling problem. In: Proceedings of the 6th parallel problem solving from nature conference, PPSN VI
Wagner HM, Whitin T (1958) Dynamic version of the economic lot size model. Manag Sci 5:89–96
Watson J-P, Beck J, Howe A, Whitley L (2003) Problem difficulty for tabu search in job-shop scheduling. Artif Intell 143(2):189–217
Wu A, Chiang D, Chang C-W (2010) Using order admission control to maximize revenue under capacity utilization requirements in MTO B2B industries. J Oper Res Soc Jpn 53(4):38–44
Zhao Z, Ball MO, Kotake M (2005) Optimization-based available-to-promise with multi-stage resource availability. Ann Oper Res 135(1):65–85
Zobolas GI, Tarantilis CD, Ioannou G (2009) A hybrid evolutionary algorithm for the job-shop scheduling problem. J Oper Res Soc 60:221–235

Chapter 4
Inventory Control

Inventory control has been the subject of intense study since the era of industrialization, due to its significant cost savings potential. Major efforts in Operations Management were carried out since the 1950s in order to optimize the costs of an industrial organization by optimizing the inventories of raw materials, work-in-progress (WIP) as well as finished products; for a detailed review of initiatives such as MRP and MRP II, or just-in-time (JIT) see Hopp and Spearman (2008). The mathematical analysis of inventory systems starts with Harris's economic order quantity (EOQ) model (Harris 1913), received a major boost in the 1950s and the 1960s, the time when computers and operations research methods started to become well known and indispensable tools for scientists, engineers and managers alike, and continues to this day, in both fundamental as well as niche aspects of inventory theory and control. Throughout this chapter, it is assumed that stocked items are non-perishable and therefore have no life-time limitations.

4.1 Deterministic Demand Models and Methods

4.1.1 The EOQ Model

The easiest models of inventory control assume that demand for a certain item to be stocked is deterministic and so we start with the presentation of classical results regarding the EOQ and economic production quantity (EPQ) models and their extensions.

The context of the problem to be solved in the EOQ model involves a company that faces continuous and constant demand for a particular product. The rate of demand then, measured in items demanded per unit time is D. The company is a retail company so it has to order the products it resells to its end-customers. There is a fixed order cost K associated with each order the company places. The purchase price of each item the company purchases is p_o and the company resells such

I. T. Christou, *Quantitative Methods in Supply Chain Management*,
DOI: 10.1007/978-0-85729-766-2_4,

Fig. 4.1 Inventory evolving in time with constant demand rate. The inventory is replenished to the quantity Q at the end of a time interval of length T. The demand rate in the figure is equal to the $\tan\varphi$ of the angle formed by the inventory plot line and the time axis, and is equal to Q/T

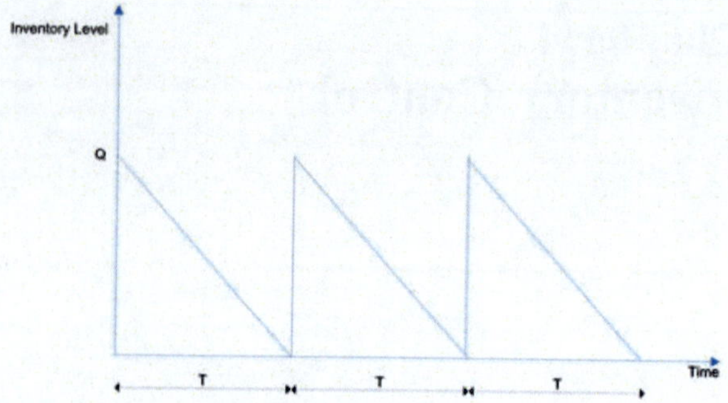

items at a unit price p_s (which should clearly be greater than p_o). On the other hand, there is a holding cost rate h associated with every item stored in the company's warehouse, representing the opportunity cost of the money invested in inventory, liabilities associated with holding and handling inventory, etc. The holding cost rate is expressed as a percentage per time unit, so the quantity $H = hp_o$ measured in currency units per unit time per stock item expresses the cost of holding one item for one-time unit in the company's warehouse. When the company places an order to its outside supplier, the order is *always fulfilled* regardless of the order size immediately, so *the order lead-time L* is *zero*. Since the company faces constant and continuous demand, demand is stationary. The problem is to minimize the total costs the company faces per unit time by selecting the time T that passes between placing consecutive orders of such an amount Q so that at the end of the next interval there will be exactly zero inventory left in the warehouse (this is known as the *Zero Inventory Property*). The fluctuations of inventory in time are shown schematically in Fig. 4.1.

From the description of this problem, we can compute the total costs to be minimized as a function of the variables Q or T, since the two variables are dependent on each other by the relation $Q = DT$ since we place an order exactly when we have no inventory, and at the end of the next interval T, we must be left exactly with zero inventory as well. The cost of purchasing a quantity Q of stock items is obviously Qp_o but it can be ignored because the purchasing cost *per unit time* becomes

$$C_p = Qp_o/T = Dp_o$$

which is a constant value and thus does not enter the optimization process.

The other two costs that must be measured are the fixed purchase cost that occurs every time an order is placed with the company's supplier, and the holding cost of holding inventory. If an order is placed every T time units, the fixed cost C_k incurred per unit time is then

$$C_k = K/T = KD/Q$$

Finally, we must measure the holding costs per unit time. Assuming at the beginning of a time interval starting at t_0 of length T an order of size $Q = DT$ has just arrived (with zero prior inventory), the inventory $I(t)$ at any time t in the interval $[t_0, t_0 + T]$ will be

$$I(t) = Q - Dt$$

as inventory is depleted from the system at the constant rate D. The total holding cost of the system within an interval of time of length T, assuming we place orders every T time unit then becomes

$$\int_0^T HI(t)\mathrm{d}t = H\int_0^T (Q - Dt)\mathrm{d}t = H\left(QT - D\frac{1}{2}T^2\right) = HT(Q - Q/2) = HTQ/2$$

(see Fig. 4.1). This quantity is equal to $HDT^2/2$ or $HQ^2/(2D)$. Therefore, the total holding cost of the system *per unit time*, assuming we place orders every T time unit becomes

$$C_I = HQ/2 = HDT/2$$

All in all then, the *total cost of the system per unit time* $C_K + C_I$ is computed as the function:

$$C_{\text{tot}}(Q) = \frac{KD}{Q} + \frac{HQ}{2} \tag{4.1}$$

Alternatively, the total cost as a function of the reorder interval T is expressed as

$$C_{\text{tot}}(T) = \frac{K}{T} + \frac{HDT}{2} \tag{4.2}$$

In order to minimize its costs then, the company needs to solve the following nonlinear optimization problem:

$$(\text{EOQ}) \min_{Q \geq 0} C_{\text{tot}}(Q) = \frac{KD}{Q} + \frac{H}{2}Q$$

The above-stated optimization problem is defined over the set $F = \mathbb{R}_+$. and assumes the value $C_{\text{tot}}(0) = +\infty$ at zero. Now, it is easy to verify that the function $C_{\text{tot}}(Q)$ is convex in all of its domain of definition, i.e. the interval $[0, +\infty)$, and its unique global minimum is attained at the point where

$$\frac{\mathrm{d}C_{\text{tot}}}{\mathrm{d}Q}(Q) = -\frac{KD}{Q^2} + \frac{H}{2} = 0$$

Solving the above equation for Q, we obtain

$$Q^* = \sqrt{\frac{2KD}{H}} \tag{4.3}$$

which is strictly greater than zero, so the constraint $Q \geq 0$ is inactive, and so does not need to be taken into account. The corresponding minimum cost per time unit becomes

$$C^*_{\text{tot}} = \frac{KD}{\sqrt{2KD/H}} + \frac{H\sqrt{2KD/H}}{2} = \sqrt{2KDH}$$

The quantity Q^* is known as the economic order quantity, hence the name of the model ("economic" here is meant in the sense of "optimal in economic terms"). The optimal time interval between placing orders is then, obviously

$$T^* = \frac{Q^*}{D} = \sqrt{\frac{2K}{HD}} \tag{4.4}$$

In the above derivation, and model, the variables T and Q were considered continuous, whereas in reality Q might be discrete (if the items in stock represent quantities of indivisible units of stock items, such as pens or pencils, etc. as opposed to, say, bulk paint in kilogram, that can be considered a continuous variable). In such a case where Q is a discrete variable, by the convexity of the function $C_{\text{tot}}(Q)$ in its continuous form, the optimal order quantity would be computed as

$$Q^{**} = \begin{cases} \lfloor Q^* \rfloor, & \text{if } \frac{KD}{\lfloor Q^* \rfloor} + \frac{H\lfloor Q^* \rfloor}{2} \leq \frac{KD}{\lceil Q^* \rceil} + \frac{H\lceil Q^* \rceil}{2} \\ \lceil Q^* \rceil, & \text{otherwise} \end{cases}$$

and the time interval T^{**} would be defined from the equation $Q = DT$.

As a practical example, consider a retail company that faces demand that can be considered constant throughout the company's planning horizon for a particular product. Even though such an assumption may at first sound absurd, in practice, there are many items that face a near-constant demand. To give an example, the demand for a drug for a chronic disease that a drug store in an isolated village faces can have a constant rate for years at a time, assuming the village's population remains the same over those years, and new occurrences of the particular disease do not happen. Let the purchase price p_o of a unit item be €10, the fixed order cost K be €200, the demand rate be $D = 20$ items per unit time and the holding cost rate h be set at 8%. Then, $H =$ €0.8/unit item/unit time, and the EOQ quantity becomes $Q^* = \sqrt{\frac{2\times200\times20}{0.8}} = \sqrt{10^4} = 100$. Since this quantity is integral, there is no need to check its floor or its ceiling. The optimal time interval between order placements becomes $T^* = Q^*/D = 100/20 = 5$ time units, and the total cost becomes $C^*_{\text{tot}} = \sqrt{2 \times 200 \times 20 \times 0.8} =$ €80/unit time.

4.1.1.1 EOQ Sensitivity Analysis

In the EOQ model, the total cost of inventory is formulated as

$$C_{\text{tot}}(Q) = KD/Q + HQ/2$$

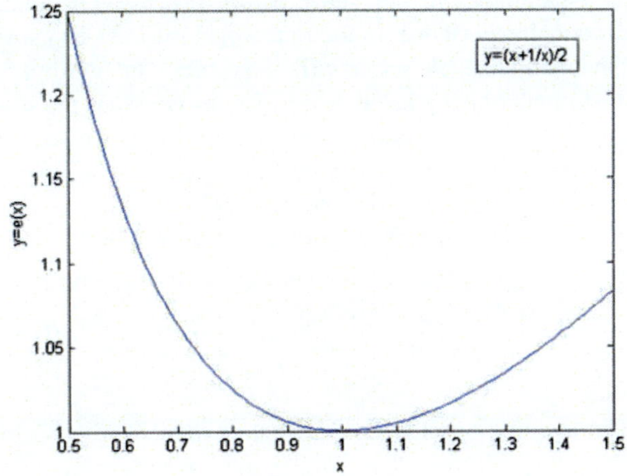

Fig. 4.2 Plot of the function $e(x)$

and the optimal cost is given by the formula

$$C^*_{tot} = \frac{KD}{\sqrt{2KD/H}} + \frac{H\sqrt{2KD/H}}{2} = \sqrt{2KDH}$$

We can write the total cost $C_{tot}(Q)$ as $KDQ^*/(QQ^*) + HQQ^*/(2Q^*)$. Now notice that, at Q^*, the two terms that form the total cost are equal:

$$KD/Q^* = HQ^*/2$$

which implies that $KD/Q^* = HQ^*/2 = C_{tot}(Q^*)$ Using the above equalities in the previous expression for the total cost, we get

$$C_{tot}(Q) = C_{tot}(Q^*)Q^*/(2Q) + C_{tot}(Q^*)Q/(2Q^*) = C_{tot}(Q^*)[Q^*/Q + Q/Q^*]$$

The last equation can also be expressed as

$$C_{tot}(Q) = C^*_{tot}e(Q^*/Q)$$

where

$$e(x) = \frac{1}{2}\left(x + \frac{1}{x}\right)$$

The function $e(x)$ defined above is plotted in Fig. 4.2. Its plot reveals why the EOQ model is rather insensitive even to significant deviations from the optimal order quantity Q^*. Indeed, even when Q deviates from Q^* by 10%, the percentage deviation of the total cost from its optimal value is <0.45%.

4.1.1.2 EOQ Under Discounts

In the above analysis, we have assumed that the cost of purchase of a single stock item unit p_o is constant and independent of the order size, and thus does not enter as a parameter in the optimization process. Of course, as everyone knows from practical experience, in the real world the cost of a single unit is not independent of the order size. A seller will very often be willing to offer a discount for good customers placing large orders. Let us consider the so-called "incremental quantity discounts" case (Hadley and Whitin 1963). In this case, the company's supplier charges a price of $p_{o,0}$ for the first q_1 units, charges $p_{o,1}$ monetary units (e.g. €) for the next q_2-q_1 items (i.e. items $q_1 + 1, \ldots, q_2$), and so on, until a last quantity size q_m after which the price remains fixed no matter how large the order size. Setting $R_j = \sum_{k=0}^{j-1} p_{o,k}(q_{k+1} - q_k)$ with $q_0 = 0$, the total cost of an order of Q units such that $q_j \leq Q \leq q_{j+1}$ becomes

$$C_P(Q) = R_j + p_{o,j}(Q - q_j), \quad j = 0, \ldots, m$$

The average purchase cost per unit then simply becomes

$$\overline{C}_P = \frac{C_P(Q)}{Q} = \frac{R_j}{Q} + p_{o,j}\left(1 - \frac{q_j}{Q}\right), \quad j = 0, \ldots, m$$

So, in the EOQ model, the total average cost per unit time when ordering a quantity of size Q with incremental discounts becomes:

$$C_{tot}(Q) = Dp_{o,j} + \frac{D(K + R_j - p_{o,j}q_j)}{Q} + \frac{h(R_j + p_{o,j}(Q - q_j))}{2} \quad j : q_j \leq Q \leq q_{j+1} \tag{4.5}$$

Optimizing the above cost function results in the following optimization problem:

$$\min_{j=0,\ldots,m, q_j \leq Q \leq q_{j+1}} C_{tot,j}(Q) = Dp_{o,j} + \frac{D(K + R_j - p_{o,j}q_j)}{Q} + \frac{h(R_j + p_{o,j}(Q - q_j))}{2} \tag{4.6}$$

which can be easily solved by the following procedure. First, solve each of the $m + 1$ independent problems $\min_Q C_{tot,j}(Q)$ for $j = 0, \ldots, m$ yielding the EOQ solution

$$Q_j^* = \sqrt{\frac{2D(K + R_j - p_{o,j}q_j)}{hp_{o,j}}}$$

Finally, set

$$C_{tot,j}^* = \begin{cases} C_{tot,j}\left(Q_j^*\right), & \text{if } q_j < Q_j^* \leq q_{j+1} \\ +\infty, & \text{otherwise} \end{cases}$$

for each $j = 0, \ldots, m$ and choose as Q^* the value of the argument of the minimal quantity $C_{tot,\, j^*}$ among the $m + 1$ values for $j = 0, \ldots, m$.

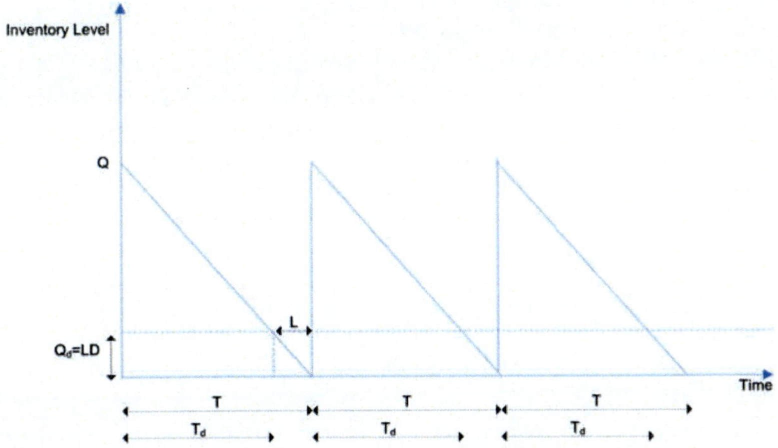

Fig. 4.3 Inventory evolution and order points in the presence of constant lead-time $L > 0$

4.1.1.3 Non-Zero Lead-Times and Backorders in the EOQ Model

The above analysis of the classical EOQ model assumed there is no lead-time between order placement and order fulfillment. When there is a positive (but constant) lead-time $L > 0$ between order placement and order fulfillment—assuming we need to avoid any stock-outs that will have to be dealt with either via backorders or even worse, as lost sales cases—the only change required in the model is to modify the time within the order cycle when an order is placed, but not the order period T^*, nor of course the order quantity Q^*. In particular, the order placement must occur prior to the end of the order period witnessed by the end of the current inventory stock, exactly because there is a positive lag L in order delivery. This (see Fig. 4.3) forces the order to take place at a time T_{d} which is L time units before the end of the current order period. In other words, T_{d} must be such that

$$T_{\mathrm{d}} = T^* - L$$

On the other hand, even though in a deterministic setting such as the one we are currently studying, it is possible to attain zero stock-outs and correspondingly zero back-orders or lost sales, it is nevertheless sometimes acceptable to allow stock-outs, but incurring a penalty whenever they occur. Assume therefore again, the case of zero lead-times, $L = 0$, and that whenever a request for an item arrives and there is no stock available on-hand, the order is "*backlogged*" so as to be fulfilled when new stock arrives at a later time. This backlogging action incurs a backorder penalty cost $P > 0$ that adds up with every unit that is in the backlog for each unit of time. The penalty P, *measured in monetary units per time unit per item*, represents the cost of keeping a customer waiting one-time unit for one item, whereas the holding cost H can be thought of as the cost of keeping one item unit per unit time waiting for a customer (Gallego 2004).

When backorders are allowed, it is possible to let the inventory in Fig. 4.1 drop below zero (accumulating backorders). The optimization question is by what level should the inventory drop below zero so as to minimize total costs per unit time, including holding, backorder and fixed ordering costs. If the allowed minimum level of the inventory is r (with $r \leq 0$), then the total average cost per time unit becomes

$$C_{\text{tot}}(T, r) = \frac{1}{T}\left[K + H\int_0^T (I(t) + r)_+ \mathrm{d}t + P\int_0^T (I(t) + r)_- \mathrm{d}t\right] \tag{4.7}$$

where $x_+ = \max(x, 0)$, and $x_- = \max(-x, 0)$, and where inventory $I(t)$ is still equal to $D(T - t)$ over the interval $[0, T]$. The two integrals in the expression of the total cost whenever backorders are allowed express holding and backorders, respectively. Evaluating the two integrals gives

$$\int_0^T (I(t) + r)_- \mathrm{d}t = -\int_{T+r/D}^T (D(T - t) + r)\mathrm{d}t = \frac{r^2}{2D}$$

$$\int_0^T (I(t) + r)_+ \mathrm{d}t = \int_0^{T+r/D} (D(T - t) + r)\mathrm{d}t = \frac{D\left(T + \frac{r}{D}\right)^2}{2}$$

Therefore, we expand $C_{\text{tot}}(T, r)$ as follows:

$$C_{\text{tot}}(T, r) = \frac{1}{T}\left[K + \frac{Pr^2}{2D} + \frac{HD\left(T + \frac{r}{D}\right)^2}{2}\right] \tag{4.8}$$

It is easy to verify that the above cost function is *jointly convex in both variables T* and r—meaning that the function f defined in $\mathbb{R}_+^* \times \mathbb{R}_-$ as $f(x) = C_{\text{tot}}(x_1, x_2)$ is convex in its whole domain of definition—so, from the first order sufficient conditions in Chap. 1, the global minimum of this function subject to the constraint $-r \geq 0$ satisfies the following:

$$\nabla C_{\text{tot}}(T^*, r^*) = \begin{bmatrix} 0 \\ -1 \end{bmatrix}\lambda^*, \qquad \lambda^* \geq 0, \quad \lambda^* \leq 0, \quad \lambda^* r^* = 0$$

If the constraint $-r \geq 0$ is inactive at r^*, the condition of course reduces to $\nabla C_{\text{tot}}(T^*, r^*) = 0$. Indeed, when $P > 0$, this is always the case as we shall verify immediately below. By expanding this equation for the partial derivative with respect to r, we arrive at the equation

$$r = -\frac{H}{H+P}DT \tag{4.9}$$

which implies that given a reorder interval T, the optimal reorder point r is not zero, but rather the fraction $H/(P + H)$ of the total demand within that interval. This fraction does tend to zero as P tends to infinity, making zero indeed the optimal reorder point when backorders are not allowed. Solving the 2×2 system of nonlinear equations $\nabla C_{tot}(T, r) = 0$ is not particularly hard. Manipulating the equation for the first derivative of the total cost function with respect to the order interval T, results in the following:

$$\frac{\partial C_{tot}(T,r)}{\partial T} = -\frac{K}{T^2} - \frac{Pr^2}{2DT^2} + \frac{HD}{2}\left[\frac{2(T+r/D)T - (T+r/D)^2}{T^2}\right] = 0 \Leftrightarrow$$

$$HD(T+r/D)(T-r/D) - \frac{Pr^2 + 2KD}{D} = 0$$

By substituting r with the expression $-HDT/(H + P)$ in the above equation, and after some algebra, we finally arrive at

$$T^* = \sqrt{\frac{P+H}{PH}}\sqrt{\frac{2K}{D}}$$

$$r^* = -\frac{H}{P+H}D\sqrt{\frac{P+H}{PH}}\sqrt{\frac{2K}{D}} = -\sqrt{\frac{H2KD}{P(P+H)}} \tag{4.10}$$

$$C^*_{tot} = \sqrt{2KHD}\sqrt{\frac{P}{H+P}}$$

The above relations are of fundamental importance in inventory control. As we shall see in the next section on inventory control under stochastic demands, similar relations hold exactly or approximately even when demand is allowed to fluctuate randomly, albeit in a stationary manner.

4.1.1.4 Multi-Item EOQ Model Coordination

Besides the many limiting assumptions of the EOQ model that however did not turn out in practice to be so severe, another very important practical issue is that of *coordinating orders*. A real world company usually manages a large base of stock keeping units (SKUs) and it is often a good practice to order different items from a given family of products that a given supplier provides together as a single order, instead of managing each individual item according to its own EOQ-based optimal reorder interval (the benefits of this practice can often be measured quantitatively as well, for example when placing an order for different items incurs a single fixed ordering cost K rather than the sum of the fixed ordering costs of each individual item in the bulk order).

It is not hard to see how far from optimality in the worst-case the total cost of managing a family of n items would be when the company is forced to order all n items according to the same time interval T. In such a case, assuming zero lead-times, no backorders, and individual order costs for each item K_i with demand rates D_i and holding costs H_i for $i = 1, \ldots, n$, the total cost of managing all n items—using (4.2)—becomes:

$$C_{\text{coord}}(T) = \sum_{i=1}^{n} \left[\frac{K_i}{T} + \frac{H_i D_i T}{2} \right] \tag{4.11}$$

This is equivalent to managing a single item with fixed order cost equal to $K_1 + K_2 + \cdots + K_n$ and holding cost $H_1D_1 + H_2D_2 + \cdots + H_nD_n$ therefore, using (4.4) we get

$$T^*_{\text{coord}} = \sqrt{\frac{2\sum_{i=1}^{n} K_i}{\sum_{i=1}^{n} H_i D_i}} \tag{4.12}$$

and the optimal cost of such a constrained coordinated system becomes

$$C^*_{\text{coord}} = \sum_{i=1}^{n} C_{\text{tot},i}(T^*_{\text{coord}}) = \sqrt{2\sum_{i=1}^{n} K_i \sum_{i=1}^{n} H_i D_i} \tag{4.13}$$

where $C_{\text{tot},i}(T)$ is the cost of managing item i by placing an order $Q_i = D_i T$ every T time units.

On the other hand, from the standard EOQ model, we know that we may optimize the cost of each individual item by ordering $Q_i^* = D_i T_i^*$ every T_i^* time units where—from (4.4)—the order interval is

$$T_i^* = \sqrt{\frac{2K_i}{H_i D_i}}$$

and the optimal cost of item i is given as $\sqrt{2K_i D_i H_i}$. Therefore the total cost of managing all items, but each item individually, is given by

$$C^*_{\text{tot}} = \sum_{i=1}^{n} \sqrt{2K_i D_i H_i} \tag{4.14}$$

The ratio of the optimal costs of the two systems is

$$r = \frac{C^*_{\text{coord}}}{C^*_{\text{tot}}} = \frac{\sqrt{2\sum_{i=1}^{n} K_i \sum_{i=1}^{n} H_i D_i}}{\sum_{i=1}^{n} \sqrt{2K_i D_i H_i}} \tag{4.15}$$

and is always greater than or equal to 1. Letting R denote the quantity

$$R = \frac{\max_{i=1,\ldots,n}\left\{\frac{K_i}{H_iD_i}\right\}}{\min_{i=1\ldots,n}\left\{\frac{K_i}{H_iD_i}\right\}}$$

it is possible to establish the following bound on r (Gallego 2004):

$$r \leq \frac{1}{2}\left[\sqrt[4]{R} + \frac{1}{\sqrt[4]{R}}\right]$$

which implies that as long as R is not too big, the relative cost increase by using a common (optimized) time interval to manage all n items remains very small.

Even further, if the order cost of placing a single order for all n items is fixed at $K = \max_{i=1,\ \ldots,\ n}\{K_i\}$, then the total cost of the coordinated system becomes

$$C'_{\text{coord}}(T) = \sum_{i=1}^{n}\left[\frac{K}{T} + \frac{H_iD_iT}{2}\right] \tag{4.16}$$

which represents a system managing a single item with order cost K and holding costs $\sum_{i=1}^{n} H_iD_i$ therefore, the optimal reorder interval must be given by

$$T'_{\text{coord}} = \sqrt{\frac{2K}{\sum_{i=1}^{n} H_iD_i}} \tag{4.17}$$

and the optimal cost becomes

$$C'_{\text{coord}}(T'_{\text{coord}}) = \sqrt{2K\sum_{i=1}^{n} D_iH_i} = \sqrt{2\max_{i=1\ldots n}\{K_i\}\sum_{i=1}^{n} D_iH_i} \tag{4.18}$$

Now, the ratio of the optimal costs of the two systems

$$r = \frac{C^*_{\text{coord}}}{C^*_{\text{tot}}} = \frac{\sqrt{2\max_{i=1,\ldots,n}\{K_i\}\sum_{i=1}^{n} H_iD_i}}{\sum_{i=1}^{n}\sqrt{2K_iD_iH_i}}$$

is no longer greater than or equal to one, and the benefits of coordination become apparent. By ordering all items at once and incurring a single fixed order cost $K = \max_{i=1,\ \ldots,\ n}\{K_i\}$, the coordinated system may have significantly lower costs than the system where items are individually managed according to their optimal individual settings.

As an example, consider the case of a family of five products, $\pi_1, \ldots, \pi_5$. Product characteristics are as follows: $H_1D_1 = 100$, $K_1 = 1{,}000$, $H_2D_2 = 50$, $K_2 = 500$, $H_3D_3 = 45$, $K_3 = 200$, $H_4D_4 = 20$, $K_4 = 200$, and finally, $H_5D_5 = 10$ and $K_5 = 100$. The optimal cost of the uncoordinated system is then given as: $\sqrt{2 \times 100 \times 1000} + \sqrt{2 \times 50 \times 500} + \sqrt{2 \times 45 \times 200} + \sqrt{2 \times 20 \times 200} + \sqrt{2 \times 10 \times 100}$ and this equals approximately 939.14. Assuming that when placing a single order containing a mix of all of the above products entails a single fixed cost of 1,000 (the maximum of the K_i), then the cost of the coordinated system becomes $\sqrt{2 \times 1000 \times (100 + 50 + 45 + 20 + 10)}$ which equals approximately 670.82, which means that the ratio r is about 0.71, representing a significant cost saving by ordering all items at the same time.

Extending the above discussion, it is not hard to imagine a *heuristic* algorithm that will attempt search to find groups (clusters) of n products with similar characteristics in terms of the ratio $K/(HD)$ from the same family of products that can be ordered together as a single order incurring only the maximum of the fixed order cost within the group so as to optimize the overall cost of managing all products in the family. Such an algorithm will iterate over the total number of groups to consider, starting from the case of a single group—just discussed above—and finally considering the case of as many groups as there are products to manage—also discussed above. The intermediate cases, where the number of groups k ranges from 2 to $n - 1$ may be handled by utilizing the polynomial-time Dynamic Programing algorithm for 1D clustering of data introduced in Sect. 1.1.3, where the data are the numbers $K_i/(H_iD_i)$ $i = 1, \ldots, n$. The output of this process would be a clustering of items that should be managed together—with common order intervals—so as to obtain as small a total cost as possible. In general, we shall often observe that the theme of *coordination* in supply chain management can achieve significant cost savings in many different settings and contexts.

4.1.2 The EPQ Model

The EPQ model is an easy extension to the EOQ model described above, applied to a production environment, rather than a retailer/distribution environment. In a production/manufacturing setting, products are usually produced in batches of certain measurable size (known as the *lot size*) simply because fixed setup costs make frequent switching of jobs in a production line prohibitively expensive. On the other hand, large inventories of finished goods—or WIP items—are not desirable since they tie money in inventory which implies opportunity costs as well as risks associated with holding inventories. The setting therefore is exactly the same in a production environment, as in a retailer/distribution environment: there is a trade-off between inventory costs and fixed production setup costs to be balanced in an economic sense. The extra modeling requirement that enters in a production environment is production orders are not fulfilled instantaneously but rather, there is a finite production rate μ—measured in products per unit time—with which the company may produce its products. Let the demand rate—also measured in products per unit time—be constant, and denoted by λ, and assume that $\lambda \leq \mu$ (otherwise the production capacity is not enough to keep up with demand). We define the *utilization coefficient* $\rho = \lambda/\mu$.

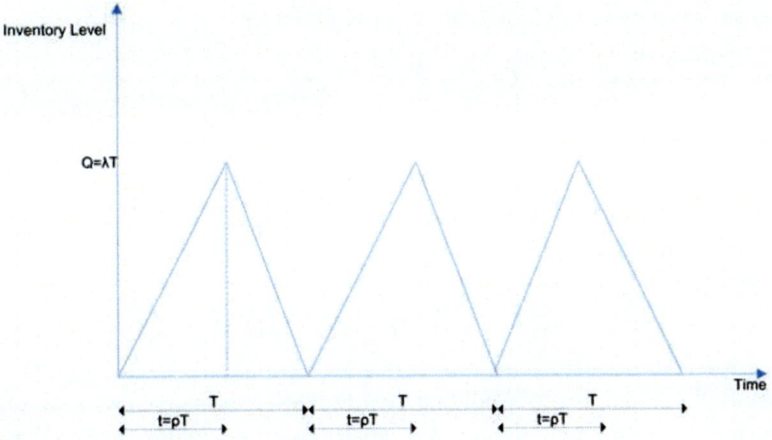

Fig. 4.4 Inventory evolution in the EPQ model

Production in the EPQ model occurs in cycles of length T to be determined, as in the EQO model. At the start and end of a cycle, inventory will be set to zero. During such a cycle, the total demanded quantity of items will be $Q = \lambda T$. On the other hand, a production run of time-length t will result in a quantity $q = \mu t$ that will be produced in that time interval, and incurs a fixed setup cost K. Starting at the beginning of a cycle then, obviously a new production run must start and continue for time t such that $\mu t = \lambda T$, or in other words, $t = \rho T$. During this time, inventory accumulates at the constant rate $\mu - \lambda$ over the interval $[0, \rho T]$, and reaches its peak value of $(\mu - \lambda)\rho T$ at time ρT. After that, inventory will drop with constant rate λ until it reaches zero at time T. A schematic of this repeating cycle is shown in Fig. 4.4.

Denoting $I(t)$ the inventory level at time t, the average inventory in the interval $[0 ,T]$ is given—as in the EOQ model—by the formula

$$\bar{I} = \frac{1}{T}\int_0^T I(t)\mathrm{d}t$$

and since $I(t)$ for t in $[0 ,T]$ is formulated as:

$$I(t) = \begin{cases} (\mu - \lambda)t\,, & t \in [0, \rho T] \\ \lambda(T - t)\,, & t \in [\rho T, T] \end{cases}$$

the average inventory in the interval [0 ,*T*] is given by

$$\frac{1}{T}\left[\int_0^{\rho T}(\mu-\lambda)t\mathrm{d}t+\int_{\rho T}^{T}\lambda(T-t)\mathrm{d}t\right]=\frac{1}{T}\Big[(\mu-\lambda)(\rho T)^2/2+\lambda T(T-\rho T)$$
$$-\lambda[T^2-(\rho T)^2]/2\Big]$$

which reduces to

$$\bar{I}=\frac{1}{2}\lambda(1-\rho)T$$

As in the EOQ model, given a holding cost rate h—measured in *monetary units per unit item per unit time*—the *average holding cost per unit time* becomes

$$C_I=\frac{1}{2}h\lambda(1-\rho)T$$

whereas the *setup cost per unit time* is obviously

$$C_K=\frac{K}{T}$$

and the *total cost of production per unit time*, as a function of the cycle length T becomes

$$C_{\text{tot}}(T)=\frac{K}{T}+\frac{h\lambda(1-\rho)T}{2} \tag{4.19}$$

which is the same formula as (4.2) but with different coefficients for the holding cost—in particular, the holding cost coefficient DH in the EOQ formula is replaced with the coefficient $h\lambda(1-\rho)$. Working exactly as in the EOQ model, the optimal cycle length T^* is now given by

$$T^*=\sqrt{\frac{2K}{h\lambda(1-\rho)}}$$

The optimal production quantity Q^* is given by

$$Q^*=\sqrt{\frac{2K\lambda}{h(1-\rho)}}$$

and the optimal average cost per unit time becomes

$$C^*_{\text{tot}}=C_{\text{tot}}(T^*)=\sqrt{2Kh\lambda(1-\rho)}$$

4.1.2.1 EPQ Model with Backorders

Using exactly the same steps as in the EOQ model, when backorders are allowed, incurring a penalty cost for shortages p per item per unit time, the EPQ model with backorders has a cost function

$$C_{\text{tot}}(T,r) = \frac{1}{T}\left[K + h\int_0^T (I(t)+r)_+ \mathrm{d}t + p\int_0^T (I(t)+r)_- \mathrm{d}t\right] \tag{4.20}$$

where $r \le 0$ is a decision variable indicating the minimum level the inventory level is allowed to drop to. As in the EOQ case, evaluating the two integrals is straightforward and results in the following expression for the total EPQ with backorders cost model

$$\begin{aligned} C_{\text{tot}}(T,r) =& \frac{K}{T} + \frac{h}{T}\left[\frac{(1-\rho)\lambda T + r}{2}\left(T + \frac{r}{\lambda(1-\rho)}\right)\right] \\ &+ \frac{p}{T}\left[\frac{(1-\rho)\lambda T^2}{2} + \left(T + \frac{r}{\lambda(1-\rho)}\right)\frac{r-(1-\rho)\lambda T}{2}\right] \end{aligned} \tag{4.21}$$

Again, this function is jointly convex in both variables (its Hessian matrix is P.D. for all $T > 0$, $r \le 0$). For any chosen cycle length T then, the optimal reorder point r is given as the solution to the equation $\partial C_{\text{tot}}(T, r)/\partial r = 0$ and solving this equation for r results in the following equation which is in complete analogy to (4.9)

$$r = -\frac{h}{h+p}\lambda(1-\rho)T \tag{4.22}$$

and substituting the above equation to the equation for the total cost of the EPQ with backorders we get the optimal EPQ cost when backorders are allowed for any given cycle length

$$C_{\text{tot}}(T) = \frac{K}{T} + \frac{ph}{2(p+h)}\lambda(1-\rho)T \tag{4.23}$$

The graph of the optimal EPQ cost with backorders allowed as a function of the cycle length T is shown in Fig. 4.5.

By setting the derivative of $C_{\text{tot}}(T)$ to zero, we obtain the optimal cycle length for the EPQ model with backorders allowed

$$T_b^* = \sqrt{\frac{2K(p+h)}{ph\lambda(1-\rho)}} \tag{4.24}$$

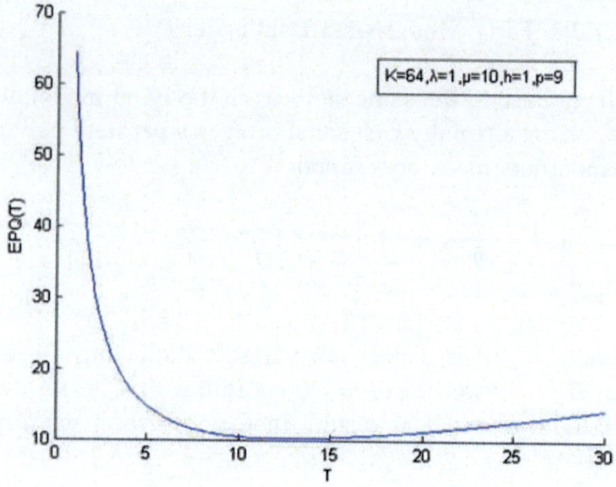

Fig. 4.5 Graph of the EPQ cost $C_{tot}(T)$ with backorders allowed

and the optimal total cost being:

$$C^*_{tot} = \sqrt{\frac{2Khp\lambda(1-\rho)}{p+h}} \tag{4.25}$$

4.2 Stochastic Demand Models and Methods

As demand in the real world can hardly be deterministic—even though in some settings, this just might be the case, as already mentioned in the previous section—controlling an inventory system effectively requires that the notion of demand uncertainty be directly taken into account in any mathematical model of the system to be controlled. This leads to problems with *deep* mathematical structure and has lead to new developments in the fields of applied probability, statistics, as well as queuing theory and optimal control.

4.2.1 Single-Period Problems: The Newsboy Problem

We start the discussion of stochastic inventory control with the easiest inventory control problem involving uncertainty: the *newsboy problem*—also known as the newsvendor problem, or the Christmas tree problem. The problem can be formulated as the determination of the optimal order quantity Q^* of newspapers a newsboy should purchase for his newsstand, so as to optimize their expected profits, when only a *single order* can be placed before demand realization the next day. Problems of this type are called single-period problems. The newsboy can purchase newspapers at a price p_o and re-sells each at the price p_s. The problem is that demand for the newspapers is not known prior to the newsboy's purchasing decision, so if the newsboy orders too few newspapers, he will face opportunity costs by loosing profits $p_s - p_o$ on each newspaper demanded but not available. Further assume an extra "goodwill" cost incurred for each lost sale p_g. On the other hand, if he orders too many newspapers he will face certain cost $p_o - p_r$for each unsold newspaper where $p_r \geq 0$ is a "salvage" value that the newsboy may be able to recoup (usually by "selling back" the unsold newspapers to the newspaper itself). Since there is uncertainty in how much the demand for the newspapers is going to be the next day, the best the newsboy can do is try to optimize his *expected profits*. Let D be a random variable denoting the demand for the newspapers of the next day, and assume it can be well approximated by a continuous variable (the conclusions also hold for the discrete case as well). Assuming the density $f(x)$ and cumulative distribution function $F(x)$ of the demand D are known, the expected profits of the next day's operation for the newsboy's stand will be given as:

$$P(Q) = p_s E[\min\{D, Q\}] - p_o Q + p_r E[(Q - D)_+] - p_g E[(D - Q)_+]$$

We can easily compute each expectation in the above sum, assuming demand can be well approximated by a continuous distribution

$$E[\min\{D, Q\}] = \int_0^Q x f(x)\mathrm{d}x + \int_Q^\infty Q f(x)\mathrm{d}x$$

$$E[(Q - D)_+] = \int_0^Q (Q - x) f(x)\mathrm{d}x$$

$$E[(D - Q)_+] = \int_Q^\infty (x - Q) f(x)\mathrm{d}x$$

where the lower limit of integration being zero is forced simply because we assume there can be no negative demand, so that $f(x) = 0$ for all $x < 0$. Substituting the above expressions in the formula for the expected profit (or loss) function of the newsboy's daily operation, we get

$$P(Q) = p_s\left(\int_0^Q x f(x)\mathrm{d}x + \int_Q^\infty Q f(x)\mathrm{d}x\right) - p_o Q + p_r \int_0^Q (Q - x) f(x)\mathrm{d}x - p_g \int_Q^\infty (x - Q) f(x)\mathrm{d}x$$

$$= (p_s - p_r)\int_0^Q x f(x)\mathrm{d}x + p_r Q F(Q) - p_o Q + (p_g + p_s) Q(1 - F(Q)) - p_g\left(\bar{D} - \int_0^Q x f(x)\mathrm{d}x\right)$$

$$= (p_s - p_r + p_g)\int_0^Q x f(x)\mathrm{d}x + p_r Q F(Q) - p_o Q + (p_g + p_s) Q(1 - F(Q)) - p_g \bar{D}$$

where $\bar{D} = \int_{-\infty}^{\infty} xf(x)\mathrm{d}x$. The above function is *concave* in Q for any density function $f(x)$ and distribution $F(x)$, as can be easily verified by computing its second derivative which is negative for all Q. To compute the optimal order quantity Q^*, we simply invoke Fermat's theorem, and set $\mathrm{d}P(Q)/\mathrm{d}Q = 0$. This yields the following *distribution independent result* for the optimal quantity Q^* to be ordered:

$$Q^* = F^{-1}\left(\frac{p_s + p_g - p_o}{p_s + p_g - p_r}\right)$$

The above result can be interpreted as follows: the optimal order quantity Q^* to be ordered in a "newsboy" single-period problem must be such that the probability that demand will be less than or equal to the quantity Q^* is equal to the fraction $(p_s + p_g - p_o)/(p_s + p_g - p_r)$. This simple result that holds irrespectively of the demand distribution plays a fundamental role in many inventory systems, single-echelon or multi-echelon. The result holds in many multi-item (Abdel-Malek and Montanari 2005; Lau and Lau 1996) or multi-period (finite horizon) settings as well, or in production environments with limited resources, requiring outsourcing (Zhang and Du 2010) or not, etc. It also holds in infinite horizon problems facing stationary demand as we shall see immediately in the following sections.

4.2.2 Continuous Review Systems

In this section, as well as the next, again we assume single-echelon inventory installations, facing independent demand, and having to make ordering decisions to place purchase orders to a supplier who can always satisfy any given order (regardless of the order size) in a fixed known lead-time $L \geq 0$. Even though the above assumptions may at first seem unrealistic, in reality they are quite often reasonable enough in practice to be of use in accurate models of the operation of single-echelon inventory systems. Further, many results developed for systems assuming constant procurement lead-time L, still hold when the lead-time is a random variable following e.g. the normal or other distributions.

Continuous review systems in particular, also known as *transaction reporting* inventory control systems, make a decision whether to order and how much to order, as soon as a demand request is materialized. Immediately upon receiving the demand order at any time t (and filling it from stock if the inventory level $I(t)$ is sufficient high, otherwise back-logging the demand so as to be filled as soon as replenishment stock arrives at the system) a decision is made whether to order more, and how much more in particular from the external supplier.

We shall study the cost structure for the continuous review systems which consists of three elements, directly corresponding to the cost structure of the deterministic EOQ model with backorders (Sect. 4.1.1.3):

1. There is a fixed ordering cost $K \geq 0$ measured in monetary units (euro, dollars, yen or whatever currency is used by the inventory system), incurred every time an order is placed, regardless of the order size
2. There is a holding cost rate h measured in monetary units per unit time per stock unit, applied to the net inventory on-hand plus on-order at every unit time, that represents opportunity and other costs associated with holding and maintaining inventory
3. Finally, there is a penalty cost rate p measured in monetary units per unit time per stock unit backordered that is applied to every back-logged stock unit every time unit that the backorder remains unfulfilled

From the above, it is clear that we shall study systems that incur *linear* holding and backorder costs.

4.2.2.1 The (r, Q) Continuous Review Policy

The rule that the (r, Q) continuous review policy implements is the following: "*immediately following a demand request, place an order of size Q if the inventory position of the system IP (on-hand inventory plus on-order minus backorders) has fallen below the reorder point r, and do nothing otherwise*".

This policy makes sense when the demand process has *continuous sample paths* because otherwise the policy cannot maintain a steady-state (once the *IP* overshoots the reorder point r, it will never reach it again).

In practice, inventory managers often solve this problem using the EOQ with backorders model, even if demand is highly uncertain during the (EOQ-imposed) reorder interval T. A standard procedure is to assume constant demand rate $D = \lambda$, where λ is an estimate of the mean rate of the demand process from past historic data, and then using the formulas (4.10) to determine T^*, and consequently to compute the quantity $Q^* = DT^*$. The reorder interval r^* may then be computed taking the lead-time into account too, and the uncertainty of demand during the lead-time, essentially by increasing appropriately the safety stock level to account for it.

Even though the above practical procedure for determining "good" parameters r and Q is very often quite adequate—especially when demand volatility during the lead-time is low—it is not difficult to establish a procedure for the exact optimization of the parameters r and Q so as to minimize total expected costs per unit time in an infinite horizon setting, when demand is assumed stationary, under some fairly general assumptions. To do that, we first need to develop a model for the total expected costs incurred by the (r, Q) policy. We shall follow the analysis provided by Zheng (1992) in his seminal paper on the properties of the continuous review (r, Q) policy.

Assume that an inventory system follows the (r, Q) policy, and faces stochastic demands that arrive at a mean rate of λ units/unit time. Define $G(y)$ to be the rate at which expected inventory (holding and backorder) costs accumulate at time

$t + L$ when the inventory position IP at time t equals y, so that $IP(t) = y$, and denotes by $D(t)$ the cumulative demand in the interval $(t, L + t)$. Since demand is assumed stationary, the stochastic process generating $D(t)$ does not change as t changes, and thus we can write D for $D(t)$. In this case, we can express the function $G(y)$ as

$$G(y) = hE[(y - D)_+] + pE[(D - y)_+] \tag{4.26}$$

where D is the random variable representing *cumulative demand in any interval of time of length L*, and $E[X]$ denotes the expectation of the random variable X. Letting $F(.)$ be the cdf of the variable D, the function $G(y)$ can be expressed as $G(y) = (h + p) \int_0^y F(x)\mathrm{d}x + p(\lambda L - y)$.

As long as the inventory position in steady-state is uniformly distributed in $(r, r + Q]$ (see discussion below), the total expected cost per unit time as a function of r and Q is given by

$$c(r, Q) = \frac{\lambda K + \int_r^{r+Q} G(y)\mathrm{d}y}{Q} \tag{4.27}$$

when demands are generated from a continuous stochastic process. When demand is a discrete random variable (as in the Poisson case for example), the exact form of the cost function becomes the following:

$$c(r, Q) = \frac{\lambda K + \sum_{y=r+1}^{r+Q} G(y)}{Q} \tag{4.28}$$

If the mean demand rate is reasonably large, and so the order quantity Q is of not too small a size, formula (4.27) can accurately be used instead of the exact formula (4.28), even when dealing with discrete demand processes, and it has the added advantage of greatly simplifying mathematical analysis. Hadley and Whitin (1963) first prove that when demand is Poisson and L is constant, the distribution of the inventory position is uniformly distributed in $\{r + 1, \ldots, r + Q\}$. Their line of argument is essentially the same as that which they used for the periodic review (r, nQ, T) policy which we present in Sect. 4.2.3.2, and for this reason we do not repeat it here. We only mention that Browne and Zipkin (1991) proved that the inventory position of a system implementing the (r, Q) policy is uniformly distributed in $(r, r + Q)$ independent of the stochastic process generating demands as long as this process is non-decreasing stochastic process with stationary increments and continuous sample paths (Serfozo and Stidham 1978).

In case of Poisson demands, the integral in Formula (4.27) can be expanded (Galliher et al. 1957; Hadley and Whitin 1963) to finally become:

$$c(r,Q) = \frac{\lambda K}{Q} + h\left(\frac{Q+1}{2} + r - L\lambda\right) + (h+p)B(r,Q)$$

$$B(r,Q) = \frac{1}{Q}(\beta(r) - \beta(r+Q)) \tag{4.29}$$

$$\beta(x) = \frac{(L\lambda)^2}{2}\bar{P}(x-1;L\lambda) - L\lambda x\bar{P}(x;L\lambda) + \frac{x(x+1)}{2}\bar{P}(x+1;L\lambda)$$

where the function $B(r, Q)$ measures the average number of backorders in an interval of length L, and $\bar{P}(x;\lambda)$ is the complementary cumulative Poisson distribution.

Zipkin (1986) first established the joint convexity of the backorders function $B(r, Q)$ in both variables regardless of the lead-time demand distribution, even with stochastic lead-times, establishing thus the convexity of the function $c(r, Q)$ when Q is also considered a *continuous* variable. He proved convexity of the function by directly showing that the Hessian of the function $B(r, Q)$ is non-negative definite (see Chap. 1). Indeed, from (4.29) the function $B(r, Q) = (\beta(r) - \beta(r + Q))/Q$ where the function $\beta(v)$ for *any* general continuous demand distribution is given by

$$\beta(v) = \int_v^\infty (t-v)\bar{F}(t)\mathrm{d}t$$

where $\bar{F}(x)$ is the complementary cumulative distribution of the lead-time demand D. From this expression, we may obtain the first derivative of $\beta(v)$ using Leibnitz's rule for differentiating a function of the form $\int_{a(x)}^{b(x)} f(t,x)\mathrm{d}t$ to get $\beta'(v) = -\int_v^\infty \bar{F}(t)\mathrm{d}t \le 0,\ \forall v \in \mathbb{R}$. Differentiating again, we get $\beta''(v) = \bar{F}(v) \ge 0,\ \forall v \in \mathbb{R}$, and then, after yet another differentiation we obtain finally, $\beta'''(v) = -f(v) \le 0,\ \forall v \in \mathbb{R}$ where $f(.)$ is the pdf of the lead-time demand D. From these, it is immediate that the function $B(r, Q)$ is non-increasing in both r and Q. Then, to establish the partial derivatives of $B(r, Q)$, the following quantities are introduced:

$$b_1(r,Q) = \beta(r) - \beta(r+Q) + Q\beta'(r+Q)$$

$$b_2(r,Q) = \beta(r) - \beta(r+Q) + Q\beta'(r)$$

$$b_3(r,Q) = \beta'(r) - \beta'(r+Q) + Q\beta''(r+Q)$$

$$b_4(r,Q) = \beta''(r) - \beta''(r+Q)$$

The Hessian matrix of $B(r, Q)$ is then given as

$$\nabla^2 B(r,Q) = \begin{bmatrix} \frac{\partial^2 B(r,Q)}{\partial r^2} & \frac{\partial^2 B(r,Q)}{\partial r \partial Q} \\ \frac{\partial^2 B(r,Q)}{\partial Q \partial r} & \frac{\partial^2 B(r,Q)}{\partial Q^2} \end{bmatrix}$$

$$= \begin{bmatrix} \frac{b_1(r,Q)+b_2(r,Q)-Qb_3(r,Q)}{Q^3} & -\frac{b_3(r,Q)}{Q^2} \\ -\frac{b_3(r,Q)}{Q^2} & \frac{b_4(r,Q)}{Q} \end{bmatrix}$$

To prove the Hessian matrix is non-negative definite is enough to prove that all principal sub-matrices determinants are non-negative (Apostol 1981). For fixed r, define the function $g(Q) = b_1(r, Q) + b_2(r, Q) - Qb_3(r, Q)$. From the way the functions $b_i(r, Q)$ $i = 1, 2, 3$ are defined, it is obvious that $g(0) = 0$, and that its derivative $g'(Q) = -Q^2\beta'''(r+Q) = Q^2 f(r+Q) \geq 0, \quad \forall r, Q \in \mathbb{R}$. Therefore, we have that $\partial^2 B(r, Q)/\partial Q^2 \geq 0$ for all $Q > 0$. Also, $\beta''(v)$ is a non-increasing function so for every $Q > 0$, $b_4(r, Q)$ is non-negative for all r, and thus $\partial^2 B(r,Q)/\partial r^2 \geq 0, \quad \forall Q > 0$. The only remaining thing is to show that the determinant of the Hessian is also non-negative, which is equivalent to showing that, assuming the reorder point r is fixed at *any arbitrary* value, the function $d(Q) = \det(\nabla^2 B(r, Q))Q^4 = (b_1(r, Q) + b_2(r, Q) - Qb_3(r, Q))b_4(r, Q) - [b_3(r, Q)]^2$ is greater than or equal to zero for all $Q > 0$. It is easy to verify that $d(0) = 0$. Then, differentiating the expression for $d(Q)$, we get $d'(Q) = f(r + Q)[b_1(r, Q) + b_2(r + Q) + Qb_3(r, Q) + Q^2 b_4(r, Q)]$, and it happens that the function $e(Q) = b_1(r, Q) + b_2(r + Q) + Qb_3(r, Q) + Q^2 b_4(r, Q)$ satisfies $e(0) = 0$, and $e'(Q) = 2(\beta'(r) - \beta'(r + Q) + Q\beta''(r)) \geq 0$ for all $Q > 0$ since the third derivative of $\beta(v)$ is always non-positive. Thus, $e(Q)$ is always non-negative, as it is non-decreasing for $Q > 0$, and $e(0) = 0$. And thus, $d(Q)$ is non-decreasing, and since $d(0) = 0$, $d(Q)$ is also non-negative *for all r, and all* $Q \geq 0$, this proves that the Hessian of $B(r, Q)$ is non-negative definite, regardless of the lead-time demand distribution!

The above fact, immediately implies that any algorithm for unconstrained convex optimization (and in fact any algorithm for unconstrained optimization reviewed in Chap. 1 of this book) will determine the unique parameters r^* and Q^* producing the global minimum of the expected total costs function $c(r, Q)$, where the parameters r, Q will be treated as continuous and then will be rounded—if needed—to the nearest integer that will provide the best value $c(r_i, Q_i)$ (one would have to examine all four integer combinations $(\lfloor r \rfloor, \lfloor Q \rfloor), (\lceil r \rceil, \lfloor Q \rfloor), (\lfloor r \rfloor, \lceil Q \rceil), (\lceil r \rceil, \lceil Q \rceil)$ to find the globally optimal integer parameter combination).

Some further analysis by Zheng of the first order necessary conditions for unconstrained optimization—treating again the variables as continuous—is interesting. In particular, given any Q, the function $c(r, Q)$ is obviously convex in r, and therefore has a unique minimizer which can be denoted as $r(Q)$. Now, for any $Q > 0$, the point $r = r(Q)$ must satisfy the condition

$$G(r) = G(r + Q)$$

since from the first order necessary conditions, we must have that $\partial c(r, Q)/\partial r = 0$ and expanding the partial derivative of c with respect to r, we have $\partial c(r, Q)/\partial r = G(r + Q) - G(r)$, from which, by setting the expression to zero we get the desired condition. This can be interpreted as follows: for a given order quantity $Q > 0$, the optimal reorder point r should be such that *the starting holding and backorder costs of a replenishment cycle should equal the ending costs*. From the implicit function theorem (Apostol 1981) we have that $r(Q)$ as a function of the order quantity Q is continuous and differentiable for all $Q > 0$. Differentiating the equation $G(r(Q)) = G(r(Q) + Q)$ we get:

$$G'(r(Q))r'(Q) = G'(r(Q) + Q)[r'(Q) + 1] \Leftrightarrow$$
$$r'(Q) = \frac{G'(r(Q) + Q)}{G'(r(Q)) - G'(r(Q) + Q)}$$

with $r(Q)$ being continuous and differentiable, and differentiating $r'(Q)$ in the above expression once again, we obtain

$$r''(Q) = \frac{G''(r(Q))[r'(Q)]^2 - G''(r(Q) + Q)[r'(Q) + 1]^2}{G'(r(Q) + Q) - G'(r(Q))}$$

Using the above expressions, one can easily show that $dr(Q)/dQ$ is in the interval $(-1, 0)$ and thus, $r(Q)$ is decreasing in Q, but that $r(Q) + Q$ is increasing in Q. Therefore, the limits of $r(Q)$ and $r(Q) + Q$ exist, and in fact are minus infinity, and plus infinity, respectively. This is because $[r(Q) + Q] - [r(Q)] = Q \to \infty$ as $Q \to \infty$, so the limit of $[r(Q) + Q]$ must be positive infinity. The function $r(Q)$ must then go to minus infinity, because otherwise, if it had a finite limit, say a, $\lim_{Q\to\infty} G(r(Q)) = G(a) < \infty = \lim_{Q\to\infty} G(r(Q) + Q)$ which is a contradiction.

Defining the function

$$H(Q) = G(r(Q)) \qquad \text{for all } Q > 0$$

and defining as $H(0) = \lim_{Q\to 0+} G(r(Q)) = G(y_0)$, where y_0 is the unique minimizer of the convex function $G(y)$, we can express the function $c(Q) = c(r(Q), Q) = \min_r c(r, Q)$ as follows:

$$c(Q) = \frac{K\lambda + \int_0^Q H(y)\mathrm{d}y}{Q}$$

By differentiating the "optimal holding and backorders" function $H(Q)$, we get $H'(Q) = G'(r(Q))r'(Q) = (G'(r(Q))G'(r(Q) + Q))/[G'(r(Q)) - G'(r(Q) + Q)]$ which is always greater than zero, thus $H(Q)$ is increasing in Q, and because $H''(Q)$ (after some algebra) can also be shown to be strictly greater than zero, it follows that $H(Q)$ is a convex increasing function of Q. The limit of $H'(Q)$as $Q \to \infty$ can also be computed using del' Hospital's rule (since both the nominator and denominator tend to infinity), so, differentiating the nominator and denominator of the formula for $dH(Q)/dQ$, we get the fraction

$$\frac{-[(p+h)F(r(Q))-p][(p+h)F(r(Q)+Q)-p]}{(p+h)(F(r(Q)+Q)-F(r(Q)))}$$

where $F(.)$ as defined before is the cumulative distribution function of the lead-time demand D in an interval of length L. This fraction, since $F(r(Q) + Q)$ tends to 1 as Q tends to infinity, and $F(r(Q))$ tends to zero as Q tends to infinity, tends to the value $hp/(h + p)$, which is the asymptotic slope of the function $H(Q)$. From the above, it follows that $c(Q)$ is also a convex function (the average of a convex function in the interval [0, Q], is also a convex function; this fact is very easy to prove from the definition of convexity and first principles of calculus, see Lemma 4.2 in Sect. 4.2.3.2).

From the above, it now becomes evident (by the convexity of $c(Q)$) that the point (r^*, Q^*) is the unique optimizer of the function $c(r, Q)$ if and only if $c(r^*, Q^*) = G(r^*) = G(r^* + Q^*)$. This means that another means for determining the globally-optimal solution (r^*, Q^*) is to solve the system of equations

$$c(r,Q) = G(r)$$
$$G(r) = G(r+Q)$$

which may or may *not* be easier than utilizing any algorithm for unconstrained optimization of Chap. 1.

4.2.2.2 The (*s*, *S*) Continuous Review Policy

The major problem with the (r, Q) policy discussed above is that it does not apply when the demands can arrive in order sizes larger than one, for example when demands are generated from a compound Poisson process (or, stuttering Poisson process). The reason is that the reorder point r can then be "overshot" at some point, and since the policy only allows for an order of size Q to be replaced, the order-up-to level $r + Q$ would no longer be reachable.

The continuous review (s, S) policy dictates the following rule for controlling the inventory system: "*after every demand, if the inventory position of the system IP has dropped below the reorder level s, place an order of size S—IP*".

It has been shown—using the notion of K-convexity introduced by Scarf (1959)—that for a single-echelon continuous review system that backlogs all unfulfilled orders, with fixed order lead-times, facing a stationary, compound renewal demand process, and a three element cost structure consisting of holding, backorder and fixed ordering costs, such that the expected rate $G(y)$ at which holding and backorder costs accumulate in time when the inventory position is at level y, has a unique global minimum, an (s, S) policy is optimal, in which the best (s, S) policy will outperform every other conceivable rule of ordering for controlling the inventory system.

Below we describe a particularly elegant, and easy to understand algorithm for finding optimal parameters for the (s, S) policy developed by Zheng and Federgruen in 1991 building on the work of Federgruen and Zipkin (1984) and Stidham

(1986). The algorithm works for *discrete* demand processes (e.g. Poisson and compound Poisson) but can be extended to the continuous case as well.

Algorithm Zheng–Federgruen (s,S) Policy Optimization

Inputs: holding and backorders cost rate function $G(y)$, expected total cost function $c(s,S)$.
Outputs: optimal parameters s^*, S^*.
Function $y = \text{findmin}(G, y_0)$
Inputs: function $G(y)$, initial point y_0
Outputs: the unique integer minimizer y^* of the function $G(.)$.

Begin

1. Set $y^* = y_0$.
2. while $G(y_0 - 1) \leq G(y_0)$ do
    a. Set $y_0 = y_0 - 1$.
3. end-while
4. while $G(y_0+1) \leq G(y_0)$ do
    a. Set $y_0 = y_0 + 1$.
5. end-while
6. Set $y^* = y_0$.
7. return y^*.

End

Begin

1. Set $y^* = \text{findmin}(G, 0)$.
2. Set $s = y^*$, $S_0 = y^*$.
3. Set $s = s - 1$.
4. while $c(s,S_0) \leq G(s)$ do
    a. Set $s = s - 1$.
5. end-while
6. Set $s_0=s$.
7. Set $c0 = c(s_0,S_0)$, $S0 = S_0$.
8. Set $S = S0 + 1$.
9. while $G(S) \leq c0$ do
    a. if $c(s,S) < c0$ then
        i. Set $S0 = S$
        ii. while $c(s,S0) \leq G(s + 1)$ do
            1. Set $s = s + 1$.
            2. Set $c0=c(s,S0)$.

iii. end-while
iv. Set $S = S + 1$.

b. end-if

10. end-while
11. Set $s^* = s$, $S^* = S0$.
12. return (s^*, S^*).

End.

The above algorithm is perhaps the fastest algorithm to date for determining the optimal parameters s^* and S^* of the (s, S) policy. It first determines the global minimum of the function $G(y)$, y^*. The initial order-up-to level S_0 is set to this value (as well as $S0$). Steps 3–6 then compute the optimal reorder point s_0 for the given order-up-to level, and set s to this value as well. Then, in the remaining steps, a search (by increments of one) for the smallest value of S that is greater than $S0$ so that it is an improvement over $c(s, S0)$ in cost (Zheng and Federgruen prove that this will happen if and only if $c(s, S0) > c(s, S)$, without any need to search for s as well). If such an S is indeed found, then the optimal reorder point s for the new $S0 = S$ is found by incrementing s—with step-size of 1—until $c(s, S0) > G(s + 1)$. The optimality of the new reorder point is based on the same logic which proves that steps 3–6 are indeed correct for the determination of the optimal reorder point s_0 given an order-up-to level S (Zheng and Federgruen 1991).

4.2.3 Periodic Review Systems

Periodic review systems review the state of the inventory position IP at equally-spaced points in time t_0, $t_0 + T$, $t_0 + 2T$, $t_0 + 3T$, … and make ordering decisions only at these points in time. The length of the time interval between reviews is known as the review period T. Clearly, as the review period $T \to 0$, a periodic review system will approach a corresponding continuous review system (exact proof of this statement can be found in Hadley and Whitin 1963). Costs are divided in two main categories, as for the continuous review systems:

- Fixed costs, including a fixed review cost incurred every time a review is performed that is independent of whether or not an order is decided, and a fixed ordering cost independent of order size that is incurred whenever an order is placed, and
- holding and stock-out inventory costs associated with maintaining inventory and backorders, respectively.

Further, it is well known (see e.g. Hadley and Whitin 1963) that an optimized continuous review system outperforms its optimized periodic review counterpart even when the periodic review system has zero review cost. The main reason for this is the extra inventory that the periodic review policies require the system to hold to protect against stock-outs in an interval of length $L + T > L$ which is the protection interval for a continuous review system.

4.2.3.1 The (*R*, *T*) Periodic Review Policy

The (*R*, *T*) policy, also known as *base-stock policy*, places orders in an inventory system using the following simple rule:

> Every *T* time units, review the Inventory Position *IP* (on-hand stock plus on-order stock minus backorders) and if it is below the reorder level *R*, order a quantity of $R - IP$ unit.

The (*R*, *T*) policy is a special case of the (*r*, *nQ*, *T*) policy to be discussed in the next section. Rao (2003) established the convexity of the long-term expected costs of a single-echelon inventory system operating under the (*R*, *T*) policy, for all demand types such that the probability of ordering at any review is convex in *T*. For continuous demand process—e.g. when demand in an interval of length *T* follows the normal distribution—the probability of ordering at any review is constant, and equal to 1, thus convex. For discrete demand processes such as the Poisson process that are routinely assumed in the related literature, the ordering probability is also convex in *T* as the reader should be able to verify by themselves easily (the ordering probability in the (*R*, *T*) policy is equivalent to the probability that demand in an interval of length *T* will be non-zero). In all such cases, the total long-term expected cost per unit time of the (*R*, *T*) policy is jointly convex in both variables, meaning that optimization of the policy can be done via any convex optimization algorithm, and the solution found is guaranteed to be the globally optimal solution minimizing long-term expected total costs per unit time. Theorem 4.7 is essentially an extension of the analysis provided by Rao.

4.2.3.2 The (*r*, *nQ*, *T*) Periodic Review Policy

The (*r*, *nQ*, *T*) policy is an extension of the (*R*, *T*) policy, in that order sizes must be quantized to be an integer multiple of some quantity *Q*. This policy—without explicit consideration of the review interval *T* as a decision parameter—was first introduced by Morse (1959) to reflect the very common practice of order quantization in the real world. Orders are quantized in the real world for a variety of reasons, including:

- Pallet sizes
- Container sizes and costs
- Material handling considerations

It is important to notice at this point that the order quantum *Q* may be due to internal reasons (e.g. because material handling dictates at the company's warehouse that all orders are received in boxes of some size to be determined) or external reasons (e.g. the supplier demands that orders are taken in number of pallet-loads, etc.) or a combination of the two. This distinction between internally determined Q and externally given base batch size Q plays an important role in determining optimal cost policies, because in the first case, the company may determine without constraints the optimal Q whereas in the second case, the company must decide on a batch size Q that must be a multiple of the externally given quantum batch size Q_b.

The policy dictates the following rule (doctrine): *every T time units, the inventory position IP of the item managed (stock on—hand, plus stock on—order minus backorders) is reviewed, and if it is below the threshold reorder level r, an order of nQ items is placed, with n being the smallest nonnegative integer that brings the IP above r.*

When the review interval T is externally fixed (typically reflecting business constraints, such as, deliveries accepted only on the first day of each month, so T is fixed to be one month and so on), a common periodic review policy in such situations is the (r, nQ) policy.

The standard cost structure for the (r, nQ, T) policy includes the following costs:

1. A fixed review cost K_r that is incurred every time a review takes place
2. A fixed order cost K_o that is incurred every time an order is placed
3. Inventory holding costs accumulating whenever stock is held
4. Inventory penalty costs accumulating whenever stock-outs occur that are however fully backlogged.

Usually, the inventory holding and penalty costs are assumed *linear*, in that a constant holding rate h is applied to each unit that is held on stock per time unit, and a constant penalty rate p is applied to each unit of stock that is backlogged per unit time. Some times, a one-time penalty fee $\hat{p}$ is incurred when an item is requested but it is out-of stock (so-called stock-out penalty). In the most general case therefore that we shall be concerned with, the function that computes the penalty costs associated with the time t for which a "backorder" remains unfulfilled will be a linear function of the form $\pi(t) = \hat{p} + pt$. Most of the analysis will assume $\pi(t) = pt$ unless we explicitly mention otherwise. We shall model the long-term expected costs per time unit incurred by the (r, nQ, T) policy and derive algorithms for the optimal r, Q and T parameters for minimizing these costs under the following additional assumptions that are also almost standard in the related literature:

- Demand for the item is a stationary stochastic process
- Demands in any time interval of length T are independent identically distributed random variables with a probability density function (pdf) $f(x, T)$ and cumulative distribution cdf $F(x, T)$ so that the mean demand is $\mu_T = \mu T$ where μ is a constant.
- The lead-time for order delivery is a constant $L \geq 0$.

The mathematical development will assume continuous demand processes but the results hold for many discrete demand processes as well, by substituting the integrals and pdfs by summations and corresponding probability mass functions, respectively.

Let $D(t)$ be a random variable denoting demand in a time interval $[t_0, t_0 + t]$ of length t. We assume that $D(t)$ is a Stochastically Increasing, Linear in t (SIL) random variable, where a random variable $X(t)$ whose distribution is parameterized by t is said to be stochastically increasing if and only if for any increasing function $g(t)$, $E[g(X(t))]$ is increasing in t, and the random variable X is said to be stochastically linear if it is both stochastically convex and concave, where $X(t)$ is stochastically convex (SCX) (respectively concave (SCV)) if for any convex (respectively concave) function $g(t)$, $E[g(X(t))]$ is convex (respectively concave). We also assume $D(t)$ such that it has stationary and independent increments.

Let $R = r + Q$ denote the least upper bound on the inventory position IP. This inventory position, immediately after a review epoch is then at a level $IP = R - X(Q)$ where $X(Q)$ is a random variable. This random variable follows the uniform distribution in the interval $[0, Q]$ in case demand is a continuous variable, and uniform in the set $\{0, \ldots, Q\}$ in case demand is discrete. This result was first given by Hadley and Whitin (1961), and we repeat their argument as it is of fundamental importance in many inventory problems (single-echelon as well as multi-echelon). The argument will be given for a Poisson demand process generating unit demands with mean rate μ. We compute the steady-state probabilities $\rho(r + j)$ that the inventory position IP of the system with reorder point r, immediately after a review is $IP = r + j, j = 1, 2, \ldots, Q$. Note that if we determine the state of the system to be defined by its IP immediately after a review, the process generating transitions between states is a Markov process discrete in space and time (because cumulative demands within review intervals and between two consecutive reviews are assumed as independent random variables). It is possible to compute the transition probabilities a_{ij} that a system in state $r + i$ in a review, will be found in state (i.e. IP) $r + j$ in the immediately succeeding review. Assume first that $j \leq i$. The demand within the review period must have been $D = i - j + nQ$ where n is some natural number greater than or equal to zero. The probability that demand is $i - j + nQ$ is $p(i - j + nQ;\ \mu T)$ where $p(x; \lambda) = p(D = x; \lambda) = e^{-\lambda}\lambda^x / x!$ is defined only for integer $x \geq 0$, and represents the *Poisson density function*. Because the events that demand D was $i - j$, $i - j + Q$, $i - j + 2Q$, ... are mutually exclusive, the transition probabilities a_{ij} must be given by the following sum:

$$a_{ij} = \sum_{n=0}^{\infty} p(i - j + nQ; \lambda T), \qquad j = 1, \ldots, i$$

If $j > i$, the in-between reviews' period demand D must have been $i - j + nQ$ where $n > 0$ is some natural number, so transition probabilities a_{ij} must be given by the following sum:

$$a_{ij} = \sum_{n=1}^{\infty} p(i - j + nQ; \lambda T), \quad j = i + 1, \ldots$$

Summing up the a_{ij} over all $i = 1, \ldots, Q$ yields:

$$\sum_{i=1}^{Q} a_{ij} = \sum_{i=1}^{j-1}\sum_{n=1}^{\infty} p(i-j+nQ;\mu T) + \sum_{i=j}^{Q}\sum_{n=0}^{\infty} p(i-j+nQ;\mu T)$$
$$= \sum_{n=0}^{\infty}\left[\sum_{k=1}^{j-1} p(nQ+Q-k;\mu T) + \sum_{k=0}^{Q-j} p(k+nQ;\mu T)\right]$$
$$= \sum_{n=0}^{\infty} p(n;\mu T) = 1$$

for all $j = 1, \ldots, Q$. Now, the steady-state probabilities $\rho(r + j)$ must satisfy the following linear system of Q equations with Q unknowns which represent "probability flow balance" in the steady-state:

$$\rho(r+j) = \sum_{i=1}^{Q} \rho(r+i)a_{ij}, \quad j = 1\ldots Q$$

and since $\sum_{i=1}^{Q} a_{ij} = 1$ for all j, we have that $\rho(r+j) = 1/Q, \quad \forall j = 1,\ldots,Q$ is the unique solution of the $Q \times Q$ system as the reader can easily verify by substituting the solution into the equations, showing that indeed the inventory position IP immediately after a review follows the uniform distribution in $\{1, \ldots, Q\}$.

Since the order replenishment lead-time L is assumed constant, assuming a review takes place at any point in time which we may arbitrarily set to zero, the *inventory level I* (which is the on-hand inventory when there is stock available and is equal to minus the total size of backorders outstanding when the system is out-of-stock, so I can take on any real value) must satisfy the following system dynamics equation:

$$I(L+t) = IP(0) - D(L+t) = R - X(Q) - D(L+t), \quad \forall t \in [0, T] \tag{4.30}$$

Equation 4.30 is also known in the literature as the "*Inventory Balance Equation*" as it relates stock arrivals and departures in time with no losses. Since $D(t)$ is stationary, so is $I(t)$.

Let $P_o(Q, T)$ denote the probability of ordering in any given review when the order quantum is Q and the time interval between reviews is T. P_o can be modeled easily because $X(Q) \sim U(0, Q)$. In particular, P_o is the probability that the demand $D(T)$ between two consecutive reviews will lead to an order being placed in the second review, which will happen if and only if $D(T) > IP(0) - r$, or equivalently, $X(Q) + D(T) > Q$. Therefore, the ordering probability is given by

$$P_o(Q,T) = \frac{1}{Q}\int_{Q}^{+\infty}\int_{0}^{Q} f(y-x,T)\mathrm{d}x\,\mathrm{d}y = \frac{1}{Q}\int_{Q}^{+\infty} [F(y,T) - F(y-Q,T)]\mathrm{d}y \tag{4.31}$$

The ordering probability when demand $D(t)$ follows the normal distribution $N(\mu t, \sigma^2 t)$ is shown in Fig. 4.6.

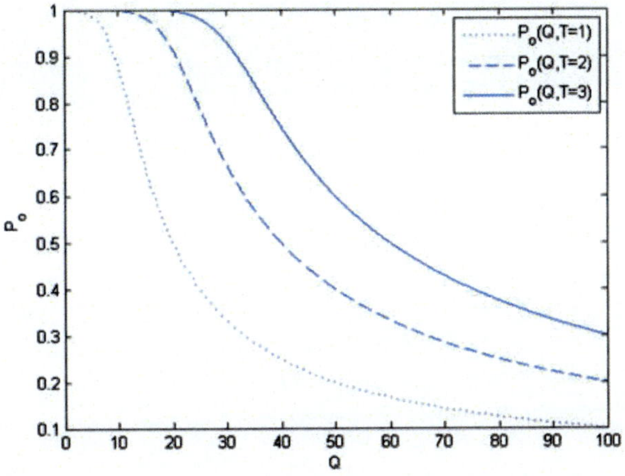

Fig. 4.6 Graph of the ordering probability $P_o(Q, T)$ as a function of Q in the (r, nQ, T) policy under normal demand $D(T) \sim N(10T, 9T)$

The shape of $P_0(Q, T)$ as a function of Q seems to be first concave, then convex. Indeed, when the demand distribution is uni-modal (so it has a unique local maximum), this is always the case. To prove this claim we need to establish a few easy lemmas.

Lemma 4.1 *Assume $f(x, T)$ is a uni-modal pdf with single maximum at $x_0(T)$, that vanishes for negative x, corresponding cdf $F(x, T)$. Then, $F(x, T)$ is convex until $x_0(T)$ then concave.*

Proof Since $F(x,T) = \int_0^x f(t,T)\mathrm{d}t$, $F_x(x,T) = f(x,T)$, $F_{xx}(x,T) = f_x(x,T)$, we have that $f_x(x, T)$ is non-negative for $x \le x_0(T)$ and non-positive for $x > x_0(T)$. Therefore,

$$F_{xx}(x,T) = \begin{cases} \geq 0, & x \leq x_0(T) \\ \leq 0, & x \geq x_0(T) \end{cases}$$

so, $F(x, T)$ is convex until $x_0(T)$ and concave afterward, as stated. QED.

Notice that most important distributions in inventory theory, such as the normal distribution, the Poisson distribution, the Erlang distribution (within appropriate parameter range), the binomial distribution and so forth, are all uni-modal.

We also need the following lemma:

Lemma 4.2 *Assume $f(x)$ is a convex (concave) function in an interval $[a,b]$. Then its average, $F(x) = \frac{1}{x-a}\int_a^x f(t)\mathrm{d}t$ is also convex (concave) function in $[a,b]$.*

Proof It is enough to prove only the convex case, and the concave case is obtained by considering the function—$f(x)$. We must prove $F(\lambda x_1 + (1-\lambda)x_2) \le \lambda F(x_1) + (1-\lambda)F(x_2)\ \forall \lambda \in [0, 1]$. We have that

$$F(\lambda x_1 + (1-\lambda)x_2) = \frac{\int_a^{\lambda x_1+(1-\lambda)x_2} f(t)\mathrm{d}t}{\lambda x_1 + (1-\lambda)x_2 - a}$$
$$= \int_0^1 f(\lambda(\mu x_1 + a(1-\mu)) + (1-\lambda)(\mu x_2 + a(1-\mu)))\mathrm{d}\mu$$

where the second equality is obtained by substituting variables $t = \mu(\lambda x_1 + (1-\lambda)x_2 - a) + a$ so that

$$\mathrm{d}t = (\lambda x_1 + (1-\lambda)x_2 - a)\mathrm{d}\mu,$$
$$t \to a \Leftrightarrow \mu \to 0 \quad t \to \lambda x_1 + (1-\lambda)x_2 \Leftrightarrow \mu \to 1$$

The integrand of the last integral obeys the convex inequality

$$f(\lambda(\mu x_1 + a(1-\mu)) + (1-\lambda)(\mu x_2 + a(1-\mu)))$$
$$\leq \lambda f(\mu x_1 + (1-\mu)a) + (1-\lambda)f(\mu x_2 + (1-\mu)a)$$

so we have that

$$F(\lambda x_1 + (1-\lambda)x_2) \leq \lambda \int_0^1 f(\mu x_1 + (1-\mu)a)\mathrm{d}\mu + (1-\lambda)\int_0^1 f(\mu x_2 + (1-\mu)a)\mathrm{d}\mu$$

and by substituting the variables μ in each integral with

$$t_i = \mu x_i + (1-\mu)a, \quad \mu \to 0 \Leftrightarrow t_i \to a \quad \mu \to 1 \Leftrightarrow t_i \to x_i \quad \mathrm{d}\mu = \mathrm{d}t_i/(x_i - a)$$

$i = 1, 2$ we get $F(\lambda x_1 + (1-\lambda)x_2) \leq \lambda \frac{\int_a^{x_1} f(t)\mathrm{d}t}{x_1 - a} + (1-\lambda)\frac{\int_a^{x_2} f(t)\mathrm{d}t}{x_2 - a}$. QED.

Lemma 4.3 *Assume that cumulative demand D(T) follows a distribution with a uni-modal pdf f(x, T) with single corresponding maximum at $x_0(T)$ and cdf F(x, T). Then $P_o(Q,T) = \frac{\mu T}{Q} - \frac{\int_Q^\infty (1-F(x,T))\mathrm{d}x}{Q}$ is concave until $x_0(T)$ and convex asymptotically as Q increases to infinity.*

Proof We have that $\mu T = \int_0^\infty xf(x, T)$\{\rm d\}$x = \int_0^\infty (1 - F(x, T))dx$ so the function can be written as $P_o(Q,T) = \frac{\int_0^Q (1-F(x,T))\mathrm{d}x}{Q} = \frac{1}{Q}\int_0^Q \bar{F}(x,T)\mathrm{d}x$ where $\bar{F}(x,T)$ is the *complementary* cumulative distribution of the demand. By Lemma 4.1, F(x, T) is convex until $x_0(T)$, then concave, so the function $\bar{F}(x,T)$ is concave until $x_0(T)$, and by Lemma 4.2, so is its average, which is the function $P_o(Q,T)$. For $Q > x_0(T)$ observe that the function can be written as $P_o(Q,T) = \frac{\int_0^{x_0(T)} \bar{F}(x,T)\mathrm{d}x}{Q} + \frac{\int_{x_0(T)}^Q \bar{F}(x,T)\mathrm{d}x}{Q - x_0(T)}\left(\frac{Q - x_0(T)}{Q}\right) = \frac{C}{Q} + F_3(Q,T)\frac{Q - x_0(T)}{Q}$ where

$C = \int_0^{x_0(T)} \bar{F}(t,T)\mathrm{d}t$ is a constant and the function $F_3(Q,T) = \frac{\int_{x_0(T)}^{Q} \bar{F}(x,T)\mathrm{d}x}{Q - x_0(T)}$ is a convex function in Q. As $Q \to \infty$, $\frac{Q - x_0(T)}{Q} \to 1$ and the function $P_o(Q, T)$ becomes asymptotically convex in Q. QED

Theorem 4.4 *Assume cumulative demand $D(T)$ follows a distribution with a unimodal pdf $f(x, T)$ with finite momentums and single corresponding maximum at $x_0(T)$ and cdf $F(x, T)$. Then, there exists a point $x_1(T)$ greater than or equal to $x_0(T)$ such that $P_o(Q, T)$ is concave for all Q in $[0,x_1(T)]$ and convex for all Q greater than $x_1(T)$. The point $x_1(T)$ is the unique root of the equation $g(Q) = \bar{F}_x(Q,T)Q^2 - 2Q\bar{F}(Q,T) + 2\int_0^Q \bar{F}(x,T)\mathrm{d}x = 0$.*

Proof We have that $P_o(Q,T) = \frac{1}{Q}\int_0^Q \bar{F}(x,T)\mathrm{d}x$. Taking the derivatives of P_o with respect to Q we get

$$\frac{\partial P_o(Q,T)}{\partial Q} = \frac{\bar{F}(Q,T)Q - \int_0^Q \bar{F}(x,T)\mathrm{d}x}{Q^2}$$

$$\frac{\partial^2 P_o(Q,T)}{\partial Q^2} = \frac{(\bar{F}_x(Q,T)Q + \bar{F}(Q,T))Q^2 - \bar{F}(Q,T)Q^2 - 2Q\left(Q\bar{F}(Q,T) - \int_0^Q \bar{F}(x,T)\mathrm{d}x\right)}{Q^4}.$$

From these equations we see that the partial second derivative of the function $P_o(Q, T)$ with respect to Q, for $Q \geq 0$ has the same sign as the function $g(Q) = \bar{F}_x(Q,T)Q^2 - 2Q\bar{F}(Q,T) + 2\int_0^Q \bar{F}(x,T)\mathrm{d}x$. But the function g satisfies $g(0) = 0$, and $g'(Q) = Q^2\bar{F}_{xx}(Q,T) = -Q^2 f_x(Q,T)$. Therefore, while $f(x, T)$ increases as a function of x, $g'(Q)$ is negative, thus in the interval $[0,x_0(T)]$ the function $P_o(Q, T)$ remains concave. On the other hand, $\forall x \geq x_0(T), g(Q) \nearrow$ because $f(x, T)$ decreases after $x_0(T)$. Also, since f is a pdf with finite momentums, $\lim_{Q\to+\infty} g(Q) = 2\int_0^\infty \bar{F}(x,T)\mathrm{d}x > 0$, so there exists a unique point $x_1(T)$ in $[x_0(T), +\infty)$ such that $g(x_1(T)) = 0$. To the left of $x_1(T)$ the function $P_o(Q, T)$ remains concave, and to the right of this point the function $P_o(Q,T)$ is convex. QED

The next corollary is useful when the distribution of the demand $D(T)$ has *compact support.*

Corollary 4.5 *Assume that the pdf $f(x, T)$ satisfies the conditions of Theorem 4.4, and also vanishes outside an interval $[a(T),b(T)]$. Then $P_o(Q,T) = 1$ for Q in $[0,a(T)]$ and $P_o(Q,T) = \frac{\int_0^\infty xf(x,T)\mathrm{d}x}{Q} = \mu_T/Q$, $Q \geq b(T)$ where μ_T is the first momentum of $f(x, T)$.*

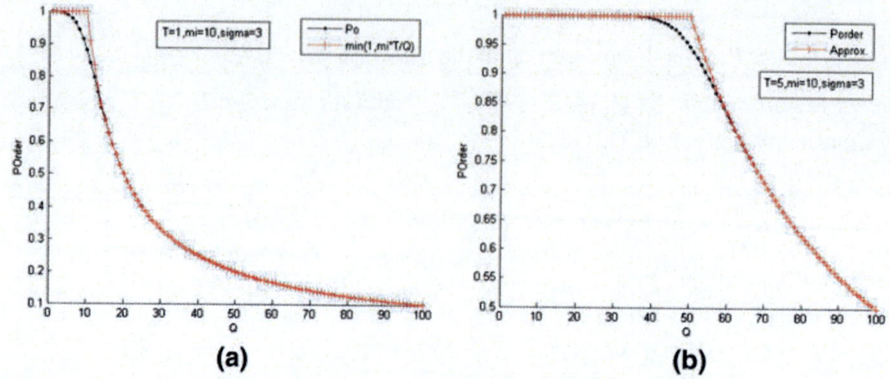

Fig. 4.7 Graphs of the ordering probability in the (r, nQ, T) policy under normal demand $D(T) \sim N(10T, 9T)$ versus the function $P_a(Q, T)$ for $T = 1$ (graph **a**) and $T = 5$ (graph **b**). $P_a(Q,T)$ approximates the true ordering probability very well almost everywhere except in a region where $\mu T/Q \sim 1$ and in particular in a region well-contained in the interval where the pdf $f(x, T)$ is strictly positive, i.e. for Q in $[\mu T\text{-}3\sigma\sqrt{T}, \mu T + 3\sigma\sqrt{T}]$

Proof Immediate from the definition of $P_o(Q,T)$ and the fact that f vanishes outside the interval $[a(T), b(T)]$. QED.

The above provides the basis for a useful approximation of the non-convex ordering probability $P_o(Q, T)$. In the continuous review counterpart of the policy (Sect. 4.2.2.1), namely the (r, Q) policy, the long-term average fixed costs are formulated as $K_o\mu/Q$ (with μ being the mean rate of demand), which is convex in Q. The following function is a reasonable approximation of the ordering probability in the (r, nQ, T) policy under the assumptions we set forth, and under the extra assumption of the uni-modal form of the pdf of cumulative demand.

$$P_a(Q,T) = \min\left\{\frac{\mu T}{Q}, 1\right\} \tag{4.32}$$

In Fig. 4.7 we plot the actual ordering probability $P_o(Q, T)$ versus the function $P_a(Q, T)$ as a function of Q for a fixed T for the case where demand in an interval of length T follows the normal distribution $N(10T, 9T)$.

We now turn our attention to the long-term expected inventory holding and backorder costs incurred per unit time, represented by the function $H(r, Q, T)$. We shall assume that no one-time fixed penalty for stock-outs occurs, so that $\hat{p} = 0$.

When the review period has length T, the expected *instantaneous* holding and backorder inventory cost per unit time at an instant t in an interval $[t_0, t_0 + T]$ starting immediately after a review (that without loss of generality we can arbitrarily set at $t_0 = 0$) is given by the expression

$$
\begin{aligned}
g(r,Q,t) &= hE\big[(I(L+t)_+\big] + pE\big[I(L+t)_-\big] \\
&= hE[r+Q-X(Q)-D(L+t)] \\
&\quad + (h+p)E[(X(Q)+D(L+t)-r-Q)_+] \\
&= h\left(r+\frac{Q}{2}-\mu(L+t)\right) + \frac{h+p}{Q}\int_{r+Q}^{+\infty}(y-r-Q)\int_0^Q f(y-x,L+t)\mathrm{d}x\,\mathrm{d}y
\end{aligned}
\tag{4.33}
$$

where $f(x, T)$ is the pdf of the cumulative demand in a time interval of length T. The second equation uses the fact that for any random variable X, $E[X_+] = E[X] + E[X_-]$ and the second equation uses the convolution of the pdf of thc random variable $Y = X(Q) + D(L + t)$. Therefore, the long-term expected inventory holding and backorder costs per unit time is the average of the function $g(r, Q, t)$ in the interval $[0, T]$ which yields:

$$
\begin{aligned}
H(r,Q,T) &= \frac{1}{T}\int_0^T g(r,Q,t)\mathrm{d}t \\
&= h\left(r+\frac{Q}{2}-\mu\left(L+\frac{T}{2}\right)\right) + (h+p)\int_{r+Q}^{+\infty}(y-r-Q)\frac{1}{Q}\int_0^Q\frac{1}{T}\int_0^T f(y-x,t)\mathrm{d}t\,\mathrm{d}x\,\mathrm{d}y
\end{aligned}
\tag{4.34}
$$

The function $H(r, Q, T)$ is jointly convex in all three variables. To prove this claim, observe that H can be written equivalently in the following form:

$$
\begin{aligned}
H(r,Q,T) &= h\left(r+\frac{Q}{2}-\mu\left(L+\frac{T}{2}\right)\right) \\
&\quad + (h+p)E[(X(Q)+D(L+t(T))-r-Q)_+]
\end{aligned}
\tag{4.35}
$$

where $t(T)$ can be considered a random variable following the uniform distribution $U(0, T)$ in the interval $[0, T]$, and $X(Q)$ is also a random variable following the uniform distribution $U(0, Q)$, therefore, $X(Q)$ is SIL in Q, $t(T)$ is SIL in T and $D(L + t)$ is SIL in t. We can state the following lemma:

Lemma 4.6 *For all* $\theta \in [0,1]$, *the random variable* $Y(\theta)$ *parameterized by* θ $Y(\theta) = X(\theta Q_1 + (1-\theta)Q_2) + D(L + t(\theta T_1 + (1-\theta)T_2)) - [\theta r_1 + (1-\theta)r_2]$ *is SCX in* θ.

Proof To prove this observe:

(i) For any $r_1, r_2 \geq 0$, $\theta r_1 + (1 - \theta)r_2$ is linear, hence SCX in θ.

(ii) Since $E[X(\theta Q_1 + (1-\theta)Q_2)] = \frac{\theta Q_1+(1-\theta)Q_2}{2}$, then $X(\theta Q_1 + (1-\theta)Q_2)$ is either SIL in θ (if $Q_1 \geq Q_2$) or SDL in θ (if $Q_1 < Q_2$). So, for any $Q_1, Q_2 \geq 0$, $X(\theta Q_1 + (1-\theta)Q_2)$ is SCX in θ.
(iii) Since $D(t)$ is SIL in t and $L + t(T)$ is SIL in T, then $D(L + t(T))$ is SIL in T. Moreover, $D(L + t(\theta T_1 + (1-\theta)T_2))$ is either SIL in θ (if $T_1 \geq T_2$) or SDL in θ (if $T_1 < T_2$). So, for any $T_1, T_2 \geq 0$, $D(L + t(\theta T_1 + (1-\theta)T_2))$ is SCX in θ.

Therefore, from (i–iii) above the algebraic sum of such terms, the required expression is SCX in θ. QED

The prior claim that $H(r, Q, T)$ is jointly convex in all three variables is now easy to prove:

Theorem 4.7 *The average inventory holding and backorders cost $H(r, Q, T)$ of the (r, nQ, T) policy is jointly convex in the variables (r, Q, T).*

Proof From (4.35) it suffices to prove that $E[(X(Q) + D(L + t(T)) - r - Q)_+]$ is convex in (r, Q, T). Let $\Pi(V) = E[(X(Q) + D(L + t(T)) - r - Q)_+]$, where $V = (r, Q, T)^T$. It is known (e.g. Rockafellar 1970) that $\Pi(V)$ is convex in V if and only if, for any $V_i = (r_i, Q_i, T_i)$, $i = 1, 2$, the function $b(\theta) = \Pi(\theta V_1 + (1-\theta)V_2)$ is convex in θ for $0 \leq \theta \leq 1$. Since, from Lemma 4.6, $X(\theta Q_1 + (1-\theta)Q_2) + D(L + t(\theta T_1 + (1-\theta)T_2)) - [\theta r_1 + (1-\theta)r_2]$ is SCX in θ, so is $[X(\theta Q_1 + (1-\theta)Q_2) + D(L + t(\theta T_1 + (1-\theta)T_2)) - [\theta r_1 + (1-\theta)r_2]]_+$. Therefore, $b(\theta)$ is convex in θ and so $\Pi(V)$ is convex in V. QED

It is important at this point to stress that joint convexity of the function $H(r, Q, T)$ in all its variables only holds under the assumptions set forth in the beginning of the section, and in particular when the lead-time L is a constant. When L is a random variable, under appropriate conditions, Silver and Robb (2008) have shown that for important demand distributions such as the Gamma and the normal distribution, the function is no longer convex in the review period T.

The long-term expected total costs for the (r,nQ,T) policy under the cost structure described above are then given by the following expression:

$$C(r,Q,T) = \frac{K_r}{T} + \frac{K_o P_o(Q,T)}{T} + H(r,Q,T) \tag{4.36}$$

It follows from the above immediately that the function $H_o(r, Q, T)$ defined as

$$H_o(r,Q,T) = \frac{K_r}{T} + H(r,Q,T) = C(r,Q,T) - \frac{K_o P_o(Q,T)}{T}$$

is a convex function in all three policy variables r, Q and T. This function is also the cost function of an alternative (r,nQ,T) policy for which all parameters and settings are the same as in the original problem, except that $K_o = 0$ so that there is no fixed ordering cost, and as such is clearly a lower bound on the value of the function $C(r, Q, T)$.

When the process-generating demands follow the normal distribution $N(\mu T, \sigma^2 T)$ Hadley and Whitin (1963) expand and calculate the integrals in formulas (4.31) and (4.34), and derive the following closed-form expression for the cost function:

$$C(r,Q,T) = \frac{K_r}{T} + \frac{K_o}{T}\left[\frac{\mu T}{Q}\Phi\left(\frac{Q-\mu T}{\sigma\sqrt{T}}\right) + \bar{\Phi}\left(\frac{Q-\mu T}{\sigma\sqrt{T}}\right) - \frac{\sigma\sqrt{T}}{Q}\phi\left(\frac{Q-\mu T}{\sigma\sqrt{T}}\right)\right]$$
$$+ h\left[\frac{Q-\mu T}{2} + r - L\mu\right]$$
$$+ (h+p)B(r,Q,T)$$

where $\phi(x), \Phi(x), \bar{\Phi}(x)$ are respectively the pdf of the standard normal distribution $N(0,1)$, its cdf, and its complementary cdf, and where $B(r, Q, T)$ is the function measuring the average number of time units of shortage incurred per time unit, i.e. the expected number of backorders at any point in time, and is given by the expression:

$$B(r,Q,T) = \frac{1}{QT}[\Xi(r,L+T) - \Xi(r,L) - [\Xi(r+Q,L+T) - \Xi(r+Q,L)]]$$

$$\Xi(r,L) = \left[\mu^2 L^3/6 - \frac{\sigma^4 r}{4\mu^3} - \mu L^2 r/2 - \frac{\sigma^2 r^2}{4\mu^2} + \frac{\sigma^2 L^2}{4} + Lr^2/2 - \frac{r^3}{6\mu} - \frac{\sigma^6}{8\mu^4}\right]\bar{\Phi}\left(\frac{r-L\mu}{\sigma\sqrt{L}}\right)$$
$$+ \left[\frac{\mu L^{5/2}\sigma}{6} - \frac{\sigma L^{3/2} r}{3} + \frac{\sigma\sqrt{L}r^2}{6\mu} + \frac{\sigma^3 L^{3/2}}{12\mu} + \frac{\sigma^3\sqrt{L}r}{4\mu^2} + \frac{\sigma^5\sqrt{L}}{4\mu^3}\right]\phi\left(\frac{r-L\mu}{\sigma\sqrt{L}}\right)$$
$$+ \frac{\sigma^6}{8\mu^4}\exp(2/\sigma^2)\bar{\Phi}\left(\frac{r+L\mu}{\sigma\sqrt{L}}\right)$$

The above expressions are valid as long as $\Phi(-\mu T/(\sigma\sqrt{T})) \sim 0$ i.e. essentially zero, so that the probability of negative demand occurring between two successive reviews is essentially zero, which will be true for all those $T \geq T_{\min} = (3\sigma/\mu)^2$.

The above convexity results allow us to state an exact algorithm for determining the optimal parameters r, Q and T of the (r, nQ, T) policy within a user-defined parameter precision for the parameters, namely ε_Q and $\varepsilon_T > 0$ (essentially *discretizing* the order quantities as well as review period to be multiples of some user-defined quanta ε_Q and ε_T).

Algorithm (r, nQ, T) Policy Optimization

Inputs: review cost K_r, ordering cost K_o, lead-time L, demand pdf $f(x, T)$, linear holding cost rate h, linear backorder penalty rate p, order search quantum ε_Q, time search quantum ε_T, minimum review period $T_{\min} \geq 0$.

Outputs: optimal parameters r^*, Q^*, T^*.

Begin

1. Set $c^* = +\infty$.
2. for $T = T_{\min}, T_{\min} + \varepsilon_T, T_{\min}+2\varepsilon_T, T_{\min} + 3\varepsilon_T, \ldots$ do
   a. for $Q = 0, \varepsilon_Q, 2\varepsilon_Q, 3\varepsilon_Q, \ldots,$ do
      i Set $r' = \arg\min_r H_o(r, Q, T)$ (solving the corresponding 1D convex optimization problem)
      ii Set $c^{Q,T} = C(r', Q, T)$, $H_{\mathrm{m}}^{Q,T} = H_o(r', Q, T)$
      iii if $c^{Q,T} < c^*$
         1. Set $c^* = c^{Q,T}$, $r^* = r'$ $Q^* = Q$, $T^* = T$
      iv. end-if
      v. if $H_m^{Q,T} > c^* \wedge H_m^{Q,T} > H_m^{Q-\varepsilon_Q,T}$ break
   b. end-for
   c. Set $H_T^* = \min_{r,Q} H_o(r, Q, T)$ (solving the corresponding 2D convex optimization problem).
   d. if $H_T^* > c^*$
   e. end-for
3. return (r^*, Q^*, T^*)

End

Theorem 4.8 *The algorithm (r, nQ, T) Policy Optimization terminates in a finite number of iterations with the optimal policy parameters within the specified accuracy.*

Proof The algorithm is guaranteed to terminate as the function $H_o(r, Q, T)$ goes to infinity as $T \to +\infty$ and it is jointly convex on (r, Q, T) and also $\lim_{Q\to\infty} H_o(r_{Q,T}, Q, T) = +\infty$ where $r_{Q,T} = \arg\min_r H_o(r, Q, T)$. Therefore, the conditions in steps 2.a.v as well as 2.d will eventually be met and the algorithm will terminate. The conditions are also sufficient: For the case of step 2.a.v, there is no point in searching for any larger Q as it is guaranteed that the cost function, being greater than the lower bound $H_o()$ will always be greater than our current incumbent value, as all other values in the range $[Q, +\infty) \times [T, +\infty)$ will yield higher costs (the lower bound is now increasing in Q). For the case of step 2.d it is obvious that at the value of T for which the condition is met, the sequence H_T^* is increasing (otherwise it would have been impossible to have found a cost value less than the lower bound) and thus, from now on the sequence $c_T^* = \min_{(r,Q)} C(r, Q, T)$ will always be above the current c^* which becomes the global optimum. QED.

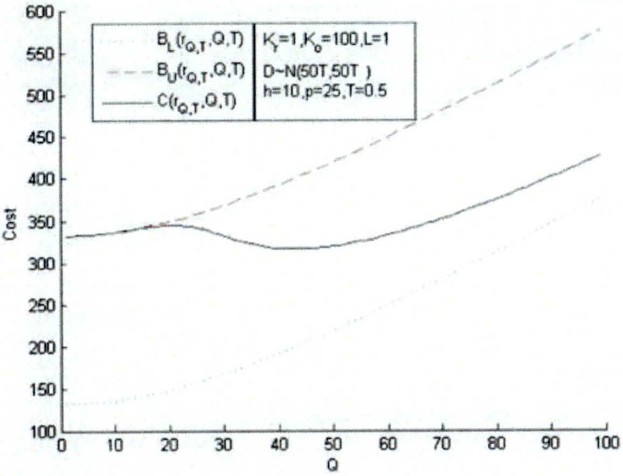

Fig. 4.8 Plot of the long-term expected total cost as a function of the batch order size Q. The review interval T is fixed at 0.5 and demand in an interval of length T follows the normal distribution with parameters $\mu_T = 50T$ and $\sigma_T = \sqrt{(50T)}$. The bounding functions $B_L(r_{Q,T}, Q, T) = H_o(r_{Q,T}, Q, T)$ and $B_U(r_{Q,T}, Q, T) = H_o(r_{Q,T}, Q, T) + K_o/T$ are also plotted

The algorithm's outer loop iterates in increasing values of the review interval length T, and the inner loop computes the optimal parameters r_T and Q_T for the given review interval T. The workings of the inner loop are shown in Fig. 4.8 where for a given T one can see that the cost function $c(Q) = C(r_{Q,T}, Q, T)$ has multiple local minima due to the non-convex ordering costs. The global optimum is determined with the help of the lower bound function $H_o(r_{Q,T}, Q, T)$.

On the other hand, Fig. 4.9 shows the sequence of values c_T^* mentioned in the proof of Theorem 4.8, being the minimum expected total cost as a function of the review interval T. The figure reveals that as T gets larger, the optimal (r, nQ, T) policy with given large T degenerates to the base-stock policy where $Q = 0$ (or $Q = 1$ for discrete demand processes). Figure 4.10 is analogous to Fig. 4.9 but for Poisson demand process.

From the graphs in Figs. 4.9 and 4.10, it becomes obvious that the control parameter T is of crucial importance for minimizing system costs even when the review cost is very small (by orders of magnitude) compared to the other system costs, i.e. fixed ordering costs, linear holding costs, linear backorder costs. For example, in the continuous demands case, if the review interval is arbitrarily—and wrongfully—set to $T = 1$, even after optimizing the parameters r and Q for $T = 1$, the system operates at more than 3% worse than optimal, and if the review interval is set to $T = 2$, the optimized for $T = 2$ system operates at more than 43.2% worse than optimal!.

Another important observation concerns the behavior of an inventory system that implements the (r, nQ, T) policy, as a function of the *spread* of the pdf of the demand. When faced with increased volatility in demand, some "empirical" advice from managers is to shorten the review interval with the implied reasoning

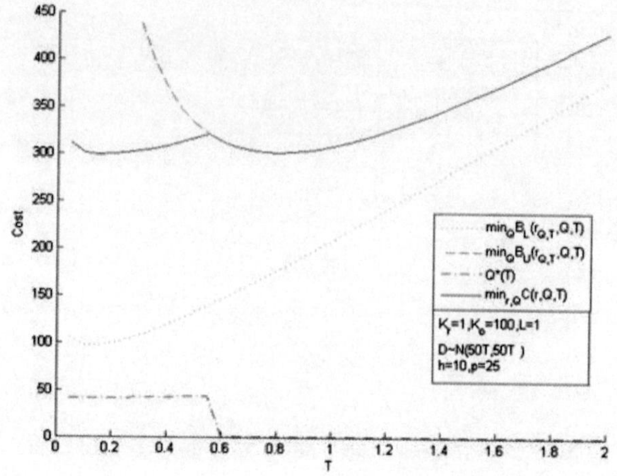

Fig. 4.9 Plot of the minimum long-term expected total cost as a function of the review interval T. Demand in an interval of length T follows the normal distribution with parameters $\mu_T = 50T$ and $\sigma_T = \sqrt{(50T)}$. The bounding functions $B_L(r_{Q,T},Q,T) = H_o(r_{Q,T},Q,T)$ and $B_U(r_{Q,T},Q,T) = H_o(r_{Q,T},Q,T) + K_o/T$ are also plotted, together with a plot of the optimal $Q^*(T)$ value for each T. Notice that the curve is bi-modal, with two local minima, the second of which (T_2) occurs when the system operates as an (R, T) policy system, as the corresponding $Q^*(T_2) = 0$

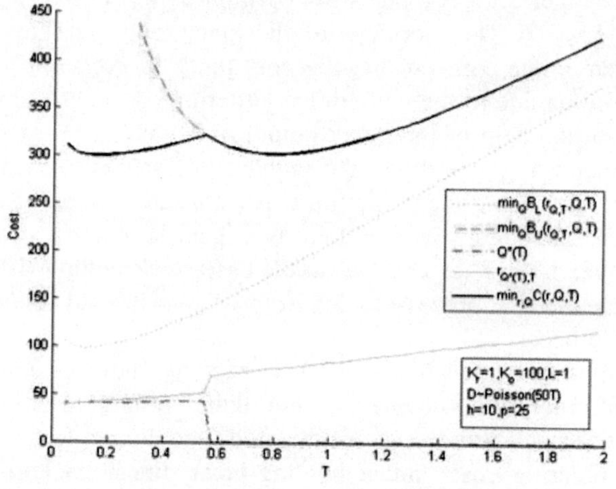

Fig. 4.10 Plot of the minimum long-term expected total cost as a function of the review interval T. Demand in an interval of length T follows the Poisson distribution with mean $\lambda_T = 50T$. The bounding functions $B_L(r_{Q,T},Q,T) = H_o(r_{Q,T},Q,T)$ and $B_U(r_{Q,T},Q,T) = H_o(r_{Q,T},Q,T) + K_o/T$ are also plotted, together with a plot of the optimal $Q^*(T)$ value for each T as well as a plot $r_{Q^*(T),\,T}$

that the fluctuations of demand should be monitored closely so as to "respond quickly" to sudden influxes, etc. Another, more mathematically-oriented advice is the opposite of the previous advice is to expand the review interval, so as to have

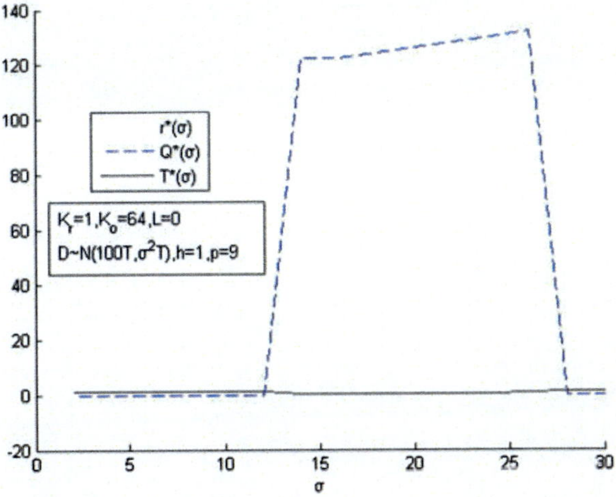

Fig. 4.11 The behavior of the optimal parameters r^*, Q^*, and T^* as a function of the demand variance for normally-distributed demands and no lead-time. Problem parameters are shown in the figure

better statistics on the demand characteristics (Silver and Robb 2008). In Fig. 4.11 we plot the optimal reorder point r^*, batch size Q^* and review interval T^*, as a function of the standard deviation σ of a demand process that follows the normal distribution $N(\mu T, \sigma^2 T)$ for an inventory system with $K_r = 1$, $K_o = 64$, $h = 1$, $p = 9$, $L = 0$ and $\mu = 100$. As can be seen from the plot, the first advice (shorten the review cycle) is valid when demand variance increases from 12 to 14 (the optimal T^* suddenly drops from 1.193 to 0.22 representing a *very significant shortening of the review period*!), only to increase—more gradually—to its prior levels to 1.137 and make the second advice sound, when σ is increased to 28. Notice that T^* again increases sharply and suddenly (as it dropped when σ increased before), and these changes are accompanied by significant changes in the other two parameters r^* and Q^*. In particular, outside the interval (14,28), the optimal policy is the (R, T) policy with $Q^* = 0$, whereas inside this interval, the optimal policy balances ordering costs with inventory holding and penalty costs by high-batch order quantities $Q^* \gg 0$.

The same erratic behavior of T^* as a function of the demand variance is observed in Fig. 4.12a, b, where for different problem settings, even though *optimal average cost monotonically increases smoothly with* σ, the optimal $T^*(\sigma)$ initially increases as demand volatility increases, but later on decreases, though not so sharply. Therefore, an increase in demand volatility does not necessarily mean that the review cycle should be shortened or extended, but it depends on the particular problem settings and value of the variance. It seems that there is no particular rule of thumb (increase or decrease the review interval) that can safely be followed regarding this particular question. Running an optimization algorithm for optimizing system costs is the sound advice again.

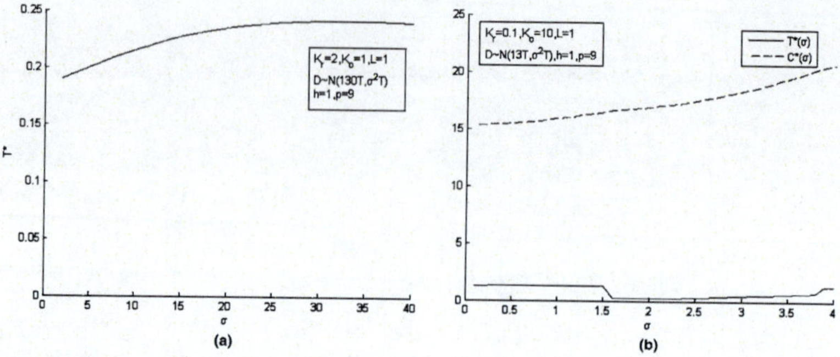

Fig. 4.12 The behavior of the optimal parameter T^* as a function of demand variance σ for normally-distributed demands with positive lead-time $L = 1$

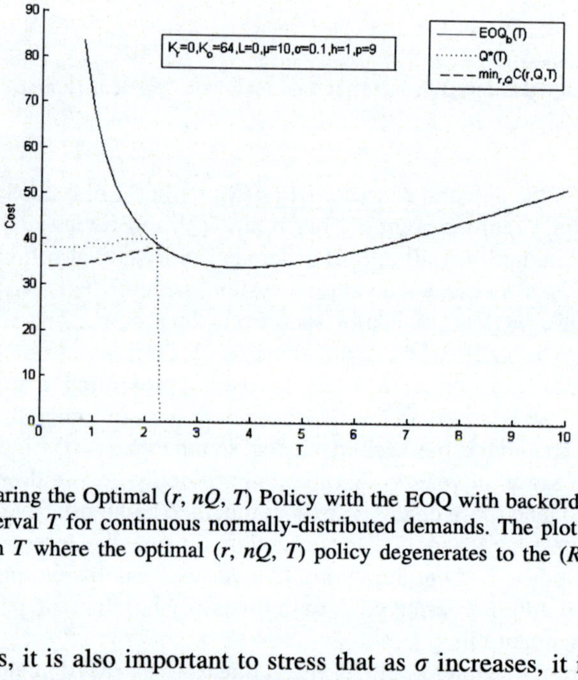

Fig. 4.13 Comparing the Optimal (r, nQ, T) Policy with the EOQ with backorders as a function of the review interval T for continuous normally-distributed demands. The plot of $Q^*(T)$ shows the exact point in T where the optimal (r, nQ, T) policy degenerates to the (R, T) policy with $Q = 0$

Nevertheless, it is also important to stress that as σ increases, it is always also possible to operate on the second locally optimal T_2 of the function $C(T) = \min_{r,Q} C(r, Q, T)$ which will result in a cost value *very close* to the globally optimal value $C(r^*, Q^*, T^*)$ (see Figs. 4.9, 4.10) when selecting the appropriate $r_{Q^*(T_2),T_2}$ and $Q^*(T_2)$. In this light, the more mathematically-oriented advice of *increasing the review interval T as volatility increases is the better option.*

In Fig. 4.13, we show how well the EOQ policy with backorders allowed approximates the (r, nQ, T) optimal policy as a function of the review interval T, for the case where demand is *nearly deterministic* ($D(T)$ follows the normal distribution $N(\mu T, \sigma^2 T)$ with $\mu = 10$ and $\sigma = 0.1$) and review costs are ignored

($K_r = 0$). As can be seen from the plot, the EOQ cost with backorders allowed (assuming a deterministic and constant demand rate $D = \mu$) is an excellent approximation of the optimal (r, nQ, T) policy for T *reasonably large and when demand is nearly deterministic*. For smaller T however, the EOQ deviates significantly from the cost of the optimal (r, nQ, T). This is due to the fact that the EOQ model assumes an order is placed every T-time units, which for small T can be easily avoided by choosing large enough Q. Indeed, for $T = 1$, the probability of ordering at the optimal $Q^*(T)$ is significantly less than 1, which makes up for the cost difference of the two models.

Externally-Imposed Quantization on the Order Size

An additional advantage of the above algorithm is that it may also be used in this exact form *without any modifications* to solve the problem where external constraints quantize the order size. In particular, assume that an externally-defined order size $Q_b > 0$ is imposed on the policy. We have already seen that the function $B_L(r, Q, T) = H_o(r, Q, T)$ is a lower bound on the function $C(r, Q, T)$. Analogously, the function $B_U(r, Q, T) = H_o(r, Q, T) + K_o/T$ is an upper bound on $C(r, Q, T)$ since $P_o(Q,T) \leq 1$ for all $Q \geq 0$, $T > 0$. Now, for any $T > 0$, the functions $B_L(r_{Q,T}, Q,T)$ and $B_U(r_{Q,T}, Q,T)$ where $r_{Q,\ T} = \arg\min_r H_o(r, Q, T)$ are convex increasing functions in Q. Convexity follows immediately from the joint convexity of the function $H(r, Q, T)$ in all three variables, and it is very easy to see that minimizing over any variable (in this case r) preserves joint convexity in the other two variables. The monotonic increase of $H(r_{Q,T}, Q,T)$ in Q is then obvious from the fact that if a function $f(x,y)$ is monotonically increasing in y, then the function $\bar{f}(x,y) = \int_0^x f(t,y)\mathrm{d}t/x$ is also monotonically increasing in y since $f(x,y_1) \leq f(x,y_2)$ for all x and all $y_1 \leq y_2$ and integrating $f(t,y_1)$ and $f(t,y_2)$ in x in the interval $[0, x]$ and dividing by x preserves the inequality.

For every $T > 0$ then, the 1D integer programing problem

$$(r, nQ, T - \text{ext}) \min_{Q,k} C(r_{Q,T}, Q, T)$$

$$\text{s.t.} \begin{cases} Q = kQ_b \\ k \in \mathbb{N}^* \end{cases}$$

has a unique finite solution $Q^*(T, Q_b)$. A straight-forward algorithm for solving the above problem will start with $k = 1$ and will continue increasing k in increments of 1 until $B_L(r_{kQ_b,T}, kQ_b, T) \geq \min_{j=1\ldots k-1}\{C(r_{jQ_b,T}, jQ_b, T)\}$. The minimizing value for the problem ($rnQT$-ext) is $Q^* = k^*Q_b$ where $k^* = \arg\min_{j=1\ldots k}\{C(r_{jQ_b,T}, jQ_b, T)\}$, exactly as the above algorithm specifies. Further, we may establish the following simple lemma:

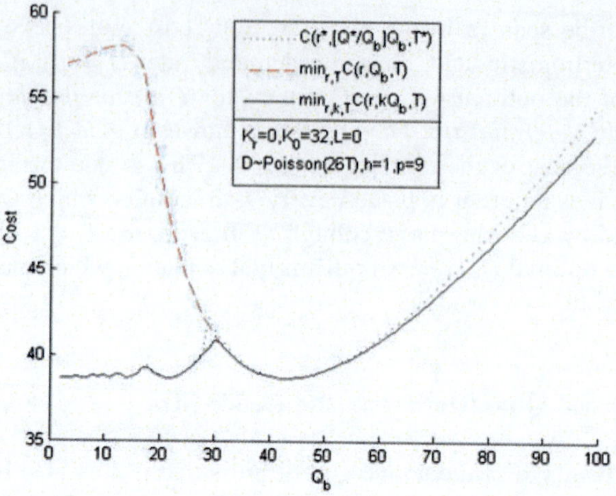

Fig. 4.14 Plot of the optimal long-term expected total cost as a function of an externally-fixed base batch size Q_b. Demand is discrete and at any interval of length T follows the Poisson distribution with mean $26T$. The *dotted line* shows the performance of the "rounding heuristic" discussed in the text, whereas the *dashed line* shows the performance of the "naïve heuristic" also discussed in the text

Lemma 4.9 *For the problem*

$$(r, nQ, T - \text{ext}\,2)\ \min_{r,Q,k,T} C(r, Q, T)$$

$$\text{s.t.} \begin{cases} Q = kQ_b \\ T > 0 \\ k \in \mathbb{N}^* \end{cases}$$

the optimal solution is located at a finite value T.

Proof For continuous demand processes, it is known that $\lim_{T\to\infty} B_i(r_{0,T}, 0, T) = +\infty,\quad i = L, U$ (Rao 2003)—similar result holds for discrete demand processes as well. Now, for both the lower and upper bounds of the total cost function it holds that $B(r_{Q,T}, Q, T)$ are convex in T approaching infinity as T grows to infinity. Therefore, the cost function $C(r, Q, T)$ is minimized at a finite value of T even under the constraints of the problem (*rnQT-ext*2). QED.

To find the optimal solution of a system controlled by the (r,nQ,T) policy under the constraint that Q must be a multiple of an externally-fixed base batch size Q_b, we only need run the algorithm for (r,nQ,T) policy optimization with $\varepsilon_Q = Q_b$. Running the algorithm for various values of Q_b we obtain the graphs in Fig. 4.14 for a Poisson demand process (the continuous counterpart of this case with normal demands is shown in Fig. 4.15). In these graphs we also compare the performance of two easy heuristics for operating the (r, nQ, T) policy under the externally-fixed base batch order size constraint:

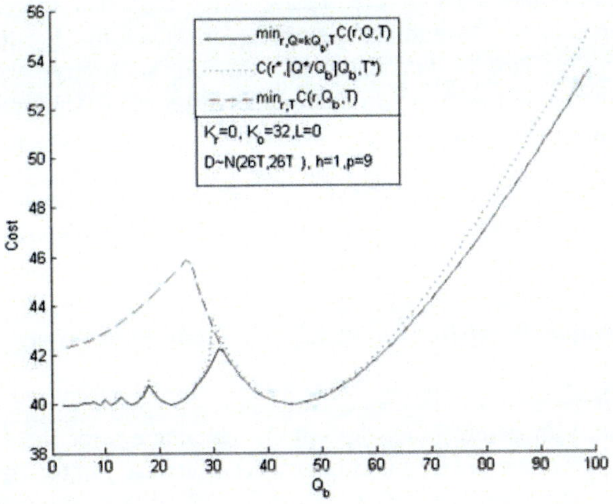

Fig. 4.15 Plot of the optimal long-term expected total cost as a function of an externally-fixed base batch size Q_b. Demand is continuous and at any interval of length T follows the normal distribution with mean and variance $26T$. The *dotted line* shows the performance of the "rounding heuristic" discussed in the text, whereas the *dashed line* shows the performance of the "naïve heuristic" also discussed in the text

1. The "naïve" heuristic, by which Q is set to the externally given base batch size Q_b, $Q = Q_b$, and then the parameters r and T are optimized for this base order size, and
2. The "rounding" heuristic, by which the unconstrained optimization problem is solved in order to determine the optimal parameters r^*, Q^* and T^*, and then Q is set to be the nearest multiple of Q_b to Q^*, so that $Q = [Q^*/Q_b]Q_b$.

The results in Fig. 4.14 show that the naïve heuristic performs very poorly when the unconstrained optimal Q^* (=43 in this case) would be large compared to Q_b, reaching gaps of up to 50% worse than the optimal constrained policy. The rounding heuristic is better, but still can perform up to 4.2% worse for some values of Q_b around 30. Other than that, the graph shows oscillations in the optimal cost of the constrained policy that reach a minimum as the values of Q_b come close to being divisors of the optimal unconstrained batch Q^*. Therefore, values of Q_b near 11 and 21 have small deviation from the optimal unconstrained policy. Once Q_b exceeds Q^*, the oscillations stop, and it is always optimal to set $Q = Q_b$, but the deviation from the optimal unconstrained policy will grow without bound to infinity as $Q_b \to \infty$. As a conclusion then, a manager managing such a system has a lot to gain by negotiating the externally-imposed base batch size Q_b to be smaller than the unconstrained optimal batch size Q^*, and in fact to negotiate a size Q_b that is as close as possible to being an exact divisor of Q^*. The same conclusions hold true in the case of continuous demands in Fig. 4.15. Notice also that the naïve heuristic performs better than the "rounding" heuristic for values of $Q_b > Q^*$ *and in fact performs optimally*. This is due to the fact that for such large values of Q_b as mentioned above Q_b is the optimal constrained value of Q. The optimization step of the naïve heuristic results in the optimal r and T values for the optimal constrained value of Q, whereas the rounding heuristic, leaves the r and T values at their setting for the optimal unconstrained value of Q, namely Q^*, which makes it sub-optimal whenever $Q_b > Q^*$.

Fast Near-Optimal Heuristics for (r, nQ, T) Policy Parameter Optimization

The only difficulty in determining the optimal parameters for the (r, nQ, T) policy—since the inventory holding and backorder costs are jointly convex in (r, Q, T)—is the non-convex ordering probability $P_o(Q,T)$ that renders the whole cost function $C(r, Q, T)$ non-convex as the previous discussion has made clear. On the other hand, we saw that the function $P_a(Q,T) = \min\{\mu T/Q, 1\}$ is an excellent approximation of the function $P_o(Q,T)$ almost everywhere except in a small "*transition region*" that is within the area over which the pdf $f(x, T)$ of the cumulative demand is positive, assuming uni-modality of $f(x, T)$.

A strategy that seems good at this point is to solve the two *convex programing problems*

$$(rnQT - P0)\ \min_{r,Q,T} B_0(r,Q,T) = \frac{K_r}{T} + \frac{K_o \mu}{Q} + H(r,Q,T)$$

and

$$(rnQT - P1)\ \min_{r,Q,T} B_1(r,Q,T) = \frac{K_r}{T} + \frac{K_o}{T} + H(r,Q,T)$$

Denote the optimal- and unique-solution of $(rnQT\text{-}P0)$ as (r_0^*, Q_0^*, T_0^*), and the solution of $(rnQT\text{-}P1)$ as (r_1^*, Q_1^*, T_1^*), and choose as near-optimal policy parameters the set of parameters that has a best value for the function $C(r, Q, T)$, i.e. choose:

$$(r^*, Q^*, T^*) = \begin{cases} (r_0^*, Q_0^*, T_0^*), & \text{if } C(r_0^*, Q_0^*, T_0^*) \le C(r_1^*, Q_1^*, T_1^*) \\ (r_1^*, Q_1^*, T_1^*), & \text{otherwise} \end{cases}$$

This heuristic requires only solving two convex unconstrained optimization problems (strictly speaking the problems have bound constraints as T and Q must both be non-negative) and constitutes therefore a polynomial-time algorithm. Its performance on large numbers of test cases with normal as well as Poisson demands has been very impressive: the heuristic always obtains the optimal

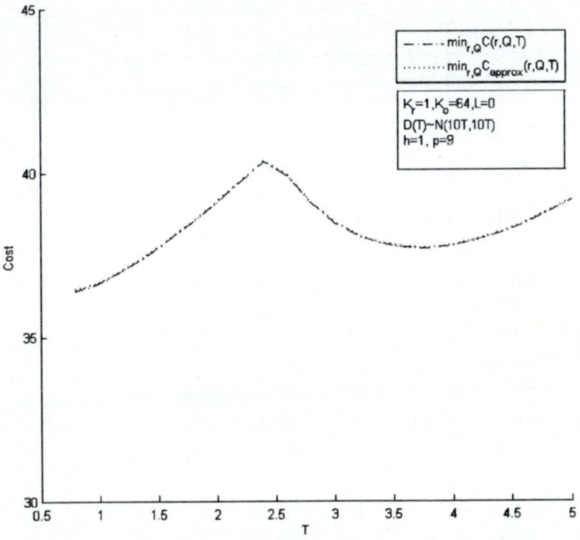

Fig. 4.16 Plot of the optimal (r, nQ, T) policy cost and its approximation C_{approx} as a function of the review interval T. The two functions are nearly indistinguishable in the plot

solution irrespective of the system parameters (fixed costs, demand rates, holding and backorder cost rates, etc.). The reason is made clear in Fig. 4.16 that shows the cost differences between the actual cost $\min_{r,Q}C(r, Q, T)$ as a function of T and the function $C_{\text{approx}}(T) = \min\{\min_{r,Q}B_0(r, Q, T), \min_{r,Q}B_1(r, Q, T)\}$. As can be seen, there exist only tiny differences between the two functions in the "transition region" before the optimal (r, nQ, T) policy degenerates to order-up-to (R, T) policy (the second "valley" in the graph). This region however never contains the globally optimal review interval T^*, so it has no effect in the approximation of the true cost function.

Table 4.1 tabulates the results of running the above heuristic, and compares with the optimal policy found by running the exact (r, nQ, T) policy optimization algorithm for the case of continuous demands following the normal distribution with $\mu = 50$, $\sigma = \sqrt{50}$, lead-time $L = 1$ and various values of the parameters K_r, K_o, h and p shown in the table. We must note that both algorithms run with an additional lower bound constraint on the value of the review interval T, so we require that $T \geq T_{\min}$ where $T_{\min}$ is chosen so that the demand in an interval of length T is with very high probability (0.997) non-negative (since demand follows $N(\mu T, \sigma^2 T)$ this requirement translates to $\mu T - 3\sigma\sqrt{T} \geq 0$ or equivalently, $T \geq T_{\min} = (3\sigma/\mu)^2$).

In Table 4.1 the first four columns specify the parameters K_r, K_o, h and p for each of the test cases. The rest of the parameters are as specified above, and with the exception of the value K_r, they are the same test cases used by Rao (2003) in his experiments with the (R, T) policy. The next four columns specify the optimal

Table 4.1 Comparison between heuristic and exact optimization of (r, nQ, T) policy parameters

Parameters				Optimal (r, nQ, T) policy				Heuristic solution				
K_r	Ko	h	p	r^*	Q^*	T^*	c^*	r	Q	T	c	Δ%
0.5	1	10	25	58.89	0	0.18	101.85	58.89	0	0.18	101.85	0,000
0.5	5	10	25	60.62	0	0.24	121.75	60.6	0	0.239	121.76	*0.008*
0.5	25	10	25	66.69	0	0.439	180.04	66.69	0	0.439	180.04	0.000
0.5	100	10	25	43.08	40.8	0.18	296.74	43.08	40.82	0.18	296.74	0.000
2	1	10	25	59.18	0	0.19	110.13	59.17	0	0.189	110.13	0.000
2	5	10	25	61.21	0	0.26	127.74	61.29	0	0.263	127.74	0.000
2	25	10	25	67.04	0	0.45	183.41	67.03	0	0.45	183.41	0.000
2	100	10	25	78.75	0	0.802	302.83	78.75	0	0.802	302.83	0.000
0.5	1	20	20	54.44	0	0.18	133.27	54.44	0	0.18	133.27	0.000
0.5	5	20	20	55.29	0	0.215	154.35	55.35	0	0.217	154.36	*0.006*
0.5	25	20	20	59.54	0	0.39	219.05	59.59	0	0.392	219.05	0.000
0.5	100	20	20	36.93	35.13	0.18	355.45	36.93	35.13	0.18	355.45	0.000
2	1	20	20	54.44	0	0.18	141.6	54.44	0	0.18	141.6	0.000
2	5	20	20	55.9	0	0.24	160.94	55.85	0	0.238	160.94	0.000
2	25	20	20	59.78	0	0.4	222.84	59.8	0	0.401	222.84	0.000
2	100	20	20	66.85	0	0.694	358.99	66.94	0	0.694	358.99	0.000
0.5	1	15	100	63.32	0	0.18	200.69	63.32	0	0.18	200.69	0.000
0.5	5	15	100	63.32	0	0.18	222.89	63.32	0	0.18	222.89	0.000
0.5	25	15	100	68.5	0	0.33	300.41	68.51	0	0.33	300.41	0.000
0.5	100	15	100	52.63	31.75	0.18	450.96	52.63	31.74	0.18	450.96	0.000
2	1	15	100	63.32	0	0.18	209.03	63.32	0	0.18	209.03	0.000
2	5	15	100	63.9	0	0.197	230.85	63.89	0	0.197	230.85	0.000
2	25	15	100	68.79	0	0.338	304.9	68.8	0	0.338	304.91	*0.003*
2	100	15	100	52.63	31.74	0.18	459.29	52.63	31.74	0.18	459.29	0.000

values for each parameter of the optimal (r, nQ, T) policy along with the optimal expected cost c^*. The final five columns specify the parameters r, Q and T found by the application of the heuristic method, along with the value $C(r, Q, T)$ and the percentage deviation of the heuristic solution from the optimal value (labeled Δ%). As is evident from the data, the heuristic essentially always finds the optimal solution.

In the following, we provide a first step toward the mathematical justification of the heuristic's performance. The functions $B_i(Q,T)$ are defined as

$$B_i(Q,T) = B_i\left(r_{Q,T}, Q, T\right) i = 0, 1.$$

Lemma 4.10 *Assume the pdf $f(x, T)$ satisfies the conditions of Corollary* 4.5 *for all T. Let $\mu = \mu_T/T$, where μ_T* is the mean of the pdf $f(x, T)$. *Then, the (r, nQ, T) policy long-run average cost function $c(Q, T) = \min_r c(r, Q, T)$ within the interval* [a(T),b(T)] *is bounded from below by the following:*

$$c(Q,T) \geq B_0(a(T),T) - K_o\mu\frac{b(T)-a(T)}{a(T)b(T)}$$
$$c(Q,T) \geq B_0(Q,T) - K_o\mu\frac{b(T)-Q}{Qb(T)}$$
$$c(Q,T) \geq B_1(Q,T) - K_o\frac{b(T)-\mu T}{b(T)T}$$
$$c(Q,T) \geq B_1(a(T),T) - K_o(1-P_o(Q,T))/T$$

and outside the interval $c(Q, T) = \min\{B_0(Q, T), B_1(Q, T)\}$.

Proof The behavior of $c(.,.)$ outside the interval $[a(T), b(T)]$ is established by Corollary 4.5. Within the interval $[a(T), b(T)]$ we have that $P_o(Q,T)$ is a decreasing function and assumes its minimum value at $b(T)$. Therefore

$$\begin{aligned} c(Q,T) &= \frac{K_r}{T} + \frac{K_o P_o(Q,T)}{T} + H(Q,T) \geq \frac{K_r + K_o\mu_T/b(T)}{T} + H(Q,T) \\ &\geq \frac{K_r + K_o\mu_T/Q}{T} + H(Q,T) + K_o\left(\frac{\mu_T/b(T)}{T} - \frac{\mu_T/Q}{T}\right) \\ &= B_0(Q,T) + K_o\frac{\mu(b(T)-Q)}{Qb(T)} \end{aligned}$$

where we use the notation $H(Q,T) = \min_r H(r, Q, T)$. This proves the second inequality. The first inequality follows by setting $Q = a(T)$. Similarly, to obtain the third inequality, we have:

$$\begin{aligned} c(Q,T) &= \frac{K_r}{T} + \frac{K_o P_o(Q,T)}{T} + H(Q,T) \geq \frac{K_r + K_o\mu_T/b(T)}{T} + H(Q,T) \\ &\geq \frac{K_r + K_o}{T} + H(Q,T) + K_o\left(\frac{\mu_T/b(T)}{T} - \frac{1}{T}\right) \\ &= B_1(Q,T) - K_o\frac{(b(T)-\mu T)}{b(T)T} \geq B_1(0,T) - K_o\frac{b(T)-\mu T}{b(T)T} \end{aligned}$$

The last inequality follows immediately from the fact that $c(Q, T) = B_1(Q, T) + K_o P_o(Q, T)/T - K_o/T$ and that $B_1(Q,T)$ is increasing in Q. QED.

Corollary 4.11 *There exists a* T_1 *large enough, so that for all* $T \geq T_1$, *and for all* Q *in the interval* $[a(T), b(T)]$, $c(Q, T) \geq B_1(0, T)$.

Proof The function $H(Q, T)$ increases without bounds in Q *and* T, so $B_1(a(T), T) \geq B_1(0, T)$ and further, the term $(b(T) - \mu T)/(b(T)T)$ goes to zero from above as T tends to infinity, therefore there exists a $T_1 > 0$ so that.

$$B_1(a(T_1),T_1) - K_o\frac{(b(T_1)-\mu T_1)}{b(T_1)T_1} \geq B_1(0,T_1).$$

and for all $T \geq T_1$, the desired inequality $c(Q, T) \geq B_1(0, T)$ holds in the interval $[a(T), b(T)]$. QED.

4.2.3.3 The (s, S, T) Periodic Review Policy

The (s,S,T) policy is the optimal policy form for an inventory system facing stochastically increasing stationary demands with independent increments (see Hadley and Whitin 1963), under the standard cost structure discussed for the (r, nQ, T) policy, namely:

1. A fixed review cost K_r that is incurred every time a review takes place
2. A fixed order cost K_o that is incurred every time an order is placed
3. Inventory holding costs accumulating whenever stock is held
4. Inventory penalty costs accumulating whenever stock-outs occur that are however fully backlogged.

The (s,S,T) policy implements the following rule:

"At equally spaced points in time t that are T time units apart, review the Inventory Position IP(t) of the system, and if IP(t) < s place an order of size S − IP(t)."

We have already seen that in the case of continuous review, deriving the analytical formulas for the expected long-term costs of the (s, S) policy or finding the optimal parameters of this policy is highly non-trivial—though from an algorithmic implementation point of view is not that hard at all. The same is the case for the periodic review policy (s, S, T); interestingly, there has been very little research on the global optimization of the (s, S, T) policy, even though some heuristics have appeared in the literature as well as in software codes, in particular in the quantitative system for business (QSB) software suite (Chang and Desai 2003), which however assume that the penalty function for a backorder outstanding for a period of time of length t is of the form $\pi(t) = \hat{p}$, i.e. each backorder incurs a fixed cost irrespective of the length of time for which the backorder remains unfulfilled.

Under the cost structure discussed above, the total expected costs for a system operating the (s, S, T) policy are given—in complete analogy with (4.36)—as follows:

$$C(s,S,T) = \frac{K_r}{T} + \frac{K_o P_o(\Delta, T)}{T} + H(s,S,T) \qquad (4.37)$$

where $P_o(\Delta, T)$ with $\Delta = S - s$ represents the probability of ordering immediately after a review takes place which is of course independent of the reorder level s and only depends on the spread Δ and the review interval T, and the function $H(s, S, T)$ pools together the expected holding and backorder costs per unit time when the system operates under the specific controls s, S and T.

We shall discuss the (s, S, T) policy model under the same assumptions that we placed in the previous section for the (r, nQ, T) policy, but *we shall restrict our attention to the case of discrete demands and in particular, Poisson demands, where order sizes are always unity*. Hadley and Whitin (1963) provided analytical formulas for computing the function $C(s, S, T)$ when the system operates under a demand pattern that is generated from a Poisson process with rate λ. Notice that in

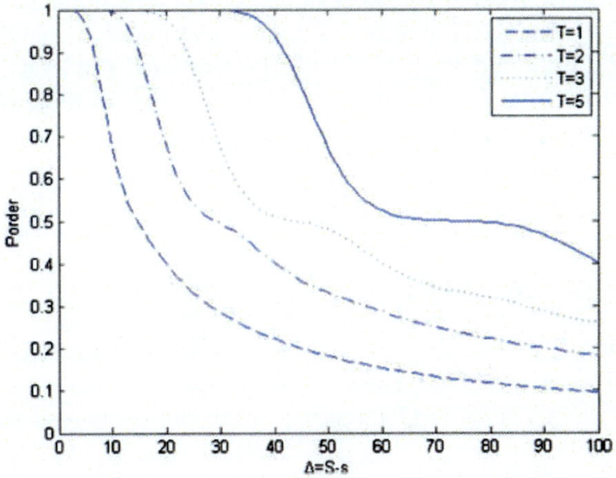

Fig. 4.17 Plot of the (*s*, *S*, *T*) policy ordering probability $P_o(\Delta, T)$ for poisson process generating demands with Rate $\lambda = 10$

such a case, the demands between reviews are indeed independent random variables regardless of the value of the review period $T > 0$. To derive the formulas, they explicitly computed the steady-state probability $\rho(s + j)$ of the system inventory position *IP* being in position $s + j\, j = 1, \ldots, \Delta = S - s$ immediately after a review. The distribution of the *IP* however is no longer uniform as it was in the (r, nQ, T) case studied in the previous section, but is a rather complicated expression:

$$\rho(s+i) = \frac{\sum_{n=0}^{\infty} p^{[n]}(S-s-i;\lambda T)}{\sum_{n=1}^{\infty}\sum_{j=1}^{S-s} np^{[n-1]}(S-s-j;\lambda T)\bar{P}(j;\lambda T)}, \quad i = 1,\ldots,S-s$$

where $p^{[n]}(x;\lambda)$ is the *n*th convolution of the pdf of the demand distribution $p(x;\lambda)$ with demand rate λ and represents the probability that exactly *x* units of stock are demanded in *n* periods—each of length *T*—and $\bar{P}(x;\lambda)$ denotes the complementary cumulative distribution of $p(x;\lambda)$. This expression is actually valid independently of the exact form of the distribution of the demand, as long as the requirements set forth previously (stationary with independent increments) hold. It is well known that the *n*th convolution of the Poisson distribution satisfies $p^{[n]}(x;y) = p(x;n\lambda)$ (see for example Hadley and Whitin 1963).

The ordering probability P_o for the (*s*, *S*, *T*) policy then is given by the following

$$P_o(\Delta, T) = \frac{1}{\sum_{n=1}^{\infty}\sum_{j=1}^{\Delta} np(\Delta - j;(n-1)\lambda T)\bar{P}(j;\lambda T)} \tag{4.38}$$

A plot of the ordering probability as a function of the spread Δ for various values of the review interval *T*, and for a given value of the Poisson demand rate $\lambda = 10$ is shown in Fig. 4.17.

Developing the expression for the long-term expected holding and backorder costs per unit time $H(s, S, T)$, requires computing the function $G(y,T)$ representing (s, S, T) policy's expected (one review period) cost of carrying inventory and backorders incurred from $L + t$ to $L + t + T$ when a system reviewed at time t is found at inventory position $IP = y$. This function depends on the review interval T, but not on the specific values of s and S. Using the expression for the steady-state probabilities $\rho(s + j)$ $j = 1, \ldots, S - s$, Hadley and Whitin developed the analytical expression for the function $G(y,T)$

$$G(y,T) = hT[y - LT - \lambda T/2] + (h+p)b(y,T) \tag{4.39}$$

where the function $b(y,T)$ measures the expected number of time units of shortage incurred in the time interval $[L{+}t, L{+}t{+}T]$ when the IP at review time t is y, and for the case of Poisson demands is given as:

$$\begin{aligned} b(y,T) = &\frac{\lambda}{2}\left[(L+T)^2\bar{P}(y+1;\lambda(L+T)) - L^2\bar{P}(y-1;L\lambda)\right] \\ &+\frac{y(y+1)}{2\lambda}[\bar{P}(y-1;\lambda(L+T)) - \bar{P}(y-1;L\lambda)] \\ &- y[(L+T)\bar{P}(y;\lambda(L+T)) - L\bar{P}(y-L\lambda)] \end{aligned}$$

Parenthetically, we note that the corresponding formula for the function $G(y,T)$ in Hadley and Whitin (1963) (p. 275 formula (5-104)) contains a typo, in that the term h in the first term of the sum making up $G(y,T)$ was omitted. In case the penalty function for the system being out-of stock is of the form $\pi(t) = \hat{p} + pt$, $\hat{p} > 0$ then yet another term of the form $\hat{p}e(y,T)$ must be added to the function $G(y,T)$, where the function $e(y,T)$ measures the expected number of backorders incurred in the time interval $[L + t, L + t + T]$ when the IP at review time t is y, and is given (Hadley and Whitin 1963) as:

$$\begin{aligned} e(y,T) = &\lambda(L+T)\bar{P}(y-1;\lambda(L+T)) - y\bar{P}(y;\lambda(L+T)) \\ &- L\lambda\bar{P}(y-1;L\lambda) + y\bar{P}(y;L\lambda) \end{aligned}$$

Under the given assumptions, the function $-G(y,T)$ is uni-modal in y for any given T, and $G(y,T)$ is in fact convex when $\hat{p} = 0$ (Stidham 1986), a property that has been widely used to devise algorithms for the optimization of the restricted (s, S) periodic review policy. A graph of $G(y,T)$ is shown in Fig. 4.18.

Notice that $G(y,T)$ has indeed a single minimum, but it is not convex. The reason for the non-convexity of $G(y,T)$ are the time-independent stock-out penalty fees $\hat{p}$ incurred. As the expected number of backorders within the interval $[L + t, L + t + T]$ is the same when the inventory position y is negative at time t (which clearly implies that $s < 0$ as well) the costs related with stock-outs $(\hat{p}e(y,T))$ are the same and are independent of y. As soon as $y \geq 0$, this term starts depending on y however, and in fact radically decreases with y, and essentially reaches zero and stays at zero for y large enough. (The latter property also holds for the standard backorder costs $(h + p)b(y, T)$).

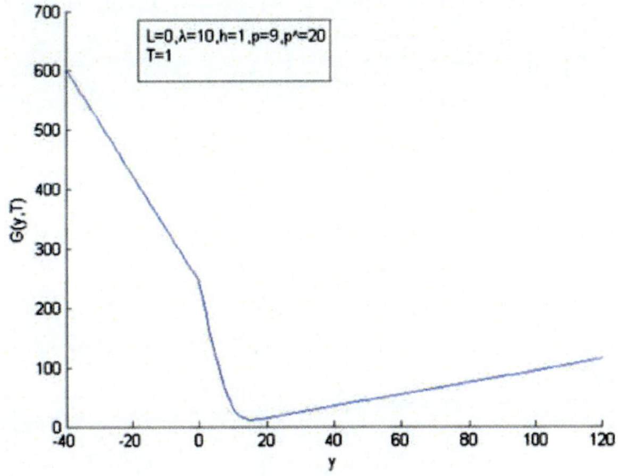

Fig. 4.18 Plot of the (*s*, *S*, *T*) policy's *G*(*y*,*T*) as a function of the inventory position *y* immediately after a review and ordering decision

Now, the total expected cost of the (*s*, *S*, *T*) policy per unit time can be expressed as follows (again, see Hadley and Whitin 1963 for the detailed derivation):

$$C(s,S,T) = \frac{K_r}{T} + \frac{K_o P_o(S-s,T)}{T} + H(s,S,T)$$
$$H(s,S,T) = \frac{1}{T}\frac{\sum_{n=0}^{\infty}\sum_{j=1}^{S-s} p(S-s-j;n\lambda T)G(s+j,T)}{\sum_{n=1}^{\infty}\sum_{j=1}^{S-s} np(S-s-j;(n-1)\lambda T)\bar{P}(j;\lambda T)} \tag{4.40}$$

Now, in a way analogous to that which we used for deriving some properties of the (*r*, *nQ*, *T*) policy in Sect. 4.2.3.2, we may define the lower and upper bounds $B_L(S,\Delta,T)$ and $B_U(S,\Delta,T)$ where $\Delta = S - s$, as follows:

$$B_L(S,\Delta,T) = K_r/T + H(S-\Delta,S,T), B_U(S,\Delta,T) = (K_r + K_o)/T + H(S-\Delta,S,T).$$

It should be obvious that for all *s*, *S*, and *T*, the following inequalities hold:

$$B_L(S,S-s,T) \le C(s,S,T) \le B_U(S,S-s,T)$$

The two bounding functions have an interesting physical interpretation: they represent the long-term expected cost per unit time of an (*s*, *S*, *T*) policy that has no fixed ordering cost, but has a review cost that is paid at each review that is K_r or $K_r + K_o$, respectively. Alternatively, they represent the long-term expected cost per unit time of a system without review costs but with fixed ordering costs K_r or $K_r + K_o$ that are forced to incur at every review regardless of the ordering decision. Another interesting property of the bounding function $B_L(S,\Delta,T)$ is that, obviously from the definition, $\lim_{\Delta\to\infty}[C(S-\Delta,S,T) - B_L(S,\Delta,T)] = 0$. Therefore, for any fixed *S* and *T*, the total average cost of the (*s*, *S*, *T*) policy always coincides with its upper bound for $\Delta = 0$ ($\Delta = 1$ for discrete demand distributions) and asymptotically reaches its lower bound as $\Delta \to \infty$.

The function $C_1(S,\Delta, T) = C(S - \Delta, S, T)$ is known to be convex in the order-up-to level S. This can be easily shown by observing that of all costs making up $C(s, S, T)$, only the holding and backorder cost function $H(s, S, T)$ depend on S and with a change of variables, can be alternatively written as:

$$
\begin{aligned}
H(S,\Delta,T) &= \frac{1}{T}\int_0^T G(S,\Delta,T,t)\mathrm{d}t \\
&= h(S - E[X(\Delta,T)]) - \lambda(L + T/2) \\
&\quad + (h+p)\int_S^\infty (y-S)\frac{1}{T}\int_0^T\int_0^\Delta f(y-x, L+t)\varphi(x)\mathrm{d}x\mathrm{d}t\mathrm{d}y
\end{aligned}
$$

where the function $G(S,\Delta, T, t)$ (different from $G(y,T)$ which is independent of Δ or S and is only meaningful at review times $t = kT$, with k being a natural number) represents the expected *instantaneous* inventory holding and backorder costs at any time t within the review period interval $[0, T]$. $X(\Delta, T) = S{-}IP$ (IP being the inventory position immediately after a review) is a stationary non-negative random variable (being the output of a stochastic clearing renewal process) obviously taking values in the interval $[0,\Delta = S - s]$ with density function $\varphi(x)$, and where, as usual, $f(x, T)$ represents the pdf of cumulative demand in a time interval of length T. The above equation is strictly valid for continuous demand distributions. The functional form of $H(S,\Delta,T)$ makes it easy to see that it is convex in S. This implies that given arbitrary Δ and T, the optimal order-up-to level S can be easily determined using any one dimensional search procedure (in the discrete case, start with any level S_0 and proceed to the left or to the right of that level decrementing or incrementing the level S by 1 each time, until the function $H(S,\Delta,T)$ stops decreasing).

In Fig. 4.19, we plot the function $C_1(\Delta,T) = \min_S C_1(S,\Delta,T)$ as a function of Δ for a fixed T, along with its bounding functions $B_{L,U}(\Delta,T) = \min_S B_{L,U}(S,\Delta,T)$.

Finding the global minimum of the function $C_1(\Delta,T)$ as a function of Δ alone, is then highly non-trivial as there are many local minima as the figure shows. Fortunately, for any given T, we can use *any* algorithm that finds the globally optimal parameters of the *restricted* (s, S) policy with fixed T, to solve this problem. The Zheng–Federgruen algorithm discussed in Sect. 4.2.2.2 is also applicable for periodic review systems. However, the Zheng and Federgruen algorithm ignores T as a variable, it ignores review costs, and even more, it has to assume that the unit of time is equal to the review period and utilizes a rough-cut end-of-period costing scheme with holding and backorder costs being measured per unit time, ignoring the variation of demand within the review interval. Therefore a small adaptation is needed in order for their algorithm to solve the problem of globally minimizing the function $C(s, S, T)$ for fixed and given T:

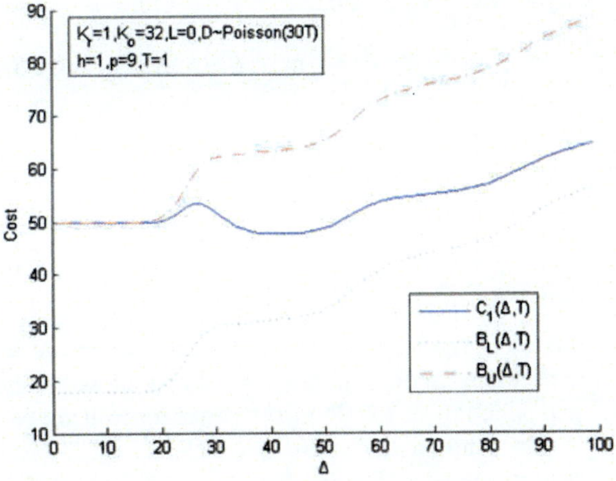

Fig. 4.19 Plot of the function $C_1(\Delta,T)$ and its bounds, B_L and B_U as a function of the spread Δ

1. The cost function $c(s, S)$ required in the Zheng–Federgruen algorithm must be $c(s, S) = C(s, S, T) - K_r/T$.
2. The function $G(y)$ in the Zheng–Federgruen algorithm needs to be defined as follows: $G(y) = G(y,T)/T$ where $G(y,T)$ are the expected holding and backorder costs incurred in the interval $[L + t, L + t + T]$ when inventory position at any review time t is y, as defined above. The division by T is required in order to account for the review period length T, since the classical Zheng–Federgruen algorithm assumes $G(y)$ measures expected holding and backorder costs per period that is the unit of time, as mentioned above.

Taking into account the above, the only issue remaining is how to determine the optimal review period T for which system costs are minimized. Fortunately, there exists an upper bound T_C on the value of T that one needs to search for. To establish this fact, we need one easy lemma—where we remind the reader that discrete demands are assumed again.

Lemma 4.12 *For any T, the following hold:*

(i) $\min_{S,\Delta\in\mathbb{N}^*} H(S,\Delta,T) = H(S_{1,T},1,T)$ *where* $S_{\Delta,T}$ is the $\text{argmin}_S H(S,\Delta,T)$ *for given* Δ *and* T.

(ii) $B_L(1,T) \le \min_{\Delta\in\mathbb{N}^*} C_1(\Delta,T) \le B_U(1,T)$

Proof The function $H(S,\Delta,T)$ represents the long-term expected total cost per unit time of a periodic review system with zero fixed costs ($K_r = K_o = 0$) and linear holding and backorder costs. For such systems, Veinott (1965) showed that the optimal inventory policy is the order-up to policy (or base-stock) policy. For any given T the optimal order-up-to level of this policy is $S_{1,T}$ and (i) follows immediately. Also, for given T, we have:

$$B_L\left(S_{1,T},1,T\right) = \min_{S,\Delta\in\mathbb{N}^*} B_L(S,\Delta,T)$$
$$\leq \min_{\Delta\in\mathbb{N}^*} C_1(\Delta,T) \leq \min_{S,\Delta\in\mathbb{N}^*} B_U(S,\Delta,T) = B_U\left(S_{1,T},1,T\right)$$

QED.

Now, it is easy to establish the following:

Theorem 4.13 $\arg\min_T \left[\min_{\Delta\in\mathbb{N}^*} C_1(\Delta,T)\right] \leq T_C$ where T_C *is the smallest* T *such that* $B_L(1,T) \geq \min_{s,S,T\leq T} C(s, S, T)$.

Proof The functions $B_{L,U}(S_{1,T},1,T)$ represent long-term expected total cost for an (R, T) base-stock periodic review policy; we have already shown the joint convexity of the cost function of (R, T) in the order-up-to level R and the period T. Rao (2003) has also shown that $\lim_{T\to\infty} H(S, \Delta = 1, T) = \infty$. Therefore, at their optimal setting it also holds that $B_{L,U}(1,T)$ are convex in T with $\lim_{T\to\infty} B_{L,U}(1, T) = \infty$. If T_C is defined as in the conditions of the Theorem, it clearly implies that the inequality $\min_{s,\,S} C(s, S, T) \geq B_L(1, T)$ holds for any $T \geq T_C$ and thus the minimizer of the function $C(T) = \min_{s,\,S} C(s, S, T)$ is smaller or equal to T_C. QED.

The above theorem implies the correctness of the following exact optimization algorithm for the determination of the globally-optimal (s, S, T) periodic review policy parameters that is based on a discretization of the review period T, according to a time-quantum ε_T. Recall that the same approach was taken in devising the algorithm for the determination of the optimal parameters for the (r, nQ, T) policy.
Algorithm (s, S, T) Policy Optimization
Inputs: review cost K_r, ordering cost K_o, lead-time L, discrete demand pdf $f(x, T)$, linear holding cost rate h, linear backorder penalty rate p, time search quantum ε_T.
Outputs: optimal parameters s^*, S^*, T^* and expected cost c^*.

Begin

1. Set $c^* = +\infty$, $B_U = +\infty$
2. for $T = \varepsilon_T, 2\varepsilon_T, 3\varepsilon_T, \ldots$ do
   a. Set (s_T, S_T) to be the optimal parameters for the restricted periodic review (s,S) policy with review interval T (by calling the Zheng–Federgruen algorithm, modified as mentioned above)
   b. Set $c_T = C(s_T, S_T, T)$
   c. if $c_T < c^*$ then
      i. Set $c^* = c_T$, $s^* = s_T$, $S^* = S_T$, $T^* = T$
   d. end-if
   e. Set $S_{1,T} = \text{argmin}_S B_L(S, 1, T)$ (by solving the discrete unconstrained 1D convex programming problem $\min_S B_L(S, 1, T)$).
   f. Set $B_{L,T} = B_L(S_{1,T},1,T)$
   g. if $B_{L,T} \geq c^*$ break.

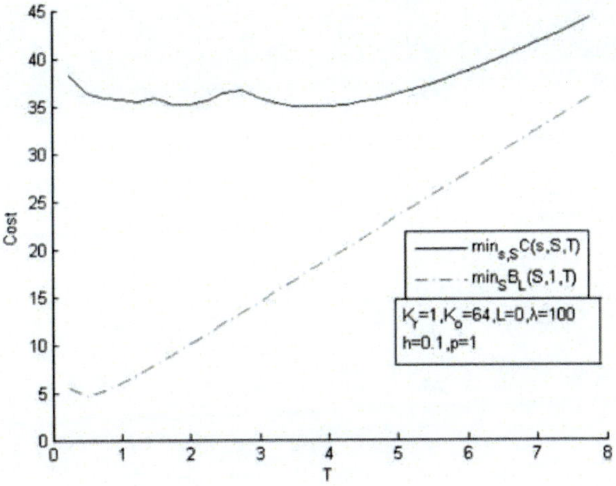

Fig. 4.20 Plot of the functions $C(T)$ and $B_L(T) = B_L(1,T)$ for an inventory system following the (s, S, T) policy with the parameters shown in the graph

3. end-for
4. return (s^*,S^*,T^*,c^*).

End.

In Fig. 4.20 we plot the application of the algorithm, showing the values $C(T)$ and $B_L(T) = B_L(S_{1,T},\Delta = 1,T)$ until the stopping conditions of the Theorem 4.13 are satisfied. The parameter ε_T was set to 0.25 to speed-up computations. On an Intel Core-2 Duo processor (4 cores) running Windows XP at 2.2 GHz with 3 GB RAM memory, the above algorithm implemented in matlab took approximately 5 min of cpu-time to run. Most of this time is spent in computing the series in the expressions (4.38) and (4.40). A note on this computation is in order. The infinite sums in the calculations required for the computation of the value $C(s, S, T)$ are computed by adding terms consecutively until five consecutive terms in the series add up to less than 10^{-9} of the current sum. Strictly speaking, this does not guarantee that the series has been computed within an accuracy of 10^{-9} or any other number, since we cannot provide an upper bound for the value of the sum of the remaining terms in the series, but for all practical purposes it has proven to be a very robust and accurate stopping criterion. To illustrate the convergence of the partial sums, see Fig. 4.21. In the figure, $s1(n) = \sum_{k=0}^{n}\sum_{j=1}^{S-s} p(S-s-j;k\lambda T)$ $G(s+j,T), s2(n) = \sum_{k=1}^{n}\sum_{j=1}^{S-s} kp(S-s-j;(k-1)\lambda T)\bar{P}(j;\lambda T)$ and $s3(n)$ is the expression for the cost $C(s, S, T)$ truncated after only the first n terms of each infinite series in the expression (4.40) have been added up. In the particular example, $s = 9$, $S = 26$, $T = 10^{-3}$, $K_r = 0$, $K_o = 10$, $L = 1$, $D(T) \sim$ Poisson$(10T)$, $h = 1$, $p = 10$. After adding up approximately 4,000 terms, the two partial sums $s1$ and $s2$ have converged according to the criterion specified above, and thus the computations for $s1$ and $s2$ stop.

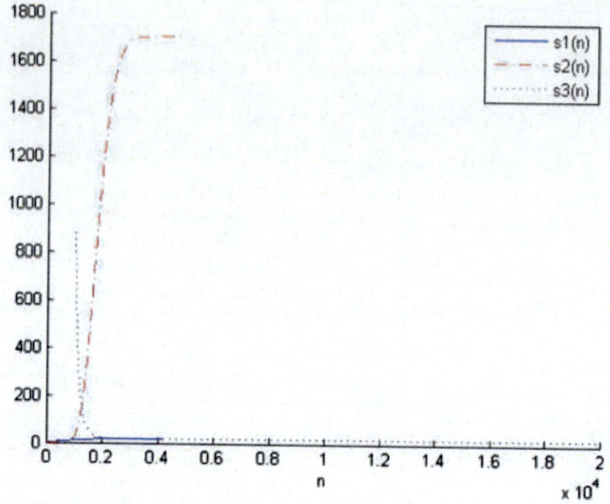

Fig. 4.21 Convergence plot of the two infinite series in the expression for the calculation of $C(s, S, T)$

The long running times associated with the exact optimization algorithm for the (s, S, T) policy demand the development of some heuristics for the fast determination of near-optimal policy parameters. Fortunately, such a heuristic is indeed possible.

Fast Determination of Near-Optimal (s,S,T) Policy Parameters

The key observation to developing a fast heuristic for the optimal determination of (s, S, T) policy parameters is a comparison between the (r, nQ, T) policy and the (s, S, T) policy. We have already seen that for large enough T, both policies reduce to the same policy, the (R, T) policy, so it is interesting to compare the two policies for smaller T values. Figure 4.22 shows exactly such a comparison, for one particular system setting. Even though there do exist observable differences between the two policies' cost for certain values of the review period T, the important thing to notice is that the difference of the two policies at the optimum setting ($T^* \approx 6.4$) is essentially nil. Further experiments comparing the two policies with a test suite used by Zheng (1992) and Rao (2003), reveal that the optimal review period of the (r, nQ, T) policy is always extremely close or completely coincides with the optimal T of the (s, S, T) policy (see Table 4.2).

Combining this observation with the fast heuristic developed in Sect. 4.2.3.2 for the (r, nQ, T) policy, leads immediately to the following algorithm:

Algorithm Heuristic (s, S, T) Policy Optimization

Inputs: review cost K_r, ordering cost K_o, lead-time L, discrete uni-modal demand pdf $f(x, T)$, linear holding cost rate h, linear backorder penalty rate p, time search quantum ε_T and maximum search distance δ_T.
Outputs: near-optimal parameters s^*, S^*, T^* and cost c^*.

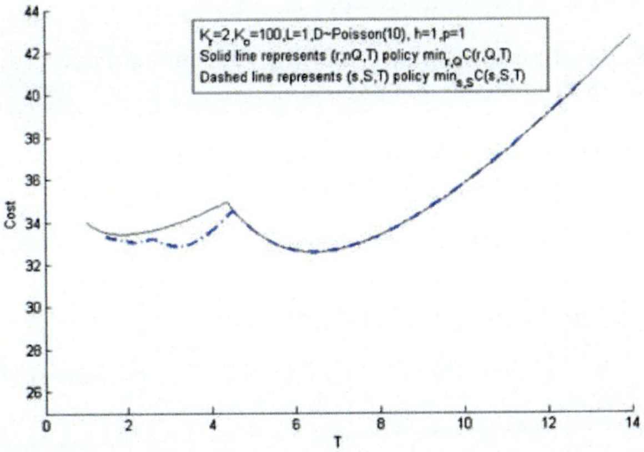

Fig. 4.22 Plot of the optimal (s, S, T) policy versus the optimal (r, nQ, T) policy as functions of the review interval T

Table 4.2 Comparison between Heuristic and exact optimization of (s, S, T) policy parameters

Parameters $L = 1$, $\lambda = 50$				(s, S, T)				Approx. optimization				
K_r	K_o	h	p	s^*	S^*	T^*	c^*	s	S	T	c	Gap%
1	1	10	25	55	58	0.16	105.32	55	58	0.16	105.32	0.000
1	5	10	25	55	61	0.25	124.5	55	61	0.25	124.5	0.000
1	25	10	25	48	67	0.16	181.01	48	67	0.16	181.01	0.000
1	100	10	25	43	79	0.19	298.16	43	79	0.19	298.16	0.000
1	1000	10	25	22	135	0.27	858.48	21	135	0.245	858.53	0.006
1	1	15	100	59	62	0.13	203.81	59	62	0.13	203.81	0.000
1	5	15	100	59	64	0.19	229.13	59	64	0.19	229.13	0.000
1	25	15	100	54	69	0.12	299.99	54	69	0.12	299.99	0.000
1	100	15	100	50	79	0.11	450.46	51	79	0.13	450.52	0.013
1	1000	15	100	42	128	0.15	1185.14	42	128	0.15	1185.14	0.000
1	1	20	20	50	53	0.14	135.2	50	53	0.14	135.2	0.000
1	5	20	20	50	55	0.21	156.36	50	55	0.21	156.36	0.000
1	25	20	20	44	60	0.18	219.71	44	60	0.18	219.71	0.000
1	100	20	20	37	67	0.19	354.95	37	67	0.19	354.95	0.000
1	1000	20	20	6	100	0.28	1015.78	5	100	0.23	1015.9	0.012

Begin

1. Set $c^* = +\infty$
2. Set $(r_0,Q_0,T_0) = \text{argmin}_{r,Q,T} B_0(r,Q,T)$ (solving the corresponding continuous un-constrained convex optimization problem considering continuous demand distribution that best approximates the discrete demand density).
3. Set $(r_1,Q_1,T_1) = \text{argmin}_{r,Q,T} B_1(r,Q,T)$ (solving the corresponding continuous unconstrained convex optimization problem considering continuous demand distribution that best approximates the discrete demand density).

4. for $T = T_0 - \delta_T$ to $T_0 + \delta_T$ step ε_T do
   a. Set $(s_T,S_T) = \text{argmin}_{s,S}C(s,S,T)$ (calling the Zheng–Federgruen algorithm), $c_T = C(s_T,S_T,T)$.
   b. if $c_T < c^*$ then
      i Set $c^* = c_T$, $s^* = s_T$, $S^* = S_T$, $T^* = T$.
   c. end-if
5. end-for
6. for $T = T_1 - \delta_T$ to $T_1 + \delta_T$ step ε_T do
   a. Set $(s_T,S_T) = \text{argmin}_{s,S}C(s,S,T)$ (calling the Zheng–Federgruen algorithm), $c_T = C(s_T,S_T,T)$.
   b. if $c_T < c^*$ then
      i. Set $c^* = c_T$, $s^* = s_T$, $S^* = S_T$, $T^* = T$.
   c. end-if
7. end-for
8. return (s^*, S^*, T^*, c^*).

End

The performance of the heuristic algorithm in terms of solution quality compares very well with that of the (s, S, T) policy optimization algorithm developed before. Some results are shown in Table 4.2 which indicates the very high quality of the heuristic solutions obtained by the algorithm, and was run with settings $\delta_T = 0.01$ and $\varepsilon_T = 0.01$. The problems, again, are taken from the paper by Rao (2003) for the (R, T) policy, and we have added an additional review cost $K_r = 1$. The last column labeled "Gap%" indicates the *percentage* deviation between the heuristic solution found by the heuristic (s, S, T) policy optimization algorithm and the exact (time-discretized) (s, S, T) policy optimization algorithm. As can be observed from the table, the maximum percentage deviation in all examples is negligible (about 0.01%) and in most cases, the algorithm does obtain the optimal solution. On the other hand, the speed-up factor obtained is *more than one hundred* (100); as an example, the heuristic optimization of the example used to create Fig. 4.20 takes <1 s on the exact same computational settings as before!

Finally, it is worth considering the performance of general-purpose randomized heuristic algorithms for global nonlinear optimization, such as the methods introduced in Sect. 1.1.1.3 of this book. We run the EA algorithm for unconstrained optimization of the function $f(s,S,T) = C(\min\{[s], [S] - 1\}, [S], |T|)$ where $[x] = round(x)$ denotes the nearest integer to the number x. The above alterations were made in order to turn the constrained mixed integer optimization problem

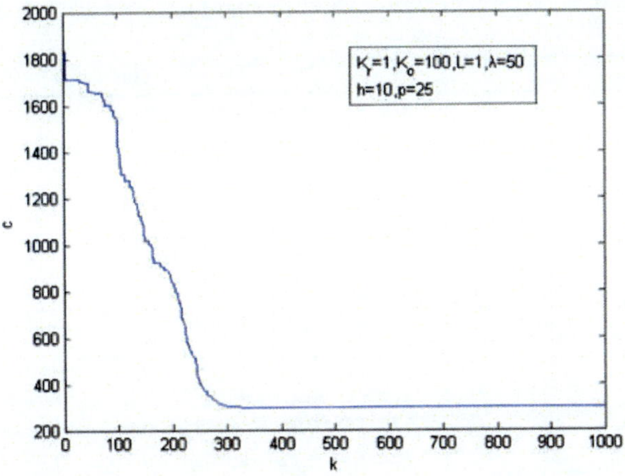

Fig. 4.23 Plot of the evolution of the EA algorithm for continuous unconstrained optimization on the function $C(\min\{[s], [S]-1\}, [S], |T|)$. The algorithm was run with user-defined parameters $\sigma = [1\ 1\ 1]^T$ and $x_0 = [0\ 1\ 1]^T$ and essentially finds the globally-optimal solution within 350 (*s*, *S*, *T*) policy evaluations

$$\min_{s,S,T} C(s,S,T)$$
$$\text{s.t.} \begin{cases} T \geq 0 \\ S - s \geq 1 \\ s, S \in \mathbb{Z} \end{cases}$$

into the unconstrained continuous optimization problem $\min_{s,S,T} f(s,S,T)$ so the EA algorithm can be applied.

The results are impressive, as can be seen in Fig. 4.23, showing the progress of the EA algorithm over 1,000 iterations. The best parameter settings found by the algorithm for the problem shown in the figure are $(s, S, T) = (43, 79, 0.1885)$, with a best cost value of 298.16, which agrees with the exact and specialized heuristic algorithms in five decimal digits as can be seen from Table 4.2! The running time of the EA algorithm is reasonable (less than a minute of cpu-time), but certainly more than the running time of the heuristic approximation algorithm developed before for the policy.

4.2.3.4 Comparing (*R*, *T*), (*r*, *nQ*, *T*) and (*s*, *S*, *T*) Periodic Review Policies

It has already been mentioned that the optimal periodic review doctrine in the existence of non-zero fixed ordering costs is the (*s*, *S*, *T*) policy. However, various research during the past three decades with periodic review systems with fixed period *T* has shown that the *differences* between these policies at their optimal settings should be relatively small, but not negligible, often exceeding

Table 4.3 Comparison between optimal (*s*, *S*, *T*), (*r*, *nQ*, *T*) & (*R*, *T*) policy costs

Parameters $L = 1$, $\lambda = 50$				(*s*, *S*, *T*)		(*r*, *nQ*, *T*)			(*R*, *T*)		
K_r	K_o	*h*	*p*	*T**	*c**	*T**	*c**	Gap%	*T**	*c**	Gap%
1	1	10	25	0.16	105.3	0.16	105.4	0.04	0.2	105.4	0.04
	5	10	25	0.25	124.5	0.25	124.6	0.06	0.3	124.6	0.06
	25	10	25	0.16	181.0	0.45	181.6	0.34	0.5	181.6	0.34
	100	10	25	0.19	298.2	0.16	299.5	0.43	0.8	301.7	1.18
	1000	10	25	0.27	858.5	0.23	859.3	0.09	2.39	865.4	0.80
	1	15	100	0.13	203.8	0.13	204.0	0.07	0.1	204.0	0.07
	5	15	100	0.19	229.1	0.19	229.5	0.14	0.2	229.5	0.14
	25	15	100	0.12	300.0	0.1	302.0	0.68	0.3	304.8	1.60
	100	15	100	0.11	450.5	0.11	451.7	0.28	0.6	464.1	3.03
	1000	15	100	0.15	1185.1	0.13	1185.8	0.06	1.78	1217.9	2.76
	1	20	20	0.14	135.2	0.14	135.3	0.06	0.1	135.3	0.06
	5	20	20	0.21	156.4	0.22	156.5	0.10	0.2	156.5	0.10
	25	20	20	0.18	219.7	0.4	220.2	0.24	0.4	220.2	0.24
	100	20	20	0.19	355.0	0.69	357.4	0.68	0.7	357.4	0.68
	1000	20	20	0.28	1015.8	0.21	1017.5	0.17	2.02	1020.4	0.45
5	1	10	25	0.25	124.6	0.25	124.6	0.00	0.25	124.6	0.00
	5	10	25	0.3	139.1	0.3	139.1	0.00	0.3	139.1	0.00
	25	10	25	0.48	190.4	0.48	190.4	0.00	0.48	190.4	0.00
	100	10	25	0.81	306.6	0.81	306.6	0.00	0.81	306.6	0.00
	1000	10	25	2.39	867.1	2.39	867.1	0.00	2.39	867.1	0.00
	1	15	100	0.19	229.4	0.19	229.5	0.00	0.19	229.5	0.00
	5	15	100	0.23	248.4	0.23	248.4	0.01	0.23	248.4	0.01
	25	15	100	0.35	316.4	0.35	316.4	0.01	0.35	316.4	0.01
	100	15	100	0.61	470.7	0.61	470.7	0.00	0.61	470.7	0.00
	1000	15	100	0.27	1204.0	0.25	1205.7	0.14	1.78	1220.1	1.34
	1	20	20	0.22	156.5	0.22	156.5	0.00	0.22	156.5	0.00
	5	20	20	0.28	172.6	0.28	172.6	0.01	0.28	172.6	0.01
	25	20	20	0.42	230.0	0.42	230.0	0.01	0.42	230.0	0.1
	100	20	20	0.7	363.1	0.7	363.1	0.00	0.7	363.1	0.00
	1000	20	20	2.02	1022.4	2.02	1022.4	0.00	2.02	1022.4	0.00

10% [see Zheng and Chen (1992) for a comparison of periodic review (*r*,*nQ*) and (*s*, *S*) policies, and Rao (2003) for a comparison between (*R*, *T*) with no review cost K_r and continuous review (*r*,*Q*)]. When explicitly taking into account the review interval *T*, in the case of lost sales, where when an order arrives and the system is out-of-stock, it is simply lost without incurring any other costs, Reis (1982) shows by simulation experiments that the cost differences between the (*R*, *T*) and (*s*, *S*, *T*) policies are very small for a very large range of values for the ordering cost K_o. This result apparently also holds for the case studied above, of full backlogging of order requests arriving during system stock-outs, as one can see in the results presented in Table 4.3. The largest percentage deviation between the optimal (*s*, *S*, *T*) and (*r*,*nQ*,*T*) policy observed is only 0.68%, implying that the

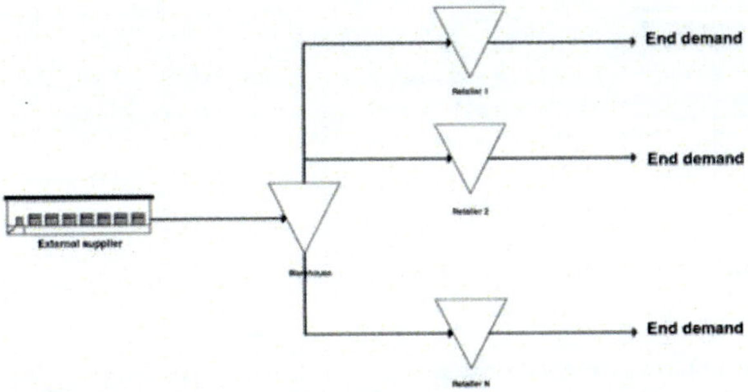

Fig. 4.24 A multi-echelon distribution system

(r,nQ,T) policy, besides offering the advantages of easier handling of orders received etc., is also very close to fully optimizing expected relevant costs. This is in sharp contrast with the previous findings mentioned above when the review period was not taken into consideration that lead to conclusions that the (r,nQ) policy may lag significantly behind the optimal (s, S) policy.

4.3 Multi-Echelon Inventory Control

The previous sections focused on so-called single-echelon inventory installations, where a single installation decides *independently* of its upstream suppliers or downstream customers how and when to order so as to minimize its total costs assuming an infinite planning horizon and assuming the demand, when stochastic, is and will be generated from the same stationary stochastic process. Indeed, examples fit well in such a framework (retail stores that are not part of a chain owning central warehouses, as well as various types—but certainly not all—of manufacturers).

However, there are cases when an organization owns several stages, or echelons, in its supply chain. An easy example would be a large retail chain, owning several warehouses, each of which feeds several near-by retail outlets of the organization. Such a situation, known as a *multi-echelon distribution system* is shown schematically in Fig. 4.24.

Alternatively, in an assembly type system, several different raw material providers feed into a final manufacturing stock-point that applies some transformation to the raw materials and produces a stock of final goods to be distributed to its end-customers (Fig. 4.25).

Interestingly (Rosling 1989), from an inventory control point of view, assembly type multi-echelon systems are equivalent to an appropriately transformed serial system such as the one shown in Fig. 4.26. Such types of systems are a central part of Supply Chain Management in organizations owning multiple stages in the supply chain. Since serial systems present the least difficulties to model and

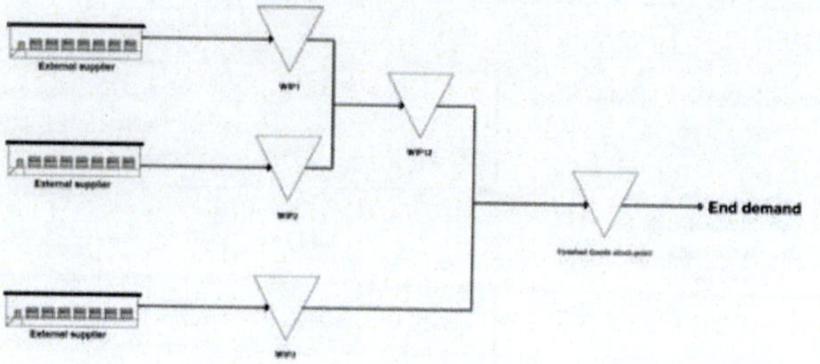

Fig. 4.25 A multi-echelon assembly system

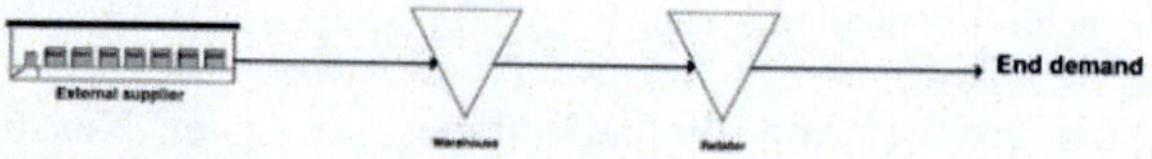

Fig. 4.26 A 2-echelon serial system. The retailer is known as echelon 1, and the warehouse is known as echelon 2

analyze mathematically, they will form the objective of the study of the following sections.

4.3.1 Serial 2-Echelon Inventory System Under Deterministic Demand

Consider the two-stage inventory control system of Fig. 4.26, and assume—as in the case of the EOQ model and its variants—that the demand the final stock-point faces has a constant rate, D. Let us call the final stock-point in the chain facing external customer demand (that could be for example the retailer store), echelon 1, and call the previous stock-point echelon 2 (that could be the warehouse), that orders stock from an external supplier that has infinite capacity, and can therefore accommodate any order of any size without any lead-time delays, so that $L_2 = 0$. Echelon 2 can accommodate any order from the downstream echelon also in zero lead-time $L_1 = 0$. Each echelon faces an inventory holding rate H_i and a fixed ordering cost K_i. Now, if the downstream echelon wishes to optimize its own costs, the EOQ model applies, and echelon 1 will order a quantity Q_1 determined by the EOQ formula (4.3). But then, even though the downstream echelon's demand is constant, the upstream echelon's demand rate is deterministic but hardly constant. In fact, demand at echelon 2 arrives only at discrete time points spaced apart according to the formula $T_1 = Q_1/D$. To facilitate the analysis of this system, Clark and Scarf (1960) introduced the concept of *echelon inventory level*, which is defined for any given stock-point within a generic supply chain network as *all physical stock at that stock-point plus all stock in transit or on-hand at any stock-point downstream minus any backorders at the most downstream stock-point.*

Therefore, when an echelon inventory level is negative, it indicates that there are backorders at the most downstream stock-point(s) and are certainly larger than the total physical stock in that echelon. It is easy to see that now, not only echelon 1, but also the upstream echelon stock has the same saw-tooth pattern in time, since it appears now that the echelon stock 2 faces the *same* constant deterministic demand rate D. Assuming no backorders are allowed, if echelon 1 orders a quantity Q_1 according to the EOQ with no backorders model, it should be obvious that the upstream stock-point 2 should order in integer multiples of that order quantity, so that $Q_2 = nQ_1$ for some natural number n. The reason is that otherwise an excess inventory will have to remain in the warehouse of echelon 2 at the end of an order cycle at the upstream stock-point, incurring extra holding costs that can be eliminated when exact multiples of the order size Q_1 are ordered by the upstream stock-point. When using echelon inventory levels however, it is necessary to modify the holding cost rates H_i of the various echelons because the same item will be counted when computing the stock level of more than one echelons. Therefore, when computing echelon costs for a stock-point i it is necessary to use holding rates

$$\hat{H}_i = H_i - \sum_{j \in P(i)} H_j$$

where $P(i)$ is the set of all predecessor stock-points of i in the supply network. With this definition of holding cost rates $\hat{H}_i$, we can formulate the problem as a *mixed integer programing problem*, minimizing the total costs in the serial chain that is now viewed as a single system:

$$(2ESS)\ \min_{n,Q_1} C_{\text{tot}}(n, Q_1) = \left(K_1 + \frac{K_2}{n}\right)\frac{D}{Q_1} + \left(n\hat{H}_2 + \hat{H}_1\right)\frac{Q_1}{2}$$

$$\text{s.t.} \begin{cases} Q_1 \geq 0 \\ n \in \mathbb{N}^* \end{cases}$$

Once the optimal Q_1^*, n^* values are found, the order quantity at the upstream echelon 2 is simply $Q_2^* = n^* Q_1^*$. Determining the minimizing point of the function $C_{\text{tot}}(n, Q_1)$ however, even though it contains a discrete variable, is not difficult, and does not require any sophisticated algorithm. In particular, notice that the function is convex and differentiable in the continuous variable Q_1 and therefore Q_1^* must satisfy the condition $\partial C_{\text{tot}}(n^*, Q_1^*)/\partial Q_1 = 0$ —since at $Q_1 = 0$, the function becomes infinite. Computing the partial derivative of C_{tot} with respect to Q_1 and solving for the optimal $Q_1^*(n)$ for a given value of n gives:

$$\frac{\partial C_{\text{tot}}(n, Q_1)}{\partial Q_1} = -\frac{(K_1 + K_2/n)D}{Q_1^2} + \left(\hat{H}_1 + n\hat{H}_2\right)/2 = 0 \Leftrightarrow$$

$$Q_1^*(n) = \sqrt{\frac{2(K_1 + K_2/n)D}{\hat{H}_1 + n\hat{H}_2}}$$

But a natural number n^* will be the global minimizer of $C_{tot}(n, Q_1)$ if and only if it is also the global minimizer of the function $c(n) = (K_1 + K_2/n)(\hat{H}_2 n + \hat{H}_1)$. The function $c(x)$ is convex and differentiable function of its argument x, and its minimizer is at $x^* = \sqrt{K_2\hat{H}_1/(K_1\hat{H}_2)}$. Therefore, the optimal value of n, n^* and the optimal order quantity Q_1^* must be determined by:

$$n^* = \arg\min\{c(\lfloor x^* \rfloor), c(\lceil x^* \rceil)\}$$

$$Q_1^* = \sqrt{\frac{2(K_1 + K_2/n^*)D}{\hat{H}_1 + n^*\hat{H}_2}}$$

4.3.2 Serial 2-Echelon Inventory System Under Stochastic Demand

When the two echelon system shown in Fig. 4.26 faces demands that cannot be assumed deterministic, modeling the problem presents serious challenges. If one attempted to optimize the total cost of such a multi-echelon system by optimizing each installation independently so as to derive an optimal ordering policy that each supplier will follow independently of the other, the result could be a highly sub-optimal system:

- The distribution of demand at higher levels of the supply chain becomes more and more complicated even for fairly simple end-user demand: when the retailer faces demands drawn from the poisson distribution, and uses an (s, S) ordering policy, then the warehouse faces demand drawn from the Erlang distribution with parameter k directly depending on $S - s$, which is a much more complex distribution.
- Lead-time of orders observed at the retailer is highly non-deterministic and depends on whether the warehouse has available stock or not; the pdf of the *lead-time demand* is no longer uni-modal even if end-consumer demand is uni-modal.
- Ordering decisions at one echelon definitely have implications on the optimal cost of the other echelons.
- If the retailer (and each installation further upstream in the supply chain) uses an (*autonomous*) optimal policy that optimizes its own costs, then even slight fluctuations in end-customer demand will result in large fluctuations in the upper levels of the supply chain: this is the bull-whip effect (Forrester 1961) that is due to the inherent instability of such supply chains (Daganzo 2003) that will be discussed briefly in the next section.

In the following *simplified* analysis (based on Axsater 2010) that illustrates the Clark and Scarf decomposition approach to multi-echelon inventory systems

(Clark and Scarf 1960), we will show that an (echelon) order-up-to policy is optimal for both echelons for a 2-echelon serial system facing stochastic demands *when there are no fixed* (review or order set-up) *costs*. We shall assume an *infinite time horizon* consisting of discrete, equally-spaced points in time $\{t_0, t_1 = t_0 + t, \ldots, t_n = t_0 + nt \ldots\}$ over which we wish to minimize expected long-term costs, and that demands are i.i.d. random variables drawn from the normal distribution that occur in the demand epochs t_i. The normality of the demands can very easily be dropped to include essentially any probability distribution as long as demands are always generated from the same stationary stochastic process.

As previously, let L_1 and L_2 denote some constant lead-times measured in time-periods (therefore being non-negative integer quantities) that each echelon experiences (generally assumed non-zero), let μ denote the average period demand and let σ denote the standard deviation of it. Let h_j denote the *holding cost per unit time and period* at installation j, and p_1 denote the shortage cost per unit time and period at the retailer (there are no penalty costs associated with warehouse backorders). Let $I_{e,j}(t)$, $I_{i,j}(t)$ denote respectively the echelon and installation *inventory level* at j just before demand for period t occurs (since demand is assumed stationary the parameter t plays no role in the analysis). Let $D(n)$ denote the random variable representing cumulative demand over n periods; and finally, let $y_1(t)$ denote the realized echelon stock Inventory Position IP_1 at installation 1 in period $t + L_2$ just before period demand occurs, and let y_2(t) denote the echelon stock Inventory Position IP_2 at installation 2 in period t just before demand occurs for that period. The following analysis also assumes that events take place at the beginning of each period in the following order: installation 2 places any order it wishes to place, then period delivery from outside supplier (with infinite capacity) arrives at installation 2, then the retailer (installation 1) places any orders it wishes to place at the warehouse (installation 2), then period delivery from the warehouse arrives to the retailer, and finally outside end-customer demand materializes at the retailer, and holding and shortage costs are evaluated at this point.

Dropping the parameter t denoting a particular time-period, when the warehouse orders, its *echelon* inventory position $IP_{e,2}$ is identical to the realized inventory position y_2. The echelon inventory level $I_{e,2}$ L_2 time-periods later will be $I_{e,2} = y_2 - D(L_2)$ where the random variable $D(L_2)$ has mean $\mu_2' = L_2\mu$, and standard deviation $\sigma_2' = \sqrt{L_2}\sigma$.

Regarding the realized echelon inventory position y_1 at the retailer after ordering in period $t + L_2$ inventory balance constraints dictate that we must have $y_1 \le I_{e,2} = y_2 - D(L_2)$. As for the *installation stock inventory level* $I_{i,1}$ measured *right after* demand in period $t + L_1 + L_2$ has materialized, it is equal to $I_{i,1} = y_1 - D(L_1 + 1)$, where the random variable $D(L_1 + 1)$ clearly has mean and standard deviation $\mu_1'' = (L_1 + 1)\mu, \sigma_1'' = \sqrt{L_1 + 1}\sigma$.

For the warehouse, it is convenient to consider ordering in an arbitrary period t, and to evaluate associated costs in period $t + L_2$ whereas for the retailer, it is convenient to consider the ordering that occurs in period $t + L_2$ and measure costs in period $t + L_2 + L_1$. We may now formulate the total costs for the 2-echelon serial system, which, in the absence of fixed costs as mentioned before, comprise of holding and back-order costs only.

The expected long-term holding costs for the warehouse in period $t + L_2$ when *echelon inventory positions* at installation i are y_i are given as:

$$C_2 = h_2 E\left[I_{e,2} - y_1\right] = h_2\left(y_2 - \mu_2'\right) - h_2 y_1$$

(no back-order penalty costs exist at the warehouse level). Similarly, the average holding and back-order costs at the retailer in period $t + L_2 + L_1$ are similarly given as

$$\begin{aligned} C_1 &= h_1 E\left[(y_1 - D(L_1+1))_+\right] + p_1 E\left[(y_1 - D(L_1+1))_-\right] \\ &= h_1\left(y_1 - \mu_1''\right) + (h_1 + p_1) E\left[(y_1 - D(L_1+1))_-\right] \end{aligned}$$

where the obvious relationship $x_+ = x + x_-$ was used.

From the above, it is clear that the total expected long-term (per period) system cost depends only on the parameters $y_i\ i = 1, 2$ so the optimal policy must be such that it results in y_i that minimizes the sum $C = C_1 + C_2$. By adding the quantity $h_2 y_1$ from C_2 and subtracting the same quantity from C_1 we obtain two new costs for installations 1 and 2 that add up again to the same total cost C as before:

$$\begin{aligned} \hat{C}_1 &= e_1 y_1 - h_1 \mu_1'' + (h_1 + p_1) E\left[(y_1 - D(L_1+1))_-\right] \\ \hat{C}_2 &= h_2\left(y_2 - \mu_2'\right) \end{aligned}$$

where $e_1 = h_1 - h_2$. Now, the new cost for the warehouse is clearly independent of the y_1 variable, but the new retailer costs *could* implicitly depend on y_2 because of the constraint $y_1 \leq y_2 - D(L_2)$. However, this is not the case, and the optimal value for the new retailer cost $\hat{C}_1$ is independent of y_2. To see why, consider the relaxation of the constrained stochastic optimization problem $\min_{y \in \mathbb{R}^2}\{\hat{C}_1(y_1) | y_1 \leq y_2 - D(L_2)\}$ where the constraint is dropped. The *relaxed* problem is essentially the classical news-boy problem analyzed in Sect. 4.2.1 and its solution is given by

$$\Phi\left(\frac{y_1^* - \mu_1''}{\sigma_1''}\right) = \frac{h_2 + p_1}{h_1 + p_1}$$

where $\Phi(x)$ is of course the standard normal distribution $N(0,1)$. Now, if the optimal solution to the relaxed problem satisfies the constraint of the original problem $y_1^* \leq y_2 - D(L_2)$ then clearly the optimal value for y_1 for the real constrained problem must be the value obtained from the solution of the news-boy problem; but if the constraint is violated, then because of the convexity of the function $\hat{C}_1(y_1)$ the optimal value for the variable y_1 is exactly equal to $y_2 - D(L_2)$ and the optimal policy that sets the inventory position y_1 to these values is an echelon stock order-up-to policy with order-up-to echelon stock level $S_{e,1} = y_1^*$.

Determining the optimal policy at the warehouse can be done along the same arguments. The only decision variable to optimize is y_2 which must be set so that the *total cost* C is minimized:

$$C = h_2\left(y_2 - \mu_2'\right) + \hat{C}_1\left(S_{e,1}\right) + \int_{y_2 - S_{e,1}}^{+\infty} \left[\hat{C}_1(y_2 - x) - \hat{C}_1\left(S_{e,1}\right)\right] \frac{1}{\sigma_2'} \phi\left(\frac{x - \mu_2'}{\sigma_2'}\right) \mathrm{d}x$$

where $\phi(x)$ denotes the pdf of the standard normal distribution $N(0,1)$. The above objective function represents yet another news-boy problem; let y_2^* denote the minimizer of this news-boy problem. The optimal policy at the warehouse—since the outside supplier in the 2-echelon system is assumed to have infinite capacity—is therefore an echelon order-up-to policy with $S_{e,2} = y_2^*$, as claimed in the beginning.

4.3.3 Stability of Serial Multi-Echelon Supply Chains

The bullwhip effect already mentioned in the previous section is the very real phenomenon where small fluctuations in the demand patterns experienced by downstream echelons in a supply chain magnify considerably as orders propagate to upstream installations so that manufacturers and raw material providers experience demand patterns that are highly erratic with long periods of no demand at all followed by small "bursts" of large-size orders; such demand patterns of course have very serious consequences on manufacturer costs, and eventually, on the retail price of goods, because manufacturers are forced to frequent setup changes, incurring significant fixed-charge costs, lead-times are likely to inflate in order to accommodate large but infrequent orders, personnel costs increase due to over-times required when demand suddenly increases above current (planned) capacity and so on.

Various researchers have given different answers regarding what causes the bullwhip effect (it is so prevalent that it is taught in business schools with the help of the classical "beer distribution" game, a business game with four players, a retailer, a wholesaler, a distributor and a manufacturer linked in a serial chain; the orders seen by the manufacturer after a few iterations, invariably become chaotic). Some of the reasons quoted include bounded rationality of the decision-makers, high lead-times, batch order sizes to minimize setup costs and so on.

However, even if computer algorithms implement optimal so-called autonomous policies that minimize the total inventory costs at each echelon—which is often the case today—so that bounded (human) rationality is not an issue, the bullwhip effect still all too often manifests itself. A control theory point of view has unveiled the systemic nature of this problem (Daganzo 2003), and we shall present in brief one main result of this theory of supply chains.

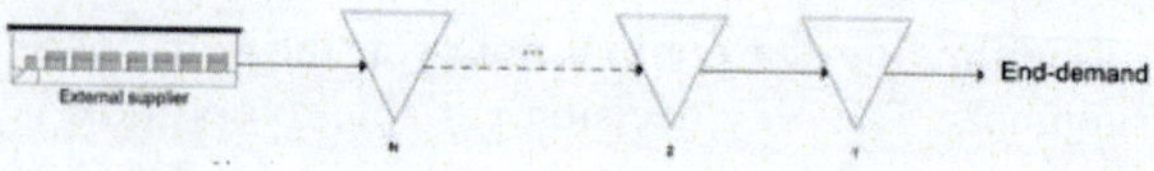

Fig. 4.27 A multi-echelon serial system

For the sake of simplicity, consider the serial multi-echelon supply chain shown in Fig. 4.27 that moves a single type of product; as is custom in multi-echelon supply chain management, higher echelon numbers on the left indicate upstream suppliers, and the retailer facing external customer demand is echelon number 1 (the external end-customer may be denoted as echelon 0). Each echelon j in the chain works as follows: at discrete demand epochs $t_n = t_0 + hn$, $n = 0,1,\ldots$ (where $h > 0$ is an arbitrary constant) an order may arrive from the echelon's unique customer, echelon $j - 1$ (or the external customers if $j = 0$), and depending on the echelon's ordering policy an order may be placed on the echelon's unique supplier, echelon $j + 1$ (unless $j = N$, the most upstream supplier in the supply chain). When an order is placed by echelon j (to its supplier), the supplier $j + 1$ immediately sends an acknowledgment to j, including the highest item number that will be sent when the order size is fulfilled a lead-time later. This item number of course always increases with each good shipped from $j + 1$ by one. The so-called "Newell curve" $N_j(t)$ is a function that returns the acknowledgment number received at time t by echelon j, and represents the cumulative flow of goods that have passed through j by time $t + L_j$, where L_j is a deterministic lead-time that echelon $j + 1$ quotes (on the assumption that $j + 1$ is not out-of-stock so that the order will have to be back-logged and serviced later). $N_j(t)$ of course is equal to the order number received by echelon $j + 1$ at time t. The cumulative item number $E_j(t + L_j)$ *expected* at echelon j by time $t + L_j$ is then related to the Newell curves by the equation

$$N_j(t) = E_j\left(t + L_j\right), \quad j = 1, 2, \ldots$$

Finally, consider the function $S_j(t)$ returning the actual arrival of physical items at echelon j by time t. If items are not delayed (so that the lead-time L_j is respected by supplier $j + 1$) obviously we must have $S_j(t) = E_j(t)$, but in the general case, we must have

$$S_j(t) = \min\left\{E_j(t), S_{j+1}\left(t - P_{j+1}\right)\right\}, \quad j = 0, 1, \ldots, N - 1$$
$$S_N(t) = E_N(t)$$

where $P_j \geq L_{j-1}$is the so-called "processing time" that represents the sum of production/handling time at echelon $j + 1$ plus the standard lead-time L_j that $j + 1$ quotes when it is not out-of-stock. S_0 and E_0 are the curves for the external end-customer. The order size placed by echelon $j > 0$ at time $t_n = t_0 + hn$ can also be expressed via the Newell curve, since $Q_j(t_n) = N_j(t_{n+1}) - N_j(t_n)$, but of course, it is the inventory policy π_j followed by each echelon j that will define the function $Q_j(t_n)$. In an *autonomous supply chain*, where the echelons base their decision on "*local*" information only (i.e. their own initial state and the history of orders placed by their unique customer), the function Q_j will be a function of the form:

$$Q_j(t_{n+1}) = Y_j(\tilde{N}_{j,n})$$

where $\tilde{N}_{j,n} = \begin{bmatrix} N_j(t_n) & N_{j-1}(t_n) & N_{j-1}(t_{n-1}) & \cdots & N_{j-1}(t_{n-p}) \end{bmatrix}^T$ is a vector gathering together the information regarding order history of the echelon's customer for a number $p > 0$ of past periods, and the echelon's own current state.

A necessary and sufficient condition for the whole supply chain to operate *without any back-order episodes ever* (i.e. stock-outs at any echelon) is the following:

$$N_j(t + P_{j+1} + L_{j+1} - L_j) \leq N_{j+1}(t), \quad \forall j = 0, 1, \ldots, N-1, \forall t \geq t_0$$

The above inequality guarantees on-time deliveries and prevents stock-outs from occurring anywhere in the supply chain. It can be interpreted as follows: to avoid any stock-outs ever anywhere in the supply chain, the lag between order times at consecutive echelons must exceed a constant $M_j = P_j + L_j - L_{j\text{-}1}$, i.e. echelons should not "overwhelm" their immediate suppliers with orders faster than the suppliers can fulfill.

Of course, the above condition is only a condition for reliability regarding stock-outs and quoted lead-times, and does not relate to stability of the supply chain operation. An inventory policy π by which echelons in the supply chain decide how much and when to order from their upstream suppliers is called *stable in the small* if the deviations from a steady-state S across the whole chain can be bounded uniformly as tightly as desired by bounding the deviations in the input. A policy π is then defined to be *stable in the large* if the associated function $Q_j(t_n)$ is such that for every $\delta > 0$ no matter how large, there exists a $\delta' > 0$ such that $(\forall n \in \mathbb{N} : Q_0(t_n) \leq \delta) \Rightarrow (\forall n \in \mathbb{N}, j = 1, 2, \ldots, N-1 : Q_j(t_n) \leq \delta')$or in other words when the most downstream echelon 0 (i.e. the external end-customer) places bounded orders at all times, *all order sizes remain bounded throughout the supply chain at all times*. Such supply chains do not exhibit the bullwhip effect. It is not difficult to confirm that the (s, S) reorder point, order-up-to policies that are optimal in the contexts of inventory cost structures that we studied in the previous chapters are in fact stable in the large, and therefore supply chains using the (s, S, T) policy autonomously—without regard for demand patterns downstream—cannot suffer the bullwhip effect! However, they are *not* stable in the small. The same results hold true for the very common (R, T) or base-stock policy and the (r,Q) and (r, nQ, T) policies studied before.

4.4 Bibliography

Several books were focused or even devoted solely on inventory control since the beginning of the industrialization era. Early examples include Harris's book on "Operations and Cost" (Harris 1915). Some of the most influential books on the topic include Arrow et al. (1958); Hadley and Whitin (1963); Silver et al. (1998); Zipkin (2000); Porteus (2002); Axsater (2010) to name a few. Inventory control research is regularly published in a number of scholarly journals, including *Operations Research, Management Science, Naval Research Logistics Quarterly, Computers & Operations Research, Manufacturing & Service Operations Management, European Journal of Operational Research, International Journal of Production Research, International Journal of Production Economics, Production Planning & Control, IIE Transactions*. Research is also published in journals such as the *Journal of the Operational Research Society, Econometrica, IEEE Transactions on Systems, Man, and Cybernetics* (*Part A*), *IEEE Transactions on Automatic Control, IEEE Transactions on Automation Science & Engineering, Advances in Applied Probability* and others.

The seminal paper on the (R, T) policy should be considered that of Rao (2003). Research on the (r,nQ) policy is very extensive. For some select publications, see Morse (1959); Galliher et al. (1957); Veinott (1965); Naddor (1975); Roundy (1985); Zheng and Chen (1992) and more recently, Larsen and Kiesmuller (2007); Li and Sridharan (2008). The paper by Larsen and Kiesmuller is among the very few that explicitly considers the review interval T as a decision variable in the optimization process. Lagodimos et al. (2010a) is another paper that explicitly considers the review interval as a decision variable for the (r, nQ, T) policy parameter optimization process.

Lagodimos et al. (2010b) is the first paper as far as we know that deals with the effects of exogenous constraints on the base batch size Q_b in the (r, nQ, T) policy parameter optimization. Another paper dealing with such effects in (s, S) policy parameter optimization is Hill (2006).

Research on the (s, S, T) policy confined in its simpler periodic (s, S) form—originally introduced by Arrow et al. (1951) is very extensive. Scarf (1959) proved the optimality of the policy in a finite horizon setting, and later, Iglehart (1963) proved optimality of the policy in the infinite horizon case under fairly general conditions, which were weakened even further by Veinott (1966). Optimization algorithms for the restricted (s, S) periodic review policy are detailed in Veinott and Wagner (1965); Johnson (1968); Bell (1970); Stidham (1977); Federgruen and Zipkin (1985); Zheng and Federgruen (1991). The algorithms for the optimization of the (s, S, T) policy in all three parameters form part of the author's current research. See Lagodimos et al. (2011) for a first treaty of the topic of including the review interval as a decision variable in the optimization of the (s, S, T) policy. The heuristic algorithms for the optimization of the (s, S, T) policy are first presented in this text.

Finally, research on multi-echelon inventory systems starts with the seminal paper of Clark and Scarf (1960), and continues to this day (Axsater 1990; Diks and de Kok 1998; and Diks and de Kok 1999) are just a few highly-cited papers in this field. Even full books (Sherbrooke 2004) are devoted solely on the topic.

4.5 Exercises

1. A vendor of seasonal items wants to purchase a number of Christmas trees for the upcoming season. Based on previous years' records, she estimates demand for Christmas trees at her shop to be a normally-distributed random variable with mean $\mu = 250$ and standard deviation $\sigma = 50$. If the purchase price for a tree is $p_o = €30$ which she then sells at $p_s = €50$, assuming she cannot recover any value at all from any unsold tree, what is the quantity of trees she should purchase in order to maximize her expected profits?
2. Implement the "(*r*, *nQ*, *T*) Policy Optimization" algorithm of Sect. 4.2.3.2 for normally-distributed demands. Run the algorithm with the following parameters: $K_r = 0$, $K_o = 100$, $L = 1$, $\mu = 10$, $\sigma = 3$, $h = 1, p = 9$, for accuracies $\varepsilon_Q = 1$, $\varepsilon_T = 0.5$, 0.1, and 0.51. How do the resulting policy parameters and average cost per unit time change when the time-quantum changes?
3. The α-service measure for a single-echelon inventory system is defined as the probability that the system at any given point in time is not in stock-out, or equivalently, the fraction of time during which the system has on-hand physical inventory level greater than zero. Show that for an inventory system facing continuous demand process and using the (*r*, *nQ*, *T*) periodic review policy, for any given *Q* and *T*, the optimal re-order point $r_{Q,T}$ that minimizes the expected long-run average cost *C*(*r*, *Q*, *T*) is such that the α-service measure of the system operating with parameters $r_{Q,T}$, *Q* & *T* is equal to $\frac{p}{h+p}$, which is a news-boy condition applicable to the (*r*, *nQ*, *T*) inventory policy. Hint: derive the first order necessary conditions for optimality of the system.
4. Show that the same result as above is true for the (*R*, *T*) policy.
5. Verify that the (*R*, *T*) and (*s*, *S*, *T*) policies are stable in the large.
6. A store faces constant demand rate $D = 50$ items/day for a particular item. If the holding cost rate is $h = 8\%$ and the purchase price of the item is $p_o = €10$ whereas the fixed setup cost of placing an order is $K = €150$ then what is the optimal order quantity Q^* that minimizes total inventory and fixed setup costs per unit time?
7. Would the answer to question no. 6 change if the supplier offered a discount of 20% for any item purchased in addition to an order of 300 items?
8. Modify the algorithm implemented in exercise no. 2 so as to optimize the (*r*, *nQ*, *T*) policy when an externally provided batch size $Q_b = 24$ is provided for a system with parameters $K_r = 1$, $K_o = 64$, $L = 0$, $h = 1, p = 10$, facing

normally-distributed demand in any interval of length T with mean $\mu_T = 50T$ and $\sigma T = 50\sqrt{T}$.

9. Design and implement a heuristic algorithm for optimal ordering coordination of a large group of items having different fixed order setup costs K_i, demand rates D_i and holding costs H_i (see discussion in Sect. 4.1.1.4). Assume that for any cluster of items that is coordinated so that all replenishment orders of items within the cluster, the single fixed order setup cost for the cluster is the maximum of the individual fixed order costs of the items in the cluster.

   (a) Compare your algorithm's performance with that of uncoordinated optimization of the EOQ of each individual item.
   (b) Under what circumstances would it be better to still prefer to individually optimize each item's order releases?

10. Show that the ordering probability $P_o(T)$ in a single-echelon inventory system facing Poisson demand with mean λT in an interval of length T that is controlled through an (R, T) policy is convex in T.
    Hint: In an inventory system governed by the (R, T) policy, $P_o(T)$ is the probability that demand in an interval of length T is strictly positive.

References

Abdel-Malek L, Montanari R (2005) An analysis of the multi-product newsboy problem with a budget constraint. Int J Prod Econ 97:296–307

Apostol TM (1981) Mathematical analysis, 2nd edn. Addison-Wesley, Reading

Arrow KJ, Harris T, Marschak J (1951) Optimal inventory policy. Econometrica 19:250–272

Arrow KJ, Karlin S, Scarf H (1958) Studies in the mathematical theory of inventory and production. Stanford University Press, Stanford

Axsater S (1990) Simple solution procedure for a class of two-echelon inventory problems. Oper Res 38(1):64–69

Axsater S (2010) Inventory control. Springer, NY

Bell CE (1970) Improved algorithms for inventory and replacement stock problems. SIAM J Appl Math 18(3):682–687

Browne S, Zipkin P (1991) Inventory models with continuous stochastic demands. Ann Appl Probab 1(3):419–435

Chang Y-L, Desai K (2003) WinQSB: decision support software for MS/OM. Wiley, Hoboken

Clark A, Scarf H (1960) Optimal policies for a multi-echelon inventory problem. Manag Sci 6(4):475–490

Daganzo C (2003) A theory of supply chains. Springer, Berlin

Diks EB, de Kok AG (1998) Optimal control of a divergent multi-echelon inventory system. Eur J Oper Res 111(1):75–97

Diks EB, de Kok AG (1999) Computational results for the control of a divergent N-echelon inventory system. Int J Prod Econ 59(1–3):327–336

Federgruen A, Zipkin P (1984) Computational issues in an infinite horizon multi-echelon inventory model. Oper Res 32(4):818–836

Federgruen A, Zipkin P (1985) Computing optimal (s, S) policies in inventory systems with continuous demands. Adv Appl Probab 17:421–442

Forrester JW (1961) Industrial dynamics. Pegasus Communications Publishing Company, Waltham
Gallego G (2004) Lecture notes on production management. Department of Industrial Engineering & Operations Research, Columbia University
Galliher HP, Morse PM, Simond M (1957) Dynamics of two classes of continuous review inventory systems. Operat Res 7(3):362–384
Hadley G, Whitin TM (1961) A family of inventory models. Manag Sci 7(4):351–371
Hadley G, Whitin TM (1963) Analysis of inventory systems. Prentice-Hall, Englewood-Cliffs
Harris FW (1913) How many parts to make at once. Factory: the Magazine of Management, 10(2):135–136,152
Harris FW (1915) Operations and cost (factory management series). A.W.Shaw, Chicago
Hill RM (2006) Inventory control with indivisible units of stock transfer. Eur J Oper Res 175:593-601
Hopp W, Spearman M (2008) Factory physics, 3rd edn. McGraw-Hill/Irwin, NY
Iglehart DL (1963) Optimality of (s, S) policies in the infinite horizon dynamic inventory problem. Manag Sci 9(2):259–267
Johnson E (1968) On (s, S) policies. Manag Sci 15(1):80–101
Lagodimos AG, Christou IT, Skouri K (2010a) Optimal (r, nQ, T) inventory control policies under stationary demand. Int J Syst Sci (to appear)
Lagodimos AG, Christou IT, Skouri K (2010b) Optimal (r, nQ, T) batch ordering with quantized supplies. Comput Oper Res (to appear)
Lagodimos AG, Christou IT, Skouri K (2011) A simple procedure to compute optimal (s,S,T) inventory policy. In: Proceedings of the international conference on challenges in statistics and operations research, Kuwait, 8–10 March 2011
Larsen C, Kiesmuller GP (2007) Developing a closed-form cost expression for a (R, s, nQ) policy where the demand process is compound generalized Erlang. Oper Res Lett 35(5):567–572
Lau HS, Lau AHL (1996) The newsstand problem: a capacitated multi-product single-period inventory problem. Eur J Oper Res 94:29–42
Li X, Sridharan V (2008) Characterizing order processes of using (R, nQ) inventory policies in supply chains. Omega 36(6):1096–1104
Morse PM (1959) Solutions of a class of discrete time inventory problems. Oper Res 7(1):67–78
Naddor E (1975) Optimal and heuristic decisions in single- and multi-item inventory systems. Manag Sci 24:1766–1768
Porteus E (2002) Foundations of stochastic inventory theory. Stanford Business Books, Stanford
Rao U (2003) Properties of the (R, T) periodic review inventory control policy for stationary stochastic demand. Manuf Serv Oper Manag 5(1):37–53
Reis DA (1982) Comparison of periodic review operating doctrines: a simulation study. In: Proceedings of the 1982 winter simulation conference
Rockafellar RT (1970) Convex analysis. Princeton University Press, Princeton
Rosling K (1989) Optimal inventory policies for assembly systems under random demands. Oper Res 37(4):565–579
Roundy R (1985) 98%-Effective integer-ratio lot sizing for one-warehouse multi-retailer systems. Manag Sci 31(11):1416–1430
Scarf H (1959) The optimality of (S,s) policies in dynamic inventory problems. Technical report no. 11, Applied Mathematics & Statistics Laboratory, Stanford University
Serfozo R, Stidham S (1978) Semi-stationary clearing processes. Stoch Process Appl 6(2):165–178
Sherbrooke CC (2004) Optimal inventory modeling of systems: multi-echelon techniques, 2nd edn. Kluwer, NY
Silver EA, Robb DJ (2008) Some insights regarding the optimal reorder period in periodic review inventory systems. Int J Prod Econ 112(1):354–366
Silver EA, Pyke DF, Peterson R (1998) Inventory management and production planning and scheduling. Wiley, Hoboken
Stidham S Jr (1977) Cost models for stochastic clearing systems. Oper Res 25(1):100–127

Stidham S Jr (1986) Clearing systems and (s, S) systems with nonlinear costs and positive lead times. Oper Res 34(2):276–280

Veinott AF Jr (1965) The optimal inventory policy for batch ordering. Oper Res 13(3):424–432

Veinott AF Jr (1966) On the optimality of (s, S) inventory policies: new conditions and a new proof. SIAM J Appl Math 14(5):1067–1083

Veinott AF Jr, Wagner HM (1965) Computing optimal (s, S) inventory policies. Manag Sci 11(5):525–552

Zhang B, Du S (2010) Multi-product newsboy problem with limited capacity and outsourcing. Eur J Oper Res 202:107–113

Zheng Y-S (1992) On properties of stochastic inventory systems. Manag Sci 38(1):87–103

Zheng Y-S, Chen F (1992) Inventory policies with quantized ordering. Nav Res Logist Q 39:285–305

Zheng Y-S, Federgruen A (1991) Finding optimal (s, S) policies is about as simple as evaluating a single (s, S) policy. Oper Res 39(4):654–665

Zipkin P (1986) Inventory service level measures: convexity and approximations. Manag Sci 32(8):975–981

Zipkin P (2000) Foundations of inventory management. McGraw-Hill, NY

Chapter 5
Location Theory and Distribution Management

This chapter deals with modeling problems related to both strategic-level decisions about the location for the fixed assets of a company such as plants, warehouses, retail stores, etc., as well as with modeling tactical and operational problems such as optimizing transportation costs a company is likely to face, especially in the manufacturing sector. Several algorithms are discussed for the models presented in this chapter that combines location with transportation issues precisely because one affects the other in very significant ways.

5.1 Location Models and Algorithms

Location theory is concerned with the optimal selection of a location or several locations to place some facility on that location so that certain costs associated with this choice are minimized.

Within the context of traditional operations management, a number of methods have been developed to guide a decision maker into making near-optimal decisions regarding the location of a facility. Depending on whether the location problems concern a service or a manufacturing facility drastically different methodologies and related algorithms are put forth. If the location problem concerns the best location for a manufacturing plant or warehouse, the major costs to be considered involve land purchasing costs and transportation costs of raw materials to the factory, and transportation costs of finished goods to the markets to be served by the plant. Therefore, such problems are often called *location/allocation problem*, because they deal with the problem of *placing* facilities on optimal locations and at the same time *allocate* demand from stores/markets to the located plants or warehouses.

On the other hand, for a service organization, the major factors that should be taken into account when considering the best location for a facility involve land/office purchase/leasing costs, location visibility so it can attract customers, and implicitly the distance between the location and its potential customers. Because

I. T. Christou, *Quantitative Methods in Supply Chain Management*,
DOI: 10.1007/978-0-85729-766-2_5, © Springer-Verlag London Limited 2012

location decisions for a service facility usually involve a very high degree of fuzziness and are hard to quantify and express in numbers, we shall only be concerned with location problems for manufacturing facilities where the major costs involve fixed setup and transportation costs measured as the distance between the location and its assigned markets.

As an introduction to the concepts that will follow, consider the following *continuous* problem: given a set S of n markets located in a d-dimensional norm vector space (normally it must suffice to set d to 2 or 3) in positions $s_{[1]}, s_{[2]}, \ldots, s_{[n]}$ the objective is to find the location of k facilities $c_{[j]}$ $j = 1, \ldots, k$ in the same d-dimensional space so as to minimize the sum of the square distances between each point $s_{[i]}$ and its closest facility $c_{[j]}$. This problem is known as the minimum sum of squares clustering problem (MSSC), because it is equivalent to grouping (clustering) the data points $s_{[i]}$ together in k clusters so that the sum of the squares of the distance between each data point and the center of the group to which it belongs is minimized, where of course, the center of a group of points G is the point $c_G = \sum_{j \in G} s_{[j]} / |G|$. The problem is continuous and unconstrained in the sense that there is no constraint on where to place each of the k facilities in space. Nevertheless, despite this seeming simplicity, the problem belongs to the class of NP-hard problems (Aloise et al. 2009), and therefore most likely there does not exist any polynomial-time algorithm that guarantees its solution. This is because, after all, the problem essentially asks to determine the optimal grouping of markets together in a minimum square distances from their center's criterion.

An early algorithm for MSSC that in fact became so prevalent that very often the problem is named after the algorithm, is the famous K-Means algorithm, and is an example of a randomized heuristic that has found an extremely wide range of applications in pattern recognition, artificial intelligence, computational intelligence and business intelligence. The algorithm starts by choosing at random k places $c_{[j]} j = 1, \ldots, k$ for the initial location of the facilities (often, they are randomly chosen from the set of points S). Then, it proceeds by iterating two steps until some convergence criterion is satisfied: in the first step, it assigns each of the n markets $s_{[i]}$ to its nearest facility $c_{[j]}$; then, in the second step, once it has assigned each market to a facility, it relocates the position of each facility to the center of the markets assigned to it. The usual convergence criterion is that in the first step of an iteration, no market "switches" assigned facility. The two steps described above are "suggested" from the fact that according to the first order necessary conditions for mathematical programming (see Chap. 1), given a fixed set of facility locations, each market must be served by the facility nearest to it, and similarly, given a set of points, the location that minimizes the sum of squares of distances of each point to that location is the center of gravity of these points. The K-Means algorithm is described in pseudo-code as follows:

K-Means Algorithm

Inputs: finite data set $S = \{s_{[1]}, \ldots, s_{[n]}\}$ of d-dimensional data points, representing the location of n markets, and k, the number of k facilities to be placed in

d-dimensional space so as to minimize the sum of the square distances between the markets and their nearest facility, and a convergence criterion for stopping the computations.
Outputs: k d-dimensional vectors representing the locations of the k facilities.

Begin

1. for $i = 1$ to k do

   a. Set the position of the ith facility $c_{[i]}$ to the location s of a market randomly chosen from S ensuring that no two facilities are placed on the same market location.

2. end-for
3. Create new array $a[1, \ldots, n]$ of integers, Create new array $A[1, \ldots, k]$ of sets of integers.
4. while convergence-criteria are not satisfied do

   a. for $i = 1$ to k do

      i. Set $A_i = \{\}$.

   b. end-for
   c. for $i = 1$ to n do

      i. Set $a[i] = \arg\min_{j=1,\ldots,k} \left\| c_{[j]} - s_{[i]} \right\|$
      ii. Set $A_{a[i]} = A_{a[i]} \cup \{i\}$.

   d. end-for
   e. for i=1 to k do

      i. if $|A_i| > 0$ then Set $c_{[i]} = \frac{1}{|A_i|} \sum_{j \in A_i} s_{[j]}$else Set $c_{[i]} =$ nil.

   f. end-for

5. end-while
6. return $\{c_1, \ldots, c_k\}$.

End

Note that it is possible that a run of the algorithm will return fewer than k points when essentially two or more clusters of a previous iteration merge in one cluster—a result which is always sub-optimal in terms of the objective function to minimize. The quality of results of the K-Means algorithm depends heavily on the initial assignment of facilities in the d-dimensional space. For this reason, the algorithm is usually run many times with different random initial placements of facilities and the best result is returned as the final answer. It has been recently demonstrated (Christou 2011) that a few runs of the K-Means algorithm can be combined in a set covering/set partitioning formulation of the MSSC problem to yield superior solutions that are usually optimal or very close to optimal (Aloise et al. 2010), at least when d is low-dimensional. As we shall see, algorithms for other related location problems, such as the p-median problem, have been the source of inspiration for algorithms for solving the MSSC problem as well, hence there is a strong link between grouping problems, corresponding models, and algorithms for their solution.

When there are constraints that must be obeyed when deciding the location of choice for, say, a facility on a map, the problem becomes known as a discrete location problem. An obvious example for a discrete location problem would be to choose between two existing land plots that an organization already owns where to place a new plant so as to expand its capacity. In the following sections, we shall be concerned with three major types of discrete location problems.

1. The p-median problem,
2. The uncapacitated facility location problem an extension of the p-median to include setup costs,
3. The capacitated facility location problem that explicitly considers the possibility of plants to reach a limit on their capacity (described in Sect. 1.2.1), and
4. The p-center problem, with applications in the public sector.

Location problems when spatial interactions are considered can be formulated as quadratic assignment problems (QAP) which are a direct extension of the linear assignment problem (LAP) studied in Chap. 1, but whose computational complexity is much harder than that of the LAP. Today, there are no good variants of any Branch-and-Bound style exact method for solving the QAP for medium-sized problems (Pardalos and Pitsoulis 2000).

5.1.1 The p-Median Problem

A fundamental model in location theory, the p-median problem is a *discrete* location problem that can be stated as follows (Kariv and Hakimi 1979; Beasley 1985; Hansen and Mladenovic 1997): consider a set of L potential locations for p facilities and a set U of locations of given markets. The p-median problem is to locate simultaneously the p facilities at locations chosen from the set L so as to minimize the transportation cost for satisfying the demand of each market, each supplied exclusively from a single facility, where each facility having *infinite capacity, may accommodate any number of markets*. The latter assumption differentiates the p-median problem from the general facility location problem formulated in Sect. 1.2.1. The p-median problem—not surprisingly—also belongs to the class of NP-hard problems (Kariv and Hakimi 1979). As a practical example, the problem obviously models the decision problem an executive would face that would have to place $p > 0$ warehouses to p of several land plots the company owns so as to the minimize long-term costs associated with transportation of goods from the warehouses to the markets the company serves. Because the company already owns all of the plots from which to choose, there are no fixed costs associated with opening a warehouse at any plot, because there are no purchase costs to consider, and it is assumed that the cost of building a warehouse is the same for any of the locations the company owns.

Formulating the p-median problem presents no modeling difficulties. Consider a finite set L of m locations, a set U of n markets, and an $n \times m$ matrix D with the distances traveled (or costs incurred) for satisfying the demand of the ith market from a facility located at the jth location in L, $\forall i \in U,\ j \in L$. The objective is to minimize the sum of these distances. The corresponding optimization problem is then formulated as

$$(p\text{-median}) \min_{x,y} c(x) = \sum_{i \in U} \sum_{j \in L} d_{ij} x_{ij}$$

$$\text{s.t.} \begin{cases} \sum_{j \in L} x_{ij} = 1, & \forall i \in U \\ \sum_{j \in L} y_j = p & \\ x_{ij} \leq y_j, & \forall i \in U, j \in L \\ x_{ij}, y_j \in B = \{0,1\}, & \forall i \in U, j \in L \end{cases}$$

The case of some locations from L not being reachable from some markets in U can be easily handled by setting the corresponding entry d_{ij} in the D matrix equal to infinity, or some very large number. The binary decision variables y_j are set to 1 if a facility should open at location j in L, and zero otherwise. The x_{ij} variables are also binary and correspond to the decision to transfer goods from the facility located at j to the market i (recall that a market will be wholly served by its nearest facility). Note that in a better formulation, the binary variables x_{ij} will be replaced by continuous variables $0 \leq x_{ij} \leq 1$ without any modeling accuracy loss. The continuous variables now represent the "fraction" of demand for market i that is served from facility at j. However, because of the nature of the objective function, and because of the fact that there is no capacity quota on any of the facilities, even though an optimal solution of the new problem may contain fractional solution values for some x_{ij}, one may obtain another optimal solution for both problems by simply setting for each i, $x_{ij} = 0$ for all j such that $x_{ij} < 1$ except one j' such that $x_{ij'} > 0$, and set $x_{ij'} = 1$ (any such j' will do). Such a formulation is better because it reduces the number of discrete variables by $|U|, |L|$, leaving only $|L|$ binary variables in the problem.

The constraints $x_{ij} \leq y_j$ guarantee that a market will not be served by a location unless that location j also has an open facility. Other than that, the first constraint guarantees that each market will be served by exactly one facility—which by the optimization criterion, will be the open facility closest to it, and the second constraint guarantees that exactly p facilities will open.

During the last 50 years, a number of fast "greedy" in nature heuristics were developed for the p-median problem. The *standard greedy algorithm* (Cornuejols et al. 1977) "opens" one facility at a time, choosing the best available location each time, given the previous assignments that have been made, where "best location" is the location that reduces the objective function the most. A greedy randomized adaptive search procedure (GRASP) algorithm for the p-median problem would utilize a *randomized greedy algorithm* that would behave exactly as the standard greedy algorithm except that in each iteration, rather than selecting the best available location for placing the next facility from set A of all currently available locations, it would randomly select from a subset B of A, of size $s = \lceil a|A| \rceil$, (where $a \in (0, 1]$ is a user-defined parameter) that comprises the s best locations. For $a \approx 0$, the randomized greedy algorithm clearly degenerates to the standard greedy algorithm, whereas for $a = 1$, it degenerates to a "purely random" assignment algorithm that simply selects p locations from L at random. Yet another variant of the randomized greedy algorithm called "*random plus greedy heuristic*" would not place in set B the s best facilities and select at random from B, but rather would select at random s of the available facilities, and then choose the best one from set B. Experiments with the above heuristics show that whereas the quality of the solutions they produce is comparable, the *random plus greedy heuristic* is much faster on average than the other two.

As the size of a p-median problem in terms of the numbers $|U|$, $|L|$, and p gets larger, many early developed heuristics (such as the ones described above) experience significant deterioration in their performance. The primary reason has to do with a property of related clustering problems known as "central limit catastrophe", which states that as the size of such problems becomes larger, there are exponentially many local minima that act as "traps" for any local search algorithm that attempts to find a solution to the problem. Correspondingly, early exact methods for the problem experience exponential increase in their running times. Several heuristic (as well as exact, Branch and Bound based) methods have been proposed in the literature lately, that do not suffer—at least to such a degree so as to be unusable—from the problems mentioned earlier. We shall describe a number of such highly successful algorithms which introduce search concepts that find applications in other areas as well.

5.1.1.1 The Alternate Algorithm

One heuristic with striking resemblance to the K-Means algorithm described above is the *Alternate* algorithm described below:

Alternate Algorithm for the p-Median Problem

Inputs: Finite index set $L = \{1, \ldots, L\}$ of locations, finite index set of markets $U = \{1, \ldots, U\}$, distance matrix D describing the distance—or cost—d_{ij} between market i and location j, number of facilities p to open, and a convergence criterion for stopping the computations.

Outputs: finite set F of indices j in L representing the locations of the p facilities.
Begin

1. Set $F = \{\}$.
2. for $i = 1$ to p do
   a. Set the position of the ith facility to a location s randomly chosen from L ensuring that no two facilities are placed on the same market location, and add the position index to the set F.
3. end-for
4. Create new array $a[1, \ldots, L]$ of integers, Create new array $A[1, \ldots, p]$ of sets of integers.
5. while convergence-criteria are not satisfied do
   a. for $i = 1$ to p do
      i. Set $A_i = \{\}$.
   b. end-for
   c. for $i = 1$ to U do
      i. Set $a[i] = \arg\min_{j \in F}\{d_{ij}\}$.
      ii. Set $A_{a[i]} = A_{a[i]} \cup \{i\}$.
   d. end-for
   e. if convergence criteria are satisfied then break.
   f. Set $F = \{\}$.
   g. for $i = 1$ to p do
      i. Set $c_i = \arg\min_{k \in L}\left\{\sum_{j \in A_i} d_{jk}\right\}$.
      ii. Set $F = F \cup \{c_i\}$.
   h. end-for
6. end-while
7. return F.

End

It should be easy for the reader to understand the similarities between the two algorithms: each algorithm first determines at random some initial locations for the placement of the facilities, and then enters a loop that iterates two sequential steps: the first step determines the group of markets that each facility should serve, and the second decides for each group the optimal location of the facility that will best serve this given group. The loop usually stops when there is no improvement in the objective function in the last iteration. It should come as no surprise that a single run of the algorithm may also return a set F of fewer than p locations, and that several runs of the algorithm starting with different random placements of facilities in various locations may be needed to obtain a good-quality solution.

The computational complexity of a *single iteration of the Alternate algorithm*, in case the second step of the loop (steps 5.f–5.h) is implemented in a naïve way is $O((|U| + |L||U|)p)$. It can be significantly reduced however, with the help of appropriate bookkeeping data structures (Whitaker 1983).

5.1.1.2 Variable Neighborhood Search-Based Algorithm

A more recent heuristic, based on the concept of variable neighborhood search could clearly outperform the Alternate algorithm in many test data sets (Hansen and Mladenovic 1997).

Variable neighborhood search (VNS) is a meta-heuristic—as are the algorithms presented in Sect. 1.1.1.3 for unconstrained global optimization—for solving combinatorial optimization problems. The basic idea is to proceed in a systematic change of neighborhood within a local search algorithm, so as to strike a good balance between the two competing themes of exploitation versus exploration of the search space. The theme of "exploring the search space" is accomplished in two ways within the context of VNS:

1. Small neighborhoods—close to the current solution—are systematically searched until a solution better than the incumbent is found
2. Large neighborhoods—far from the current solution—are explored partially, in a "probing" manner so that a new solution is selected at random and then, a systematic local search within the (variable) neighborhood of that new solution is performed.

The algorithm remains at the same solution until a new incumbent solution is found, at which point it jumps to the new incumbent solution point. Neighborhoods are usually prioritized in such a way that neighborhoods increasingly far from the current one are explored, so that a process of exploitation of the current best solution and its surrounding space is followed by an exploration process where new areas are probed for promising regions of high-quality solutions.

Therefore, the VNS meta-heuristic resembles the simulated annealing algorithm constituting a "shaking process" where the degree of shaking is carefully controlled, exactly as the SA algorithm also does. The trade-off between these two processes is controlled through a few user-defined parameters.

The method requires a distance function between any two solutions x_1 and x_2 represented by the index *sets* of locations from L where facilities are to be placed. The distance between x_1 and x_2 is defined simply as $d(x_1, x_2) = |x_1 - x_2|$ and it is very easy to verify that $d(.,.)$ is a metric function in the space of all elements of 2^U that have exactly p points. The metric defines the notion of neighborhood as follows: given a parameter k, the neighborhood $N_k(x)$ of a solution x is the set of all subsets y of L containing exactly p elements such that $d(x, y) = k$. The algorithm in pseudo-code is as follows:

VNS Algorithm for the p-Median Problem

Inputs: Finite index set $L = \{1, \ldots, L\}$ of locations, finite index set of markets $U = \{1, \ldots, U\}$, distance matrix D describing the distance—or cost—d_{ij} between market i and location j, number of facilities p to open, a base algorithm B for computing an initial heuristic solution to the p-median problem, user-defined parameter $k_{\max}$.
Outputs: finite set F of indices j representing the locations of the p facilities.

Begin

1. Set $x = x^* = B()$.
2. Set $k = 1$.
3. while $k \leq k_{\max}$ do

   a. for $m = 1$ to k do

      i. Set a variable gin to a randomly selected location from $L - x$.

      ii. Set $gout = \arg\min_{j \in x} \left\{ \sum_{u \in U} \min_{i \in x \cup \{gin\} - \{j\}} d_{ui} \right\}$

      iii. Set $x = x \cup \{gin\} - \{gout\}$.

   b. end-for
   c. Set $f = c(x), f* = c(x*)$.
   d. repeat

      i. Set $D = c(x)$.
      ii. for each j in x do

         1. Find the best position l from $L - x$ to move facility located at j into.
         2. Set $\delta = c(x\text{U}\{l\} - \{j\}) - c(x)$.
         3. if $\delta \leq 0$ then Set $x = x\text{U}\{l\} - \{j\}, f = c(x)$.

      iii. end-for
      iv. Set $\Delta = c(x) - D$.

   e. until $\Delta = 0$.
   f. if $f < f^*$ then

      i. Set $f^* = f$.
      ii. Set $x^* = x$.
      iii. Set $k = 1$.

   g. else

      i. Set $f = f^*$.
      ii. Set $x = x^*$.
      iii. Set $k = k+1$.

   h. end-if

4. end-while
5. Set $F = x^*$.
6. return F.

End

5.1.1.3 Combining GA and Path Re-linking

More recently, a hybrid heuristic combining GA ideas, Tabu search, and path re-linking has been proposed for the p-median problem (Resende and Werneck 2004), and later extended for the more general uncapacitated facility location problem; this heuristic has provided impressive results for both location problems, and for this reason we present it in this section.

The algorithm works by maintaining a pool of solutions—as do genetic algorithms—and at each iteration a solution is created using some randomized heuristic algorithm that is subsequently refined using some local search procedure. The refined solution competes, based on their solution quality, with the pool of elite solutions maintained throughout the iterations for selection for recombination. The refined solution and the selected solution are then combined via a mechanism called Path Re-linking which we describe next, in order to produce a new better solution which will be candidate for entering the elite pool, the criterion for entering the elite pool always being high solution quality. After a pre-set number of such major iterations, the final elite pool may be post optimized via some local search procedure and the best solution in terms of low objective function value is returned.

The path-relinking method accepts as input two solutions x_1 and x_2 different from each other, and returns a set of new solutions that can be considered as "intermediate" points on a path from x_1 to x_2. In particular, path-relinking creates an ordering of set $D = x_2 - x_1$ and in $|D|$ consecutive iterations, it opens a facility in the next location dictated by the ordering of D and closes a facility from the set $x_1 - x_2$ so as to minimize the cost of the objective function. After the $|D|$ consecutive iterations, solution x_1 has been transformed into solution x_2. The $|D|$ intermediate solutions are returned.

In pseudo-code, the algorithm is as follows:

Hybrid Algorithm for the p-Median Problem

Inputs: Finite index set $L = \{1, \ldots, L\}$ of locations, finite index set of markets $U = \{1, \ldots, U\}$, distance matrix D describing the distance—or cost—d_{ij} between market i and location j, number of facilities p to open, a set of base algorithms B for computing heuristic solutions of the p-median problem, a local search procedure localSearch():, a selection procedure select(): $2^{2\wedge U} \times 2^U \to 2^U$, a path re-linking method relink(): $2^U \times 2^U \to 2^U$, a method add(): $2^{2\wedge U} \times 2^U \to 2^{2\wedge U}$ for adding solutions to a pool of solutions, a post-optimization procedure postOptimize(): $2^{2\wedge U} \to 2^U$, an integer parameter $esize > 0$, and an integer parameter $k_{\max} > 0$
Outputs: finite set s of indices j representing the locations of the p facilities.

Begin

1. Set $S = \{\}$, *elite* = {}
2. for $i = 1$ to *esize* do

   a. Apply the random plus greedy heuristic method to produce a solution s.
   b. Set *elite* = *elite* U $\{s\}$.

3. end-for
4. for $i = 1$ to $k_{\max}$ do
   a. Apply the random plus greedy heuristic method to produce a solution s.
   b. Set s = localSearch(s).
   c. Set s' = select(*elite*, s).
   d. if ($s' \neq$ nil) then
      i. Set s' = relink(s,s').
      ii. Set *elite* = add(*elite*, s').
   e. end-if
   f. Set *elite* = add(*elite*, s).
5. end-for
6. Set s = postOptimize(*elite*).
7. return s.

End

Resende and Werneck (2004) reported excellent quality results with the following implementations for the input procedures required by algorithm Hybrid:

- The localSearch() procedure implements a fast interchange heuristic that proceeds in iterations, and within each iteration, the "*best move*" for moving an already open facility from a location l_1 to another yet unused location l_2 in L is determined. If this best move reduces the objective function value, it is actually made, and a new iteration begins, otherwise the iterations stop and the current solution is returned (see Resende and Werneck 2003).
- The procedure select(e, s) selects a solution e from the elite pool with probability proportional to the cardinality of the set $e-s$, known as the *symmetric difference between e and s*. The reason is that it would be more likely to be profitable for the path re-linking procedure to be applied between two solutions that are as dissimilar as possible.
- The procedure add(e, s) adds to the pool e the solution s if the pool has not reached its maximum size *esize* yet. Otherwise, if the pool is full, the solution s is added only if it is better than the currently worse solution in the pool, in which case, the worse solution is removed from the pool.
- Finally, the process postOptimize(e) performs path re-linking between any possible pair of solutions in the elite pool e, and the best solution found after all *esize**(*esize* − 1)/2 path re-linking calls have been executed is returned as the output of the Hybrid algorithm.

5.1.1.4 Optimal Combination of Heuristic Base Algorithms

Note that similar to MSSC, the p-median problem is also a clustering (grouping) problem that essentially asks for the optimal partitioning of a set of markets into p groups that each will be served from the same facility to open in one of the L available locations. In fact, both problems belong to a generic class of problems defined in Christou (2011) as intra-cluster criterion-based clustering problems (IC^3). A clustering problem is said to belong in the (IC^3) class if and only if the problem is to find a partition C of a finite set S of n data-points into a predetermined number p of disjoint partition blocks $C_1, C_2, \ldots, C_p$ such that $C_i \cap C_j = \emptyset \; \forall i \neq j,\; \cup_{i=1}^{p} C_i = S$ so as to minimize a cost function of the form $c(C) = \sum_{i=1}^{p} c(C_i)$. The form of the cost function indicates that the clustering objective can be decomposed in the costs of each cluster and the cost of each cluster C_i depends only on the cluster itself and is otherwise independent of how the rest of the data-points in $S - C_i$ are partitioned into blocks. It is easy to justify why the p-median problem belongs to the (IC^3) class. The reason is that for any allocation/assignment of the p facilities into p of the available facility locations in L, say $l_1, l_2, \ldots, l_p$, the total cost of the assignment is the sum of the costs of each market to its nearest facility, which can be written as $\sum_{i=1}^{p} c(C_i)$, where $C_i = \left\{ i \in U | l_i = \arg\min_{j \in L} d_{ij} \right\}$ and $c(C_i) = \sum_{k \in C_i} d_{kl_i}$—with the standard convention that $c(C_i)$ is zero if C_i is the empty set. Also note that the following *monotone clustering property* (MCP) holds: $C_i \subseteq C_j \Rightarrow c(C_i) \leq c(C_j)$. To see why, consider without loss of generality a non-empty cluster C, and a cluster $C' = C \cup \{u\},\; u \in U - C$. Let l in L denote the best location for placing a facility to serve the markets in C, and let $l' \in L$ denote the best location among the locations in L for placing a facility to serve the markets in C'. Clearly, it must hold that $c(C) = \sum_{i \in C} d_{il} \leq \sum_{i \in C} d_{il'} \leq \sum_{i \in C} d_{il'} + d_{ul'} = \sum_{i \in C'} d_{il'} = c(C')$ which proves the MCP for the p-median problem.

Now, besides the standard MIP formulation of the p-median problem given by the (*p-median*) model, we may also view any clustering problem as a very large set partitioning problem (see Chap. 1), possibly with additional side constraints. In particular, the whole class of (IC^3) problems can be modeled as follows:

$$(IC^3) \quad \min_x \sum_{i=1}^{N} c_i x_i$$

$$\text{s.t.} \begin{cases} Ax = e \\ x^T e = p \\ x \in \mathbb{B}^N \end{cases}$$

where $A = [a_1, \ldots, a_N]$ is a matrix of dimensions $|U| \times N$ with $N = \sum_{i=1}^{|U|} \binom{|U|}{i} = 2^{|U|} - 1$ being the total number of subsets of the set U; the columns of A, a_i are the indicator vectors of each partition of the data set U, and $c_i = c(a_i)$ is the cost corresponding to forcing all markets in a_i to be served from the same—optimally chosen from L—facility. The vector x contains N binary decision variables, the ith component indicating whether the ith subset of all subsets of set U will be included in the solution (and of course, e is a column vector of N ones, and $B = \{0,1\}$). The constraints $Ax = e$ require that each market u in U will belong to *exactly one cluster*, and be served by the facility that will be opened for that cluster only, and the constraint $x^T e = p$ requires that exactly p facilities are placed among the locations in L.

Clearly, attempting to solve the (IC^3) problem as-is, is hopeless, since the number of columns of matrix A renders its size intractable even for modest values for the size of set U. But it is possible to use this formulation in order to simply optimally combine the results of some heuristic solutions, such as those produced by several runs of the Alternate algorithm or some other fast heuristic algorithm. Indeed, rather than including all possible subsets of set U in the columns of matrix A, we could only include in a restricted matrix A_B the subsets produced by a number of runs of some heuristic algorithm, and then use any algorithm capable of solving set partitioning problems with side constraints to solve the resulting restricted set partitioning problem. In fact, instead of solving the restricted set partitioning problem, we may solve the corresponding restricted set covering problem with side-constraint that follows:

$$(\mathrm{SCP}_R) \quad \min_x \quad \sum_{i=1}^{q} c([A_B]_i)x_i$$

$$\text{s.t.} \begin{cases} A_B x \geq e \\ \sum_{i=1}^{q} x_i = k \\ x_i \in \mathrm{B} \quad i = 1 \ldots q \end{cases}$$

Matrix A_B is a sub-matrix of A, and has only q columns, each corresponding to one subset of U produced by one of the runs of a base-heuristic algorithm. If the base algorithm was run n times, then $q \leq np$, where the inequality is due to the fact that the same subset of markets in U may be produced by more than one run of the base algorithm. Solving the (SCP_R) problem is of course much easier than solving the full problem (IC^3), but on the other hand, the optimal solution x^* of an (SCP_R) problem instance has no guarantee of being the optimal solution to (IC^3); in fact, there is no guarantee that x^* will be a feasible solution to (IC^3) either, because x^* may contain clusters that overlap. But in such a case, it is easy to convert x^* into a solution x^{**} that is a feasible solution to (IC^3), and in fact, is at least as good a solution as solution $\hat{x}$ of the restricted partitioning problem below:

$$(\mathrm{SPP}_R) \quad \min_x \quad \sum_{i=1}^{q} c([A_B]_i)x_i$$

$$\text{s.t.} \begin{cases} A_B x = e \\ \sum_{i=1}^{q} x_i = k \\ x_i \in \mathrm{B} \quad i = 1, \ldots, q \end{cases}$$

Indeed, to produce solution x^{**} consider each market u that appears in more than one cluster in solution x^*, and for each cluster C containing the market u, create a new cluster C' that contains all markets in C except u (this new cluster C' is not part of the clusters in A_B, but is certainly represented in once of the columns of the original matrix A). Then, add the market u to the cluster C among the newly created ones that incurs the least cost when adding the market to it. After this process terminates, a new set of clusters represented by a solution x^{**} indicating vectors in A is produced is a feasible solution to (IC^3), and by the MCP that holds for the p-median problem discussed above, the solution x^{**} must have a cost at most equal to the cost of x^*.
The above naturally lead to the following *ensemble algorithm*:

EXAMCE Algorithm for the p-Median Problem

Inputs: Finite index set $L = \{1, \ldots, L\}$ of locations, finite index set of markets $U = \{1, \ldots, U\}$, distance matrix D describing the distance—or cost— d_{ij} between market i and location j, number of facilities p to open, a set of base algorithms B for computing heuristic solutions of the p-median problem, any procedure Rm_dup() that removes duplicates in a clustering solution to produce a feasible solution for the IC^3 problem, a function Expand():$2^U \to 2^{2^\wedge U}$ that expands the set of solutions by taking as input a subset C of set U and returning a set of sets that are "neighbors" of C, and a function Local(): $2^{2^\wedge U} \to 2^{2^\wedge U}$ that implements any local search algorithm for improving the cost function of the p-median problem starting from an input initial (clustering) solution.
Outputs: finite set F of indices j representing the locations of the p facilities.

Begin

1. Apply the base algorithms in B to produce an initial set S_B of clusters.
2. Repeat

   a. Set $N = |S_B|$
   b. Set A_B to be the matrix whose columns are the membership indicator vectors of the clusters of S_B.
   c. Solve problem (SCP_R) to produce solution x^*.
   d. Set $C' = \text{Rm\_dup}(x^*, S_B)$.
   e. Set $C'' = \text{Local}(C'')$.
   f. Set $c'' = c(C'')$.
   g. Set $C''' = C'$ U C''.
   h. Set $C^4 = \{\}$.
   i. for each C in C''' do

      i. Set $C^4 = C^4$ U Expand(C).

   j. end-for
   k. Set $C^5 = C^4$ U C''.
   l. Add C^5 to S_B.

3. until no improvement in the cost c'' is made.
4. Set $F = \{\}$.
5. for each C in C'' do
   a. Set $j = \arg\min_{l \in L} \sum_{i \in C} d_{il}$.
   b. Set $F = F \cup \{j\}$.
6. end-for
7. return F.

End.

The EXAMCE algorithm as described above represents a family of algorithms for grouping problems in the (IC^3) class satisfying the MCP, and appropriate instantiations of the algorithm have been successfully used for the MSSC problem—as mentioned already—as well as the VRP to be discussed later in this chapter. The results of the EXAMCE algorithm shown in the next section were obtained using the following settings:

- The Rm_dup() procedure effectively implements the strategy for duplicates removal discussed above.
- The base algorithms B are all instances of the Alternate algorithm starting from a different random initial placement of facilities to locations in L, with the extra modification that in step 5.g.i of the Alternate algorithm, only a small randomly selected subset of the set L (comprising neighbors of the currently assigned facility location l for the cluster) is considered.
- The same algorithm (Alternate) is used as the function Local().
- The function Expand(C) works as follows: for each cluster g in set C, the τ nearest neighboring markets of the location assigned for g that are not already in g are progressively added to the set g to produce τ new clusters $g_1^+, g_2^+, \ldots, g_\tau^+$ and similarly, the τ of the markets in g that are the farthest from the location assigned for g are incrementally removed from g to produce sets $g_1^-, \ldots, g_\tau^-$ (if $\tau > |g|$, then τ is reset to $|g| - 1$). The output is the set $\{g_1^+, \ldots, g_\tau^+, g_1^-, \ldots, g_\tau^-\}$.

5.1.1.5 Comparison of Heuristic Algorithms for the p-Median Problem

In Table 5.1 we compare the solution qualities of three of the algorithms discussed above: the VNS algorithm (Sect. 5.1.1.2), the pop-star program implementing the hybrid GA and path re-linking algorithm, and the EXAMCE algorithm, which was run with a base set B of 150 different instances of the Alternate algorithm starting with a solution computed by a single run of the *random plus greedy* heuristic mentioned earlier. EXAMCE uses the SCIP open-source state-of-the-art constraint integer programming library to solve the (restricted) set covering problems it requires to solve, and since these problems

Table 5.1 Comparison between VNS, POPSTAR, and EXAMCE on the p-median problem

Problem	P	VNS	POPSTAR	EXAMCE
FL1400	10	**101248.13**	101249.54	103569.11
	20	**57856.32**	57857.94	57904.48
	30	44086.53	**44013.47**	44044.93
	40	35005.82	**35002.51**	35026.01
	50	29176.45	**29090.22**	29125.76
	60	25176.47	25166.91	**25161.11**
	70	22186.14	22126.02	**22120.02**
	80	19900.66	19876.38	**19872.90**
	90	18055.94	**17988.59**	**17988.59**
	100	**16551.20**	16552.21	16552.76
	150	**12035.56**	12036.69	12081.34
	200	9362.99	9360.95	**9359.30**
	250	7746.96	**7744.05**	7747.74
	300	6628.92	6624.36	**6619.50**
	350	5739.28	**5725.12**	5725.49
	400	5045.84	5016.04	**5006.75**
	450	4489.93	4476.67	**4468.43**
	500	4062.86	**4047.19**	4049.11

have a relatively small number of columns, the response time of the solver is in the order of a few seconds even on low-end commodity workstations. The test set we ran the algorithm on was used by Hansen and Mladenovic (1997) in their study of the performance of the VNS heuristic on the p-median problem. The data set is the FL1400 data set from the TSPLIB, which describes the locations of 1,400 data points in a two-dimensional plane. We assume that the set of markets U is the same as the set L and both sets comprise the data points described in the corresponding data set, so $|U| = |L| = 1,400$. The distance between a market m and a potential location l is simply the Euclidean distance between the two points in two-dimensional space.

The results show that although the pop-star heuristic is considered the state-of-the-art algorithm for facility location problems in general, the EXAMCE algorithm manages to find comparable quality solutions, and in seven cases, it outperforms pop-star, which was run with its default parameter settings. Pop-star is better six times, and VNS wins four times in total (in one case, EXAMCE and pop-star both find the best-known solution for the problem). The boldface entries indicate the best solution for a given problem instance.

5.1.2 The Uncapacitated Facility Location Problem

As already mentioned, the uncapacitated facility location problem is the same as the p-median problem, except that fixed setup costs f_j for the placement of a facility in *any* of the candidate locations $j \in L$ are explicitly accounted for.

These fixed setup costs usually represent land purchase costs or differing construction costs in different locations. The problem can be formulated as a MIP problem as follows:

$$\text{(UFL)} \min_{x,y} c(x) = \sum_{i \in U} \sum_{j \in L} d_{ij} x_{ij} + \sum_{j \in L} f_j y_j$$

$$\text{s.t.} \begin{cases} \sum_{j \in L} x_{ij} = 1, & \forall i \in U \\ \sum_{j \in L} y_j = p & \\ x_{ij} \le y_j, & \forall i \in U, j \in L \\ x_{ij}, y_j \in B = \{0,1\}, & \forall i \in U, j \in L \end{cases}$$

The Hybrid GA and path re-linking heuristic described in Sect. 5.1.1.3 for the p-median problem applies very well to this problem too, with minimal adaptations.

The EXAMCE algorithm (Sect. 5.1.1.4) that combines solutions in a set covering/set partitioning optimization context may also work as-is; of course, the cost function $c(C)$ of a cluster C (a subset of the markets set U) will have to be modified as follows:

$$c(C) = \min_{j \in L} \left\{ f_j + \sum_{i \in C} d_{ij} \right\}$$

The reason that EXAMCE still applies is that the problem is still in the (IC^3) class, and satisfies the monotone clustering property. To see this, consider a non-empty cluster C of markets to be served by a single facility, and consider another cluster $C' = C \cup \{u\}, u \in U - C$. The optimal cost of these clusters satisfies

$$c(C') = \min_{j \in L} \left\{ f_j + \sum_{i \in C'} d_{ij} \right\} = f_{j'} + \sum_{i \in C'} d_{ij'} \ge f_{j'} + \sum_{i \in C} d_{ij'} \ge \min_{j \in L} \left\{ f_j + \sum_{i \in C} d_{ij} \right\} = c(C).$$

The Alternate algorithm described in Sect. 5.1.1.2 also needs to be modified to work for the uncapacitated facility location problem, since determining the best location to serve a group of markets now requires the consideration of the fixed cost of each location, according to the new definition of a cluster's cost above. This modification however, is also, the only thing required for the algorithm to run.

5.1.3 The Capacitated Facility Location Problem

If we take into account the fact that a plant does not have infinite capacity, and that it may therefore be unable to serve the demands from any arbitrary number of markets, the previous uncapacitated facility location model is no longer sufficient as a problem formulation. In this case, we are given as before the fixed setup costs f_j for opening a facility in location $j \in L$, we are given the costs d_{ij} for satisfying a unit of demand in market i from location j. We are also given the capacity c_l a plant in location l would have, as well as the demand for the particular single product we are considering in each market u in U, say r_u.

The major decision variables for our problem are again, binary variables y_j indicating whether to open ($y_j = 1$) or not ($y_j = 0$) a facility at location j, and there are some further continuous decision variables x_{ij} specifying the fraction of the total demand for each market i that will have to be satisfied from location j (which will be zero of course if the location does not have a plant in the solution). The problem can now be formulated as follows:

$$\text{(CFL)} \min_{x,y} \sum_{j\in L} f_j y_j + \sum_{i\in U}\sum_{j\in L} d_{ij} r_i x_{ij}$$

$$\text{s.t.} \begin{cases} \sum_{j\in L} x_{ij} = 1, & \forall i \in U \\ \sum_{i\in U} r_i x_{ij} \leq c_j y_j, & \forall j \in L \\ 0 \leq x_{ij}, & \forall i \in U, j \in L \\ x_{ij} \leq 1, & \forall i \in U, j \in L \\ y_j \in B = \{0,1\}, & \forall j \in L \end{cases}$$

Note that the above formulation is slightly different from the formulation of Sect. 1.2.1 in, but the two formulations are equivalent. The first constraint specifies that each market's demand must be fully served. The second constraint specifies that the total demand served from a facility in any location j cannot exceed the facility's capacity (which will be zero if no facility is opened at the particular location). The final constraints dictate the bounds of the variables x_{ij} and the integrality of the variables y_j.

The optimal solution of the model (x^*, y^*) not only determines the optimal location of the facilities to open, but also how much fraction of demand each market will get from each of the open facilities, via the variables x^*. The actual supply each facility will provide will then be $s_j^* = \sum_{i\in U} r_i x_{ij}^*$, and the allocation of this supply to markets will be obviously $r_i x_{ij}^*$ for each i in U.

In the following, we shall briefly describe a Lagrangian relaxation-based heuristic for the solution of (CFL) (for more details the reader may consult Ghiani et al. 2004). By deleting the constraints associated with demand satisfaction for each market i in U, and penalizing the solutions by multiplying the degree of violation of the constraints by the Lagrange multiplier λ_i we get the following Lagrangian problem:

$$\text{(CFL-L)} \min_{x,y,\lambda} L(x,y,\lambda) = \sum_{j\in L} f_j y_j + \sum_{i\in U}\sum_{j\in L} d_{ij} r_i x_{ij} + \sum_{i\in U} \lambda_i \left[\sum_{j\in L} x_{ij} - 1\right]$$

$$\text{s.t.} \begin{cases} \sum_{i\in U} r_i x_{ij} \leq c_j y_j, & \forall j \in L \\ 0 \leq x_{ij} \leq 1, & \forall i \in U, j \in L \\ y_j \in B = \{0,1\}, & \forall j \in L \end{cases}$$

Problem (CFL-L) has the very nice property that for any given (fixed) Lagrangian vector λ of multipliers it can be decomposed into $|L|$ independent sub-problems, where the form of the j-th sub-problem is the following:

$$\text{(CFL-Lj)} \min_{w,z}\left[\sum_{i\in U}\left(r_i d_{ij}+\lambda_j\right)w_i+f_j z\right]$$

$$\text{s.t.}\begin{cases}\sum_{i\in U} r_i w_i \le c_j z \\ 0\le w_i\le 1, \qquad \forall i\in U \\ z\in\{0,1\}\end{cases}$$

This form is particularly easy to solve: for the option $z = 0$ (corresponding to not placing a facility in location j), the optimal solution requires $w = 0$ as well because of the first constraint, and the total cost of the trivial solution $(w,z) = (0,0)$ is zero. For the option $z = 1$, the problem becomes a continuous version of the binary knapsack problem studied in Sect. 1.2.1 [we can easily turn the problem in the maximization standard format for knapsack problems by considering as objective function coefficients the values $e_i = -(r_i d_{ij} + \lambda_j)$], and the optimal solution of this continuous knapsack problem is given by the following greedy algorithm:

Greedy algorithm for continuous knapsack problem

Inputs: Array c_j of costs of n items $j = 1, \ldots, n$, array v_j of the volume of each item $j = 1, \ldots, n$, and total knapsack capacity C.

Outputs: Optimal solution value z^* of the problem $\min_x\left\{z=\sum_{j=1}^{n} c_j x_j \mid \sum_{j=1}^{n} v_j x_j\le C,\right.$

$\left.0\le x_j\le 1 \forall j=1,\ldots,n\right\}$

Begin

1. Sort the array $g_i = -c_i/v_i$ $i = 1, \ldots, n$ in decreasing order, and reorder the arrays c_i and v_i according to the order in which the g values appear.
2. Set $z = 0$, $v = 0$, $i = 1$.
3. while $v \le C$ do
   a. Set $x_i = \min\{C - v, v_i\}/v_i$.
   b. Set $v = v + v_i x_i$.
   c. Set $z = z + c_i x_i$.
   d. Set $i = i + 1$.
4. end-while
5. return z.

End

Given the solutions of the independent sub-problems (CFL-Lj) for a given Lagrange multiplier vector λ, a *feasible solution for the original* (*CFL*) *problem* may be constructed by ordering the optimal values $\underline{z}_j$ of (CFL-Lj) in increasing order, and placing facilities in the locations dictated from that sorted list, until all demands for all markets in U are satisfied. In other words, first open a facility in the location l that had the smallest $\underline{z}_l$ value, and continue until no more facilities are needed to open. This procedure is optimal assuming the multipliers λ had been chosen "correctly". To obtain the "right" set of multipliers, in the spirit of the multi-commodity network flow algorithms of Sect. 1.1.2.2, it can be proved that it suffices to start with any set of values $\lambda^{[0]}$ for the vector λ and solve to optimality the problem (CFL-L) as we just described, and then update the multipliers in the k-th iteration according to the equation

$$\lambda_i^{[k+1]} = \lambda_i^{[k]} + \beta_k s_i^{[k]}, \quad k = 0, \ldots$$

where

$$s_i^{[k]} = \sum_{j \in L} x_{ij}^{[k]} - 1, \quad \forall i \in U$$

$$\beta_k = \frac{\min\{\bar{z}^{[j]}, j = 1, \ldots, k\} - \underline{z}^{[k]}}{\sum_{i \in U} \left(s_i^{[k]}\right)^2}$$

The notation $\underline{z}^{[k]}$ denotes the optimal solution value of the (CFL-L) problem in the kth iteration (with Lagrange multipliers $\lambda^{[k]}$), and $\bar{z}^{[k]}$ represents the value of the solution of the problem (CFL) found in the kth iteration by the heuristic method described above. These iterations in k may stop when a convergence criterion is satisfied (usually, when the value β_k becomes sufficiently small, or when in two successive iterations, identical solutions are obtained for the placement of facilities in L).

5.1.3.1 Multi-commodity Multi-Echelon Capacitated Facility Location

The Capacitated Facility Location model is capable of capturing many intricacies of the real-world location decision problems facing decision makers, but still fails to capture some aspects that are often fundamental:

- Plants usually do not serve customers directly but rather ship products to warehouses, each of which has a finite volume
- Warehouses store many (if not all) different types of products the manufacturing plants make within their storage facilities

Shi et al. (2004) describe the following real-world multi-commodity multi-echelon facility location problem. An (ordered) set of L plants each producing some or all of K different products must ship their products to W different warehouses that are to be built in locations chosen from a (ordered) set J of possible locations. The warehouses will then serve a (again, ordered) set I of different markets, each market i having demand w_{ik} for product $k \in K$. Each product k has volume s_k and the total volume capacity of a warehouse to be located in the jth site in J is q_j. The supply at the lth plant in L of product k is given and denoted as v_{lk}. The unit shipping costs from plant l to warehouse j of product k are denoted as c_{ljk} and the unit shipping costs from warehouse j to market i of product k are denoted as d_{jik}. As before, there are fixed costs f_j associated with building a warehouse at site $j \in J$. The problem is to find the best locations for the W warehouses so as to minimize the total set up and shipping costs of operating the warehouses so that demand for all products is fully met at all markets, and each market i gets its supplies for each product k from a unique warehouse.

The decision variables for the problem, include the binary decision variables $y_j, j \in J$ being 1 iff a warehouse opens at the jth location in J, the binary variables x_{jik} being 1 iff an open warehouse at location j serves the ith market in I with the kth product in K, and also the continuous variables u_{ljk} denoting the amount of product k that will be shipped from plant l to warehouse j.

The problem is modeled as follows:

$$(\text{MEMCFL}) \min_{x,y,u} z = \sum_{j\in J}\sum_{k\in K}\sum_{l\in L} c_{ljk}u_{ljk} + \sum_{i\in I}\sum_{j\in J}\sum_{k\in K} d_{jik}w_{ik}x_{jik} + \sum_{j\in J} f_j y_j$$

subject to:

$$\sum_{j\in J} x_{jik} = 1, \quad \forall i \in I, \quad \forall k \in K$$

single-supplier constraints

$$\sum_{i\in I}\sum_{k\in K} s_k w_{ik} x_{jik} \le q_j y_j, \quad \forall j \in J$$

warehouse volume capacity constraints

$$\sum_{i\in I} w_{ik}x_{jik} = \sum_{l\in L} u_{ljk}, \quad \forall k \in K, \quad \forall j \in J$$

flow conservation at each warehouse for each product

$$\sum_{j\in J} u_{ljk} \le v_{lk}, \quad \forall k \in K, \quad \forall l \in L$$

plant capacity constraints

$$\sum_{j\in J} y_j = W$$

number of warehouses to open constraint

$$\begin{aligned} & x_{jik} \in B = \{0,1\}, \quad \forall j \in J,\ i \in I, k \in K \\ & y_j \in B = \{0,1\}, \quad \forall j \in J \\ & u_{ljk} \geq 0, \quad \forall l \in L, \quad \forall j \in J, \quad \forall k \in K \end{aligned}$$

Due to the large number of binary decision variables and the coupling constraints relating the decisions regarding the routing of different products to warehouses and then to their final destinations, i.e. the markets in I, standard Branch and Bound or Branch and Cut algorithms fail to produce feasible solutions of the above MIP for problem settings of five plants, ten warehouse locations to choose from 100 possibilities, ten markets, and as few as three different products.

However, it is possible to use the Lagrangian relaxation ideas described before that lead to natural decompositions of the problem to obtain high quality solutions. If the two equality constraints in the formulation of the (MEMCFL) model are relaxed so that the single-supplier constraint and the product conservation constraints at the warehouse are relaxed in a Lagrangian penalty multiplier method sense, then the following relaxed problem is obtained, where λ_{ik} represent the multipliers for the single-supplier constraint, and θ_{jk} represent the multipliers for the product conservation at the warehouse constraints:

$$\begin{aligned} (\text{MEMCFL} - \text{L}) \ \min_{x,y,u,\lambda,\theta} L(x,y,u,\lambda,\theta) = & \sum_{j\in J}\sum_{k\in K}\sum_{l\in L} c_{ljk}u_{ljk} \\ & + \sum_{i\in I}\sum_{j\in J}\sum_{k\in K} d_{jik}w_{ik}x_{jik} + \sum_{j\in J} f_j y_j \\ & + \sum_{k\in K}\sum_{j\in J}\theta_{jk}\left[\sum_{i\in I} w_{ik}x_{jik} - \sum_{l\in L} u_{ljk}\right] \\ & + \sum_{i\in I}\sum_{k\in K}\lambda_{ik}\left[1 - \sum_{j\in J} x_{jik}\right] \end{aligned}$$

$$\text{s.t.} \begin{cases} \sum_{i\in I}\sum_{k\in K} s_k w_{ik} x_{jik} \leq q_j y_j, & \forall j \in J \\ \sum_{j\in J} u_{ljk} \leq v_{lk}, & \forall k \in K, \forall l \in L \\ \sum_{j\in J} y_j = W \\ x, y \in B, u \geq 0 \end{cases}$$

Now, *given the Lagrange multipliers* λ and θ, the relaxation of the product flow conservation at the warehouses decouples warehouses from plants, and the (MEMCFL-L) *problem becomes the super-position of two independent problems* that can be solved separately (and in parallel), the plant-to-warehouse transportation problem (*P–W*), and the warehouse to markets (*W*–M) capacitated facility location problem that is a combination of the standard uncapacitated facility location problem and the standard capacitated facility location problems extended in the multi-commodity domain:

$$(P-W)\ \min_u \sum_{j\in J}\sum_{k\in K}\sum_{l\in L}\left(c_{ljk}-\theta_{jk}\right)u_{ljk}$$

$$\text{s.t.}\begin{cases}\sum_{j\in J} u_{ljk}\le v_{lk}, & \forall l\in L, \quad \forall k\in K\\ u_{ljk}\ge 0, & \forall j\in J\forall l\in L, \quad \forall k\in K\end{cases}$$

and

$$(W-M)\ \min_{x,y} \sum_{i\in I}\sum_{j\in J}\sum_{k\in K}\left(d_{jik}w_{ik}-\lambda_{ik}+w_{ik}\theta_{jk}\right)x_{jik}+\sum_{j\in J} f_j y_j$$

$$\text{s.t.}\begin{cases}\sum_{i\in I}\sum_{k\in K} s_k w_{ik} x_{jik}\le q_j y_j\,, \quad \forall j\in J\\ \sum_{j\in J} y_j = W\\ x,y\in B\end{cases}$$

The (*P*–*W*) problem is a simple LP that is very easy to solve even for large-scale problem instances. The second hybrid multi-commodity capacitated facility location problem is more challenging, but, can be solved using the Lagrangian relaxation method described above for solving the problem (CFL) following the same steps and adding the constraint on the number of warehouses *W* to open.

Shi et al. found that a nested partitions method (see Sect. 1.2.2.5) obtains comparable quality results to those obtained using Lagrangian relaxation in a short fraction of the time required by the Lagrangian relaxation methods.

5.1.4 The p-Center Problem

The p-center problem bears some resemblances to the *p*-median problem discussed before, but there are also some major differences: the problem is to determine the optimal location of *p* facilities by choosing from a set *L* of potential locations so that the maximum distance of any market from a given set of market locations *U* to the nearest facility is minimized. In this sense, the problem is a so-called mini-max problem, and finds application in domains such as location of public facilities; for example, determining the optimal placement of police stations or ambulance stations in a city or district, etc. Nevertheless, the problem also finds application within a supply chain management or logistics context, when it is desired to ensure minimal service levels to all customers, or to ensure some level of "fairness" of customer service among the retail stores in a consumer chain and so on.

In the above version of the problem, we can formulate a combinatorial optimization model as follows: given a set of potential locations L, a set of markets U, a distance matrix D of dimensions $|U| \times |L|$ between each market i in U and each location j in L, and a number p of facilities to open, solve the problem

$$\min_{\pi \in 2^L, |\pi| = p} \left\{ z = \max_{i \in U} \left\{ d_{iP_\pi(i)} \right\} \right\} \quad P_\pi(i) = \arg\min \left\{ d_{ij} | j \in \pi \right\}$$

This problem can also be formulated, in the more convenient MIP form:

$$(\text{p-center}) \min_{x,y,z} z$$

$$\text{s.t.} \begin{cases} \sum_{j \in L} y_j \le p \\ \sum_{j \in L} x_{ij} = 1, & \forall i \in U \\ 0 \le x_{ij} \le y_j, & \forall i \in U,\ j \in L \\ \sum_{j \in L} d_{ij} x_{ij} \le z, & \forall i \in U \\ y_j \in B = \{0, 1\}, \quad \forall j \in L \end{cases}$$

The above (classical) MIP model deserves some discussion. Clearly, the binary variables y_j are the variables representing the decision to open (1) or not (0) a facility at location j in L. The auxiliary (continuous) variables x_{ij} will be non-zero if and only if the distance from market i to location j is minimal among the locations with an open facility (thanks to the third constraint of the model). We may view the variables x_{ij} as the fraction of time market i receives service from location j. Obviously, the first constraint ensures that no more than p facilities are opened in the optimal solution. The second constraint ensures that only one facility will serve each market in U, whereas the fourth constraint ensures that the objective measures the maximum distance between any market and its closest facility. (Note that, similar to the tighter formulation of the p-median problem discussed above, it is not necessary for the x_{ij} variables to be binary, since if in an optimal solution of the (*p-center*) model some x_{ij} turn out to have fractional values, then for each i, we can set all x_{ij} for which $x_{ij} > 0$ to zero, except one such variable, which we set to one, and the solution remains optimal). The formulation therefore overall requires $|L|$ binary variables,$|U||L| + 1$ continuous variables, and $2|U|(|L| + 1) + 1$ constraints, of which $|U||L|$ constraints are just bound constraints.

Recently, Elloumi et al. (2004) presented an alternative formulation of the above version of the p-center problem that converts the problem in to a set covering problem, which has proved before to yield fruitful decompositions in multi-commodity network flow problems (see the last sub sections of Sect. 1.2). Let $D_{\min}$ and $D_{\max}$ denote the minimum and maximum values of the elements of the distance matrix D. Let the *different* values of the elements d_{ij} of matrix D be sorted in ascending order, so that $D_{\min} = D_0 < D_1 < \cdots < D_K = D_{\max}$. The problem can be formulated in terms of the above values as follows:

$$(p\text{-center-SC}) \min_{y,z} \left[D_0 + \sum_{k=1}^{K} (D_k - D_{k-1}) z_k \right]$$

$$\text{s.t.} \begin{cases} \sum_{j \in L} y_j \leq p \\ z_k + \sum_{j \in L: d_{ij} < D_k} y_j \geq 1, & \forall i \in U, k = 1 \ldots K \\ z_k \in B = \{0,1\}, & \forall k = 1 \ldots K \\ y_j \in B = \{0,1\}, & \forall j \in L \end{cases}$$

There are $|L| + K$ binary variables and $|U|K + 1$ constraints. The decision variables y_j—as before—denote whether a facility should be placed at location j in L or not. The role of the K variables z_k is more subtle: z_k can be set to zero iff it is possible to choose p facilities and have all markets in U be served by a facility that is at a distance less than or equal to D_{k-1}. This, in turn, implies that in an optimal solution if $z_k = 0$ for some k, then $z_j = 0$ for all $j > k$ as well, and vice versa, if $z_k = 1$ for some k in the optimal solution, then $z_j = 1$ for all $j < k$ in the solution. These are consequences of the positivity of the coefficients in the objective function and the second set of constraints in the (*p-center-SC*) model. Indeed, note that in an optimal solution of this model, if $z_k = 0$ for some k , then it must be the case that all markets can be served from a distance less than D_k. When $z_k = 1$, the objective "jumps" from D_{k-1} to D_k and keeps jumping to higher D values as z_j stays at 1 with increasing j.

Now, consider the binary programming minimization model MM(δ) (Meyer 1992) defined as follows:

$$\text{MM}(\delta) = \min_{y} \left\{ \sum_{j \in L} y_j \middle| \sum_{j: d_{ij} \leq \delta} y_j \geq 1, \quad y_j \in \{0,1\}, \quad \forall i \in U, j \in L \right\}$$

where we define MM(δ) = $+\infty$ when the optimization problem defining MM(δ) is infeasible. The optimization problem defining MM is clearly a set covering problem. Since a feasible solution for MM(D_k) is also feasible for MM(D_{k+1}), clearly the inequality holds MM(D_k) $\geq$ MM(D_{k+1}) for all $k = 0, \ldots, K$. Let k^* be the smallest integer in the set $\{0, 1, \ldots, K\}$ such that MM(D_{k^*}) $\leq p$, and let y^* denote an optimal solution to MM(D_{k^*}). Then, the corresponding optimal values of the z variables, say z^*, will satisfy $(z^*)_j = 1$ for all $j = 1, \ldots, k^*$ and all $(z^*)_j = 0$ for all $j > k^*$. Therefore, to solve the model (*p-center-SC*) one has to solve up to $K + 1$ set covering problems defined by the minimization model MM(D_k) $k = 0, \ldots, K$, so as to determine the optimal value k^*, and then assign the values of the variables z^* as discussed above. Elloumi et al. (2004) prove that the lower bound provided by the optimization of the LP relaxation of the model (*p-centers-SC*) is greater than or equal to the lower bound provided by the optimal solution of the LP relaxation of the model (*p-centers*), and that in fact, solving the $K + 1$ Set Covering Problems defining the minimization model MM(D_k) can be done very fast using the standard state-of-the-art commercial MIP solvers, so that even large instances of this version of p-centers problem containing more than 1,400 markets and equal number of locations can be solved within seconds of cpu time on commodity workstations. They also provide a polynomial time heuristic

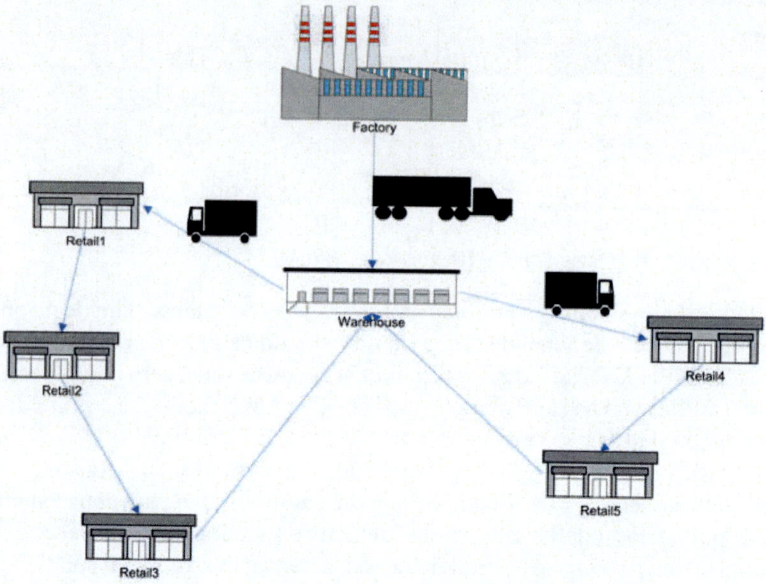

Fig. 5.1 Routing of a vehicle fleet in a multi-echelon structure

algorithm that provides excellent quality solutions due to the very good lower and upper bounds their algorithm is able to compute for the solution of the (p-center-SC) problem.

5.2 Distribution Management: Models and Algorithms

Distribution companies have as their mission statement to supply retail outlets or direct customer orders from warehouses, and less often, directly from the stock-points of manufacturing plants producing the final goods for their customers. In doing so, they have to manage fleets of trucks or small vans and in the case of island nations, even boats and ferries that carry the goods from the place of origin to the required destinations.

Having a number of origins of demand and destinations, and a fleet of vehicles to carry the demanded goods, the issue of managing this fleet so as to make optimal use of it arises naturally. In Fig. 5.1 we illustrate the issue of scheduling the available vehicles in a small fleet so as to pick up goods from given demand origins and transfer them to given destination points. The above give rise to some important families of problems in distribution management; in the remainder of this chapter, we shall examine the VRP family, which deals with the problem of building vehicle routes so as to satisfy demands at minimum cost subject to several other possible constraints such as time windows of arrival at each location, etc.

5.2.1 The Vehicle Routing Problem

Being part of the *short-haul freight transportation management*, the routing and scheduling of vehicles represents a very important component of every distribution company. The vehicle routing problem (VRP) is the problem of determining a set of minimum cost tours (in this context called routes) originating and terminating at a given central stock-point (depot) so that a corresponding number of vehicles can serve a set of given customers with the given demands.

In a standard mathematical formulation of the problem, the VRP is defined over a directed graph $G(V,E)$. Customers are located on the nodes of the graph, so that every customer is represented by exactly one node in the graph, and we are given the starting node $s \in V$ representing the base of the vehicles in the fleet. Each arc $(i,j) \in E$ has a cost c_{ij} representing usually the distance between locations i and j. The objective is to determine a set of routes starting and ending at s, so that all nodes are *served* by exactly one route once, and so that the total cost of the routes being the sum of costs of all arcs traversed is minimized. Clearly, when there is only a single vehicle in the fleet, the above description of the VRP expresses the traveling salesman problem TSP formulated in Sect. 1.2.1 so that the VRP is of at least the same complexity as the TSP problem, and is therefore an NP-hard problem. Several additional constraints are usually imposed when formulating the VRP, dictated from the actual real-world problem that it attempts to model:

- The number of vehicles in the fleet may be given or may be a decision variable for the problem
- Total demand transported by each vehicle may not exceed its capacity
- Duration of each route must not exceed an upper limit equal to the duration of a work shift
- Customers require to be served within predefined time windows.

Other constraints sometimes encountered in practice include precedence constraints that are essential when the vehicles must first visit some supply nodes where they pick up goods and then transfer them to demand (destination) nodes in the route. We shall not deal with problems where any node in G other than the central depot s can be a supply node for the vehicles.

We shall describe a set partitioning modeling approach to the VRP subject to time-window constraints (VRPTW) that simultaneously optimizes the number of vehicles to use. This modeling approach leads to an exact method for solving the VRPTW problem utilizing the concepts of column generation for solving the LP relaxation of the set partitioning model within a Branch and Bound framework that has a lot in common with the methods discussed in Sect. 1.2.2.4. In the process, we shall discuss fast dynamic programming-based algorithms for the shortest path problem with resource constraints (SPPRC) which was formulated and briefly discussed in Sect. 1.1.2.2.

In formulating the VRPTW, an extra parameter t_{ij} representing the time it takes to serve customer i and then traverse the arc (i,j) is introduced. An immediate implication of this "merging together" of the travel time t_{ij} between nodes i and j, and the service time of customer i, is that a node cannot be visited by a vehicle route *unless* it also serves the customer at that node. In the (rare) case where this is not so, and routes are allowed to "cross", the time t_{ij} along an arc only denotes travel time, and a new parameter, say st_i is needed to indicate *service time of customer i*. Each customer i has *non-negative demand* d_i and has an associated time-window $[a_i, b_i]$ during which it accepts service from a vehicle (a vehicle may be scheduled so that it arrives at any node i before time a_i but will have to wait until the start of the time window). The total demand of the customers to be served in a route must be less than the vehicle's capacity, $Q > 0$.

We define a *feasible route r* satisfying the $k > 0$ customers n_j $j = 1, \ldots, k$ to be a path in the graph G $\{\langle s,n_1\rangle, \langle n_1,n_2\rangle, \ldots, \langle n_k,s\rangle\}$ where $\langle i,j\rangle$ denotes a path [usually, but not necessarily in all models, comprising a single arc (i,j)] connecting node i to node j that satisfies the following two properties:

1. $\sum_{j=1}^{k} d_j \le Q$ (capacity constraint)
2. Define the *service start time* T_i for any customer $i = 1,\ldots, k$ given the route r as follows:

$$T_i = \begin{cases} 0, & i = 0 \\ \max\{a_i, T_{i-1} + t_{<n_{i-1},n_i>}\}, & i > 0 \end{cases}$$

where $t_{\langle i,j\rangle} = \sum_{e\in\langle i,j\rangle} t_e$. Then, for route r to be feasible, the T_i must satisfy $T_i \le b_i$ for all $i = 1, \ldots, k$. Let R denote the set of all such feasible routes.

Now, the VRPTW problem may be formulated as a set partitioning problem:

$$(VRPTW) \min_x \sum_{i=1}^{|R|} \sum_{(i,j)\in r_i} c_{ij} x_i$$

$$\text{s.t.} \begin{cases} Ax = e \\ x \in B^{|R|}, \quad B = \{0,1\} \end{cases}$$

where e is a column vector of $|R|$ ones, and $A = \begin{bmatrix} a_1, & a_2, & \ldots, & a_{|R|} \end{bmatrix}$ is a matrix whose columns a_i are the indicator vectors of the nodes in $V-\{s\}$ participating in the ith feasible route, r_i. The decision variable x_i therefore corresponds to the decision (value set to one) to include the ith feasible route in the solution of the VRPTW problem or not (value set to zero). Constraints on the maximum number of routes to be created—usually due to vehicle fleet size limitations—can be easily incorporated into the model by adding the side constraint $e^T x \le m$, where m is the total number of vehicles in the fleet.

5.2.1.1 Exact Methods for the Vehicle Routing Problem with Time Windows

Given the very large number of columns of matrix A, Desrochers et al. (1992) designed and implemented a Branch and Price type algorithm (see Sect. 1.2.2.4) using a column generation method for solving the LP relaxation of the VRPTW model for the standard model where routes cannot "cross", for which they pose the extra restriction that the costs c_{ij} of all arcs in E are non-negative integers. Let γ_{ir} denote the number of times node $i \in V$ is visited in the feasible route represented in the rth column of the matrix A. Also, define X_d to denote the total number of routes in the solution, and z to denote the total distance traveled by all routes, i.e. the total cost of the solution x. Desrochers, Desrosiers and Solomon work with the following set covering problem instead of the original VRPTW:

$$(VRPTW - SC) \min_{x,X_d,z} \sum_{i=1}^{|R|} \sum_{e \in r_i} c_e x_i$$

$$\text{s.t.} \begin{cases} \sum_{r=1}^{|R|} \gamma_{jr} x_r \geq 1, & \forall j \in V - \{s\} \\ \sum_{r=1}^{|R|} x_r = X_d \\ \sum_{i=1}^{|R|} \sum_{e \in r_i} c_e x_i = z \\ x \in B^{|R|}, z, X_d \in \mathbb{N} \end{cases}$$

The first constraint in the problem together with the binary nature of the variables $x_i = 1, \ldots, |R|$ make (VRPTW-SC) a set covering problem instead of set partitioning. The intermediate two constraints introducing variables X_d and z are "auxiliary" constraints and variables only added because they help the specific algorithm for solving this problem, as we shall see. In the following, as already hinted at, it is assumed that all feasible routes r generated are paths of the form $\{(s,n_1)\ ,\ (n_1,n_2),\ \ldots,\ (n_k,s)\}$ so that *the only nodes visited in the route are the customers that must be served by the route.*

To solve the LP relaxation of the (VRPTW-SC) model in an LP-based Branch and Bound framework, observe that the dual variables π_j, $j \in V - \{s\}$ associated with the first set of constraints in the model, and the dual variables π_d and π_z associated with the last two constraints of the model are enough to compute the reduced cost of a route $r = \{(s, n_1), (n_1, n_2), \ldots, (n_k, s)\} \in R$, which is simply

$$\bar{c}_r = \sum_{i=0}^{k} \left[(1 - \pi_z) c_{n_i n_{i+1}} - \pi_{n_i} \right]$$

where we use the convention $n_0 = n_{k+1} = s$. Looking at the above equation, it makes sense to define the *reduced cost of an arc* to be

$$\bar{c}_{ij} = (1 - \pi_z)c_{ij} - \pi_i \quad \forall (i,j) \in E$$

The algorithm that solves the LP relaxation of the (VPPTW-SC) model, starts with a small set of columns corresponding to a trivial solution in which there is a dedicated vehicle for each node (customer) in G, and after the (trivial) optimal solution to this restricted set covering problem is found, the algorithm solves a dedicated *sub-problem* to find any reduced cost columns (routes) that should enter the solution. This sub-problem turns out to be a SPPRC (corresponding to the duration consumed along each arc) defined over the same graph G with the same arcs having the same duration, but with arc costs equal to their reduced costs $\bar{c}_{ij}$ defined above. A pseudo-polynomial time algorithm for the SPPRC problem is discussed in the next section. Once the solution of the SPPRC fails to produce a route of negative total (reduced) cost, the current LP solution is optimal. Otherwise, the negative reduced cost routes corresponding to the paths found by solving the SPPRC are added as extra columns to the restricted LP problem and the new augmented LP is solved again. Once the optimal solution of the LP relaxation at the root node of the Branch and Bound tree is found, the B&B algorithm proceeds with a specialized branching strategy:

- The algorithm first checks whether to branch on the number of vehicles: if X_d is fractional, say $X_d = v$, the algorithm creates two branches: the extra inequality $X_d \leq \lfloor v \rfloor$ is added to the left sub-problem of the current node, and the extra inequality $X_d \geq \lceil v \rceil$ is added to the right sub-problem.
- Next, at each branch, if the solution of the LP relaxation is still fractional, if the objective function value is also fractional (so that $z = u$ is fractional), the cut $z \geq \lceil u \rceil$ is added to the current sub-problem without destroying the structure of the SPPRC sub-problem that needs to be solved for the column-generation step of the problem.
- Finally, when the solution of a node is such so that both z *and* X_d are integer, but one or more of the x variables are not, branching is performed on the arcs of the sub-problem network. In particular, when fractional column values x_r exist, for each arc (i,j) in G with a fractional flow, an arc-score is computed that depends on the flow value of the arc as well as the number of fractional routes in which the arc participates, and the arc with the highest score is chosen to be branched on as follows (remember that the problem formulation in (VRPTW-SC) does not directly include the arc flows as variables): to fix the flow of an arc (i,j) at zero, the arc is simply removed from the sub-problem network, and any column containing this arc has its cost heavily penalized so that it is priced out in the re-optimization of the LP at the next level of the B and B tree; to fix the value of an arc (i,j) at one, it simply suffices to remove all arcs (i,n) and (l,j) in E where $n \neq j$ and $i \neq l$ form the sub-problem graph, and heavily penalize the cost of any route that contains such arcs, as before. In the re-optimization of the LPs at the next level in the Branch and Bound tree, the SPPRC problems generate new columns as needed that obey the requirements $x_r = 0$ or $x_r = 1$, respectively.

Experimenting with this algorithm, the authors realized that the issue of multiple coverage of a node by more than one route never arises, and this should be obvious because their algorithm is an exact method guaranteeing the optimal solution to the VRPTW problem, and in the optimal solution, having two vehicles *serving* the same customer can never be optimal (the cost may be further reduced by simply removing the duplicate customer from all but one of the routes serving them).

We now turn our attention to solving the SPPRC problem with time windows as the resource constraint which plays the role of the sub-problem in the column generation approach to solving the (VRPTW-SC) model.

5.2.1.2 An Algorithm for Solving the Shortest Path Problem with Resource Constraints and Time Windows

The shortest path problem with resource constraints *and* time windows is defined over a *directed* graph $G(V,E)$ on which real costs $c_{ij} \in \mathbb{R}$ exist on each edge (i,j) in E as well as a *non-negative duration* t_{ij} that represents the duration of traversing the arc (i,j). There also exist real demands $d_j \in \mathbb{R}$ on each node j in V. The objective of the shortest path problem with resource constraints and time windows (SPPRCTW) is to find the minimum cost path from a source node s to a terminal node f so that each node i in the path is visited within a specified interval $[a_i,b_i]$ (a_i or b_i may be negative or positive infinity respectively, in which case the interval becomes open on the left or the right), and that the partial sums of all demands from s to any node j in the path to f remains less than or equal to a given capacity constant Q.

To analyze the problem, define the cost $C_j(d,t)$ as the minimum cost of the partial path going from node s to the node $j \in V$ having accumulated total demand d and ready to leave node j at time t or later. The DP equations then hold for the cost $C_j(d,t)$:

$$C_s(0,0) = 0$$
$$C_j(d,t) = \min_{(i,j)\in E} \{C_i(d',t') + c_{ij} | t' + t_{ij} \le t, d' + d_j \le d, t' \in [a_i, b_i]\}$$
$$\forall j \in V, \quad \forall d : d_j \le d \le Q, \quad \forall t \in [a_j, b_j]$$

This problem is NP-hard (Desrochers et al. 1992), but as mentioned already, there are pseudo-polynomial time algorithms for solving it. The pulling algorithm of the same authors *assumes that all data are integer valued* and maintains two sets of labels for each node in the graph. The first set of labels includes the labels associated with feasible paths from s to the node and therefore defines primal solutions that constitute an upper bound $\bar{C}_j(d,t)$ on the value of the optimal path to that node, whereas the second set of labels includes labels $\underline{C}_j(d,t)$ associated with lower bounds on the cost of a path ending at node j in state (d,t), where "*state*" is defined as the two-dimensional vector that represents the partial sum of demands summed over the nodes of the path ending at node j and also the earliest time at which the visitor (vehicle) may leave from node j. This algorithm has a computational complexity $O\left(Q^2\left[\sum_{i\in V}(b_i+1-a_i)\right]^2\right)$:

The Pulling Algorithm for SPPRCTW

Inputs: Directed weighted graph $G(V,E)$ with arc weights c_{ij} arc durations t_{ij} for each arc (i,j) in E, node demands d_j and time windows $[a_j,b_j]$ for each node j in V, vehicle capacity Q, and initial node s.
Outputs: Optimal paths from s to every node in the network satisfying all resource and time-window constraints in all nodes in the path from source to each destination.

Begin

/* initialization */

1. for each $j \in V - \{s\}$ do
   a. Set $\bar{P}_j = \emptyset$.
   b. for each $d = 0$ to Q do
      i. for each $t = a_j$ to b_j do
         1. Set $\underline{C}_j(d,t) = -\infty, \quad \bar{C}_j(d,t) = +\infty$.
         2. Set $\bar{P}_j = \bar{P}_j \cup \{(d,t)\}$.
      ii. end-for
   c. end-for
   d. Set $P_j = \emptyset, R_j = \emptyset$.
2. end-for
3. Set $\underline{C}_s(0,0) = \bar{C}_s(0,0) = 0$, $P_s = \{(0,0)\}$, $\bar{P}_s = \emptyset$, $R_s = \emptyset$.

/* search for next state */

4. Set $W = \bigcup_{j\in V} \bar{P}_j$.
5. if $W = \emptyset$ GOTO 14.
6. Find the minimum in lexicographic order state vector of any node j (d,t) in W.

/* update state information and labels */

7. Set $\underline{C}_j(d,t) = \min\limits_{(i,j)\in E}\{\underline{C}_i(d',t') + c_{ij} | t' + t_{ij} \le t, t' \in [a_i,b_i], \quad d' + d_j \le d\}$
8. Set $\bar{C}_j(d,t) = \min\limits_{(i,j)\in E}\{\bar{C}_i(d',t') + c_{ij} | t' + t_{ij} \le t, t' \in [a_i,b_i], \quad d' + d_j \le d\}$.
9. Set $i = \arg\min\limits_{i:(i,j)\in E}\{\bar{C}_i(d',t') + c_{ij} | t' + t_{ij} \le t, t' \in [a_i,b_i], \quad d' + d_j \le d\}$.
10. if $\underline{C}_j(d,t) = \bar{C}_j(d,t)$ then
    a. Set $P_j = P_j \cup \{(d,t)\}$, $\bar{P}_j = \bar{P}_j - \{(d,t)\}$.
    b. Set $R_j = R_i \cup \{(i,j)\}$.
11. else
    a. Set $P_j = P_j - \{(d,t)\}$, $\bar{P}_j = \bar{P}_j \cup \{(d,t)\}$.

12. end-if
13. GOTO 4.
14. return $R = \bigcup_{j \in V} \{R_j\}$.

End.

5.2.1.3 Heuristic Methods for Solving the Vehicle Routing Problem

Note that the VRPTW problem is such that the ideas developed for solving problems in IC^3 class discussed in Sect. 5.1.1.4 are also applicable. Thus, in an adaptation of the spirit of the EXAMCE algorithm for the p-median problem, one may generate heuristically a restricted set R' of high-quality feasible routes for the VRPTW ensuring that all nodes are covered by at least one of the generated routes, and then solve the corresponding set covering problem:

$$(\text{VRPTW-SC2}) \min_x \sum_{i=1}^{|R'|} \sum_{e \in r_i} c_e x_i$$

$$\text{s.t.} \begin{cases} A_{R'} x \geq e \\ x_i \in B = \{0,1\}, \quad \forall i = 1, \ldots, |R'| \end{cases}$$

where $A_{R'}$ is the appropriate sub-matrix of matrix A in the VRPTW model. If in the optimal solution x^* of (VRPTW-SC2) there are nodes served by more than one route, then any procedure that removes the node appearances from all routes containing them except one, will result in a new solution that is feasible for the VRPTW, with better (or equal) cost than the cost of x^*.

To generate high-quality feasible routes covering all customers, after creating a dedicated route for each customer (node), one may attempt to iteratively select and merge pairs of routes to create a new single route that will serve all customers in the individual routes, while observing the feasibility constraints for the routes (this is known as the *savings heuristic*, a more detailed discussion of which can be found in Ghiani et al. 2004).

The Cluster First Route Second Heuristic

Another heuristic—appropriate when there is an upper limit on the number m of vehicles to use—that can be used to construct (some) high-quality routes is the following Cluster-First-Route-Second heuristic: First, nodes are partitioned into disjoint subsets $V_k \subseteq V - \{s\} : \bigcup_{i=1}^{m} V_k = V - \{s\}$ so that each subset satisfies the vehicle capacity constraint $\sum_{v \in V_k} d_v \leq Q$ (being associated with a single vehicle). The partitioning is done according to some clustering criterion based on the distance between the nodes—possibly, by applying the p-median problem with $p = m$.

Then, a single feasible route is constructed for all nodes in $V_k \cup \{s\}$ for each $k = 1, \ldots, m$ using any algorithm for the TSP, and the resulting tour of the nodes of $V_k \cup \{s\}$ starting and ending at s, is modified if needed to ensure compliance with the time-window constraints of the problem.

5.2.2 The Tankering Problem

Another transportation-related planning problem that offers the potential for large savings (especially in the airline industry) is that of determining the optimal locations along an airplane's route to refuel—although the problem also appears within the context of refueling a vehicle following a (relatively long-haul) route determined by the solution of a VRP problem above—the problem also applies to the maritime sector, and is also relevant because the ship's tanks have ample capacity to accommodate fuel for many trips. Fuel prices are known to fluctuate widely from city to city or between countries, and when an airplane's (or vehicle's) route has several legs (see discussion in Sect. 3.4.2 and above) choosing to "carry" extra fuel (so-called *tankering fuel*) in one flight so the airplane does not have to refuel on the next (more expensive) airport can result in significant savings. It has been estimated that by 1999, a major US airline operator was able to save more than a *few million dollars annually* after purchasing a software decision support tool to help planners decide how much fuel an aircraft should carry in order to avoid paying high-fuel prices.

Mathematically, the problem can be stated as a constrained nonlinear optimization problem. Let $I = (1, 2, \ldots, n)$denote the sequence of indices of cities that comprise an aircraft's route from its current location (l_0) until the *first* location (l_n) that offers fuel at a price p_n *lower* than the current location's price p_0. The tankering problem (TP) need only determine the fuel to be purchased at locations $l_0, \ldots, l_{n-1}$ since it would be optimal—from a fuel cost point of view—to arrive at airport l_n with (essentially) empty tanks. Let $p_i > 0$ denote the price per gallon of fuel at location l_i and assume the airplane has arrived at its current location l_0 with fuel level f_0. Let $c_i : \left[f_i^{\min}, f_i^{\max}\right] \to \left[0, f_i^{\max}\right]$be a function returning the fuel consumption of the particular aircraft (not just type of aircraft) flying the leg$(l_i \to l_{i+1})$. Clearly, it must hold

$$x - \bar{f} \le c_i(x) \le x, \quad \forall x \in \left[f_i^{\min}, f_i^{\max}\right]$$

where $\bar{f}$ is an upper bound on the fuel level that the particular aircraft is allowed to land with (independent of where it is landing), and $f_i^{\min}, f_i^{\max}$ are the minimum and maximum fuel levels that the aircraft is allowed to *take off with* for the particular flight. The function $c_i(x)$ certainly depends on x because when carrying more fuel, an aircraft (or vehicle) also burns more fuel during its trip, but this relation is *not highly nonlinear*. In other words, the function $c_i(x)$ can often be approximated by a piece-wise linear function of the form

$$c_i(x) = c_{i,0} + \sum_{j=1}^{k-1} c_{i,j}\left(f_{i,j+1} - f_{i,j}\right) + c_{i,k}\left(x - f_{i,k}\right), \quad x \in \left[f_{i,k}, f_{i,k+1}\right]$$

where $\bigcup_{j=1}^{N-1} \left[f_{i,j}, f_{i,j+1}\right] = \left[f_i^{\min}, f_i^{\max}\right]$ is a *partitioning* of the space of acceptable fuel levels for the particular flight leg i and $c_{i,k} \in (0, 1]$ $k = 1, \ldots, N-1$ are small constants.

The tankering problem can therefore be formulated as follows, where x denotes the decision variables of how much fuel to purchase at each airport $l_0, \ldots, l_{n-1}$.

$$(\text{TP}) \min_{x,f} \sum_{i=0}^{n-1} p_i x_i$$

$$\text{s.t.} \begin{cases} x_i + f_i \le f_i^{\max}, & i = 0, \ldots, n-1 \\ x_i + f_i \ge f_i^{\min}, & i = 0, \ldots, n-1 \\ f_i = x_{i-1} + f_{i-1} - c_{i-1}(x_{i-1} + f_{i-1}), & i = 1, \ldots, n \\ x \ge 0, 0 \le f \le \bar{f} e \end{cases}$$

The variables f_i denote the fuel level with which the aircraft lands at airport l_i $i = 1, \ldots, n$, and e, as usual is an n-dimensional column vector of ones.

The objective function directly measures the total fuel costs to fly to the last city in the sequence, l_n. The first two constraints ensure that the fuel level at each leg is within acceptable levels, and the third set of constraints is the "fuel balance" constraint, maintaining that the fuel level when arriving at the destination is the fuel level when taking off minus the fuel consumed during the flight.

Despite the *nonlinear* nature of the problem, in practice usually it can be solved rather easily, due to its low dimensionality. Indeed, the number n of flight legs required to plan for before a lower price airport is encountered is very small (typically between 3 and 8). Also, the number of intervals $[f_{i,j}, f_{i,j+1}]$ required to partition the acceptable fuel interval $\left[f_i^{\min}, f_i^{\max}\right]$ so as to have a linear fuel consumption function within each interval is also very small (in the order of $N < 5$). A standard method in Integer Programming allows us to model a piece-wise linear function with binary variables: the function $c_i(x)$ can be modeled using $N - 1$ binary variables $y_{i,j}$ taking the value 1 iff $x \in \left[f_{i,j}, f_{i,j+1}\right)$ and N continuous variables $\lambda_{i,j}$ as follows:

$$c_i(x) = \sum_{j=1}^{N} \lambda_j C_{i,j}$$

$$\lambda_{i,1} \le y_{i,1}$$

$$\lambda_{i,j} \le y_{i,j-1} + y_{i,j}, \quad j = 2, \ldots, N-1$$

$$\lambda_{i,N} \le y_{i,N-1}$$

$$\sum_{j=1}^{N} \lambda_{i,j} = 1$$

$$\sum_{j=1}^{N-1} y_{i,j} = 1, \quad i = 0, \ldots, n-1$$

$$y_{i,j} \in B = \{0, 1\}, \quad j = 1, \ldots, N-1$$

$$\lambda_{i,j} \ge 0, \quad j = 1, \ldots, N$$

where $C_{i,j} = c_i(f_{i,j})$ for each $i = 0, \ldots, n-1$ and $j = 1, \ldots, N$. Thus the (TP) can be modeled as a MIP with $n(N-1)$ binary variables and $2n + N$ continuous variables, and $n(N + 6)$ constraints (excluding box constraints). For typical values of n and N the problem is usually rather easy to solve directly in any modern solver.

When N or n becomes large (which may be the case in maritime transportation planning), a dynamic programming (DP) approach to the TP can solve the problem usually much faster. As is usual when DP is applied, we discretize the search space by discretizing the fuel level f_i that the aircraft is allowed to have *when arriving at airport* l_i. Consider an arbitrary discretization of the interval $[0,\bar{f}]$ into k_i discrete fuel levels $\left(0 = g_{i,1} < g_{i,2} \ldots < g_{i,k_i} = \bar{f}\right)$ and consider the restricted (as opposed to relaxed) problem whereby the aircraft is required to land at airport l_i with a fuel level equal to $g_{i,j}$ for some j in $\{1,2, \ldots, k_i\}$. Now, the minimum cost refueling plan so as to arrive at destination l_r where $r = 1,2, \ldots, n$ is governed by the following DP equation:

$$z_r\left(g_{r,i}\right) = \min_{j=1\ldots k_{r-1}} \left[z_{r-1}\left(g_{r-1,j}\right) + p_{r-1}x \middle| d(x) = g_{r,i}, f_{r-1}^{\min} \leq g_{r-1,j} + x \leq f_{r-1}^{\max}\right]$$
$$i = 1 \ldots k_r,\ z_0\left(g_{0,1} = f_0\right) = 0, \quad k_0 = 1$$

The function d is defined as $d(x) = g_{r-1,j} + x - c_{r-1}\left(g_{r-1,j} + x\right)$ so solving the equation $d(x) = g_{r,i}$ is trivial since it is a piece-wise linear function. When the equation $d(x) = g_{r,i}$ results in a value x that is not in the interval $\left[f_{r-1}^{\min} - g_{r-1,j}, f_{r-1}^{\max} - g_{r-1,j}\right]$, the alternative of reaching location l_{r-1} with fuel level $g_{r-1,j}$ and then arriving at l_r with fuel level $g_{r,i}$ is obviously infeasible and thus discarded. The optimal fueling plan for the entire set of flights from l_0 to l_n is then the value $z_n(0)$. The details of the algorithm that implements the DP equation for the discretized version of the TP are left as an exercise for the reader.

It is interesting to note at this point that the US airline operator mentioned above, obtained large savings by only solving a *one-step tankering problem*, where the problem is restricted to deciding how much fuel to purchase at an airport when the next airport in the aircraft's route is more expensive. The problem in this case, reduces to a problem of the form

$$(1-\mathrm{TP})\ \max x$$
$$\text{s.t.} \begin{cases} (p_1 - p_0)x \geq p_0\left(c_0(f_0 + x) - c_0\left(\max\left\{f_0^{\min}, f_0\right\}\right)\right) \\ f_0^{\min} \leq f_0 + x \leq f_0^{\max} \end{cases}$$

Since the function $c_0()$ is monotonically increasing in its domain of definition, the problem is trivial to solve without resorting to an LP solver. Devising an algorithm to solve the (1-TP) model is left as an exercise for the reader.

Also note that in a more complete version of the problem, some consideration could be given to the issue of wear and tear of the aircraft's tires during landing due to the extra weight caused by the tankering fuel, but due to the many orders of magnitude in costs between fuel prices and tire stress, it is hardly ever taken into account.

5.3 Integrated Location and Distribution Management

Facility location and vehicle routing are problems that belong to fundamentally different decision making levels of an enterprise, since the first category is clearly of strategic importance due to the extremely one-shot high costs that it entails, whereas the second category comprises part of the standard day-to-day operations. Clearly, once made, strategic decisions cannot easily be "undone", whereas tactical- or operational-level decisions are relatively easy to "correct" or "modify" when the need arises. Nevertheless, the two problems are also fundamentally linked: once facilities and warehouses are *located*, each market to be served will have to be *allocated* to one (and sometimes more than one) facility, and then goods from upstream echelons in the supply chain will have to be *physically transferred* to the most downstream echelons (retailers) facing the end-customer demand, which gives rise to scheduling the available vehicle fleet for such pick-ups and deliveries. Sub-optimal decisions in locating plants or warehouses therefore will reverberate high costs for a very long time in terms of high-fuel costs for vehicles transferring the goods, high costs for maintaining a larger fleet of vehicles than would be otherwise necessary, high personnel costs, and so on. Sub-optimal location decisions will also translate into higher lead-times for transferring goods between installations, and this will have consequences on the inventory costs required to maintain the desired service levels.

Overall therefore, it becomes obvious that location decisions are of fundamental importance not only because of the high fixed costs associated with purchasing land plots and erecting a manufacturing facility, but also because of the high costs that will result from operating the "placed" supply chain network even when the network operates optimally. The ideal therefore would be to simultaneously optimize all associated costs when designing a supply chain taking into consideration fixed charge costs for different locations where to place factories, warehouses, or retail stores, transportation costs of the products to be offered through the supply chain, costs of purchasing, operating and maintaining an optimally sized fleet of vehicles and their crews, and of course, the costs of holding inventories at all echelons in the supply chain, all under significant uncertainty regarding the values of many problem parameters:

- Product demand at each market
- Fuel prices
- Personnel costs including union regulations, etc.
- Fleet depreciation and so on...

Finding the optimal solution to the above "total" problem seeking to optimize the whole supply chain as a single system is currently out of reach, and it is for this reason that each problem (location, distribution management, inventory management, planning, forecasting, etc.) is usually treated separately, as an independent sub-problem that must be solved within the constraints imposed by the solution chosen for higher level decision making problems. At the top of this decision-making hierarchy are usually the location problems studied in this chapter. For this reason, the more factors and related costs that location models take into account, the more likely it is that decision problems at the tactical and operational level will be less constrained from reaching the true globally optimal performance. In that regard, some organizations implement the following methodology when solving strategic location problems: the location problem is solved using one or more of the methods discussed in Sect. 5.1 and then markets are allocated to placed facilities. Then, the simultaneous vehicle fleet sizing and routing problem is solved again to determine the implications of the location decisions on the operational level. If the results are not satisfactory, the location problem is re-optimized having additional constraints arising from the operational-level concerns. These two steps are repeated until they converge to a satisfactory solution.

5.4 Bibliography

Location problems have been extensively studied in the *Operations Research and Optimization community*. The journal *Location Science* was devoted exclusively to the study of location problems from a quantitative point of view. It has since been incorporated to the journal *Computers and Operations Research*. Other journals where problems related to location science and optimization are discussed include *Management Science*, *Transportation Science*, *INFORMS Journal on Computing*, *European Journal of Operational Research*, *Mathematics of Operations Research*, *Annals of Discrete Mathematics*, *Discrete Applied Mathematics*, *Mathematical Programming*, *RAIRO Operations Research*, *Journal of Algorithms*, *Journal of Heuristics*, and so on. Several books have been devoted solely on the subject as well, e.g. see Love et al. (1988).

Early work on algorithms for the p-median and facility location problems can be found in Kuehn and Hamburger (1963), Maranzana (1964), Erlenkotter (1978), Kariv and Hakimi (1979). Beasley (1985) offers an exact algorithm for the p-median, and around that time, the fast heuristic of Whitaker (1983) was published. Later papers include Megiddo and Supowit (1984), Moreno et al. (1990) and Captivo (1991). More recently, Hansen and Mladenovic (1997) presented a VNS heuristic for the p-median, Ghosh (2003) presented general neighborhood search heuristics for the uncapacitated facility location problem, Hoefer (2003) presents an experimental comparison, and Resende and Werneck (2003) present the complicated details of the implementation of a fast swap-based heuristic for the p-median problem based on Whitaker's work Whitaker (1983).

However, the fastest heuristic—that produces some of the best-known quality results available today—for the uncapacitated facility location problem remains Resende and Werneck (2004); the paper describes a fast hybrid heuristic for the p-median problem, but as the authors show in a successor paper the algorithm can be applied with very little customization to the uncapacitated facility location problem as well.

An easy introduction to distribution management is given in Ghiani et al. (2004). The linking between location and distribution management is nicely illustrated in Laporte (1988). Wren (1981) discusses the VRP in the context of the public transport sector. Golden and Assad (1988) provide an in-depth treatment of VRP problems illustrating the major methods known until then.

A long series on important papers for VRP with or without resource constraints and with or without time-window constraints came from the GERAD research center in Montreal, Canada. Indicatively, we mention Desrosiers et al. (1984), Desrochers and Soumis (1988), Desrochers and Soumis (1989), Desrochers et al. (1992), and references therein. A more recent paper on the same theme is Yunes et al. (2005). A comprehensive survey on Tabu search-based heuristics for the VRPTW can be found in Braysy and Gendreau (2002).

An integrated approach to production planning and distribution management is presented in Bilgen and Gunther (2009), where the authors, building on the work of Christou et al. (2007) and others, present a joint MIP optimization model that simultaneously optimizes production and transportation costs in different scenarios where full or less than full truckloads are allowed when vehicles start their routes to deliver finished goods from plants to distribution centers.

5.5 Exercises

1. Implement the standard greedy heuristic for the *p*-median problem and test its performance on the TSPLIB problem FL1400. How does it compare to the solutions recorded in Table 5.1.
2. Implement the Alternate algorithm for the *p*-median problem and test its performance on the TSPLIB problem FL1400. How does it compare to the solutions recorded in Table 5.1.
3. Can you propose any modifications to the Alternate algorithm so that it is applicable as a heuristic for the uncapacitated facility location problem? Justify your answer.
4. Design an algorithm for the Shortest Path Problem with Time-Windows and resource constraints on an acyclic graph, using as starting point the Pulling algorithm described in Sect. 5.2.1.2. How can such an algorithm exploit the fact that the input graph is acyclic? Implement the algorithm.
5. The chinese postman problem (CPP) defined over a *weighted-edge directed* graph $G(V, E, W)$, is to determine a minimum-cost route traversing all arcs and edges of the graph *at least once* (respecting directionality of course).

   (a) Model the problem as a MIP.
   (b) Propose and implement a heuristic for solving the CPP.

References

Aloise D, Desphande A, Hansen P, Popat P (2009) NP-hardness of Euclidean sum-of-squares clustering. Mach Learn 75(2): 245–248

Aloise D, Hansen P, Liberti L (2010) An improved column generation algorithm for minimum sum of squares clustering. Math Program Ser A, 20 April 2010, published online

Beasley JE (1985) A note on solving large p-median problems. Eur J Oper Res, 21(2):270–273

Bilgen B, Gunther H-O (2009) Integrated production and distribution planning in the fast moving consumer goods industry: a block planning application. OR Spectrum, 18 June. 2009, first published online

Braysy O, Gendreau M (2002) Tabu Search heuristics for the vehicle routing problem with time windows. Top 10(2):211–237

Captivo EM (1991) Fast primal and dual heuristics for the p-median location problem. Eur J Oper Res, 52(1):65–74

Christou IT (2011) Coordination of cluster ensembles via exact methods. IEEE Trans Pattern Anal Mach Intell 33(2):279–293

Christou IT, Lagodimos AG, Lycopoulou D (2007) Hierarchical production planning for multi-product lines in the beverage industry. J Prod Plan Control 18(5):367–376

Cornuejols G, Fisher M, Nemhauser GL (1977) On the uncapacitated location problem. Ann Discret Math, 1:163–177

Desrochers M, Soumis F (1988) A reoptimization algorithm for the shortest path problem with time windows. Eur J Oper Res, 35(2):242–254

Desrochers M, Soumis F (1989) A column generation approach to the urban transit crew scheduling problem. Trans Sci 23(1):1–13

Desrochers M, Desrosiers J, Solomon M (1992) A new optimization algorithm for the vehicle routing problem with time-windows. Oper Res 40(2):342–354

Desrosiers J, Soumis F, Desrochers M (1984) Routing with time-windows by column generation. Networks, 14(4):545–565

Elloumi S, Labbe M, Pochet Y (2004) A new formulation and resolution method for the p-center problem. INFORMS J Comp 16(1):84–94

Erlenkotter D (1978) A dual-based procedure for uncapacitated facility location. Oper Res 26(6):992–1009

Ghiani G, Laporte G, Musmanno R (2004) Introduction to logistics systems planning and control. Wiley, Chicester

Ghosh D (2003) Neighborhood search heuristics for the uncapacitated facility location problem. Eur J Oper Res, 150:150–162

Golden BL, Assad AA (1988) Vehicle routing: methods and studies. North-Holland, Amsterdam

Hansen P, Mladenovic N (1997) Variable neighborhood search for the p-median. Locat Sci 5(4):207–226

Hoefer M (2003) Experimental comparison of heuristic and approximation algorithms for uncapacitated facility location. Lect Notes Comput Sci, 2647

Kariv O, Hakimi SL (1979) An algorithmic approach to network location problems: part 2, the p-medians. SIAM J Appl Math, 37:539–560

Kuehn AA, Hamburger MJ (1963) A heuristic program for locating warehouse. Manag Sci, 9(4):643–666

Laporte G (1988) Location-routing problems. In: Golden BL, Assad AA (eds) Vehicle routing: methods and studies. North-Holland, Amsterdam

Love RF, Morris JG, Wesolowsky GO (1988) Facilities Location. North-Holland, Amsterdam

Maranzana FE (1964) On the location of supply points to minimize transportation costs. Operat Res Quart 15:261–270

Megiddo N, Supowit KJ (1984) On the complexity of some common geometric location problems. SIAM J Comput 13(1):182–196

Meyer RR (1992) Lecture notes on integer programming. Dept of Computer Sciences, University of Wisconsin

Moreno J, Rodrigez C, Jimenez N (1990) Heuristic cluster algorithm for multiple facility location-allocation problem. RAIRO Oper Res 25(1):97–107

Pardalos PM, Pitsoulis LS (2000) Nonlinear assignment problems: algorithms and applications. Kluwer, Dordrecht

Resende MGC, Werneck RF (2003) On the implementation of a swap-based local search procedure for the p-median problem. In: Proceedings of the 5th Workshop on algorithm engineering and experiments, pp. 119–127, SIAM

Resende MGC, Werneck RF (2004) A hybrid heuristic for the p-median problem. J Heuristics 10(1):59–88

Shi L, Meyer RR, Bozbay M, Miller AJ (2004) A nested partitions method for solving large-scale multi-commodity facility location problems. J Syst Sci Syst Eng 13(2):158–179

Whitaker R (1983) A fast algorithm for the greedy interchange for large scale clustering and median location problems. INFOR, 21:95–108

Wren A (1981) Computer scheduling of public transport urban passenger vehicle and crew scheduling. North-Holland, Amsterdam

Yunes TH, Moura AV, deSouza CC (2005) Hybrid column generation approaches for urban transit crew management problems. Transp Sci 39(2):273–288

Chapter 6
Epilogue

With this short chapter, we come to the end of a journey into modeling and solving problems related to the field broadly known as Supply Chain Management. The focus of this book has been on *quantitative methods*: rigorous modeling & analysis of the problem at hand, and systematic development of efficient algorithms for solving it. Starting with a review of the major ideas behind linear, nonlinear and combinatorial optimization, we discussed methods for demand forecasting, advanced planning and scheduling techniques, inventory control under deterministic or stochastic demand, facility location theory and algorithms, and distribution management. The problems we studied have all been cast as optimization problems, and exact or heuristic approaches to solving them were presented. Important developments in the field of optimization algorithm design as well as in the fields of computer architecture and hardware design have made it possible to solve NP-hard problems arising in Supply Chain Management that were completely out-of-reach just 10 years ago.

But just as algorithm and computer design has improved dramatically during the past few years, so has the sheer size and complexity of the problems of today. Very often, problems that have been relatively easy to solve by now, when slightly modified to take into account a real-world consideration or constraint that was originally left out of the formulation of the problem, become extremely difficult to solve. For example, when randomness cannot be cast aside from the problem formulation, even modeling of the problem may present extreme difficulties, let alone devising efficient algorithms for solving it. This is the case for example with modeling allocation of inventories in divergent multi-echelon inventory systems facing stochastic demands. Determining in such cases the form of the optimal policy—if one exists—is a highly non-trivial task. Randomness also entails the possibility of many alternative scenarios that must be taken into account when deciding which strategy to pursue for a particular problem, and all too often the number of alternative scenarios to consider blows up to such an extent that considering all possibilities necessary in order to draw safe conclusions becomes impossible even if a modern highly parallel super-computer is available.

I. T. Christou, *Quantitative Methods in Supply Chain Management*,
DOI: 10.1007/978-0-85729-766-2_6, © Springer-Verlag London Limited 2012

The above-mentioned difficulties in solving certain classes of problems by no means imply that quantitative methods are not suitable for certain real-world situations. Whenever a problem is intractable in its full form, the obvious practice is to carefully consider only the most important factors that determine the problem solution. Devising a manageable model has thus been a major theme of this book. The second theme of the book has been to demonstrate that for the most important problems in Supply Chain Management, even when determining the exact optimal solution to a problem is not possible, devising efficient heuristics that can produce high-quality near-optimal solutions in a reasonable amount of time on commodity computers is still possible and is very likely to be possible in the foreseeable future as problem sizes increase. The major driving force behind this capability has been (a) the development of highly efficient *exact methods* for linear and combinatorial optimization implemented in state-of-the-art commercial software packages (cplex®, gurobi®) and even free and open-source packages (scip, clp, etc.), and (b) the development of highly efficient *meta-heuristics* for local and global search such as those mentioned in Chap. 1.

Nevertheless, there are also new trends and problem areas in Supply Chain Management that are emerging, and we mention a few of them. These are not always areas where quantitative methods are likely to play a major role.

1. *RFID technology*: the widespread adoption of the RFID technology for tracking inventories, pallets, semi-finished products, finished goods etc. is likely to lead to serious costsavings for the manufacturing and logistics sectors alike due to highly visible monitoring of the flow of goods, vehicle fleets and their drivers, as well as the *business processes* driving these flows. Newly available middleware for enabling RFID tags, readers, gates, and accompanying application-level software will make the adoption process of RFID technology as important as the Electronic Data Interchange (EDI) format had been in the past for conducting business-to-business transactions.
2. *Reverse and green logistics*: recycling used materials is of great importance to a sustainable development for the twenty-first century. Estimating carbon footprints of manufacturing and logistics processes and eventually goods will be an important activity in the coming years, as will the field of optimizing the process by which raw materials from used products that have finished their useful lifecycle are salvaged so as to reuse them without entailing high costs for the environment.
3. *Global sharing of information*: information has already been more and more readily available throughout the supply chain. Upstream manufacturers often know in real-time the actual demand for their products at the retail stores they supply. This has formed the basis for the Vendor Managed Inventory initiatives that often met with significant success in the past decade. Optimizing decisions under global information is an already active area of research in various subfields of Supply Chain Management.
4. In some sectors, it is already possible, or will soon be possible to *jointly optimize previously (artificially) separated sub-problems to obtain better solutions to the*

overall problem than the super-position of the two sub-problems alone can provide. Examples from the transportation sector include *integrated* air line crew pairing and crew assignments generation, whereas from the manufacturing sector, an easy example would be *integrated facility location and routing*. These developments have been made possible through the use of sophisticated algorithms (sometimes based on Lagrangian relaxation ideas or column generation ideas) for very large-scale optimization using coarse-grain decomposition and parallelization to solve many nearly independent sub-problems and then *coordinate and merge* the sub-problem solutions to form a final overall optimal solution.
5. *Social network analysis* will likely be important in forecasting trends in product demand, in combination with traditional time-series analysis methods, and data mining techniques applied to social networks which will be an important tool for marketing and product managers.
6. Finally, the emerging fields of *business intelligence and computational intelligence* will likely form an important ingredient of next generation decision support systems that will have to make decisions based on huge amounts of aggregated information; such tools will aid the decision maker and analysts even in the initial stages of formulating the right model to solve.

Having said all that, if by reading this book, the reader decides to look for more information or research on any of the topics mentioned in its pages, the book will have achieved its purpose in full.

About the Author

Dr. Ioannis T. Christou holds a Dipl. Ing. Degree in Electrical Engineering from the National Technical University of Athens, Greece, an MBA degree from Athens University of Economics & Business and the National Technical University of Athens, and an M.Sc. and Ph.D. in Computer Sciences from the University of Wisconsin at Madison, Madison, WI, USA. He has held senior posts at TransQuest Inc., and Delta Technology Inc., has been an MTS at Lucent Technologies Bell Labs, USA, and an area leader in Data and Knowledge Engineering at Intracom S.A., Greece. He has consulted various private and public sector companies in Greece on a variety of business intelligence & SCM and IT related issues, and has developed large-scale MIS systems for ΔEH S.A. (public electricity utility company of Greece), 3-E S.A., Velti S.A., GAP S.A., and others. He developed the LOTO bid-line generation system for Delta Air Lines, and the Fraud-Detection system for lotteries and online games of chance for Intralot S.A. He has taught "software engineering" and "computer programming laboratory" as an adjunct Assistant Professor at the Computer Engineering & Informatics Department of the University of Patras, Greece; "production systems" at the Production and Management Engineering Department of the Democritus University of Thrace, Greece; "business management for engineers" & "information systems modeling" at the Information Networking Institute of Carnegie-Mellon University, Pittsburgh, PA, USA; and "modern methods for network optimization" at the Doctorate School of Aalborg University, Aalborg, Denmark. Dr. Christou is currently an Associate Professor at Athens Information Technology, Athens Greece, where he teaches graduate-level courses on "systems analysis and design", "logistics and supply chain management", and "network optimization", and an adjunct professor at Carnegie-Mellon University, Pittsburgh, PA, USA. His research work has appeared in IEEE Transactions on Pattern Analysis and Machine Intelligence, Mathematical Programming, Interfaces, Journal of Global Optimization, International Journal of Production Research, Production Planning & Control, International Journal of Systems Science, Computers and Operations Research, and other high impact-factor journals and conferences. He is a member of IEEE, the ACM, and the Technical Chamber of Greece.

I. T. Christou, *Quantitative Methods in Supply Chain Management*,
DOI: 10.1007/978-0-85729-766-2,

Index

I. T. Christou, *Quantitative Methods in Supply Chain Management*,
DOI: 10.1007/978-0-85729-766-2,